Marine Biodiversity

Developments in Hydrobiology 183

Series editor

K. Martens

Marine Biodiversity

Patterns and Processes, Assessment, Threats, Management and Conservation

Edited by

H. Queiroga[1], M.R. Cunha[1], A. Cunha[1], M.H. Moreira[1], V. Quintino[1], A.M. Rodrigues[1], J. Serôdio[1] & R.M. Warwick[2]

[1]*Universidade de Aveiro, Departamento de Biologia, CESAM - Centro de Estudos do Ambiente e do Mar, Campus Universitário de Santiago, 3810-193 Aveiro, Portugal*
[2]*Plymouth Marine Laboratory, West Hoe, Plymouth, PL1 3DH, United Kingdom*

Reprinted from Hydrobiologia, Volume 555 (2006)

Springer

Library of Congress Cataloging-in-Publication Data

A C.I.P. Catalogue record for this book is available from the Library of Congress.

ISBN 1-4020-4321-x

Published by Springer,
P.O. Box 17, 3300 AA Dordrecht, The Netherlands

Cover illustration: Logo of the Symposium

Printed on acid-free paper

Printed in the Netherlands

TABLE OF CONTENTS

Hydrobiologia (2006) 555:ix
H. Queiroga, M.R. Cunha, A. Cunha, M.H. Moreira, V. Quintino, A.M. Rodrigues, J. Serôdio & R.M. Warwick (eds),
Marine Biodiversity: Patterns and Processes, Assessment, Threats, Management and Conservation
DOI 10.1007/s10750-005-1100-x

Foreword

This volume presents some of the contributions given to the 38th European Marine Biology Symposium, held in Aveiro, Portugal, between 8th and 12th September, 2003. The Symposium was convened under the theme of Marine Biodioversity, and subdivided into four subthemes: Patterns and processes, Assessment, Threats, and Management and conservation.

The Symposium attracted 305 scientists from 26 countries, who presented 88 oral communications and 238 posters. In line with the tradition of the EMBS, most of contributions were made by students, with more than 150 participants enrolled in this category. Such participation of enthusiastic young researchers, which gives the opportunity to forge new friendships and to stimulate the exchange of scientific ideas, provides at the same time a link between generations of marine biologists. Therefore, we believe that the 38th EMBS was a proud heir, and at the same time a worthy trustee, of the EMBS philosophy.

The reasons for selecting the theme need not to be justified. Understanding the functioning of the Marine Ecosystems is the first step to measure and predict the influence of Man, and to find solutions for the enormous array of problems, at different scales, we face today. We hoped that the Symposium could present the current research and concerns about Marine Biodiversity in Europe. This Proceedings Volume is organised according to the four subthemes of the Symposium. In order to provide additional structure and facilitate navigation by the interested reader, we divided each of the subthemes into sections. These sections do not parallel taxonomic groups, but rather issues commonly perceived as relevant by scientists concerned with the study, protection and management of Marine Biodiversity.

The Symposium was hosted by the University of Aveiro, and organised by members of the Department of Biology. Financial support for the meeting was provided by the Fundação Luso-Americana para o Desenvolvimento and by the Fundação Calouste Gulbenkian, while Fundação para a Ciência e Tecnologia sponsored the publication of the Proceedings. We thank all our sponsors for their kind support. An EMBS without the Yellow Submarine contest is not an EMBS. The Yellow Submarine was organised by graduation and post-graduated students of the Department of Biology, who also helped in various tasks during the meeting. Efficient secretariat support was provided by GrupUNAVE.

We also wish to thank the many colleagues that helped us with the task of refereeing the manuscripts that were submitted for publication, who have given their time freely. All manuscripts were reviewed by at least two internationally renowned scientists. It is not possible to acknowledge here, individually, the 121 referees involved with all the manuscripts that were submitted, but we believe that their effort strongly contributed to enhance the quality of the papers in this volume. Whatever effect these papers shall have on future research on Marine Biodiversity, a great part is indebted to them. A word of recognition in due to Carla Domingues, who provided substantial help during the final stages of editing the manuscripts included in this volume.

Finally, we hope that this Proceedings Volume brings back to the participants of the 38th EMBS many happy memories of the Aveiro meeting.

Henrique Queiroga
Marina Ribeiro Cunha
Ângela Cunha
Maria Helena Moreira
Victor Quintino
Ana Maria Rodrigues
João Serôdio
Richard M. Warwick

Guest Editors

Hydrobiologia (2006) 555:xi–xv

H. Queiroga, M.R. Cunha, A. Cunha, M.H. Moreira, V. Quintino, A.M. Rodrigues, J. Serôdio & R.M. Warwick (eds),
Marine Biodiversity: Patterns and Processes, Assessment, Threats, Management and Conservation
DOI 10.1007/s10750-005-1131-3

List of participants

Australia
Cruz Motta, Juan José
Dempster, Tim
Schreider, Maria

Belgium
De Mesel, Ilse
Derycke, Sofie
Franco, Maria
Hoste, Eveline
Remerie, Thomas
Steyaert, Maaike
Van Hoey, Gert
Vanaverbeke, Jan
Vandendriessche, Sofie
Vanquickelberghe, Véronique
Verween, Annick
Wittoeck, Jan
Zintzen, Vincent

Canada
Cusson, Mathieu
Pech, Daniel

Croatia
Arko-Pijevac, Milvana
Cetinic, Ivona
Fafandel, Maja
Kirineié, Marin
Kovacic, Marcelo
Kraus, Romina
Micic, Milena
Semencic, Lorena
Travizi, Ana

Finland
Bonsdorff, Erik
Kraufvelin,Patrik

France
Bachelet, Guy
Barnay, Anne-Sophie
Bhaud, Michel
Blanchet, Hugues
Dauvin, Jean-Claude
Degre, Delphine
Dolan, John
Grall, Jacques
Hily, Christian
Le Loc'h, François
Raffin, Coralie
Sorbe, Jean-Claude
Thiebaut, Eric

Germany
Lorenz, Andrea
Reiss, Henning
Schünemann, Henrike

Greece
Nicolaidou, Artemis

Ireland
Crowe, Tasman
Davenport, John
O'Connor, Nessa
Rajasekharan, Maya
Sinkanin, Christina

Italy
Addis, Piero
Balocchi, Cristiana
Basilone, Walter
Bava, Simone
Beltrano, Anna Maria
Benedetti-Cecchi, Lisandro
Bulleri, Fabio
Bussotti, Simona
Campo, Davide
Cannizzaro, Leonardo
Caroppo, Carmela
Casu, Daniela
Cattaneo-Vietti, Riccardo
Cavallo, Rosa Anna
Ceschia, Carlo
Chiantore, Mariachiara

Como, Serena
Consoli, Pierpaolo
Corti, Serena
Cossu, Andrea
De Vito, Doris
Denitto,Francesco
Falace, Annalisa
Fanelli, Giovanni
Ferlizzi, Antonio
Focard, Silvano
Fossi, Soania
Franco, Anita
Franzoi, Piero
Fraschetti, Simonetta
Garaventa, Francesca
Giacobbe, Salvatore
Giordano, Daniela
Gomei, Marina
Gravili, Cinzia
Guidetti, Paolo
Jedrzejczak, Marcin Filip
Leonardi, Maracella
Maggio, Teresa
Mangialajo, Luisa
Merello, Stefania
Milanese, Martina
Moscatello, Salvatore
Murenu, Matteo
Pellizzato, Francesca
Prato, Ermelinda
Relini, Giulio
Rubino, Fernando
Terranova, Margherita Sirna
Virgilio, Massimiliano
Vitale, Sergio
Vizzini, Salvatrice

Lithuania
Gasiunaite, Zita
Kubiliute, Jurate
Pilkaitytė, Renata
Razinkovas, Arturas

Mexico
Amaya, Ana Maria
Ardisson, Pedro
Signoret, Giséle

New Zealand
Norkko, Joana

Norway
Christie, Hartvig
Ellingsen, Kari Elsa
Eriksen, Veslemoy
Gray, John
Hagen, Nils
Moy, Frithjof E.
Olsgard, Frode

Panama
Fortunato, Helena

Poland
Michalowska, Monika
Normant, Monika
Radziejewska, Teresa
Szaniawska, Anna

Portugal
Abreu, Cristina
Adão, Helena
Amaral, Maria José
Azevedo, Isabel
Barcelos e Ramos, Joana
Boaventura, Diana
Brotas, Vanda
Cabral, Henrique
Callado, António J.
Campos, Bruno
Cancela da Fonseca, Luís M.
Cardoso, Margarida
Cardoso, Patrícia
Carreras, Marta
Carvalho, Liliana
Carvalho, Cátia
Chainho, Paula
Coelho, João Pedro
Coelho, Susana
Costa, Ana Luisa
Costa, Filipe
Costa, José Lino
Couto, André Daniel
Cruz, Sónia
Curz, Teresa
Cunha, Ângela
Cunha, Marina
Cúdria, João da Silva
Diekmann, Onno
Diogo, Hugo
Duarte, Pedro

Engelen, Aschwin
Faustino, Claúdia
Fernandes, Joana
Ferreira, Susana Margarida
Ferreira, Eduardo
Ferreira, Filipe
França, Susana
Freitas, Rosa
Gameiro, Carla
Gamito, Sofia
Gaspar, Miguel J. Baptista
Henriques, Vera
Hinzmann, Mariana
Joaquim, Sandra
Leitão, Francisco
Lillebo, Ana Isabel
Magalhães, Catarina
Marques, Joana
Martins, Irene
Matias, Miguel Graça
Matzen da Silva, Jjoana
Meireles, Sónia
Melo Rodrigues, Margarida
Meneses Martins, Gustavo
Mical, Joana
Monteiro, Carla
Moreira, Maria Helena
Neto, João Miguel
Nunes, Alexandra
Pardal, Miguel Angelo
Pastorinho, Ramiro
Patrício, Joana
Pearson, Gareth
Peralta, Nelson
Pereira, Sílvia
Pereira, Tadeu
Pereira, Ana Luísa
Pereria, Paula Maria
Perrin, Cecile
Pinheiro, Ana Cheila
Pinheiro, Ana
Queiroga, Henrique
Queiroz, Nuno
Quintino, Vitor
Ramos, Sandra
Ravara, Ascenção
Ribeiro, Joaquim Miguel
Rodrigues, Ana
Rodrigues, Clara
Sá, Rita
Sá Pinto, Alexandra Isabel
Sampaio, Leandro
Santos, Miguel
Santos, Patrícia
Sardo, Margarida
Seabra, Maria Inês
Serôdio, João
Serrão, Ester
Silva, Gilda
Silva, Helena
Silva, Susana
Silvert, William
Sobral, Dinah
Sousa, Ana
Subida, Dulce
Teixeira, Céllia
Tempera, Fernando
Vasconcelos, Paulo
Vasconcelos, Rita
Veiga, Pedro
Vieira, Sónia
Vinagre, Catarina
Weber, Gerhard Michael
Wise, Laura Simões

Russia
Plyuscheva, Maria

Singapore
O'Riordan, Ruth

Slovenia
France, Janja

Spain
Besteiro, Celia
Carabel, Sirka
Cuesta, José A.
De Pablo, Maria Jesús
Estrada, Marta
Frutos, Inmaculada
García Calvo, Beatriz
Godinez-Dominguez, Enrique
Guillén Nieto, Juan
Pallas, Alicia
Parapar, Julio
Ramos Esplá, Alfonso A.
Rodríguez Castro, Jorge
Rufino, Marta
Sabah Mazzetta, Sandra

Varela de Tuesta, Maria Mercedes
Verisimo, Patricia

Sweden
Asplund, Maria
Bergström, Lena
Eriksson, Susanne
Flach Lundgren, Elsina
Gullström, Martin
Jonsson, Anna
Nagel, Katja
Norkko, Alf
Norling, Karl
Raberg, Sonja
Stal, Johan
Svensson, Carl-Johan

The Netherlands
Amaro, Teresa
Besse Lototskaya, Anna
Bult, Tammo
Cardoso, Joana Ferreira
Holmes, Sebastian
Hummel, Herman
Jansen, Jeroen
Loff, José Francisco
Rossi, Francesca
Van der Meer, Jaap

Ukraine
Shulman, Georgy

United Kingdom
Aboim, Maria Ana
Aranda da Silva, Ana
Burrows, Michael
Chesman, Belinda S.
Colson, Isabelle
Dashfield, Sarah
Davis, Martin
Di Carlo, Giueseppe
Dodd, Jane
Gibson, Robin
Gubbay, Susan
Hawkins, Tony
Hilario, Ana
Hiscock, Keith
Hughes, Roger Neville
Jackson, Juliette
Jones, Malcolm Benjamin
Kendall, Mike
Lee, James
Mieszkowska, Nova
Moffat, Angela
Moschella, Paula
Needham, Hazel
Poloczanska, Elvira
Richardson, Kirsten
Rose, Craig
Ryland, John S.
Somerfield, Paul
Warwick, Richard
Widdicombe, Steve
Wilding, Thomas

United States of America
Berkenbush, Katrin
Dunbar, Stephen G.
Whitlatch, Robert

Sponsors

universidade de aveiro theoria poiesis praxis

FUNDAÇÃO CALOUSTE GULBENKIAN

FUNDAÇÃO LUSO-AMERICANA

FCT Fundação para a Ciência e a Tecnologia
MINISTÉRIO DA CIÊNCIA E DO ENSINO SUPERIOR Portugal

Patterns and Processes

Hydrobiologia (2006) 555:3–18
H. Queiroga, M.R. Cunha, A. Cunha, M.H. Moreira, V. Quintino, A.M. Rodrigues, J. Serôdio & R.M. Warwick (eds),
Marine Biodiversity: Patterns and Processes, Assessment, Threats, Management and Conservation
DOI 10.1007/s10750-005-1101-9

The relationship between phytoplankton diversity and community function in a coastal lagoon

Pedro Duarte[1,*], Maria Filomena Macedo[2] & Luís Cancela da Fonseca[3]
[1]*Center for Modelling and Analysis of Environmental Systems, Faculty of Science and Technology, Fernando Pessoa University, Praça 9 de Abril 349, 4249-004 Porto, Portugal*
[2]*Department of Conservação e Restauro, Faculty of Sciences and Technology, New University of Lisbon, P-2825-114 Monte de Caparica, Portugal*
[3]*IMAR/Lab. Mar. Guia and FCMA – DCB, University of Algarve, Campus de Gambelas, 8005-139, Faro, Portugal*
*(*Author for correspondence: E-mail: pduarte@ufp.pt)*

Key words: Phytoplankton diversity, production, coastal lagoons

Abstract

The decrease of biodiversity related to the phenomena of global climate change is stimulating the scientific community towards a better understanding of the relationships between biodiversity and ecosystem functioning. In ecosystems where marked biodiversity changes occur at seasonal time scales, it is easier to relate them with ecosystem functioning. The objective of this work is to analyse the relationship between phytoplankton diversity and primary production in St. André coastal lagoon – SW Portugal. This lagoon is artificially opened to the sea every year in early spring, exhibiting a shift from a marine dominated to a low salinity ecosystem in winter. Data on salinity, temperature, nutrients, phytoplankton species composition, chlorophyll *a* (Chl *a*) concentration and primary production were analysed over a year. Modelling studies based on production-irradiance curves were also conducted. A total of 19 taxa were identified among diatoms, dinoflagellates and euglenophyceans, the less abundant group. Lowest diversities (Shannon–Wiener index) were observed just before the opening to the sea. Results show a negative correlation ($p < 0.05$) between diversity and chlorophyll *a* (Chl *a*) concentration (0.2–40.3 mg Chl *a* m^{-3}). Higher Chl *a* values corresponded to periods when the community was dominated by the dinoflagellate *Prorocentrum minimum* (>90% of cell abundance) and production was maximal (up to 234.8 mg C m^{-3} h^{-1}). Maximal photosynthetic rates (P_{max}) (2.0–22.5 mg C mg Chl a^{-1} h^{-1}) were higher under lower Chl *a* concentrations. The results of this work suggest that decreases in diversity are associated with increases in biomass and production, whereas increases correspond to opposite trends. It is suggested that these trends, contrary to those observed in terrestrial and in some benthic ecosystems, may be a result of low habitat diversity in the water column and resulting competitive pressure. The occurrence of the highest photosynthetic rates when Chl *a* is low, under some of the highest diversities, suggests a more efficient use of irradiance under low biomass–high diversity conditions. Results suggest that this increased efficiency is not explained by potential reductions in nutrient limitation and intraspecific competition under lower biomasses and may be a result of niche complementarity.

Introduction

Biodiversity changes at various temporal and spatial scales (Krebs, 1994). The former may be as large as evolutionary time scales and as small as seasonal or even shorter time scales. The latter may range from latitudinal to local diversity gradients. Given the important shifts observed in biodiversity and the long-term effects of global change, it is important to understand the impact of these changes on ecosystem functioning and ecosystem services.

Biodiversity (hereafter referred as diversity) is a measure of community structure, whether it is expressed merely as species richness or with a specific index. Production is a measure of community function. Therefore, relating diversity with production is one of the several ways to relate community structure with community function. The relationship between these two parameters has been a topic of much debate over the years, mostly in terrestrial ecology. In spite of all this debate, it is not yet a matter of consensus among the scientific community (e.g. Huston et al., 2000). One question that may be asked about these two parameters, assuming that they are related, is "Which is the cause and which is the effect?". According to some classical ecology textbooks (Krebs, 1994), production may hardly be the cause, since some of the most productive ecosystems have a low diversity. Recent studies on grassland ecosystems suggest that more diverse communities are more productive, because of niche complementarity (Hector et al., 1999; Tilman et al., 2001). This leads to the concept of 'overyielding', when polycultures exhibit higher production than monocultures, due to positive synergies between different species, as in the presence of nitrogen-fixing plants. In such a case, diversity would be the cause of higher production.

When production of different terrestrial ecosystems is compared, from grasslands to rain forests, it is apparent that as diversity increases towards tropical forests, ecosystems have larger gross and net areal productions. But when these values are related to biomass standing stock (the P/B ratio), the opposite seems to be the rule (Fig. 1) (Whittaker & Likens, 1975). Following the same authors, when areal production of continental shelf ecosystems and upwelling zones are compared with production of open ocean ecosystems, the formers exhibit much higher values than the latter, whilst the opposite is true for the P/B ratio. As in terrestrial ecosystems, those with higher biomass standing stocks exhibit higher production but lower P/B ratios than the ones with lower biomass densities (Fig. 1). However, whereas large biomass standing stocks are generally associated with higher diversity in terrestrial ecosystems, the opposite seems to be the rule in pelagic marine ecosystems, where it is generally reported that increased production is associated with decreased diversity (Pearl, 1988; Krebs, 1994). There seems to be a relatively scarcity of field and experimental data relating diversity, production and the P/B ratios in different marine ecosystems, whereas this is a very active field of research in terrestrial ecosystems, with manipulative experiments in current usage (e.g. Hector et al., 1999; Tilman et al., 2001). One of the few works where phytoplankton diversity and production were analysed together is that of Agard et al. (1996). These authors found some empirical evidence to confirm the dynamic equilibrium model of Huston (1979, 1994). According to this model, diversity is reduced by competitive exclusion under conditions of high production and low levels of disturbance, or where production is too slow to allow recovery from mortality. Diversity is therefore maximised at 'intermediate' disturbance and production levels.

Changes in species composition and diversity may produce changes in community level parameters, like phytoplankton growth rate and those parameters regulating the photosynthetic response to irradiance or other limiting factors. It is important to understand how these changes are reflected in ecosystem functioning and ecosystem services. The relationship between photosynthetic rate and irradiance (P–I) is of utmost importance in phytoplankton production studies. The knowledge of the dynamics of the P–I parameters over the annual cycle can be used to estimate primary production over seasonal scales. It may also help to understand some of the mechanisms controlling photosynthesis and operating from the species to the ecosystem level (Macedo et al., 2001).

However, at the present state of knowledge, it is very difficult to relate these parameters with community structure. According to Banse (1982), phytoplankton growth rate changes allometrically with cellular carbon. However, the parameters regulating this allometric relationship are higher for diatoms than for dinoflagellates, predicting higher growth rates for the former than for the latter, when cells exhibit similar carbon contents. Gallegos (1992) observed in the estuary of the Rhode River (Maryland, USA) that the parameters of the P–I curves were higher when phytoplankton blooms were dominated by the diatom

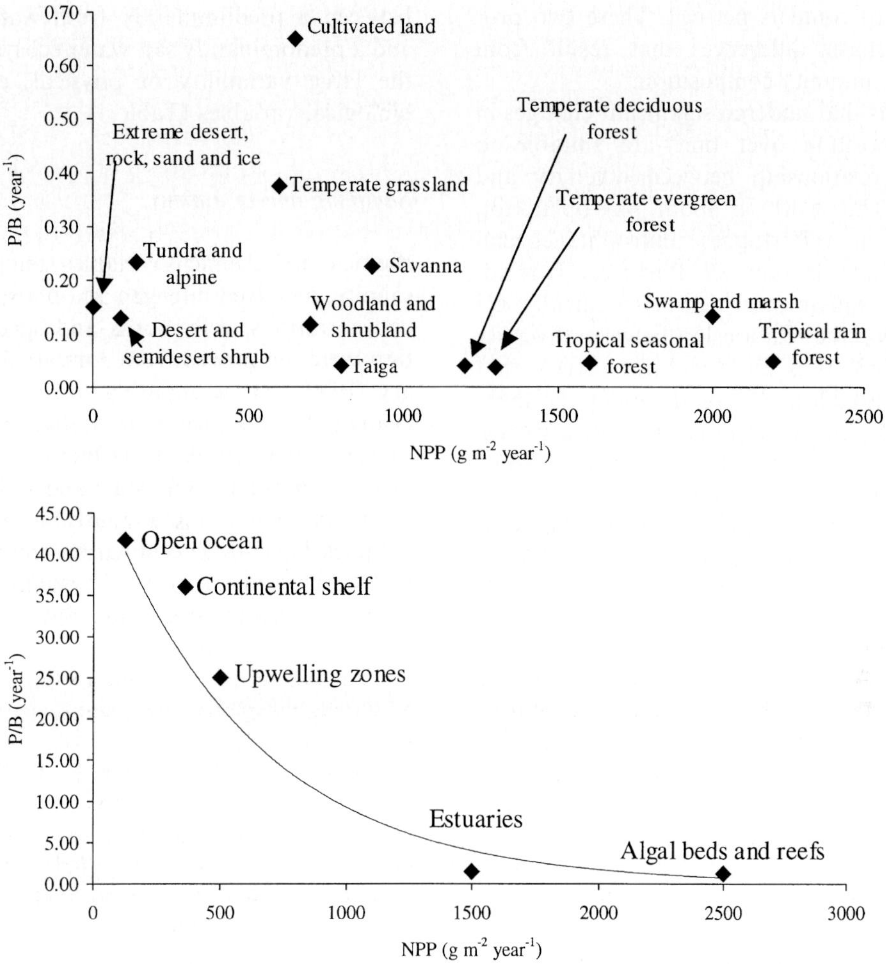

Figure 1. Relationship between areal net primary production and the P/B ratio obtained from data presented in Whittaker & Likens (1975).

Thalassiosira pseudonana Hasle & Heimdal and lower when the blooms were dominated by dinoflagellates. Shaw & Purdie (2001) observed in the UK coastal waters of the North Sea that the October peak in the parameters of the P–I curves coincided with a period in which dinoflagellates accounted for a high proportion of phytoplankton biomass. In a study conducted in Santo André coastal lagoon Macedo et al. (2001) obtained P–I curves every month for a period of 13 months together with phytoplankton species composition and cell counts. A significant Arrhenius type relationship was obtained between light saturated photosynthesis (P_{max}) and temperature when blooms were dominated by the dinoflagellate *Prorocentrum minimum* (Pavillard) Schiller. None of the previous authors related phytoplankton diversity with the P–I curve parameters. Moreover, none separated the effects of species composition from the effects of other environmental variables that may contribute to photoacclimation and photoadaptation of phytoplankton cells, leading to differences in the P–I curve parameters. According to Pahl-Wostl & Imboden (1990) photoresponse has typical time scales between a few minutes and a few hours and corresponds to the time it takes for photosynthesis to reach a steady state response to light. Photoacclimation occurs at time scales of several hours to days and corresponds to changes in cell composition, as chlorophyll *a* (Chl *a*) contents per cell. These two processes may mask differences that result from changes in community composition.

Ecosystems that undergo significant changes in species composition over time are suitable to analyse the relationship between diversity and production. This work is about phytoplankton communities in a Portuguese eutrophic coastal lagoon (St. André lagoon, SW Portugal). It is not based on an experimental design specifically defined to analyse the relationship between diversity and production. It is an exploratory study that may help to establish hypothesis about the mentioned relationship to be tested in future works. Therefore, the objective of this work is to get some insight into the following question:

"How is phytoplankton diversity related to phytoplankton community production and photosynthetic rates in a coastal lagoon?"

Methods

The sampling and analytical methods used in this study have been described elsewhere (Macedo et al., 1998, 2001). Therefore, only a brief description will be provided here.

Study area

Santo André Lagoon (38° 05′ N, 8° 47′ W) is a shallow (average annual depth of about 1 m, with a maximum of 5 m in autumn) land-locked coastal system located on the southwest coast of Portugal (Fig. 2). The lagoon is connected with the sea only in two periods: during about one month in March–April, by a man-made channel, and occasionally when seawater overpasses the dunes. In the first situation, low salinity water and sediments are exported and colonisation by marine species occurs. After the lagoon is closed, salinity progressively decreases and organic matter accumulates leading to summer dystrophy (Cancela da Fonseca et al., 1989). The lagoon receives freshwater from six small rivers forming a drainage basin of about 96 km^2. The lagoon can be stratified or vertically mixed, depending on the prevailing environmental conditions (Bernardo, 1990). Fishing is the main economic activity in Santo André Lagoon, although it is also used for recreation. The shifting between a predominantly fresh water ecosystem and a predominantly salt water ecosystem explains the large variability of physical, chemical and biological variables (Table 1).

Sampling and treatment

Physical and chemical variables (temperature, pH, salinity, dissolved nitrogen and phosphorus) Chl *a* concentrations, cell counts and species composition were monitored from January 1998 to January 1999, on a monthly basis (13 sampling campaigns) at one sampling station (Fig. 2). Water samples for phytoplankton biomass, species composition, inorganic nutrients and P–I experiments were collected simultaneously at 0.5 m depth. Samples for P–I determination were collected in the morning and kept in the dark for about 4 h before the incubations (see below).

Chemical analyses

Inorganic nutrient analyses (nitrate, nitrite, ammonia and phosphate) were performed according to the methods described in A.P.H.A. (1992) and Parsons et al. (1984). Total available inorganic carbon was determined in the water samples prior to incubation from pH (pH Meter ESD model 69) and alkalinity measurements according to Parsons et al. (1984). Samples for Chl *a* and phaeopigments (Phae) were filtered onto 0.45 μm membrane filters. Pigments were extracted in 90% acetone and analysed fluorometrically by the method of Yentsch & Menzel (1963) as modified by Holm-Hansen et al. (1965).

Species determination

Samples for species determination and enumeration were preserved with Lugol's solution (Throndsen, 1978) for about 6 months. Phytoplankton cells were counted by the Utermöhl technique in an Olympus IX70 light inverted microscope (Hasle, 1978), using the classification scheme of Drebes (1974), Dodge (1975) and Hasle et al. (1996). Phytoplankton diversity was calculated using the Shannon–Wiener function for each sampling occasion.

Table 1. Main characteristics of St. André lagoon (average ranges from Bernardo, 1990)

Average area	150 ha
Average depth	90–280 cm
Max. depth	225–540 cm
Salinity	1.9–23.5 psu
Temperature	9.5–28.6 °C
Phosphate P-P0$_4$	0.05–3.8 μmol l^{-1}
Nitrate N-N0$_3$	0.2–75.3 μmol l^{-1}
Ammonia N-NH$_4$	1.4–22.3 μmol l^{-1}
Chl *a*	1.8–61.9 mg m^{-3}
Macrophytes	94–438 g AFDW m^{-2}
Sediment org. matter	6.5–16.6%

P–I experiments

Samples were incubated in the laboratory at the same temperature measured in the field, at the time of sampling and under variable irradiance. Light was provided by 1500 W tungsten halogen lamps. Heat produced by the lights was dissipated using a cold water flow system. Irradiance (0–1445 μmol quanta m^{-2} s^{-1}) was measured by a LI-COR underwater cosine quantum sensor (model LI–192SA) and attenuation was achieved by means of grey PVC nets. Preservation of the spectral characteristics was verified by spectral analysis (see Macedo et al., 1998).

Photosynthetic rates were measured at different irradiances by the standard ^{14}C incubation technique (Steemann Neilsen, 1952) and following the ICES CM 1996/L:3 recommendations. Water samples were placed in 60 ml Winkler bottles and inoculated with 1 ml NaH^{14}CO$_3$ with 10 μCi (371.88 kBq cm^{-3}) (^{14}C Centralen). A dark bottle was used as blank.

P–I parameters were calculated from the photosynthesis and irradiance using the Eiler & Peeters (1988) model (1).

$$P(I) = \frac{I}{aI^2 + bI + c}\left[\mathrm{mg\,C(mg\,Chl}\,a)^{-1}\mathrm{h}^{-1}\right] \quad (1)$$

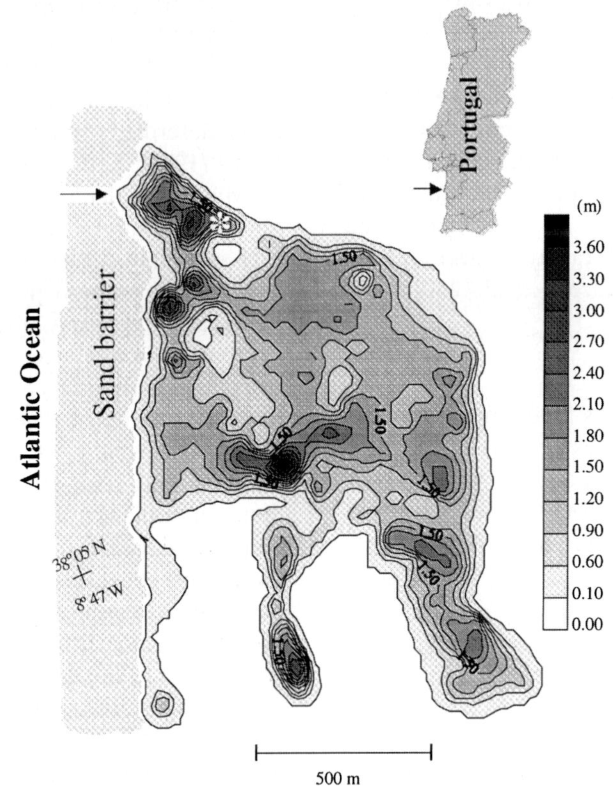

Figure 2. Santo André Lagoon bathymetry. The white asterisk marks the position of the sampling station. The arrow on the upper left corner shows the place where the artificial channel is opened between the Lagoon and the sea (see text).

where:
$P(I)$ – Light limited photosynthetic rate;
I – Irradiance (μmol quanta $m^{-2}\,s^{-1}$);
a, b and c – Adjustment parameters.

By differentiating this function, the parameters α (initial slope), P_{max} (light saturated photosynthesis) and I_{opt} (optimal irradiance) can be expressed as a function of a, b, and c:

$$\alpha = \frac{1}{c}\left[\mathrm{mgC(mgChl}a)^{-1}\,\mathrm{h}^{-1}\,\mu\mathrm{mol\,quanta}^{-1}\,\mathrm{m}^2\,\mathrm{s}\right] \quad (2)$$

$$I_{opt} = \sqrt{\frac{c}{a}}\left[\mu\mathrm{mol\ quanta\ m}^{-2}\,\mathrm{s}^{-1}\right] \quad (3)$$

$$P_{max} = \frac{1}{b + 2\sqrt{ac}}\left[\mathrm{mg\,C(mg\,Chl}\,a)^{-1}\,\mathrm{h}^{-1}\right] \quad (4)$$

A variant of this model combined with an Arrhenius temperature limitation function was also used following Duarte (1995):

$$P(I,t) = \frac{I.\exp(d - (e/t))}{aI^2 + bI + c}\left[\mathrm{mg\,C(mg\,Chl}\,a)^{-1}\,\mathrm{h}^{-1}\right] \quad (5)$$

Where, $P(I, t)$ is Light and temperature limited photosynthetic rate; d and e are parameters of the Arrhenius function; and t is Temperature (°C).

A simplification of this model was made in order to reduce the number of parameters to be estimated, by dividing both the numerator and the denominator by exp(d) (Macedo et al., 2001):

$$P(I,t) = \frac{I.\exp(-e/t)}{a'I^2 + b'I + c'} \quad (6)$$

Again, by differentiating this function, the parameters α, P_{max} and I_{opt} can be expressed as a function of a, b, and c. The solution for I_{opt} is as shown before (Eq. (3)) and the solutions for the former two parameters are shown below (Duarte, 1995):

$$\alpha = \frac{\exp(-e/t)}{c} \quad (7)$$

$$P_{max} = \frac{\exp(-e/t)}{b + 2\sqrt{ac}} \quad (8)$$

Thirteen P–I curves were fitted with Eq. (1) (one for each sampling campaign), using the Quasi-Newton non-linear least-squares regression technique (Statistica software). For some campaigns it was possible to obtain a good fit with Eq. (6) as well (see below) (Macedo et al., 2001). Linear regressions (type II) between observed and predicted values were used to verify the fitting equation. For each curve, the slope of the regression line was checked for significant differences from one and the y-intercept was checked for significant differences from zero. The significance of these differences is an indication of a poor fit to observed data (Keller, 1989). Also, analysis of variance was used to test for the significance of the variance explained by the regression line. All statistical analyses were done for a 95% confidence level. P–I curves considered in this study were only those for which all tests confirmed the quality of the obtained fit.

Mathematical simulations

Daily average primary production was calculated for each of the sampling occasions using the above mathematical relationships (1 and 6) integrated over depth and over time with parameters described in Macedo et al. (2001), simulated light intensity data over a 24-h period, with the equations described in Brock (1981) and Portela & Neves (1994), and measured *in situ* temperature.

Macedo et al. (2001) were able to fit Eq. (1) to all obtained datasets (a different parameter set for each sampling occasion) and Eq. (6) only to those datasets (a common parameter set for a total of seven sampling occasions) where the dinoflagellate *Prorocentrum minimum* was the dominant species (> 54% cell abundance). Using Eq. (1) or Eq. (6) for those periods of *P. minimum* dominance should yield similar results, whereas the opposite is true for the remaining periods (cf. – Results). Considering that phytoplankton species dominance shifted between diatoms and dinoflagellates, calculating photosynthetic rates with both equations for all sampling occasions allows us to obtain estimates of expected photosynthetic rates in the case of diatom and dinoflagellate dominance with prevailing light and temperature conditions. Since diatom dominance coincided with higher diversities, comparing obtained results may give some insight into the relationship between diversity and production.

Results

Chl *a* data for St. André lagoon is shown in Figure 3, for years 1984, 1985 (January–December), 1986 (January–May), 1998 and 1999 (January). Phytoplankton cell counts (Figure 3) were directly correlated with Chl *a* ($p < 0.05$).

The results presented in Figure 4 show that during the January 98–January 99 period, lowest phytoplankton diversity was observed in winter, whereas higher values were observed in spring or late summer. Although the correlation between Chl *a* and diversity was not significant ($p > 0.05$), it is apparent that minimum diversity corresponds to some of the highest Chl *a* values, whereas the opposite is true for maximum diversity results (Figs. 3, 4).

In Figure 5, the percentage of different phytoplankton groups over the sampling period is shown. Dinoflagellates and diatoms were the dominant groups. Lower diversities coincided with periods of dinoflagellate dominance, whereas higher diversities coincided with periods of diatom dominance. Dinoflagellate proportion, after the arc sin transformation (Underwood, 1981), is negatively correlated with diversity ($p < 0.05$). A total of 19 taxa were identified among diatoms, dinoflagellates and euglenophyceans, the less abundant group (Macedo et al., 2001).

Temporal variability of P_{max} (Eq. (4), cf. – values reported in Table II of Macedo et al. (2001)) and maximal volume integrated production (MaxProd) – the product of P_{max} and chlorophyll concentration – showed different patterns (Fig. 6). Both parameters exhibited a high variability. P_{max} showed higher values between May and August, with another maximum in October. Minimum values occurred in winter months. MaxProd reached maximum values in some winter months (February and March 1998 and January 1999). However, the lowest value occurred also in winter (January 1999). There is a negative correlation between P_{max} and Chl *a* concentration and the opposite between MaxProd and Chl *a* ($p < 0.05$). P_{max} is directly correlated solely with temperature. MaxProd is directly correlated with nitrogen and the proportion of dinoflagellates ($p < 0.05$). Highest P_{max} values coincided with some of the highest diversities (Figs. 4, 6). Cellular Chl *a*, obtained from the ratio Chl *a* /cell counts, and cellular P_{max}, obtained from the Production/Chl *a* cell counts ratio, are shown in Figure 7. There is a significant correlation between both variables ($p < 0.05$).

Temporal variability of α is shown in Figure 8. No significant correlation was found between the initial slope and any other parameter or variable, except a negative correlation with inorganic phosphorus ($p < 0.05$). However, peaks in α coincide with peaks in diversity (cf. Fig. 4).

Figure 9 depicts the parabolic relationship between diversity and MaxProd. A linear relationship results in a much lower R^2 – 0.436.

In Figure 10 a tree clustering, obtained from the Pearson correlation coefficient, displays two main groups of variables:

(i) An upper group with diversity, equitability, salinity, species richness, P_{max}, temperature, cellular P_{max}, cellular Chl *a*, diatom abundance and α, the initial slope of P–I curves, at a much larger linkage distance;
(ii) A lower group with Chl *a*, cell numbers, pH, MaxProd, the proportion of diatoms, dinoflagellate abundance, the proportion of dinoflagellates, inorganic nutrients and the ratio between nutrient and Chl *a* concentrations.

In Figure 11 the results of the mathematical simulations described above (cf. – Methodology – Mathematical simulations) are presented. The comparisons between both data sets by a one-way ANOVA did not reveal any significant differences ($p > 0.05$).

Discussion

Considering the high variability of Chl *a* data at temporal scales considerably smaller than the sampling intervals depicted in Figure 3, it is clear that the results available are insufficient to adequately describe Chl *a* dynamics. However, it is apparent that the 1998–1999 data are well within the ranges observed in previous works with maximum average Chl *a* concentration reaching ca. 60 mg m^{-3} (Cancela da Fonseca, 1989; Cancela da Fonseca et al., 1989; Bernardo, 1990). All data series show peaks in winter months. Data from 1984, 1985 and 1998 also present peaks in August or September. According to Bernardo (1990), higher Chl *a* concentrations are associated with

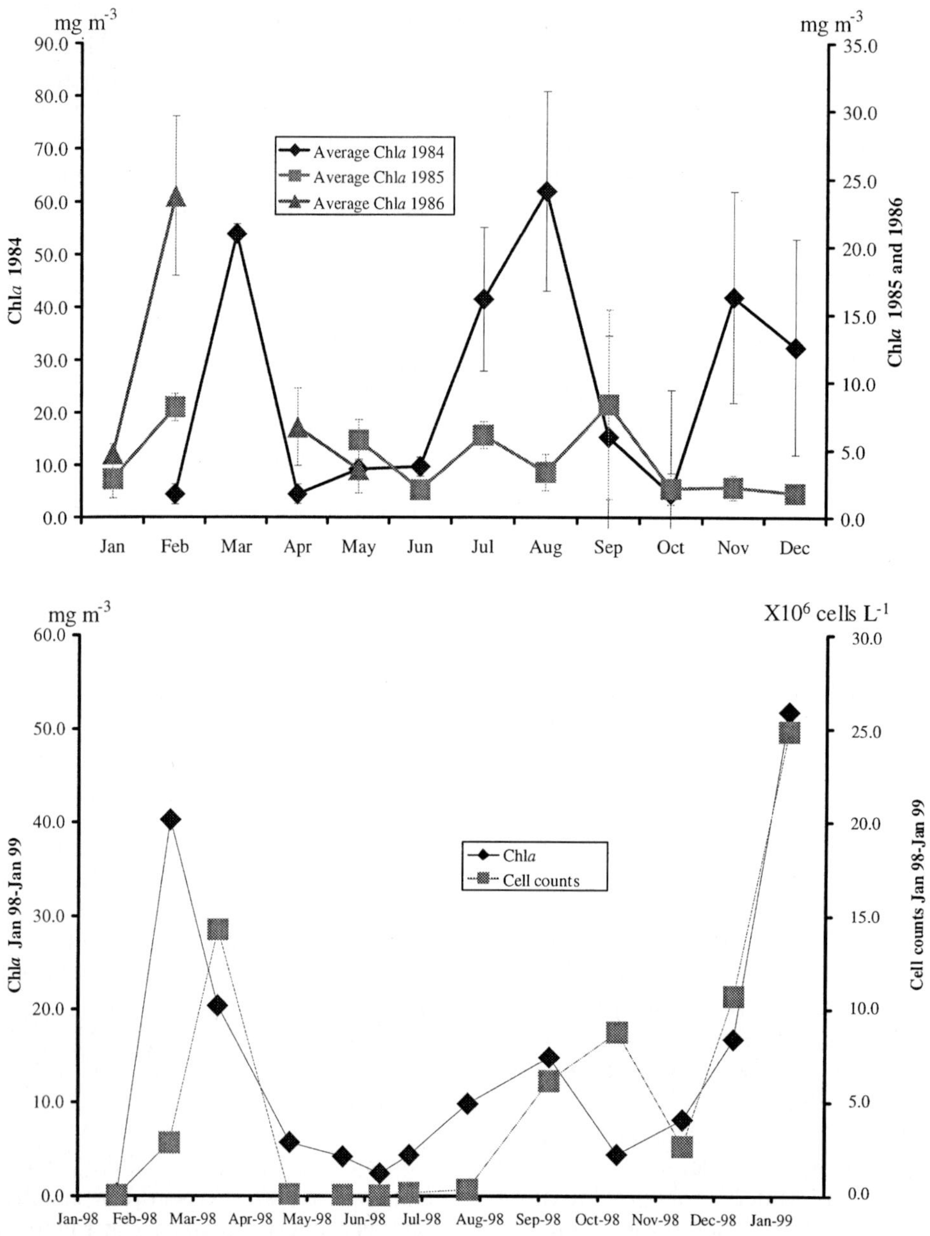

Figure 3. Chl *a* concentrations from several sampling campaigns in St. André lagoon. Data from 1984, 1985 and 1986 was taken from Cancela da Fonseca (1989), Cancela da Fonseca et al. (1989) and Bernardo (1990). Data from 1998–1999 was taken from Macedo et al. (2001). For the period 1998–1999 phytoplankton cell counts are also shown. The values for the period 1984–1986 were based on ca. 17 sampling points over the whole lagoon (also shown the 95% confidence limits). 1998–1999 data was from one sampling location (cf. – Fig. 2).

high nutrient inputs by runoff in winter months and internal nutrient recycling in summer months.

There is a relatively scarcity of data on phytoplankton species richness. In most studies only major taxonomic groups are listed, or proportions of dominant species given. Considering the number of phytoplankton species listed for some coastal ecosystems; e.g. the Rhode river estuary (Maryland, USA) (Gallegos, 1992), the Elbe (Germany), the Shelde (Belgium/The Netherlands) and the Girond (France) estuaries (Muylaert & Sabbe, 1999), the Pearl River estuary (China)

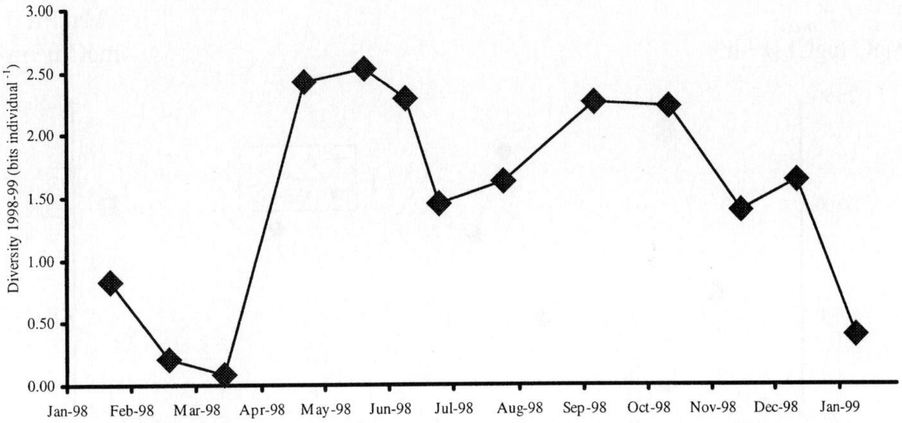

Figure 4. Diversity calculated by the Shannon–Wiener function.

(Huang et al., 2004) and the Bras de Port solar salterns in Santa Pola (Spain) (Estrada et al., 2004), species richness in Santo André lagoon (19 taxa identified) is comparable to the lower values reported – 29 species for the Girond estuary, 18 species for the Rhode river estuary and between 10 and 32 for the Bras de Port solar salterns. However, it is noteworthy that in all these studies the number of samples was much larger than in the present work. The low species richness in Santo André lagoon may partly be explained by the frequent overwhelming dominance of *Prorocentrum minimum* (up to >90% cell counts) (Macedo et al., 2001) and the alternating periods of low/high salinity (cf. – Methods, Study Area). If the Shannon diversity index is used for comparison, instead of species richness, the range reported in this study (0.08–2.53 bitts individual^{-1}) (Fig. 4)

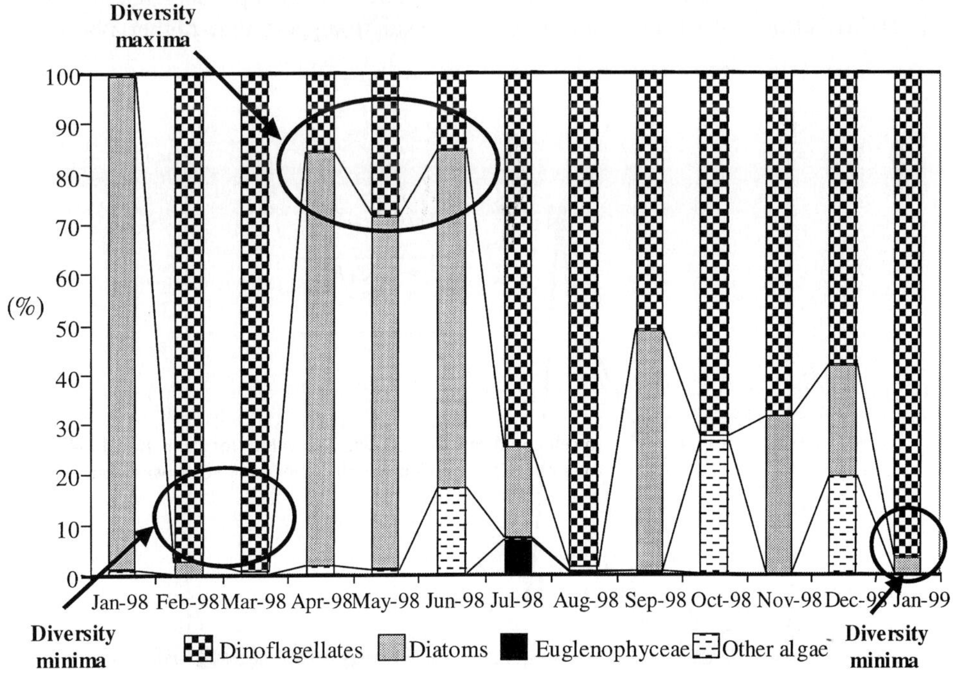

Figure 5. Percentage of phytoplankton groups over the period 1998–1999. Also shown minimum and maximum phytoplankton diversity periods (Adapted from Macedo et al., 2001).

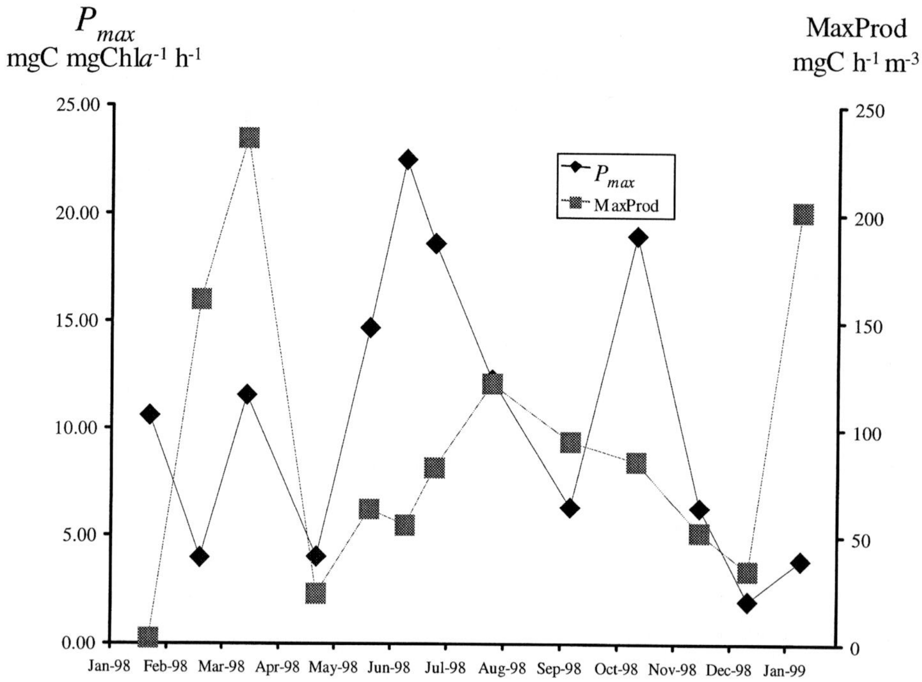

Figure 6. Light saturated photosynthesis ($P_{\max}$) and maximal volume integrated production (MaxProd) (see text).

includes the range reported in Estrada et al. (2004) (0.5 – ca. 2.6 bitts individual^{-1}), the average value reported for the Pearl river estuary (2.47 bitts individual^{-1}) (Huang et al., 2004) and the value reported for the Sado estuary (Portugal) by Peneda et al. (1980) – 1.1 bitts individual^{-1}.

The two groups depicted in Figure 10 (cf. – Results) suggest that higher photosynthetic rates

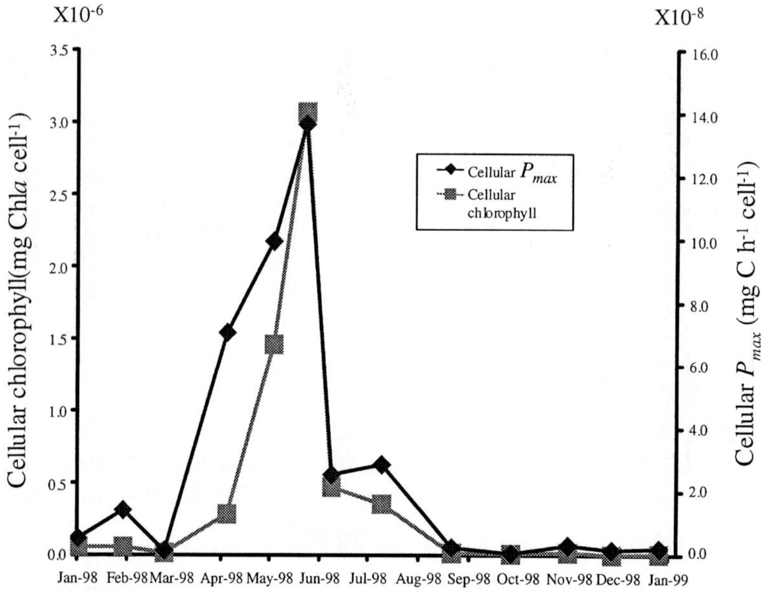

Figure 7. Cellular light saturated photosynthesis ($P_{\max}$) and cellular Chl *a* contents (see text).

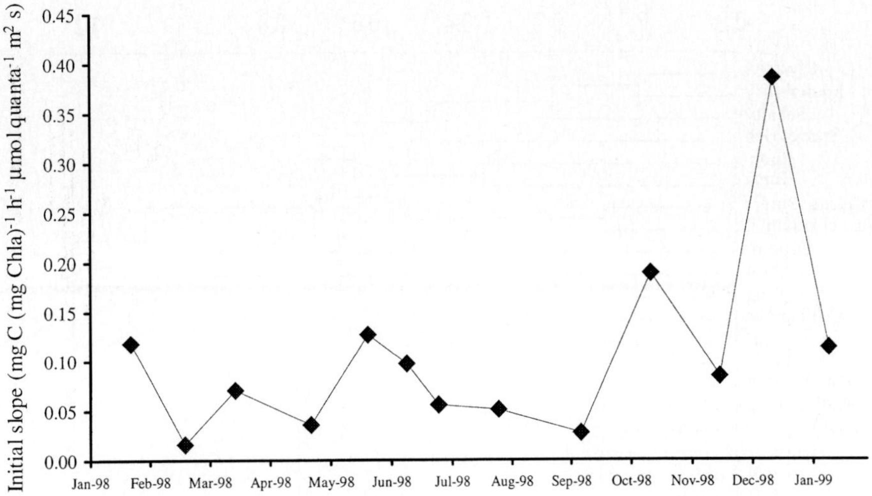

Figure 8. Initial slope (α) of the P–I curves for the study period (see text).

are associated with lower nutrient concentrations and higher temperatures, whereas higher MaxProd values are associated with higher nutrient loads and higher nutrient/Chl *a* ratios. From these results, it is apparent that higher P_{max} values are not related to release from nutrient limitation. In this work only nitrogen and phosphorus were considered. However, it is expectable that in winter periods, when MaxProd is higher, runoff transports all potentially limiting nutrients to the lagoon. If this is the case, then higher P_{max} values may be explained mostly by temperature as suggested by the positive and significant correlation referred above (cf. – Results). This is an expected result since P_{max} is known to be a function of the enzymatic processes in photosynthesis and therefore it is temperature dependent (Eppley, 1972; Harrison & Platt, 1980; Davison, 1991). The negative correlation between P_{max} and Chl *a* concentration and the opposite between MaxProd and Chl *a* concentration ($p < 0.05$) are expected, since the former is calculated from a ratio where Chl *a* is the denominator and the latter is calculated from a product by Chl *a* (cf. – Results). Generally, when Chl *a* is higher, intra and interspecific competition for light and/or nutrients is more likely to occur

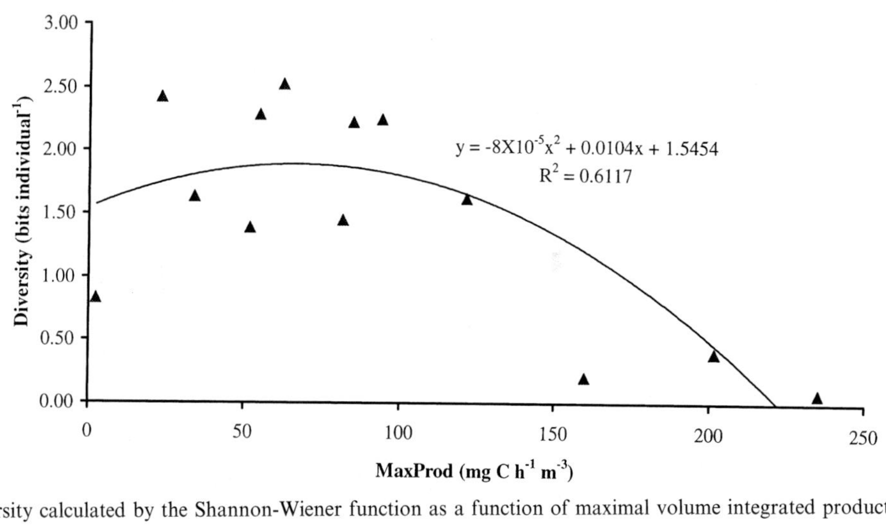

Figure 9. Diversity calculated by the Shannon-Wiener function as a function of maximal volume integrated production – MaxProd (see text).

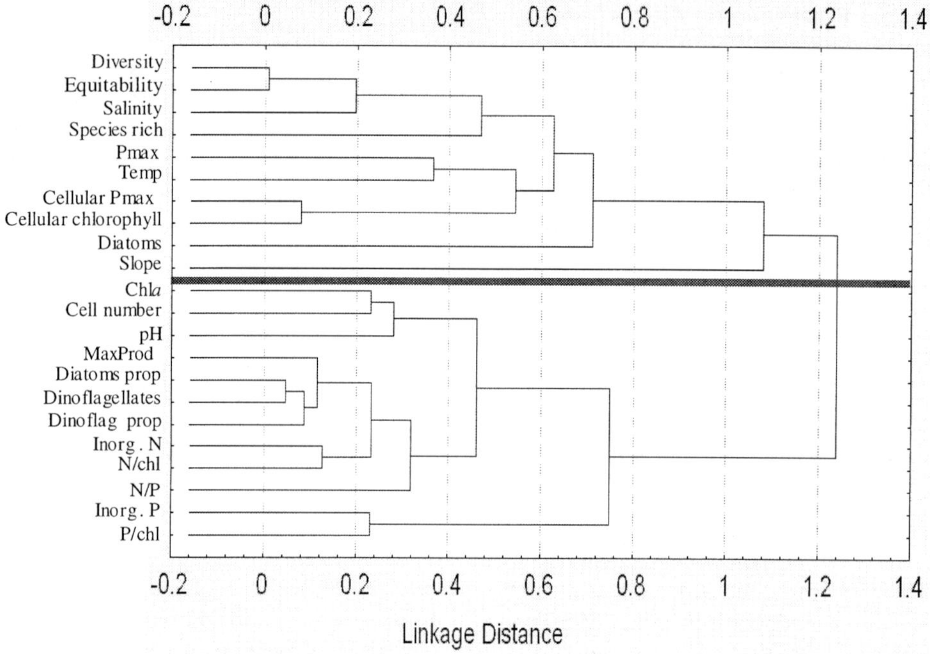

Figure 10. Cluster analysis using the Pearson correlation coefficient and the weighted pair-group average amalgamation scheme. The horizontal line separates two main groups of variables (see text).

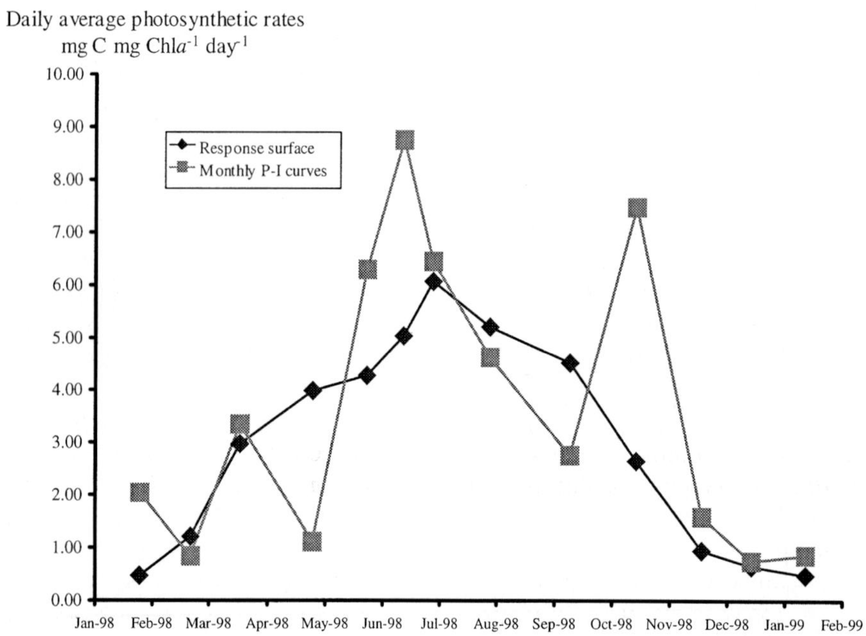

Figure 11. Daily average photosynthetic rates predicted with the P–I curves for each monthly phytoplankton sample, and with the surface response relating photosynthetic rates with irradiance and temperature, for those samples when *Prorocentrum minimum* abundance was larger than 54% of cell counts. The parameters of the P–I curves and of the response surface are described in Macedo et al. (2001) (see text).

reducing P_{max}. Further, low Chl *a* values and corresponding phytoplankton biomasses are not likely to result in high MaxProd, since this is a volume integrated value (cf. – Results).

The coincidence between diversity and α maxima (cf. – Results and Figs. 4, 8) and the fact that α is associated with higher P_{max} and diversity (cf. Results and Fig. 10), suggests that phytoplankton assemblages with higher diversity may also be more efficient at low light levels than lower diversity assemblages.

The absence of significant differences between the two simulated datasets in Figure 11 (cf. – Results) is not surprising because it is apparent that the curve based on the response surface smoothes out the larger variability of the curve calculated with monthly estimates of the photosynthetic parameters. Therefore, it is apparent that long-term (seasonal) average estimates of daily productivity are similar. However, short-term estimates may differ by more than 100%. This demonstrates the importance of having data on the temporal variability of the P–I curve parameters. P_{max} values estimated by the monthly P–I curves are much larger than maximum values obtained with the response surface obtained for the dinoflagellates, and occur in some of the higher diversity periods (Fig. 4). Therefore, it may be speculated that the observed differences were not due solely to temperature effects, already accounted for, but also to community diversity. If this is the case, then niche complementarity may be the explanation.

The results presented here on diversity and production are in contradiction to those of Hector et al. (1999) and Tilman et al. (2001), on grassland communities. Whilst these authors suggest that more diverse communities exhibit larger areal production, the results of this work reveal higher maximal volume integrated production under lower diversity. If niche complementarity is at work in St. André lagoon phytoplankton communities, its effect is reflected not on production but on photosynthetic rates. One might then ask the following question: "Why more productive phytoplankton assemblages are low in diversity?"

The coexistence of several phytoplankton species under a few limiting resources has been known as the 'paradox of plankton'. It has been explained by the non-equilibrium nature of phytoplankton communities (Krebs, 1994). The model of Huisman et al. (1999) suggested that the coexistence of several phytoplankton species may be explained by the internal dynamics of competitive interactions, capable of generating chaos and opportunities for several species to coexist under a number of limiting resources lower than the number of competing species. According to these authors, if the number of limiting factors increases (different factors for different species), there is more room for more species to coexist. The recent modelling study of Yamamoto & Hatta (2004) provides theoretical evidence for the importance of pulsed nutrient supply in increasing phytoplankton diversity. These authors found that nutrient pulses with 'intermediate frequency' (corresponding to a period of 9 days) maximised the survival of modelled species, in line with the 'Intermediate Disturbance Hypothesis' (IDH) (Connell, 1978). These studies may help to justify the diversity of plankton communities, from internal dynamics and/or external forcing, but they do not explain why more diverse communities produce less biomass in spite of being more efficient.

In one of the rare works where phytoplankton diversity and production were analysed together, Agard et al. (1996) found empirical evidence to support Huston's dynamic equilibrium hypothesis (Huston, 1979, 1994) (cf. – Introduction) – species richness of Caribbean phytoplankton appeared to be maximized under intermediate conditions of disturbance and primary production. On one hand, low production reduces recovery from mortality and may therefore reduce species diversity. On the other hand, higher production may lead to lower diversity through competitive exclusion. The parabolic relationship between diversity and MaxProd obtained in the present work agree with those findings (Fig. 9).

Before trying to explain the observed patterns in Santo André lagoon it is important to recognize that comparing the low diverse and more productive winter phytoplankton assemblages with those observed in spring and summer in St. André lagoon is a bit like comparing two different ecosystems. In winter, the lagoon is predominantly fresh water, whereas in spring and summer the opposite is true. In winter, large nutrient inputs due to rainfall, may give opportunity for some species tolerant to low salinity to reach and maintain high biomasses. Therefore, competitive exclusion is more likely to occur. Later in the year, the lower nutrient concentrations may limit biomass growth and production. Furthermore, the number of limiting nutrients is likely to increase, generating adequate conditions for the coexistence of more species, according to Huisman's hypothesis. In summer, when rainfall only rarely occurs, nutrient pulses are limited to phosphorus release from the sediments under episodes of bottom anoxia (Bernardo, 1990).

In terrestrial vegetation, although limiting nutrients may be less than coexisting species (Krebs, 1994), niche diversity may be larger, since soil heterogeneity and the plant canopy itself may provide more environmental diversity than the relatively homogeneous water column environments and therefore more opportunities for more species. This may help to explain the differences observed between terrestrial and pelagic ecosystems, concerning the relationship between diversity and production. In fact, similar differences for similar reasons are likely to occur between pelagic and some benthic ecosystems. Algal beds and reefs are generally associated with high diversity and their areal production is among the highest in marine environments (Fig. 1).

From the results discussed so far, the following conclusions may be drawn regarding the St. André lagoon phytoplankton communities:

(1) Phytoplankton communities with lower diversity are dominated by dinoflagellates, exhibit higher cell numbers, Chl *a* concentrations and production, in conjunction with higher nitrogen and phosphorus concentrations, than communities with higher diversity.
(2) The latter are dominated by diatoms, exhibit the highest photosynthetic rates and efficiencies, related with higher Chl *a* cell contents, water temperature, salinity, diversity, species richness and equitability.
(3) These highest photosynthetic rates are not correlated with either inorganic P or inorganic N. Temperature may explain part of the observed results as well as niche complementarity.
(4) Seasonal changes in photosynthetic parameters do not seem to have a major impact on community production averaged over large time scales, but have a major impact at daily time scales.

Furthermore, the following hypothesis may be defined:

(1) Higher nutrient loads in a relatively homogeneous water column may reduce the number of limiting nutrients increasing competitive pressure and leading to high dominance. Low nutrient loads are more likely to result in limitation by more nutrients.
(2) If different species are limited by two or three different nutrients (one per species) there should be more opportunity for more diversity to develop due to non-equilibrium oscillations within the community, following Huisman et al. (1999). More diverse communities are probably more efficient in utilizing irradiance energy due to niche complementarity.

These hypotheses may be tested by experimental designs similar to those employed in terrestrial ecology (Hector et al., 1999; Tilman et al., 2001). These designs imply measuring community production after the random addition of different species and have been criticised by Huston et al. (2000) among other things, due to the fact that random species addition does not mimic either natural or human-caused processes. Ideally, experiments should compare production and photosynthetic rates of different realistic species assemblages, with similar salinity and temperature tolerances, testing simultaneously for the effects of Chl *a* concentration, that is clearly related to photosynthetic rates (see above), and nutrient additions.

Acknowledgements

The authors wish to thank Maria João Guerreiro and two anonymous reviewers for their helpful comments.

References

A.P.H.A., W.W.A. & W.E.F., 1992. Standard Methods for the Examination of Water and Wastewater, 18th edn. Washington.

Agard, J. B. R., R. H. Hubbard & J. K. Griffith, 1996. The relation between productivity, disturbance and the biodiversity of Caribbean phytoplankton: applicability of Huston's dynamic equilibrium model. Journal of Experimental Marine Biology and Ecology 202: 1–17.

Banse, K., 1982. Cell volumes, maximal growth rates of univellular algae and ciliates, and the role of ciliates in the marine pelagial. Limnology and Oceanography 27: 1059–1071.

Bernardo, J. M., 1990. Dinâmica de uma lagoa costeira eutrófica (Lagoa de Santo André). Ph.D. Dissertation, University of Lisbon, 322 pp.

Brock, T. D., 1981. Calculating solar radiation for ecological studies. Ecological Modelling 14: 1–19.

Cancela da Fonseca, L. M. Q., 1989. Estudo da influência da "abertura ao mar" sobre um sistema lagunar costeiro: A Lagoa de Santo André. Ph.D. Dissertation, University of Lisbon, 355 pp.

Cancela da Fonseca, L. M. Q., A. M. Costa & J. M. Bernardo, 1989. Seasonal variation of benthic and fish communities in a shallow land-locked coastal lagoon (St. André, SW Portugal). Scientia Marina 53: 663–669.

Connell, J., 1978. Diversity in tropical rain forests and coral reefs. Science 199: 1304–1310.

Davison, I. R., 1991. Environmental effects on algal photosynthesis: temperature. Journal of Phycology 27: 2–8.

Dogde, J. D., 1975. The Prorocentrales (dinophyceae). II. Revision of the taxonomy within the genus *Prorocentrum*. Botanical Journal of the Linnean Society 71: 103–125.

Drebes, G., 1974. Marines Phytoplankton. Eine Auswahl der Helgol der Planktonalgen (Diatomeen, Peridieen) Georg Thieme-Verlag, Stuttgard, 123 pp.

Duarte, P., 1995. A mechanistic model of the effects of light and temperature on algal primary productivity. Ecological Modelling 82: 151–160.

Eilers, P. H. C. & J. C. H. Peeters, 1988. A model for the relationship between light intensity and the rate of photosynthesis in phytoplankton. Ecological Modelling 42: 199–215.

Eppley, R. W., 1972. Temperature and phytoplankton growth in the sea. Fisheries Bulletin 70: 1063–1084.

Estrada, M., P. Henriksen, J. M. Gasol, E. O. Casamayor & C. Pedrós-Alió, 2004. Diversity of planktonic photoautotrophic microorganisms along a salinity gradient as depicted by microscopy, flow cytometry, pigment analysis and DNA-based methods. FEMS Microbiology Ecology 49: 281–293.

Gallegos, C. L., 1992. Phytoplankton photosynthesis, productivity, and species composition in a eutrophic estuary: comparasion of bloom and non-bloom assemblages. Marine Ecology Progress Series 81: 257–267.

Platt, T., 1980. Variations in assimilation number of coastal marine phytoplankton: Effects of environmental co-variates. Journal of Plankton Research 2: 249–260.

Hasle, G. R., 1978. The inverted microscope method. In Sournia, A. (ed.), Monographs on Oceanographic Methodology, 6. Phytoplankton Manual. UNESCO, Paris: 148–150.

Hasle, G. R., E. E. Syvertsen, K. A. Steidinger & K. Tangen, 1996. Identifying marine diatoms and dinoflagellates. In Tomas C. R. (ed.), Identifying Marine Phytoplankton. Academic Press, pp. 387–584.

Hector, A., B. Schmid, C. Beierkuhnlein, M. C. Caldeira, M. Diemer, P. G. Dimitrakopoulos, J. A. Finn, H. Freitas, P. S. Giller, J. Good, R. Harris, P. Hogberg, K. Huss-Danell, J. Joshi, A. Jumpponen, C. Korner, P. W. Leadley, M. Loreau, A. Minns, C. P. Mulder, G. O'Donovan, S. J. Otway, J. S. Pereira, A. Prinz & D. J. Read, 1999. Plant diversity and productivity experiments in european grasslands. Science 286: 1123–1127.

Holm-Hansen, O., C. J. Lorenzen, R. W. Holmes & J. H. D. Strickland, 1965. Fluorometric determination of chlorophyll. Journal du Conseil, Conseil permanent International pour l'Exploration de la Mer 30: 3–15.

Huang, L., W. Jian, X. Song, X. Huang, S. Liu, P. Qian, K. Yin & M. Wu, 2004. Species diversity and distribution for phytoplankton of the Pearl River estuary during rainy and dry seasons. Marine Pollution Bulletin 49: 588–596.

Huisman, J., V. Opstveen & F. J. Weissing, 1999. Critical depth and critical turbulence: two different mechanisms for the development of phytoplankton blooms. Limnology and Oceanography 44: 1781–1787.

Huston, M. A., 1979. A general hypothesis of species diversity. American Naturalist 113: 81–101.

Huston, M. A., 1994. Biological Diversity: The Coexistence of Changing Landscapes. Cambridge University Press, Cambridge, 681 pp.

Huston, M. A., L. W. Arssen, M. P. Austin & B. S. Cade, 2000. No consistent effect of plant diversity on productivity. Science 289: 1255a.

ICES C. M., 1996/L:3. Biological Oceanography Committee. Report of the working group on phytoplankton ecology. Ref: C + E + Env: pp. 28–30.

Keller, A. A., 1989. Modelling the effects of temperature, light and nutrients on primary productivity: an empirical and mechanistic approach compared. Limnology and Oceanography 34: 82–95.

Krebbs, C. J., 1994. Ecology: The Experimental Analysis of Distribution and Abundance, 4a edn. Harper Collins College Publishers, 801 pp.

Macedo, M. F., J. G. Ferreira & P. Duarte, 1998. Dynamic behavior of photosynthesis-irradiance curves determined from oxygen production during variable incubation periods. Marine Ecology Progress Series 165: 31–43.

Macedo, M. F., P. Duarte, P. Mendes & J. G. Ferreira, 2001. Annual variation of environmental variables, phytoplankton species composition and photosynthetic parameters in a coastal lagoon. Journal of Plankton Research 23: 719–732.

Muylaert, K. & K. Sabbe, 1999. Spring phytoplankton assemblages in and around the maximum turbidity zone of estuaries of the Elbe (Germany), the Schelde (Belgium/The Netherlands) and the Gironde (France). Journal of Marine Systems 22: 133–149.

Pahl-Wostl, C. & D. M. Imboden, 1990. DYPHORA - a dynamic model for the rate of photosynthesis of algae. Journal of Plankton Research 12: 1207–1221.

Parsons, T. R., Y. Maita & C. M. Lalli, 1984. A Manual of Chemical and Biological Methods for Seawater Analysis . Pergamon Press, N.Y.

Pearl, H. W., 1988. Nuisance phytoplankton blooms in coastal, estuarine, and inland waters. Limnology and Oceanography 33: 823–847.

Peneda, M. C., M. M. Cruces, J. L. Biscaya, M. C. Santos, 1980. Preliminary evaluation of physico-chemical and biological data collected during a yearly cycle in the Sado estuary. In Actual Problems of Oceanography in Portugal. Junta Nacional de Investigação Científica e Tecnológica and NATO Marine Sciences Panel Lisbon Portugal. pp. 171–188.

Portela, L.I. & R. Neves, 1994. Modelling temperature distribution in the shallow Tejo estuary. In Tsakiris, G. & M. A. Santos (eds), Advances in Water Resources Technology and Management. Balkema, Rotterdam: 457–463.

Shaw, P. J. & D. A. Purdie, 2001. Phytoplankton photosynthesis-irradiance parameters in the near-shore UK coastal waters of the North Sea: temporal variation and environmental control. Marine Ecology Progress Series 216: 83–94.

Steeman Neilsen, E., 1952. Inactivation of the photochemical mechanism in photosynthesis as a means to protect cells against to high light intensities. Physiologica Plantarum 15: 161–171.

Tilman, D., P. B. Reich, J. Knops, D. Wedin, T. Mielke & C. Lehman, 2001. Diversity and productivity in a long-term grassland experiment. Science 294: 843–845.

Throndsen, J., 1978. Preservation and storage. In Sournia, A. (ed.), Monographs on Oceanographic Methodology, 6. Phytoplankton Manual. UNESCO, Paris: 69–74.

Underwood, A. J., 1981. Techniques of analysis of variance in experimental marine biology and ecology. Oceanography and Marine Biology Annual Review 19: 513–605.

Whittaker, R. H. & G. E. Likens, 1975. The biosphere and man. In Lieth, H. & R. H. Whittaker (eds), Primary Productivity of the Biosphere. Springer-Verlag, Berlin: 305–328.

Yamamoto, T. & G. Hatta, 2004. Pulsed nutrient supply as a factor inducing phytoplankton diversity. Ecological Modelling 171: 247–270.

Yentsch, C. S. & D. W. Menzel, 1963. A method for the determination of phytoplankton chlorophyll and phaeophytin by fluorescence. Deep Sea Research 10: 221–231.

Hydrobiologia (2006) 555:19–30
H. Queiroga, M.R. Cunha, A. Cunha, M.H. Moreira, V. Quintino, A.M. Rodrigues, J. Serôdio & R.M. Warwick (eds),
Marine Biodiversity: Patterns and Processes, Assessment, Threats, Management and Conservation
DOI 10.1007/s10750-005-1102-8

Ascendency as ecological indicator for environmental quality assessment at the ecosystem level: a case study

J. Patrício[1,*], R. Ulanowicz[2], M. A. Pardal[1] & J. C. Marques[1]
[1]*IMAR- Institute of Marine Research, Department of Zoology, Faculty of Sciences and Technology, University of Coimbra, 3004-517, Coimbra, Portugal*
[2]*Chesapeake Biological Laboratory, Center for Environmental and Estuarine Studies, University of Maryland, Solomons, Maryland, 20688-0038, USA*
*(*Author for correspondence: E-mail: jpatricio@ci.uc.pt)*

Key words: network analysis, ascendency, eutrophication, estuary

Abstract

Previous studies have shown that when an ecosystem consists of many interacting components it becomes impossible to understand how it functions by focussing only on individual relationships. Alternatively, one can attempt to quantify system behaviour as a whole by developing ecological indicators that combine numerous environmental factors into a single value. One such holistic measure, called the system 'ascendency', arises from the analysis of networks of trophic exchanges. It deals with the joint quantification of overall system activity with the organisation of the component processes and can be used specifically to identify the occurrence of eutrophication. System ascendency analyses were applied to data over a gradient of eutrophication in a well documented small temperate intertidal estuary. Three areas were compared along the gradient, respectively, non eutrophic, intermediate eutrophic, and strongly eutrophic. Values of other measures related to the ascendency, such as the total system throughput, development capacity, and average mutual information, as well as the ascendency itself, were clearly higher in the non-eutrophic area. When the whole-system properties of the three areas were compared, however, the values associated with the intermediate eutrophic area turned out to be the lowest, which possibly could be attributed to the unstable nature of this area. The current study provided an example of how the measures arising out of network analysis might lead to an improved understanding of the system functioning and of the eutrophication process itself.

Introduction

Ecology can be defined as the scientific study of the relationships between organisms and their environment; and, in general, can be approached from two directions: (1) as reductionism, wherein each relationship is considered by itself and the results are assembled afterwards; and (2) as holism, whereby the system is considered in its entirety and a search is undertaken to reveal properties at the system level (Jørgensen, 1997).

Previous studies have shown that an ecosystem consists of so many interacting components that it becomes impossible ever to understand how it functions by examining the component relationships in isolation. Often, when individual components of ecosystems are studied via reductionism, the reconstructed ensemble will behave differently than the sum of the parts.

To obviate such problems, one might attempt to describe phenomenologically at least part of the reality of ecosystems structure by developing ecological indicators that combine numerous environmental factors into a single value, in the hope that such an index will be useful in connecting empirical research, modelling and management (Patten et al., 2002 a, b). It is envisioned that such indicators will provide synoptic information about the state of the ecosystems they represent. Most often, they address an ecosystem's structure or functioning and encompass specific aspects or components, such as nutrient concentrations, the pattern of water flows, the diversity of vertebrates or macroinvertebrates, plant diversity, etc. Sometimes, however, they are intended to address the ecological integrity of the system as a whole.

One such holistic measure derives from the analysis of networks of trophic exchanges and is called the system 'ascendency'. Ulanowicz (1980) defines ascendency as an index that quantifies both the level of system activity and the degree of the organization with which it processes material in autocatalytic fashion. The level of activity is measured by the sum of the magnitudes of all the trophic exchanges occurring in the system, or what is called the 'total system throughput' (TST.) The organisation of the flow structure is captured by the average mutual information (AMI) inherent in how the flows are put together (Rutledge et al., 1976.) Ascendency varies jointly as (is the product of) both of these network characteristics (see below.)

Although ascendency is a rather abstract concept, it possesses manifold attributes that are useful in a number of practical circumstances – for example, quantifying the status of an ecosystem (Baird & Ulanowicz, 1989; Halfon et al., 1996; Wolff et al., 1996; Heymans & Baird, 2000; Ray & Ulanowicz, 2000), measuring the effects of perturbations on it (Baird & Heymans, 1996; Almunia et al., 1999), assessing its health and integrity (Constanza, 1992), and comparing ecosystems one with another (Baird & Ulanowicz, 1989; Heymans & Baird, 1995; Baird et al., 1991; Monaco & Ulanowicz, 1997). Recently, the organizational factor in the ascendency has been shown to be equal to the logarithm of the number of effective trophic levels inherent in the system (Zorach & Ulanowicz, 2003.)

An enormous number of studies have shown unambiguously that most European and North American estuaries are affected to some degree by organic pollution and nutrient discharges, often to an extent that gives rise to eutrophication and its linked effects upon resident biota (Diaz & Rosenberg, 1995; Norkko & Bonsdorff, 1996; Flindt et al., 1997; Marques et al., 1997; Weaver et al., 1997; Raffaelli et al., 1998; Cloern, 2001). Using ascendency, it becomes possible to determine quantitatively whether a disturbed system has grown or shrunk, developed or regressed. Furthermore, the process of eutrophication can be defined in terms of its network attributes as any increase in system ascendency (due to a nutrient enrichment) causes a rise in total system throughput which more than compensates for a concomitant fall in the mutual information (Ulanowicz, 1986).

The aim of this study was to test whether this network definition of eutrophication properly tracks changes in community structure along a gradient of eutrophication existing in the south arm of the Mondego estuary (Portugal), a small and well described temperate intertidal estuary (eg. Marques et al., 1997, 2003; Pardal et al., 2000; 2004; Cardoso et al., 2002).

Material and methods

Study area

The Mondego estuary, situated along the western coast of Portugal, is bifurcated into a northern and southern arm, each exhibiting very different hydrological characteristics (Fig. 1). The northern arm is deeper, while the southern arm is silted up, especially in upstream areas, which causes most of the freshwater discharge to flow through the northern arm. This siltation diverts most of the freshwater discharge into the northern arm. As a consequence, the water circulation in the southern arm is dependent mainly on tidal flushing and on a relatively small input of freshwater from the Pranto River, the flow of which is controlled artificially by a sluice.

Macroalgal blooms of *Enteromorpha intestinalis* (Linnaeus) Link and *Enteromorpha compressa* (Linnaeus) Greville have been observed with regularity in the Mondego over the last twenty years (Flindt et al., 1997; Marques et al., 1997, 2003; Lillebø et al., 1999; Pardal et al., 2000, 2004; Martins et al., 2001; Cardoso et al., 2002; Dolbeth et al., 2003). Nevertheless, such macroalgal blooms may not occur in exceptionally rainy years. This is most probably due to the resulting long periods

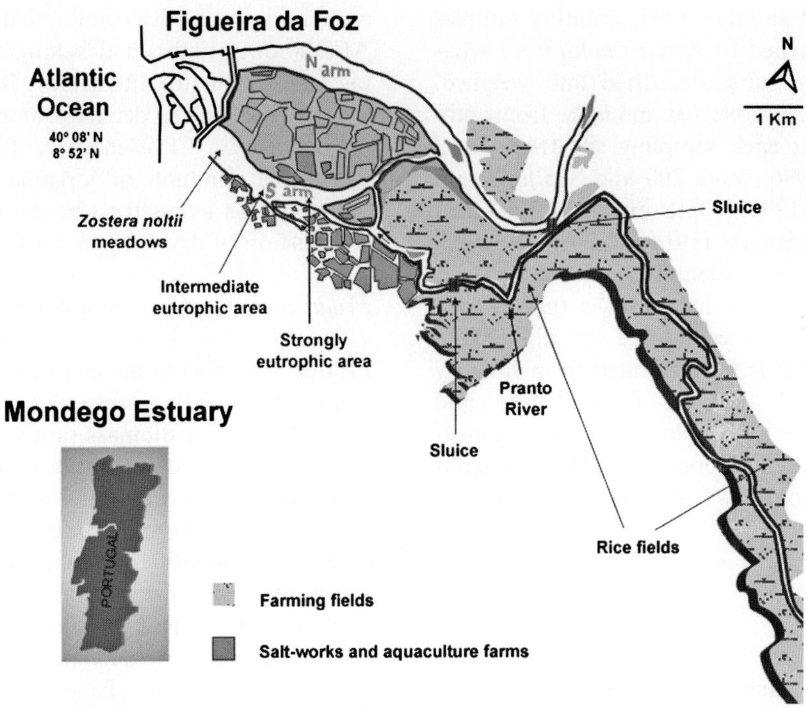

Figure 1. Mondego estuary: location of the sampling stations along a spatial gradient of eutrophication.

that salinity remains below the tolerance limit of macroalgae, coupled with a limitation of phosphorous induced by a heavy nitrogen discharge from the Pranto River (Martins et al., 2001).

Sampling was conducted in three areas in the southern arm of the Mondego estuary that represent different stages along a spatial gradient of eutrophication (Marques et al., 1997, Lillebø et al., 1999; Pardal et al., 2000, 2004; Cardoso et al., 2002; Dolbeth et al., 2003) (Fig. 1): (a) A non eutrophic area (*Zostera noltii* Hornem. beds), (b) an intermediate eutrophic area (*Z. noltii* absent, although residual roots can still be found in the sediment, and the occasional formation of abundant macroalgae mats) and (c) a strongly eutrophic area (macrophyte community totally absent for at least a decade and strong, regularly occurring blooms of *Enteromorpha* spp.).

Methods

Food webs of the ecosystem in the three areas were constructed using the 'Ecopath with Ecosim' software package, which assists the user in casting a balanced carbon budget for each trophic group. The core routine of Ecopath/Ecosim centres around the Ecopath program of Polovina (1984), which has been extended to apply to non-steady-state systems (Christensen & Pauly, 2000). It no longer assumes a steady state but instead calculates parameters on the assumption of mass balance over an arbitrary period – usually one year. When applied, Ecopath derives model parameters on the basis of two master equations, one of which describes the production term and the other which ensures the balance of energy over each compartment. For further details of the Ecopath/Ecosim package see Christensen & Pauly (2000) or visit <http://www.ecopath.org/>.

Sampling program and laboratory treatment

Chlorophyll a, detritus, macroalgae, macrophyte and macrofauna were sampled fortnightly (February 1993–January 1994), during low tide, at each of the three areas. All biological material was identified and separated into the lowest possible taxa (for more details concerning the technical procedures see Pardal et al., 2000, 2002). Between

March 1996 and January 1997, monthly samples of epiphytes attached to *Zostera noltii* were separated from their substrate, dried and weighed. Zooplankton were collected monthly from subsurface waters at each sampling site from April 1995 to April 1996, using 200 and 335 μm mesh nets (Azeiteiro, 1999). Data on fish were taken monthly from January 1991 to December 1992. The captured fish were identified and weighed (wet weight), and the dominant species in the stomach contents were analyzed (Jorge et al., 2002). Finally, wading birds were counted from January 1996 to January 1998 at fortnight-tide and monthly low-water to provide an accurate census across the three areas (Lopes et al., 2002). Seagulls were counted monthly, from November 1993 to July 1994.

Compartments

A different number of ecosystem compartments was identified in each system (Table 1): 43 in the *Zostera* meadows, 36 in the intermediate eutrophic area and 34 in the strongly eutrophic area.

Biomass

Chlorophyll a was estimated according to standard procedures (Strickland & Parsons, 1968) and values were transformed into phytoplankton biomass using a conversion factor taken from Anderson & Williams (1998) and assuming an average depth of 0,5 m over the sampling area. Epiphytes consisted only of the material attached to the aerial part of *Zostera noltii*. Plants and macrofauna were dried at 70 °C for 72 h and weighed. The ash free dry weight (AFDW) of biomass was assessed after combusting samples for 8 h at 450 °C (Pardal et al., 2000, 2002). The abundance of each zooplankton taxon was estimated by multiplying the observed number of that taxon by the average AFDW of an individual belonging to it. The weights of all taxa were summed to arrive at the annual average standing stock. Sixty-two species of fish were observed and were grouped according to their ecological and trophic characteristics. The biomass corresponding to each group was determined by multiplying its wet weight by a conversion factor taken from Jørgensen et al. (1991). The observed density of each bird species was multiplied by the average AFDW of an individual belonging to that taxon (as taken from the literature). Bacterial biomass was assigned to the detritus compartment, as recommended by Christensen & Pauly (1992). Finally, the amount of organic matter in the sediment was assessed to be the weight lost after combustion of dry samples for 8 h at 450 °C.

Production, consumption and diet composition

Production refers to the increase of tissue within a compartment over a given period. Whenever possible, Production/Biomass ratios (P/B), previously calculated for local populations (Allen, 1971), were used. When this was not feasible, values taken from the literature were utilized. Special care was exercised to identify values coming from similar Portuguese estuarine systems.

Consumption is the intake of food by a group over a given interval of time. It was entered into Ecopath as the ratio of consumption to biomass (Q/B). Q/B values for birds and fish were taken from the literature. For the other heterotrophic compartments, the production/biomass ratio was entered into the program to estimate indirectly the Q/B ratio (Hostens & Hamerlynck, 1994).

In a trophic model, such as those constructed using the Ecopath, it is predation that links the different groups into a system. Consumption for one group becomes mortality for another, making information on predation paramount to understanding the dynamics of ecosystems. Unfortunately, quantitative information on diet composition is sparse. Diet information for almost all the compartments identified here had to be obtained from the literature (e.g. Hughes, 1969; Costa, 1982; Pihl, 1985; Zajac, 1986; Sprung, 1994; Ansell et al., 1999; Azeiteiro et al., 1999; Cunha et al., 2000; Pardal et al., 2000). Initially, all the prey items of each compartment of macrofauna and fishes were listed, along with their corresponding percentages of occurrence. Each observed dietary item was then assigned to an ecologically similar species or group of species as identified above. Finally, the percentage of occurrence in the diet was assumed to be proportional to the fraction that its biomass comprised the total biomass of the group. The diets of wading birds and gulls were obtained directly

Table 1. System compartments along the three study areas

Compartments	*Zostera* meadows	Intermediate eutrophic area	Strongly eutrophic area
(1) Phytoplankton	*	*	*
(2) *Enteromorpha* spp (A)	*	*	*
(3) *Ulva* sp (A)	*	*	*
(4) *Gracilaria* sp (A)	*	*	*
(5) *Fucus* sp (A)		*	*
(6) *Zostera noltii* (M)	*		
(7) Epiphytes	*		
(8) Zooplankton	*	*	*
(9) *Hydrobia ulvae* (G)	*	*	*
(10) *Gibula umbilicalis* (G)	*		
(11) *Littorina* spp (G)	*	*	
(12) *Melita palmata* (Am)	*	*	*
(13) *Ampithoe valida* (Am)	*	*	*
(14) *Echinogammarus marinus* (Am)			*
(15) *Corophium multisetosum* (Am)		*	
(16) *Scrobicularia plana* (B)	*	*	*
(17) *Cerastoderma edule* (B)	*	*	*
(18) *Modiolus barbatus* (B)	*		
(19) *Cyathura carinata* (I)	*	*	*
(20) *Idotea chelipes* (I)	*	*	*
(21) *Sphaeroma hookeri* (I)	*		
(22) *Carcinus maenas* (D)	*	*	*
(23) *Crangon crangon* (D)	*	*	*
(24) *Alkmaria romijni* (P)	*	*	*
(25) *Capitella capitata* (P)	*	*	*
(26) *Heteromastus filiformis* (P)	*	*	*
(27) *Hediste diversicolor* (P)	*	*	*
(28) *Diopatra neapolitana* (P)	*		
(29) *Nephtys hombergii* (P)	*	*	
(30) *Lumbrineris impatiens* (P)	*		
(31) Other macrofauna detritivores	*	*	*
(32) Other macrofauna predators	*	*	*
(33) Oligochaets	*	*	*
(34) Microalgae and detritus feeders (F)	*	*	*
(35) Zooplankton consumers (F)	*	*	*
(36) Endofauna consumers (F)	*	*	*
(37) Macrofauna predators (F)	*	*	*
(38) *Trigla lucerna* (F)	*		
(39) *Pomatoschistus minutus* (F)	*		
(40) *Larus ridibundus* (Gu)	*	*	*
(41) *Larus fuscus* (Gu)	*	*	*
(42) *Charadrius alexandrinus* (W)	*	*	*
(43) *Charadrius hiaticula* (W)	*	*	*
(44) *Pluvialis squatarola* (W)	*	*	*
(45) *Calidris alpina* (W)	*	*	*
(46) Detritus	*	*	*

A = macroalgae, M = macrophyte, G = gastropoda, Am = amphipoda, B = bivalvia, I = isopoda, D = decapoda, P = polychaeta, F = fish, Gu = gull, W = wader.

from an analysis of their droppings (Cabral et al., 1999; Moreira, 1995).

Captures

A complete network requires estimates of the rates of export from the system, including the harvests of economically important species. In the present work, the harvests of the bivalve *Scrobicularia plana* da Costa and the polychaete *Hediste diversicolor* (Müller) (the only two species of economic importance) were considered so small as to be negligible.

System indices

Estimates of consumption, production and respiration generated by Ecopath with Ecosim to calculate annual AFDW budgets for each heterotrophic compartment during one year were imported into the NETWRK 4.2a software (Ulanowicz, 1999). The structures of trophic levels and cycling for each network were analyzed, and the overall system properties were calculated using algorithms described by Ulanowicz (1986), Kay et al. (1989) and Monaco & Ulanowicz (1997). Ulanowicz (1986) has characterized the developmental status of an ecosystem network in terms of a set of information-theoretic whole-system indices applied to the flow structure (see Table 2). Such global measurements include:

(1) The aforementioned total system throughput (TST), which is the sum of the magnitudes of all flows occurring in the system. Including consumptions, exports, respirations and flows into detritus. TST is a surrogate for the size of an ecosystem in the same sense that the extent of an economy may be reckoned in terms of the gross domestic product (Kay et al., 1989).
(2) The average mutual information (AMI), which represents the average amount of constraint exerted upon an arbitrary amount of mass as it flows from any one compartment to the next (Ulanowicz, 1997). It also measures the overall degree by which one compartment communicates unambiguously with any other (Rutledge et al., 1976.) Ulanowicz (1986) has suggested that both the number of trophic compartments and the extent of trophic specificity (the relative lack of trophic niche overlap) are embodied in the AMI of the flow connection between compartments. A network with high mutual information is a system with many nodes (compartments) of comparable size that are connected with each other (Baird et al., 1991). System development then becomes any increase in the AMI, which reflects increasing internal constraint within the ecosystem. Such constraints are thought to arise as autocatalytic feedback loops reinforce and increment their component pathways at the expense of other non-participating members.
(3) The ascendency (A), which is the product of (1) by (2). The ascendency is the key index that characterizes the degree of system development and maturity. It incorporates both growth and development into the same index and, simultaneously, measures network's potential for competitive advantage over other network configurations (Ulanowicz, 1986).
(4) The development capacity (C), which functions as a mathematical upper bound on the ascendency. Capacity is measured by the diversity of the flows (calculated using the Shannon-Wiener formula), as normalized by the total system throughput (Ulanowicz & Norden, 1990). The capacity is so named, because it represents the scope of the system for further development.

Table 2. System indices algorithms. T_{pq}, all possible transfers; p and q, arbitrary system component or environment; T_{ij}, trophic exchanges from *taxon i* to *taxon j*; $(n+1)$, import; $(n+2)$, export or dissipation. A dot as subscript indicates summation over that index

System indices	Algorithms
Total system throughput, TST	$\mathrm{TST} = \sum_{p,q} T_{pq}$
Average mutual information, AMI	$\mathrm{AMI} = \sum_{i,j} \frac{T_{ij}}{T_{..}} \log\left[\frac{T_{ij}T_{..}}{T_{i.}T_{.j}}\right]$
Ascendency, A	$A = \sum_{i}\sum_{j} T_{ij} \log\left[\frac{T_{ij}T_{..}}{T_{i.}T_{.j}}\right]$
Development capacity, C	$C = \sum_{i,j} T_{ij} \log\left[\frac{T_{ij}}{T_{..}}\right]$
Overhead, Φ	$\Phi = -\sum_{i,j} T_{ij} \log\left(\frac{T_{ij}^2}{\sum_k T_{kj} \sum_i T_{il}}\right)$
Specific overhead, Φ/TST	$\Phi/\mathrm{TST} = -\sum_{i,j=0}^{n+2} \frac{T_{ij}}{T_{..}} \log\left[\frac{T_{ij}^2}{T_{i.}T_{.j}}\right]$

(5) The system overhead (Ø), which is the (non-negative) amount by which the capacity, *C*, exceeds the ascendency, *A*. (Baird et al., 1991). The overhead is generated by structural ambiguities deriving from multiplicities in the system inputs, exports, dissipations and internal exchanges (functional redundancy). It quantifies the system's residual "freedom" and represents its potential for recovery or innovative restructuring.

(6) The specific overhead of the system (Ø/TST), which is the total flexibility of the system calculated on a per- unit- flow basis. It consists mostly of pathway redundancy, but in open systems it is also augmented by multiplicities in the external inputs and outputs.

Results

Table 3 summarizes the ecological statistic and indices for the three estuarine networks. The sum of consumptions, exports, respiration, production and flow to detritus was always higher in the *Zostera* meadows, followed by the strongly eutrophic area and, finally, by the intermediate eutrophic area. The annual rate of net primary production presented a similar behaviour, clearly related with the primary producers' dynamic in each of the studied areas.

Computed values for the total system throughput, development capacity, average mutual information and ascendency (Table 4) were all clearly higher in the non-eutrophic area, and were followed by those for the strongly eutrophic system. The rankings in specific overhead (Table 4) mirrored those in the redundancy (which comprises the largest component of Ø/TST), with the highest values being calculated for the intermediate eutrophic area.

The energy flow networks pertaining to the non-eutrophic and the strongly eutrophic areas were aggregated into their canonical trophic forms (Fig. 2), otherwise known as the 'Lindeman spine'. Regarding the trophic analysis, the *Zostera* meadows presented one more trophic level than those counted in the strongly eutrophic chain, however this area exhibited lower transfer efficiency at the first trophic level (14.8%).

Concerning the magnitude of recycling activity, the overall percentage of cycled matter, as indicated by the Finn cycling index, increased along with the degree of eutrophication (Table 4). The total number of cycles (Table 4) was the highest in the *Zostera* meadows (74517), followed by the intermediate eutrophic area (15009) and the fewest were counted in the strongly eutrophic area (9164). With regards to the major routes of recycling identified as a result of this study (Fig. 3), it becomes evident that material was flowing over a more complicated web of cycles in the *Zostera* community. It is also evident that the keys species, implicated in the cycle process, changed from one community to the other.

After all, comparing the behaviour of the ascendency with other ecological indicators (Table 5), showed that the heterogeneity (as computed using the Shanon–Wiener index) and the

Table 3. Summary of ecological statistic/ indices for the three estuarine networks

Statistic/indices	*Zostera* meadows	Intermediate eutrophic area	Strongly eutrophic area
Sum of consumption (g AFDW m^{-2} y^{-1})	700.67	297.25	525.72
Sum of exports (g AFDW m^{-2} y^{-1})	1707.51	34.01	169.61
Sum of respiration (g AFDW m^{-2} y^{-1})	2322.55	297.32	612.67
Flow to detritus (g AFDW m^{-2} y^{-1})	2092.06	195.09	522.40
Sum to production (g AFDW m^{-2} y^{-1})	2151	223	494
Net primary production (g AFDW m^{-2} y^{-1})	2014.83	165.58	391.065
Total primary production/total respiration	0.868	0.557	0.638
Total biomass/Total system throughput	0.045	0.061	0.076
Total biomass (no detritus) (g AFDW m^{-2})	304.18	49.91	139.24
Omnivory index	0.105	0.110	0.202

Table 4. Network analysis ecosystem indices for the three areas

Information indices	*Zostera* meadows	Intermediate eutrophic area	Strongly eutrophic area
Total System Throughput (g AFDW m−2 y−1)	10852	1154.8	2612.5
Development Capacity (g AFDW m−2 y−1; bits)	39126	5695.2	10831
Ascendency (%)	42.3	30.4	36.7
Overhead on imports (%)	12.3	8.2	6.2
Overhead on exports (%)	1.3	1.5	2.5
Dissipative overhead (%)	17.7	22.1	19.9
Redundancy (%)	26.4	37.8	34.6
Average Mutual Information (bits)	1.525	1.498	1.522
Specific Overhead	2.080	3.432	2.623
Connectance indices			
Overall connectance	1.672	2.431	2.11
Finn cycling index	5.75E-02	0.2045	0.1946
Total number of cycles	74517	15009	9164

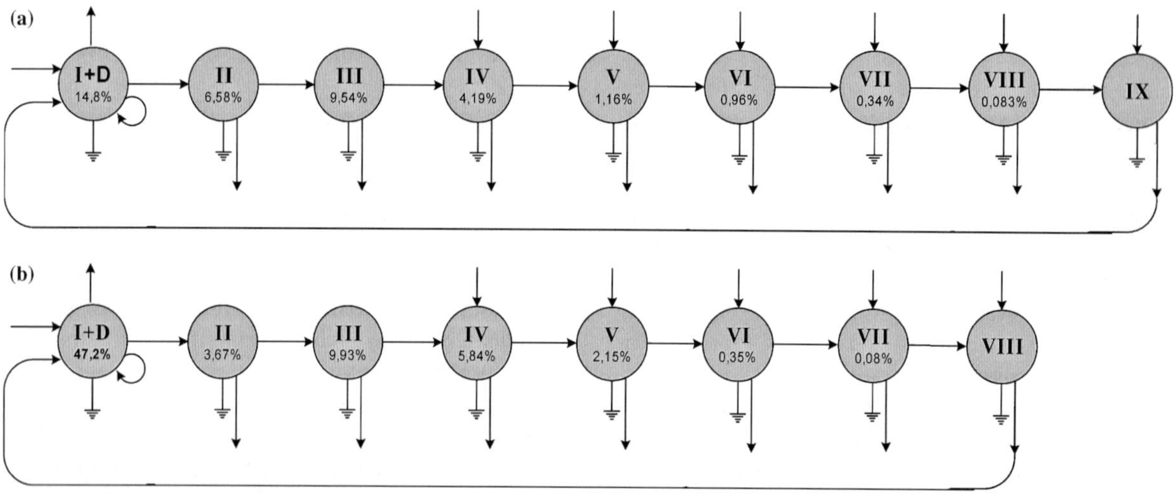

Figure 2. Qualitative representation of linear food chains: (a) *Zostera noltii* meadows and (b) strongly eutrophic area. Flows out of compartment boxes represent exports, flows arriving to compartment boxes represent exogenous system inputs and flows out of the bottom represent respiration. *Level* I + D corresponds to the association of autotrophs (level 1) and Detritus (non-living compartment). The heterotrophic compartments are divided in *levels* II-IX (in a) or *levels* II-VIII (in b), according to their diets.

specific exergy (a thermodynamic measure indicating the amount of work the system can perform on a per- unit- mass basis), both increased with greater eutrophication. On the other hand, species richness, ascendency and total exergy declined with the degree of eutrophication.

Discussion

A long-term study in the Mondego estuary has indicated that years of low precipitation have been associated with reductions in turnover rates and with increases in water column stability, salinity and light penetration (Martins et al., 2001). These changes in habitat conditions favoured the initiation of macroalgal blooms, which then served to depress the previously dominant macrophyte communities (Marques et al., 1997; Martins et al., 2001; Cardoso et al., 2002; Dolbeth et al., 2003). In the intermediate and strongly eutrophic areas, primary production is largely the result of such macroalgal blooms (Marques et al., 1997). As a consequence, production in these two systems

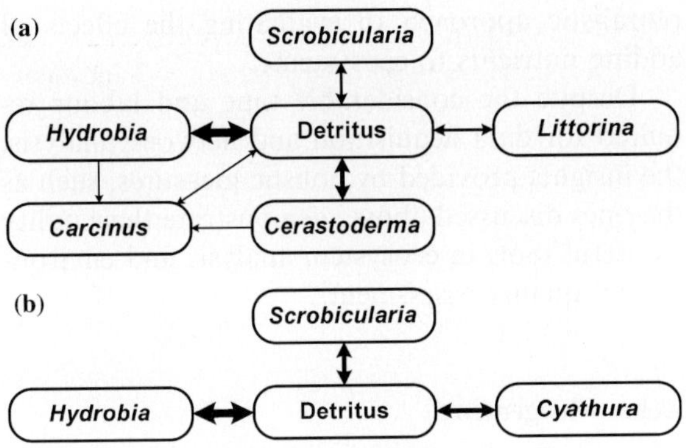

Figure 3. Cycling diagrams: (a) Diagram of those compartments involved in 91% of the cycled flow (g AFDW m^{-2} y^{-1}) in *Zostera noltii* meadows, (b) The compartments involved in 92% of the cycled flow (g AFDW m^{-2} y^{-1}) in the strongly eutrophic ecosystem.

appears as a strong pulse during the course of the blooms, but remains at very low levels for the rest of the year (Dolbeth et al., 2003). The short duration of the abundant primary production in these areas averages over the year to a significantly lower annual rate of net primary production. Odum (1969) had suggested that less-impacted systems (e.g. *Zostera* beds) should exhibit higher rates of net system production – a fact that is consistent with the results of the current study.

When the whole-system properties of the three areas were compared, an interesting pattern emerged. The measures associated with the intermediate eutrophic area did not fall between those calculated at the ends of the gradient. Rather, the intermediate eutrophic system exhibited the lowest values for ascendency, AMI, TST and development capacity and the highest values for redundancy, Ø/TST and FCI (Table 4). That is, by all indications it seemed to be the *most disturbed* of the three systems. Because ascendency is scaled by the flows of material in a system, it is likely to be dominated in the *Zostera* meadows and the strongly eutrophic site by the primary producers – seagrasses and macroalgal mats, respectively. At the intermediate site there is little macroalgal material and no seagrasses, resulting in a lower value for ascendency. A second, and not mutually exclusive, explanation is that the non-disturbed and most disturbed sites host relatively stable communities – one dominated by seagrasses and fine sediments, and the other by macroalgal mats and coarser material. When the seagrasses are lost, however, there is a coarsening of the sediments, which makes it very difficult for seagrasses to re-invade. The reason for the depressed values found in the intermediate eutrophic area, therefore, appears to lie in its unstable nature (Marques et al., 2003).

The canonical form reveals how the ecosystem is functioning in terms of its embedded, distinct trophic levels. The *Zostera* beds (Fig. 2a) are seen to have an additional trophic level beyond those visible in the strongly eutrophic chain, implying that this community possesses a more developed web with additional top consumers. The unimpacted community, however, presented a lower transfer efficiency at the first trophic level, probably because the macrophyte production usually cannot be consumed directly, but needs first to be decomposed (Lillebø et al., 1999). The intermediate eutrophic area was the most effective one in transferring material (8.9%) at the second trophic level.

Although the Finn cycling index, increased along with the degree of eutrophication, the structure of cycling changed dramatically between the unimpacted and the eutrophic systems. The total number of cycles decreased along the gradient, due to the tendency for a larger number of cycles to be found among systems possessing more

Table 5. Ecological indicators along the eutrophication gradient

Ecological indicators		*Zostera* meadows	Intermediate eutrophic area	Strongly eutrophic area
Biodiversity	Species diversity	1.80	1.51	1.21
	Heterogeneity	0.99	1.46	1.52
Specific Exergy		85.70	150.29	165.42
Exergy		25364	4789	8547
Ascendency		42.3	30.4	36.7

compartments at higher trophic levels. Looking at the major routes of recycling, it becomes evident that material was flowing over a more complicated web of cycles in the *Zostera* community. Odum (1969) has suggested that mature ecosystems recycle a greater percentage of their constituent material and energy than do pioneer or disturbed communities. Hence, according to Odum, the progressive increase in the FCI would suggest a maturation of the ecosystem. It has been observed, however, that disturbed systems also often exhibit greater degrees of recycling. The speculation is that such increase in cycling in disturbed systems is the homeostatic response that maintains in circulation resources which before the perturbation had been stored as biomass in the higher organisms (Ulanowicz, 1984; Ulanowicz & Wulff, 1991). This latter scenario seems consistent with the present results.

In the light of these results, the network definition of eutrophication (Ulanowicz, 1986) does not appear to accord with the gradient in eutrophication in the Mondego estuarine ecosystem. Rather, it would seem more accurate to describe the effects of eutrophication process in this ecosystem in terms of a disturbance to system ascendency caused by an intermittent supply of excess nutrients that, when coupled with a combination of physical factors (e.g. salinity, precipitation, etc), causes *both* a decrease in system activity *and* a drop in the mutual information of the flow structure. Even though a significant rise in the total system throughput does occur during the period of the algal bloom and does at that time give rise to a strong increase of the system ascendency, the longer-term, annual picture suggests instead that the non-bloom components of the intermediate and strongly eutrophic communities were unable to accommodate the pulse in production. The overall result was a decrease in the annual value of the system TST and, as a consequence, of the annual ascendency as well.

Comparing the behaviour of the ascendency with the more traditional ecological indicators revealed that the heterogeneity and the specific exergy both increased monotonically with greater eutrophication (Marques et al., 1997). Species richness, ascendency and total exergy; however, declined, as expected, with the degree of eutrophication. These results argue in favour of using a pluralistic approach to evaluating the effects of adding nutrients to ecosystems.

Despite the considerable time and labour required for data acquisition and network analysis, the insights provided by holistic measures, such as the ones discussed above, demonstrate their utility as useful tools in ecosystem analysis and environmental quality assessment.

Acknowledgements

The present study was carried out in the scope of the research project DYNAMOD (POCTI/M6S137431/2001) and supported by FCT (Portuguese National Board of Scientific Research) through a grant SFRH/BD/820/2000, and by the IMAR-Institute of Marine research /FLAD Grants program. The second author received support through a grant from the National Science Foundation's Biocomplexity Program (Contract No. DEB–9981328).

References

Allen, P. M., 1971. Relation between production and biomass. Journal of Fisheries Research Board of Canada 28: 1573–1581.

Almunia, J., G. Basterretxea, J. Aristegui & R. E. Ulanowicz, 1999. Benthic-pelagic switching in a coastal subtropical lagoon. Estuarine, Coastal and Shelf Science 49: 363–384.

Anderson, T. R., P. J. le & B. Williams, 1998. Modelling the seasonal cycle of dissolved organic carbon at station E1 in the English Channel. Estuarine, Coastal and Shelf Science 46: 93–109.

Ansell, A. D., C. A. Comely & L. Robb, 1999. Distribution, movements and diets of macrocrustaceans on a Scottish sandy beach with particular reference to predation on juvenile fishes. Marine Ecology Progress Series 176: 115–130.

Azeiteiro, U. M., J. C. Marques & P. Ré, 1999. Zooplankton annual cycle in the Mondego river estuary (Portugal). Arquivos do Museu Bocage III: 239–264.

Baird, D. & R. E. Ulanowicz, 1989. The seasonal dynamics of the Chesapeak Bay ecosystem. Ecological Monographs 59: 329–364.

Baird, D., J. M. MacGlade & R. E. Ulanowicz, 1991. The comparative ecology of six marine ecosystems. Philosophical Transactions of the Royal Society of London 333: 15–29.

Baird, D. & J. J. Heymans, 1996. Assessment of ecosystem changes in response to freshwater inflow of the Kromme River Estuary, St Francis Bay, South Africa: a network analysis approach. Water SA 22: 307–318.

Cabral, J. A., M. A. Pardal, R. J. Lopes, T. Múrias & J. C. Marques, 1999. The impact of macroalgal blooms on the use of the intertidal area and feeding behaviour of waders (Charadrii) in the Mondego estuary (West Portugal). Acta Oecologica 20(4): 417–428.

Cardoso, P. G., A. I. Lillebø, M. A. Pardal, S. Ferreira & J. C. Marques, 2002. The effect of different primary producers on Hydrobia ulvae population dynamics. A case study in a temperate intertidal estuary. Journal of Experimental Marine Biology and Ecology 277(2): 173–195.

Christensen V. & D. Pauly, 1992. A guide to the Ecopath II program (version 2.1). ICLARM Software. pp. 6–72.

Christensen V., C. J. Walters & D. Pauly, 2000. Ecopath with Ecosim: a User's Guide, October 2000 Edition. Fisheries Centre, University of British Columbia, Vancouver Canada and ICLARM Penang, Malaysia 130 pp.

Cloern, J. E., 2001. Our evolving conceptual model of the coastal eutrophication problem. Marine Ecology and Progress Series 210: 223–253.

Constanza, R., 1992. Toward an operational definition of ecosystem health. In Constanza, R. B. G. Norton, & B. D. Haskell (eds.) Ecosystem health: new goals for environmental management. Island Press, Washington D.C: 269 pp.

Costa M.J., 1982. Contribution à l'étude de l'écologie des poissons de l'estuaire du Tage (Portugal). Thèse P. Grade docteur ès-sciences. Université Paris VII, 256 pp.

Cunha, M. R., M. H. Moreira & J. C. Sorbe, 2000. The amphipod *Corophium multisetosum* (Corophiidae) in Ria de Aveiro (NW Portugal).II.Abundance, biomass and production. Marine Biology 137: 651–660.

Diaz, R. J. & R. Rosenberg, 1995. Marine benthic hypoxia: a review of its ecological effects and the behavioural responses of benthic macrofauna. Oceanography and Marine Biology:anAnnual Review 33: 245–303.

Dolbeth M., M. A. Pardal, A. I. Lillebø, U. Azeiteiro & J. C. Marques, 2003. Short- and long- term effects of eutrophication on the secondary production of an intertidal macrobenthic community. Marine Biology 143: 1229–1238.

Flindt, M. R., L. Kamp-Nielsen, J. C. Marques, M. A. Pardal, M. Bocci, G. Bendoricchio, S. N. Nielsen & S. E. Jørgensen, 1997. Description and comparation of the three shallow estuaries: Mondego river (Portugal), Roskield Fjord (Denmark) and the lagoon of Venice (Italy). Ecological Modelling 102: 17–31.

Halfon, E., N. Schito & R. E. Ulanowicz, 1996. Energy flow through the Lake Ontario food web: conceptual model and attempt at mass balance. Ecological Modelling 86: 1–36.

Heymans, J. J. & D. Baird, 1995. Energy flow in the Kromme estuarine ecosystem, St. Francis Bay, South Africa. Estuarine, Coastal and Shelf Science 41: 39–59.

Heymans, J. J. & D. Baird, 2000. A carbon flow model and network analysis of the northern Benguela upwelling system, Namibia. Ecological Modelling 126: 9–32.

Hostens, K. & O. Hamerlynck, 1994. The mobile epifauna of the soft bottoms in the subtidal Oosterscheld Estuary: structure, function and impact of the storm-surge barrier. Hydrobiologia 282/283: 479–496.

Hughes, R. N., 1969. A study of feeding in *Scrobicularia plana*. Journal of the Marine Biological Association of the United Kingdom 49: 805–23.

Kay, J. J., L. A. Graham & R. E. Ulanowicz, 1989. A detailed guide to network analysis. In Wulff, F. J. G. Field, & K. H. Mann (eds.) Network Analysis in Marine Ecosystems: Methods and Applications. Springer- Verlag, Heidelberg: 15–61.

Jorge, I., C. C. Monteiro & G. Lasserre, 2002. Fish community of Mondego estuary: space-temporal organisation. In Pardal, M. A. J. C. Marques, & M. A. Graça (eds.) Aquatic Ecology of the Mondego River Basin. 2 Global Importance of Local Experience, Imprensa da Universidade de Coimbra: 199– 219.

Jørgensen, S. E., 1997. Integration of Ecosystem Theories: A Pattern (2nd ed.). Kluwer Academic Publishers, The Netherlands 388 pp.

Jørgensen, S. E., S. N. Nielsen & L. A. Jørgensen, 1991. Handbook of Ecological Parameters and Ecotoxicology. Elsevier Science Publishers, Amsterdam, The Netherlands 1263 pp.

Lillebø, A. I., M. A. Pardal & J. C. Marques, 1999. Population structure, dynamics and production of Hydrobia ulvae (Pennant) (Mollusca: Prosobranchia) along an eutrophication gradient in the Mondego estuary (Portugal). Acta Oecologica 20(4): 289–304.

Lopes, R. J, J. A. Cabral, C. Pacheco, T. Múrias & J. C. Marques, 2002. Status and habitat use of waders in the Mondego estuary. In Pardal., M. A. J. C. Marques, & M. A. Graça (eds.) Aquatic Ecology of the Mondego River Basin. 2 Global Importance of Local Experience, Imprensa da Universidade de Coimbra: 219–230.

Marques, J. C., M. A. Pardal, S. N. Nielsen & S. E. Jørgensen, 1997. Analysis of the properties of exergy and biodiversity along an estuarine gradient of eutrophication. Ecological Modelling 102: 155–167.

Marques, J. C., S. N. Nielsen, M. A. Pardal & S. E. Jørgensen, 2003. Impact of eutrophication and river management within a framework of ecosystem theories. Ecological Modelling 166(1–2): 147–168.

Martins, I. I., M. A. Pardal, A. I. Lillebø, M. R. Flindt & J. C. Marques, 2001. Hydrodynamics as a major factor controlling the occurrence of green macroalgal blooms in a Eutrophic estuary. A case study on the influence of precipitation and river management. Estuarine, Coastal and Shelf Science 52: 165–177.

Monaco, M. E. & R. E. Ulanowicz, 1997. Comparative ecosystem trophic structure of three U.S.mid-Atlantic estuaries. Marine Ecology Progress Series 161: 239–254.

Moreira F. M., 1995. A utilização das zonas entre-marés do estuário do Tejo por aves aquáticas e suas implicações para os fluxos de energia na teia trófica estuarina. PhD Thesis, University of Lisbon.

Norkko, A. & E. Bonsdorff, 1996. Rapid zoobenthic community responses to accumulations of drifting algae. Marine Ecology Progress Series 131: 143–157.

Pardal, M. A., J. C. Marques, I. Metelo, A. I. Lillebø & M. R. Flindt, 2000. Impact of eutrophication on the life cycle, population dynamics and production of Ampithoe valida (Amphipoda) along an estuarine spatial gradient (Mondego estuary, Portugal). Marine Ecology Progress Series 196: 207–219.

Pardal, M. A., Marques, J. C. Graça, M. A. (eds.) 2002. Aquatic Ecology of the Mondego River Basin. Global Importance of Local Experience, Imprensa da Universidade de Coimbra 576 pp.

Pardal, M. A., P. G. Cardoso, J. P. Sousa, J. C. Marques & D. Raffaelli, 2004. Assessing environmental quality: a novel approach. Marine Ecology Progress Series 267: 1–8.

Patten B. C., B. D. Fath, J. S. Choi, S. Bastianoni, S. R. Borret, S. Brandt-Williams, M. Debeljak, J. Fonseca, W. E. Grant, D. Karnawati, J. C. Marques, A. Mosere, F. Müller, C. Pahl-Wostl, R. Seppelt, W. H. Steinborn & Yu. M. Svirezhev, 2002 a. Complex Adaptive Hierarchical Systems – chapter 3. In Costanza, B. & S. E. Jorgensen (eds.), Understanding and Solving Environmental Problems in the 21st Century, Elsevier: 41–94.

Patten, B. C., B. D. Fath, J. S. Choi, S. Bastianoni, S. R. Borret, S. Brandt-Williams, M. Debeljak, J. Fonseca, W. E. Grant, D. Karnawati, J. C. Marques, A. Mosere, F. Müller, C. Pahl-Wostl, R. Seppelt, W. H. Steinborn & Yu. M. Svirezhev, 2002 b. Complex Adaptive Hierarchical Systems – chapter 4. In Costanza, B. & S. E. Jorgensen (eds), Understanding and Solving Environmental Problems in the 21st Century, Elsevier: 95–99.

Pihl, L., 1985. Food selection and consumption of mobile epibenthic fauna in shallow marine areas. Marine Ecology Progress Series 22: 169–179.

Polovina, J. J., 1984. Model of a coral reef ecosystem I. The ECOPATH model and its application to French Frigate Shoals. Coral Reefs 3: 1–11.

Raffaelli, D., J. A. Raven & L. J. Poole, 1998. Ecological impact of green macroalgal blooms. Oceanography and Marine Biology: an Annual Review 36: 97–125.

Ray, S., R. E. Ulanowicz, N. C. Majee & A. B. Roy, 2000. Network analysis of a benthic food web model of a partly reclaimed island in the sundarban mangrove ecosystem, India. Journal of Biological Systems 8(3): 263–278.

Rutledge, R. W., B. L. Basorre & R. J. Mulholland, 1976. Ecological stability: an information theory viewpoint. Journal of Theoretical Biology 57: 355–371.

Sprung, M., 1994. Macrobenthic secondary production in the intertidal zone of the Ria Formosa- a lagoon in southern Portugal. Estuarine, Coastal and Shelf Science 38: 539–558.

Strickland, J. D. & T. R. Parsons, 1968. A practical hand-book of seawater analysis. Bulletin of Fisheries Research Board of Canada 167: 1–311.

Ulanowicz, R. E., 1980. An hypothesis on the development of natural communities. Journal of Theoretical Biology 85: 223–245.

Ulanowicz, R. E., 1984. Community measures of marine food networks and their possible applications. In Fasham, M. J. R. (ed.) Flows of Energy and Materials in Marine Ecosystems. Plenum, London: pp. 23–47.

Ulanowicz, R. E., 1986. A phenomenological perspective of ecological development. In Poston, T. M. & R. Purdy (eds.) Aquatic Toxicology and Environmental Fate. Vol. 9, ASTM STP 921. American Society for Testing and Materials, Philadelphia, USA: 73–81.

Ulanowicz, R. E. & F. Wulff, 1991. Comparing ecosystems structures: the Chesapeake Bay and the Baltic Sea. In Cole, J., G. Lovett & S. Findlay (eds.), Comparative Analyses of Ecosystems: Patterns, Mechanisms and Theories. Springer Verlag: 140–166.

Ulanowicz, R. E., 1997. Ecology, the Ascendent Perspective. Columbia University Press, New York 201 pp.

Ulanowicz, R. E., 1999. NETWRK 4.2a: A package of computer algorithms to analyse ecological flow networks. Solomons, MD, US.

Ulanowicz, R. E. & J. S. Norden, 1990. Symmetrical overhead in flow and networks. International Journal of Systems Science 21(2): 429–437.

Weaver, M. J., J. J. Magnuson & M. K. Clayton, 1997. Distribution of littoral fishes in structurally complex macrophytes. Canadian Journal of Fisheries and Aquatic Sciences 54: 2277–2289.

Wolff, M., H. J. Hartmann & V. Koch, 1996. A pilot trophic model for Golfo Dulce, a fjord-like tropical embayment, Costa Rica. Revista de Biologia Tropical 44(3): 215–231.

Zajac, R. M., 1986. The effects of intra-specific density and food supply on growth and reproduction in an infaunal polychaete, *Polydora ligni* Webster. Journal of Marine Research 44: 339–359.

Zorach, A. C. & R. E. Ulanowicz, 2003. Quantifying the complexity of flow networks: How many roles are there?. Complexity 8(3): 68–76.

Hydrobiologia (2006) 555:31–40
H. Queiroga, M.R. Cunha, A. Cunha, M.H. Moreira, V. Quintino, A.M. Rodrigues, J. Serôdio & R.M. Warwick (eds),
Marine Biodiversity: Patterns and Processes, Assessment, Threats, Management and Conservation
DOI 10.1007/s10750-005-1103-7

Phytoplankton seasonality in a highly stratified karstic estuary (Krka, Adriatic Sea)

I. Cetinić[1,]*, D. Viličić[1], Z. Burić[1] & G. Olujić[2]
[1]*Division of Biology, Faculty of Science, University of Zagreb, Rooseveltov trg 6, PO Box 333, Zagreb, Croatia*
[2]*Hydrographic Institute of the Republic of Croatia, Split, Croatia*
*(*Author for correspondence: E-mail: icetinic@botanic.hr)*

Key words: phytoplankton, estuary, Krka, Adriatic Sea, seasonality

Abstract

The abundance of phytoplankton and the composition of nutrients were analysed at three stations in the highly stratified karstic Krka Estuary (east Adriatic coast), in the period March 2000–February 2001. The phytoplankton assemblages were analysed by cluster analysis (average linkage), applied to the Bray–Curtis dissimilarity index computed on density data. Seven groups were identified, which can be combined into two major groups. The spring period was influenced by high riverine water inflow and characterised by the development of cocolithophorids and autotrophic nanoplankton, and small diatoms in winter. In the summer–autumn period domination of one large group occurred, comprising dinoflagellates (micro and nano) and colonial diatoms. Three small groups were also present in that period, each with a specific phytoplankton composition. Variance explained by the Canonical Correspondence Analysis (CCA) of the time-position data was low but corresponded with clustering of samples. CC analysis revealed temperature and salinity as the most important indicators of riverine water impact that influenced species composition and phytoplankton seasonality. The nutrient regime was highly influenced by river/seawater exchange, except in the case of orthophosphates, which showed no correlation with salinity.

Introduction

Phytoplankton is the main representative of primary production in estuarine ecosystems (Williams, 1981). A number of factors define the role of phytoplankton in estuarine production such as salinity, temperature, light (influenced by turbidity), nutrients, water dynamics and the configuration of the water basin. Also, phytoplankton composition influences various processes such as nutrient recycling, grazing, particle sinking, and food webs.

Previous studies of the Krka estuary concerning the phytoplankton composition (Viličić et al., 1990; Bakran-Petricioli et al., 1999) identified 96 and 136 taxa of microphytoplankton respectively, dominated by marine microphytoplankton. However, there is no explanation of the influence of the hydrographic conditions on the composition and seasonality of the phytoplankton in the Krka estuary.

The aim of the study is to gain more insight into the factors influencing phytoplankton composition and seasonality in this Adriatic salt-wedge estuary.

Study area

The estuary of the karstic river Krka is a salt-wedge, highly stratified estuary (Žutić & Legović, 1987) located in the central part of the eastern Adriatic coast. The estuary is 22 km long, relatively narrow except for two wider parts, Prokljan lake and Šibenik harbour (Fig. 1). It extends from the base the of Skradin waterfalls (travertine barriers) to the mouth at station E5. The sharp halocline is characterised by a salinity gradient greater than 1%/cm. The density structure in the water column comes mainly through salinity difference, while the temperature difference may be neglected (Legović, 1991; Legović et al., 1991a). The yearly nutrient input into the estuary is poor; 55 × 10^6 N mol/yr, 1.8 × 10^6 P mol/yr and 36.4 × 10^6 Si mol/yr (Gržetić et al., 1991). In the upper reach of the estuary, decomposing freshwater phytoplankton, which develops in Visovac Lake, situated above the waterfalls, is the main nutrient source (Viličić et al., 1989; Legović et al., 1996). The second nutrient source is of anthropogenic origin in the lower reach (Šibenik harbour) (Legović et al., 1994). Eutrophication in the upper reaches of the estuary can provoke occasional hypoxic conditions (Legović et al., 1991c).

Materials and methods

Seventy-five water samples were taken in March 2000, then in April, July, June, October and February 2001, at three sampling stations in the Krka estuary; E3 (in the upper estuary), E4a (middle estuary) and E5 in the mouth of the estuary. Station E4a is located in Šibenik harbour, subject to anthropogenic influences (Fig. 1).

Samples were taken, depending of the sampling station and on the thickness of the upper brackish layer, at 0, 1, 1.5, 2, 3, 4, 5, 6, 10, 20 and 40 m. All parameters were measured in aliquots of the same sample. Samples for phytoplankton analysis were preserved in 2% neutralised formaldehyde (final concentration). After sedimentation for 24 h, 50 ml subsamples were analysed, and cell counts obtained by the inverted microscope method (Utermöhl, 1958). Microplankton cells (longer then 20 μm) were counted at magnifications of 200 and 400×. Nanophytoplankton cells (2–20 μm) were counted in 20 randomly selected fields of vision, under magnifications of 400 and 1000×.

Nutrients (phosphate, nitrate, nitrite, ammonium and silicate) and oxygen were measured following the standard procedures (Strickland & Parsons, 1972).

Salinity and temperature were measured using a CTD probe (Sea Bird Electronic Inc, USA). Sea-

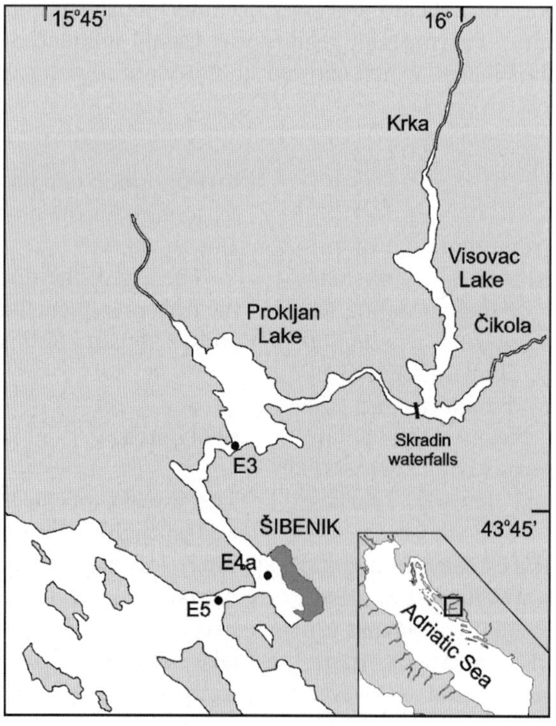

Figure 1. Krka River and estuary with stations E3, E4a and E5.

surface temperature was measured with an inverted thermometer.

Two multivariate methods were applied to analyse the data. Cluster analysis (average linkage) was used to determine the phytoplankton assemblages, applied to the Bray–Curtis dissimilarity index computed on the density data. A logarithmic transformation (log $(x+1)$) was used on the phytoplankton density data to obtain the normal distribution.

Canonical Correspondence Analysis (CCA; Ter Braak, 1995) was used to display environment–phytoplankton relationships.

Results

The highest temperatures were observed in June, in the 2–3 m layer, and the lowest, at the surface in February. Salinity was highest in the marine layer in July and October, and lowest in the brackish layer during April and February (Fig. 2).

Maximum phosphate concentration was observed in the middle estuary in June. The highest values for silicate, nitrate and nitrite were determined in February (at stations E3 and E4a). The highest ammonium concentration was determined at the mouth of the estuary in June. Oxygen saturation was highest in the surface at station E3 in June, and the lowest near the bottom layer in October (Table 1).

Maximum abundances of phytoplankton, both microphytoplankton (1.25×10^{7} cells l^{-1}) and nanophytoplankton (1.16×10^{6} cells l^{-1}), were observed in Šibenik harbour (station E4a), in March and October. High abundance was also observed at station E5 at the same time (Fig. 3).

The total number of determined taxa was 139. In the species data matrix, we list only taxa present in more then two samples throughout the study (78 taxa).

Cluster analysis of the phytoplankton assemblage (mainly identified to species level) identified seven groups of different size (Fig. 4). For every group of phytoplankton, the dominant species are defined, based on abundance and frequency (Table 2). Group A was composed of marine coccolithophorids and smaller colonial diatoms (freshwater and marine), and characterised the spring group (10 samples). Samples from group B

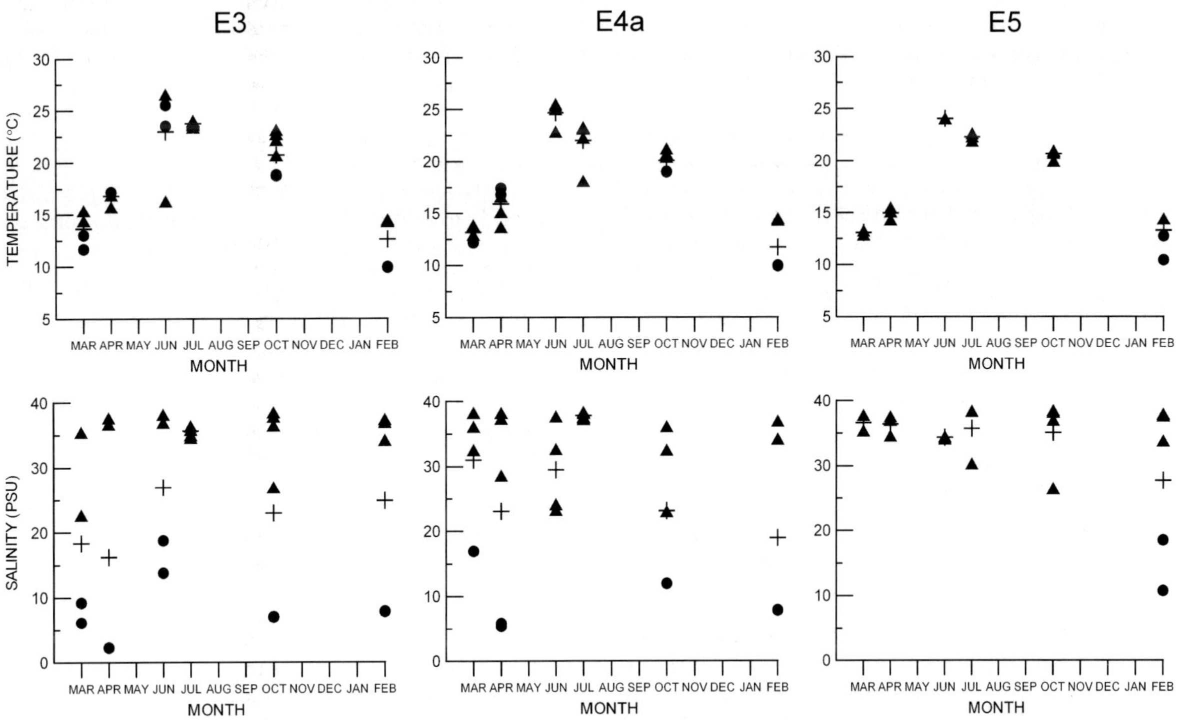

Figure 2. Temperature and salinity values measured at stations E3, E4a and E5. + denotes the average value from the monthly data set, dots represent values measured above halocline and triangles values measured below halocline.

Table 1. Maximum (Max), minimum (Min), average, standard deviation and median for observed orthophosphate (PO_4), silicate (SiO_4), nitrate (NO_3), nitrite (NO_2), ammonium (NH_4) concentrations and dissolved oxygen saturation

Station	E3	E4	E5	E3	E4	E5
	$PO_4^{3-}/\mu M$			$SiO_4/\mu M$		
Max	0.21	0.51	0.23	21.41	21.14	20.73
(Sample No)	(E3OCT20)	(E4JUN1)	(E5JUN0)	(E3MAR2)	(E4FEB2)	(E5FEB0)
Min	0.01	0.02	0.01	2.26	0.27	0.19
(Sample No)	(E3JUL20)	(E4JUL20)	(E5JUL5)	(E3OCT5)	(E4OCT4)	(E5MAR2)
Average	0.06	0.15	0.09	10.85	8.12	4.63
Standard deviation	0.04	0.12	0.07	6.22	7.26	5.65
Median	0.05	0.12	0.08	10.89	5.72	2.42
	$NO_3^-/\mu M$			$NO_2^-/\mu M$		
Max	40.51	38.12	28.19	1.01	0.41	0.31
(Sample No)	(E3FEB2)	(E4FEB2)	(E5FEB0)	(E3OCT20)	(E4FEB2)	(E5FEB0)
Min	0.16	0.07	0.07	0.03	0.01	0.02
(Sample No)	(E3JUL10)	(E4JUL10)	(E5JUL0)	(E3JUL10)	(E4JUL20)	(E5JUL5)
Average	10.06	9.66	5.70	0.20	0.21	0.17
Standard deviation	11.42	11.61	7.42	0.19	0.13	0.11
Median	5.16	4.50	3.03	0.14	0.17	0.19
	$NH_4^+/\mu M$			Dissolved oxygen % saturation		
Max	1.49	1.41	1.94	125.58	125.13	114.75
(Sample No)	(E3JUN1.5)	(E4JUL2)	(E5JUN2)	(E3JUN3)	(E4JUL2)	(E5JUN0)
Min	0.24	0.41	0.38	77.12	88.16	89.27
(Sample No)	(E3APR4)	(E4JUN2)	(E5FEB3)	(E3OCT20)	(E4MAR40)	(E5OCT3)
Average	0.71	0.81	0.95	98.66	103.80	101.20
Standard deviation	0.37	0.29	0.43	11.55	10.29	6.16
Median	0.68	0.77	0.91	97.25	104.41	99.58

Sample names are designated by stations (E3–E5), months (FEB, MAR, APR, JUN, JUL, OCT) and depths (0, 1, 1.5, 2, 3, 4, 5, 6, 10, 20, 40). For example, E3APR4 designates sample taken at E3, in April, from the depth of 4 m.

(12 samples) were taken during the month of April and are dominated by freshwater dinoflagellates, *Gymnodinium* sp, with larger abundances of nanoplankton representatives from chlorophytes, cryptophytes and dinoflagellates.

Group C predominated in summer and autumn, it was composed of colonial diatoms, some larger dinoflagellates, and a nano fraction composed of cryptophytes, smaller dinoflagellates and coccolithophorids.

Exceptions were samples taken in July at greater depths, which were incorporated into group F (5 samples), and dominated by marine diatoms and some freshwater pennate diatoms. Two more groups were defined in the summer–autumn period: group E, with two samples, dominated by *Thalassionema nitzschioides* Grun. and *Dactyliosolen fragilissimus* (Bergon) Hasle, and group D at station E3, in deeper layers, dominated by *Pseudonitzschia* spp. Marine diatoms and freshwater diatoms, as well as smaller sized chlorophytes and cryptophytes, dominated group F – the winter group (14 samples).

The CCA sample biplot (Fig. 5) reveals separation along axis 1 into two bigger groups. Axis 1 explains 6.3% of the variance in species data (eigenvalue (e.v.) = 0.257), while axis 2 explains 4.6% (e.v. = 0.186). The left side of the ordination is occupied by winter and spring samples, represented by groups A, B and D, except for the sample E4MAR40. The right side of the ordination is occupied by the summer–autumn samples (groups C, D, E, and G) with the exception of sample E3OCT2. The length of the environmental variable arrows shows the correlation with the ordination axes, and therefore the relative power

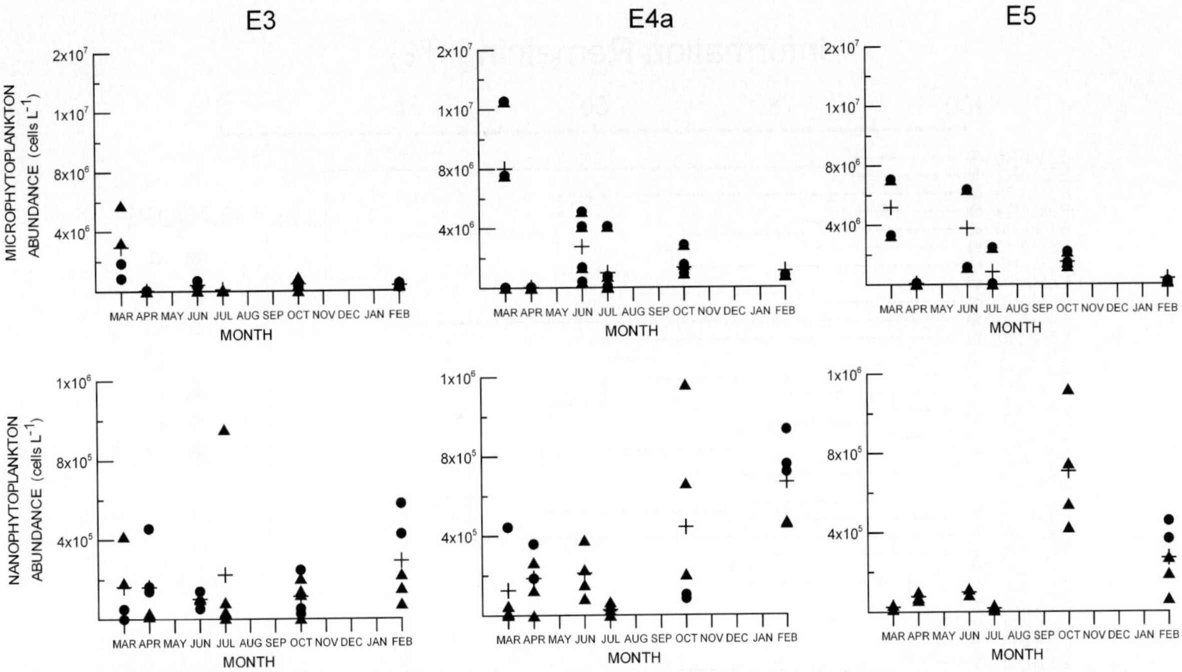

Figure 3. Microphytoplankton and nanophytoplankton abundances measured on stations E3, E4a and E5. + denotes the average value from the monthly data set, dots represent values measured above halocline and triangles values measured below halocline.

of each variable to explain the position of the sample within the ordination, while direction of the arrow shows the gradient of that environmental variable. The CCA diagram demonstrates a significant positive correlation of temperature ($r=0.803$), and strong negative correlation of nitrates ($r=-0.790$) with the first CCA axis. The strongest correlation of the second CCA axis is with salinity ($r=0.461$). Orthophosphate, ammonium and oxygen saturation shows low correlation with both of the CCA axis.

Discussion

Nutrient content is low, in accordance with results from the Krka estuary published earlier (Gržetić, 1990; Legović et al., 1994). The maximum phosphate concentration of 0.51 μM, recorded in Šibenik harbour (station E3), was due to the anthropogenic input and higher than the value recorded earlier (0.4 μM). Mean concentrations of nitrates and silica decreased from the upper to the lower estuary and these values are lower (nitrates by 60%, silicate by 80%) than those recorded before.

The highest values of nitrate and silicate concentrations occurred during the winter/spring period, and their negative correlation with salinity points to a higher river influence during that period. Phosphorus shows no significant correlation with salinity ($r=-0.04$, $p=0.01$), or temperature ($r=0.208$, $p=0.05$), supporting the fact that estuarine phosphorus is not from one source only but is the product of the decomposition of organic material in the upper reach (Legović et al., 1991b), as well as of anthropogenic origin (Šibenik sewage outfall).

Despite the low variance explained by first two axes in CCA, analysis points to similar seasonal-based sample assemblages as can be seen in cluster analysis.

Group B. with the dominant freshwater *Gymnodinium* sp., as can be seen at the CCA biplot, is mostly associated with high silicate, which also points to higher river input during April (Fig. 6). In addition, most of the samples from the winter group (F) and from the spring group (A), especially one from the upper reach of the estuary, are situated on the higher silicate gradient, indicating an important role in fresh-water/river influence on the composition of the winter–spring phytoplankton community in the estuary. Such a higher river input and conse-

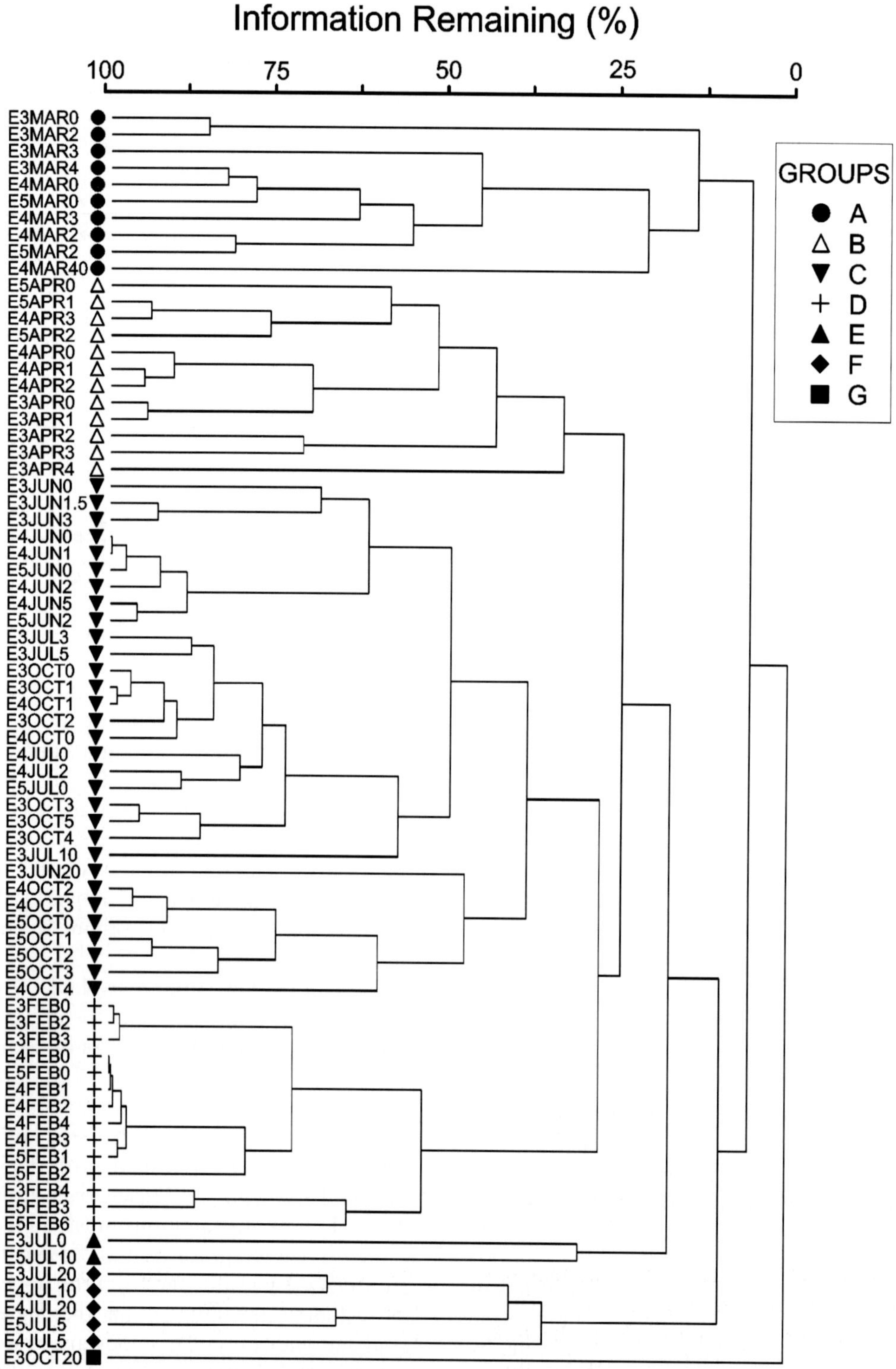

Figure 4. Dendrogram classification of phytoplankton samples produced by cluster analysis (average linkage) using Bray–Curtis distance. Seven sample groups (A, B, C, D, E, F and G) were selected after inspection of the dendrogram, and symbol coded. For abbreviation of the sample names, see Table 1.

Table 2. Dominant species of seven groups (A–G) identified by cluster analysis based on phytoplankton density data in Krka estuary

Species	Mean cell volume μm^3/cell	ORIGIN Freshwater (F), Marine (M)	GROUPS						
			A	B	C	D	E	F	G
Asterionella formosa Hassal	580	F	+					+	
Cocconeis scutellum Ehrenberg	816	M	+						
Leptocylindrus minimus Gran	680	M	+						
Licmophora sp.	ND	M	+						
Skeletonema costatum (Grev.) Cleve	1030	M	+						
Chaetoceros socialis Laud	3140	M	+			+			
Calyptrosphaera oblonga Lohm.	2740*	M	+	+					
Rhabdosphaera tignifer Schiller	2500	M	+	+					
Diatoma elongatum (Lyngb.) Agardh.	600	F		+				+	
Gymnodinium sp.	1350	F		+		+			
Chlorophyta $>10<20$ μm	300	F/M		+		+		+	
Cryptophyta $>10<20$ μm	165	F/M		+	+			+	
Dinoflagellates $>10<20$ μm	ND	M		+	+				
Cerataulina pelagica (Cleve) Hendey	18740*	M				+			
Chaetoceros danicus Cleve	2200	M			+				
Leptocylindrus danicus Cleve	680	M			+				
Pseudonitzschia spp.	2000	M			+				+
Dactyliosolen fragilissimus (Bergon) Hasle	6000	M				+	+		
Thalassiosira sp.	ND	M			+			+	
Prorocentrum micans Ehrenb.	3100*	M			+				
Prorocentrum minimum (Pav.) Schiller	1310*	M			+				
Scripsiella sp.	ND	M			+				
Coccolithophorids <10 μm	79	M			+				
Dinobryon spp.	650	F				+			
Amphiprora sp.	ND	F				+			
Chaetoceros compressus Laud.	3760	M				+			
Guinardia striata (Stolt.) Hasle	15000	M				+			
Thalassionema nitzschioides Grun.	2000	M				+	+		
Chaetoceros affinis Laud.	2300*	M					+		
Chaetoceros sp.	2000	M					+		
Gyrodinium sp.	ND	M					+		
Asterionellopsis glacialis (Castr.) Round	560	M						+	
Chaetoceros decipiens Cleve	3900	M						+	
Chaetoceros simplex Ostenf.	700	M						+	
Nitzschia longissima (Breb.) Ralfs.	80*	M						+	

Origin (freshwater, marine) and mean cell volume are shown. Cell volume marked with * are taken from Viličić (1985); ND – not defined.

quently higher nutrient load, as well as temperature increase, led to the highest microphytoplankton abundance determined along the Krka estuary in March. In spite of that riverine influence, the composition of microphytoplankton during the winter–springtime period in groups A and B is dominated mostly by typical marine winter–spring species such as coccolithophorids *Calyptrosphaera oblonga* Lohm. and *Rhabdosphaera tignifer* Schiller, and smaller marine colonial diatoms. Dominant diatoms *Chaetoceros compressus* Laud. and *Guinardia striata* (Stolt.) Hasle were also present during the spring period in Mali Ston bay, southern Adriatic Sea (Viličić et al.,

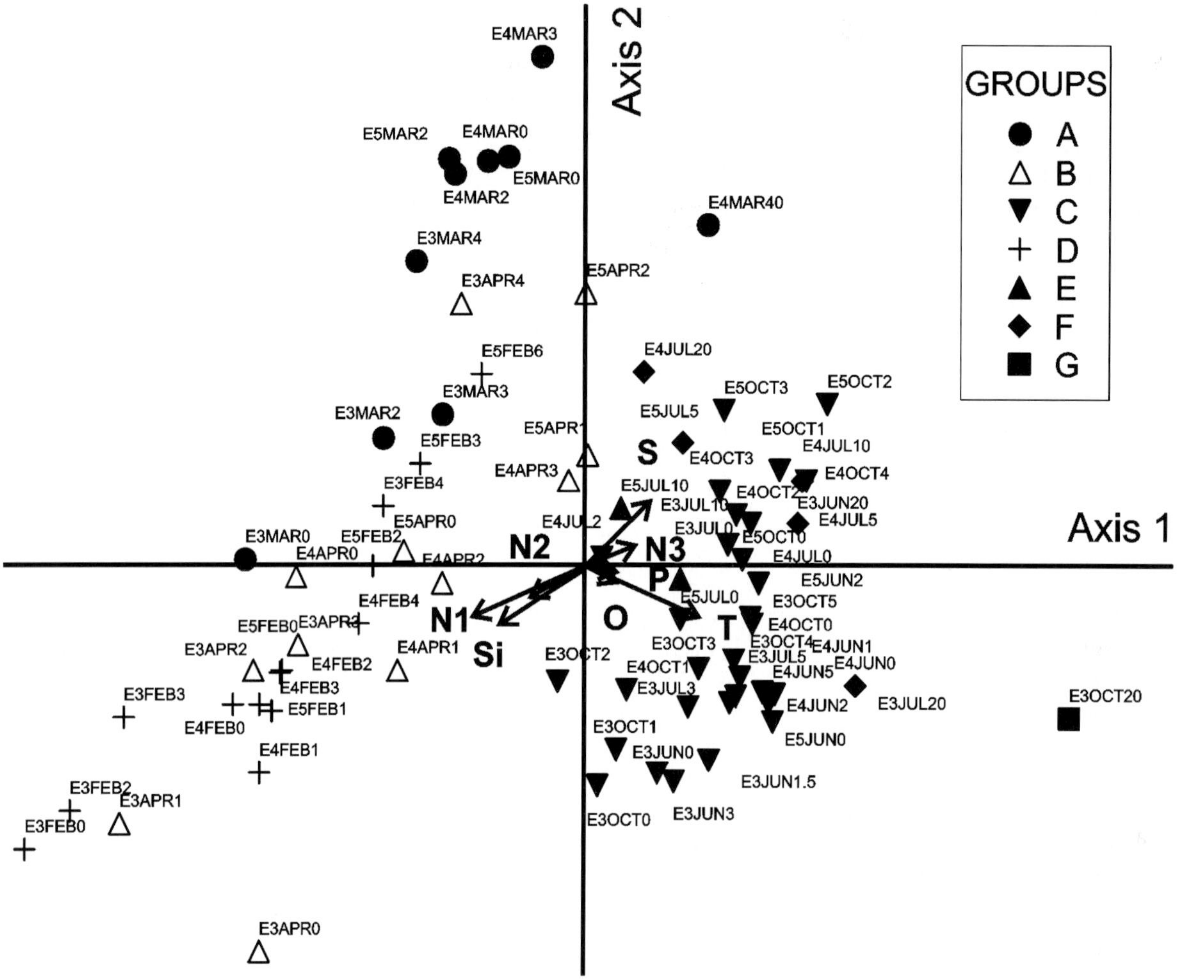

Figure 5. Ordination diagram displaying first and second axis of CCA. Samples are designated, depending on the cluster analysis affiliation, as different symbols as shown in the legend. For abbreviation of the sample names, see Table 1. The environmental variables are displayed as arrows (T – temperature, S – salinity, P – phosphate, O – dissolved oxygen saturation, N_3 – ammonia, N_2 – nitrite, N_1 – nitrate, Si – silica).

1998). Diatoms *Dactyliosolen fragilissimus* and *Cerataulina pelagica* (Cleve) Hendey were found to form localised diatom blooms in the Gulf of Trieste during the same period, after strong freshwater input (Malej et al., 1995). Regarding the nanophytoplankton composition, the presence of chlorophytes in the winter–springtime period, supports the evidence of high river influence.

The Summer–autumn period in Krka estuary was characterised by one major group C and groups E, F, G as the minor ones. In the CCA biplot samples from stations E4a and E5 from that period are linked to the higher phosphate gradient, due to the stronger anthropogenic influence in Šibenik harbour. The summer phytoplankton was composed mainly of marine species.

Group C characterised the period of summer and autumn, with diatoms such as *Chaetoceros danicus* Cleve, *Leptocylindrus danicus* Cleve and some other species of the genus *Chaetoceros*, many dinoflagellates, such as *Prorocentrum micans* Ehrenb., *Prorocentrum minimum* (Pav.) Schiller and *Scrippsiella* sp., were present. A similar situation has been observed in Gulf of Trieste, where a micro fraction of the phytoplankton community in June was dominated by dinoflagellates (Malej et al., 1995).

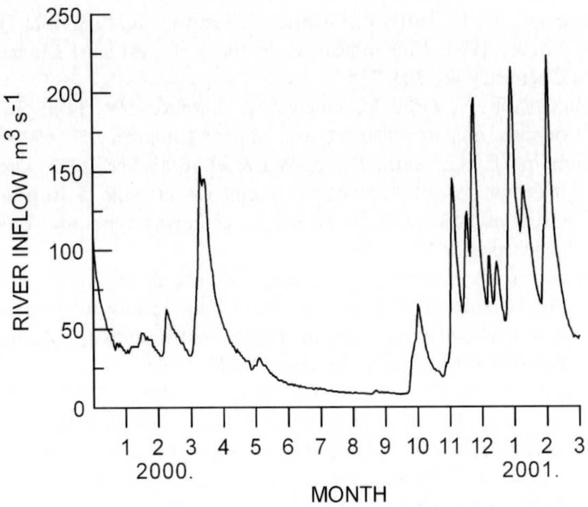

Figure 6. River Krka inflow measured during the sampling period.

The highest and the lowest nanophytoplankton abundances were present in the same period. Nanophytoplankton encountered in those samples mostly comprises dinoflagelates and cryptophytes. Highest abundances of nanophytoplankton are associated with the halocline, as was noticed in previous studies (Denant et al., 1991), because of the accumulation of inorganic and organic material along the halocline. Also, the highest picoplankton biomass in the summer–autumn period in Krka estuary was detected at the halocline (Moreira-Turcq et al., 2001). The lowest abundances of nanophytoplankton were encountered in the deeper layer (E4JUL5, E5JUL10), possibly due to low nutrient accumulation or more intensive microzooplankton grazing pressure bellow the halocline (Kršinić, 1990).

The lowest abundances of nanophytoplankton at station E3 are encountered in group G, representing part of the summer–autumn period in the Krka estuary. The separation of sample E3OCT20, the only representative in group G in the CCA biplot as well as in the cluster dendrogram, can possibly be explained by hypoxia, which usually appears in the autumn (Legović et al., 1991c). In that sample from near the bottom, we determined the lowest oxygen saturation (77.12%), and higher nitrite (1.01 μM) and phosphate (0.209 μM) concentrations. Due to the decomposition of the freshwater phytoplankton (group F with *Diatoma elongatum* (Lyngb.) Agardh. and *Asterionella formosa* Hassal) which comes from Visovac Lake, and regeneration of nutrients, the marine phytoplankton blooms usually occur in the estuary (Legović et al., 1991b). In addition, the presence of the diatom *Pseudonitzschia* spp. as a dominant species in group G, points to a higher level of eutrophication.

Higher nitrite values may appear during the summer–autumn in the Chesapeake Bay (Officer et al., 1984), and higher phosphate in the Ebro River estuary (Sierra et al., 2002) indicating microbiological decomposition of decaying phytoplankton.

Phytoplankton communities in the Krka estuary could be defined in relation to seasonality (winter–spring, summer–autumn) and eutrophication level (either anthropogenic or of natural origin).

Although variance explained in CCA biplot was low, the separation of samples into two bigger groups with temperature as the most important factor, showed a resemblance to the temperature dominated regime in the eastern Adriatic sea, as has been noticed before (Jasprica & Carić, 2001). In estuaries, salinity is the most important factor, but the answer to the possible temperature-dominated regime in the Krka estuary can perhaps be found in geomorphology. A similar interpretation has been offered for the Sado estuary, Portugal (Lemaire et al., 2002). In the Krka estuary, a brackish surface layer permanently makes up about 10% of the water column. In the wider part of the deeper estuary, the seawater exchange time is 50–100 days in winter, and 250 days in summer (Legović, 1991). In such conditions, the phytoplankton community in Krka estuary more resembles the typical coastal marine community under the influence of occasional freshwater inflow (due to riverine input or heavy rain) than an estuarine community. Phytoplankton in the Krka estuary is mostly marine, with during high river input. Freshwater is responsible for some freshwater species coming into the estuary from the freshwater accumulation above the estuary (Visovac Lake), the higher nutrient input, which provokes higher surface abundances, and occasionally phytoplankton blooms. Some of mentioned nutrients such as ammonia originated from the marine zooplankton or microbiological metabolic activity. Phosphate had no correlation with salinity, possibly because the input was not from one origin only, and not connected with seasonal changes governed by temperature or salinity.

Acknowledgements

This study was financially supported by Croatian Ministry of science and technology, (project no. 0119–144). We are thankful to colleagues from the Ruđer Bošković Institute for their collaboration during the research.

References

Bakran-Petricioli, T., D. Petricioli & D. Viličić, 1999. Taxonomic composition and seasonal distribution of microphytoplankton in the Krka river estuary. Natura Croatica 7: 307–319.

Denant, V., A. Saliot & R. F. C. Mantoura, 1991. Distribution of algal chlorophyll and carotenoid pigments in a stratified estuary: the Krka river, Adriatic Sea. Marine Chemistry 32: 285–297.

Gržetić, Z., 1990. Basic hydrologic and chemical properties of Krka estuary (in Croatian). Ph.D. Dissertation, Institute Ruder Boskovic, Zagreb: 162 pp.

Gržetić, Z., R. Precali, D. Deggobis & A. Škrivanić, 1991. Nutrient enrichment and phytoplankton response in an Adriatic karstic estuary. Marine Chemistry 32: 313–332.

Jasprica, N. & M. Carić, 2001. Planktonic diatoms and their relation to environmental factors at three stations in the southern Adriatic, Mediterranean Sea. In Jahn, R., J. P. Kociolek, A. Witkowski, P. Compere (eds), Lange-Bertalot Festschrift: Studies on Diatoms. Gantner Ruggell, Berlin: 513–536.

Kršinić, F., 1990. The planktonic ciliates in the Krka river estuary. In National Park Krka, status of research and problems of ecosystem protection. Croatian Ecological Society, pp. 259–262.

Legović, T., 1991. Exchange of water in a stratified estuary with application to Krka (Adriatic Sea). Marine Chemistry 32: 121–135.

Legović, T., Z. Gržetić & V. Žutić, 1991a. Subsurface temperature maximum in a stratified estuary. Marine Chemistry 32: 163–170.

Legović, T., D. Viličić, D. Petricioli & V. Žutić, 1991b. Subsurface phytoplankton bloom in a stratified estuary. Marine Chemistry 32: 361–374.

Legović, T., D. Petricioli & V. Žutić, 1991c. Hypoxia in a pristine stratified estuary (Krka, Adriatic Sea). Marine Chemistry 32: 347–359.

Legović, T., V. Žutić, Z. Gržetić, G. Cauwet, R. Precali & D. Viličić, 1994. Eutrophication in the Krka estuary. Marine Chemistry 46: 203–215.

Legović, T., V. Žutić, D. Viličić & Z. Gržetić, 1996. Transport of silica in a stratified estuary. Marine Chemistry 53: 69–80.

Lemaire, E., G. Abril, R. De Wit & H. Etcheber, 2002. Distribution of phytoplankton pigments in nine European estuaries and implications for an estuarine typology. Biogeochemistry 59: 5–23.

Malej, A., P. Mozetič, V. Malačič, S. Terzić & M. Ahel, 1995. Phytoplankton responses to freshwater inputs in a small semi-enclosed gulf (Gulf of Trieste, Adriatic Sea). Marine Ecological Progressive Series 120: 111–121.

Moreira-Turcq, P. F., G. Cauwet & J. M. Martin, 2001. Contribution of flow cytometry to estimate picoplankton biomass in estuarine systems. Hydrobiologia 462: 157–168.

Officer, C. B., R. B. Biggs, J. L. Taft, L. E. Cronin, M. A. Tyler & W. R. Boynton, 1984. Chesapeake Bay anoxia: origin development and significance. Science 223: 22–27.

Sierra, J. P., A. Sanchez-Arcilla, J. Del Gonzalez Ro, J. Flos, E. Movellan, C. Mosso, R. Martnez, M. Rodilla, S. Falco & I. Romero, 2002. Spatial distribution of nutrients in the Ebro estuary and plume. Continental Shelf Research 22: 361–378.

Strickland, J. D. H. & T. R. Parsons, 1972. The Practical Handbook of Seawater analysis. Bulletin of Fisheries Research Board of Canada 167: 310.

Ter Braak, C. J. F., 1995. Ordination. In Jongman, R. H. G., C. J. F. Ter Braak O. F. R. van Tongeren, (eds), Data Analysis in Community and Landscape Ecology. Cambridge University Press, pp. 91–173.

Utermöhl, H., 1958. Zur Vervollkommnung der quantitativen Phytoplankton Methodik Mitteilungen. Internationale Vereiningung fuer Theoretische und Angewandte Limnologie 9: 1–38.

Viličić, D., 1985. An examination of cell volume in dominant phytoplankton species of the middle and southern Adriatic waters. Internationale Revue der Gesamten Hydrobiologie 70: 829–843.

Viličić, D., T. Legović & V. Žutić, 1989. Vertical distribution of phytoplankton in a stratified estuary. Aquatic Science 51: 31–46.

Viličić, D., D. Petricioli & N. Jasprica, 1990. Seasonal phytoplankton distribution in the Krka estuary and Visovac lake (in Croatian). In. National Park Krka, status of research and problems of ecosystem protection. Croatian Ecological Society, pp. 317–330.

Viličić, D., N. Jasprica, M. Carić & Z. Burić, 1998. Taxonomic composition and seasonal distribution of microphytoplankton in Mali Ston Bay (eastern Adriatic). Acta Botanica Croatica 57: 1–10.

Williams, P. J., 1981. Primary productivity and heterotrophic activity in estuaries. In Martin, J. M., J. D. Burton & D. Eisma (eds) River inputs to Ocean Systems. UNESCO: p 384.

Žutić, V. & T. Legović, 1987. A film of organic matter at the freshwater/seawater interface of an estuary. Nature 328: 612–614.

Hydrobiologia (2006) 555:41–48
H. Queiroga, M.R. Cunha, A. Cunha, M.H. Moreira, V. Quintino, A.M. Rodrigues, J. Serôdio & R.M. Warwick (eds),
Marine Biodiversity: Patterns and Processes, Assessment, Threats, Management and Conservation
DOI 10.1007/s10750-005-1104-6

Factors controlling phytoplankton blooms in a temperate estuary: nutrient limitation and physical forcing

Renata Pilkaitytė & Artūras Razinkovas*
Coastal Research and Planning Institute, Klaipėda University, H. Manto 84, LT-92294, Klaipėda, Lithuania
*(*Author for correspondence: E-mail: art@corpi.ku.lt)*

Key words: phytoplankton, temperate estuary, nutrient enrichment bioassays

Abstract

A combination of enclosure nutrient enrichment experiments and historical data analysis was used to identify the factors controlling seasonal dynamics and competition of the phytoplankton community in the Curonian lagoon (Southeast Baltic Sea). Experiments using different nutrient (N, P and Si) manipulations were performed in 10-l enclosures for 48 h. Changes in chlorophyll *a* concentrations, inorganic nutrient concentrations, and plankton cell density were monitored. Results revealed that phytoplankton development in the lagoon is strongly affected by ambient physical factors (wind, temperature). Nutrient limitation, however, also plays an important role in seasonal succession mechanisms showing quite distinct seasonal development patterns. Based on the data, available phytoplankton seasonal succession in the Curonian lagoon could be described as composed by three phases corresponding to different domination and regulatory mechanisms.

Introduction

In aquatic environments, phytoplankton growth may be controlled by the supply of limiting nutrients, usually nitrogen or phosphorus. It is generally accepted that phosphorus limits freshwater ecosystems, while phytoplankton in marine waters tends to be nitrogen- (Hecky et al., 1993; Wetzel, 2001), or even iron- (Hutchins, 1995; Kawaguchi et al., 1997) limited. However, the evidence that marine systems are in fact nitrogen-limited is not nearly as strong as the evidence that freshwaters are phosphorus-limited. On the other hand, due to anthropogenic impact the phytoplankton in freshwaters increasingly tends to be nitrogen-limited (Sommer, 1989a; Aldridge et al., 1993). The Curonian lagoon is a brackish water estuarine system that receives inputs from the Baltic Sea and freshwater runoff. Therefore, the lagoon may be N- or P-limited or may shift between these states depending on the volume of inputs from marine vs. terrestrial sources.

The aim of this study was to delineate factors controlling phytoplankton development in a coastal lagoon and their interaction at distinct periods of the seasonal succession. Enclosure nutrient enrichment experiments were conducted to test the seasonal limitation patterns, while data analysis of historical time series was employed to elucidate the role of ambient physical factors (temperature and wind).

Materials and methods

This study was carried out in the Curonian lagoon – a shallow transitory freshwater basin, connected to the south-eastern Baltic Sea through the narrow Klaipėda strait (Fig. 1). The area of the lagoon is 1584 km^2; its mean depth is 3.8 m, with maximum depths of 5.8 m in the southern region and 10–12 m in the artificially deepened Klaipėda strait harbour area. The southern and central portions of the lagoon contain freshwater

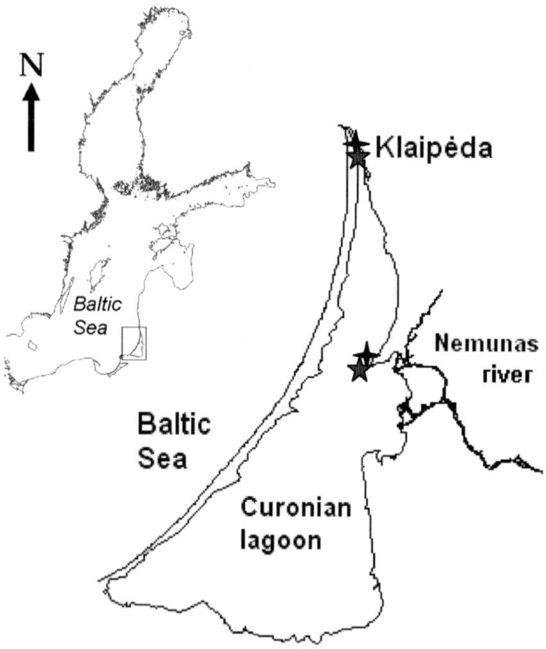

Figure 1. Experimental site locations in the Curonian lagoon. Stars indicate enclosure experimental sites, stars – MRC nutrient monitoring stations.

due to discharge from the Nemunas and other smaller rivers, while the salinity in the northern part varies from 0 to 8 PSU, depending on wind activity affecting brackish water inflow from the Baltic Sea. Western and Southern winds predominate in the Curonian lagoon. Average wind speed is 1–5 m s^{-1} (55%). However, wind speeds exceeding 15 m s^{-1} occur in autumn and winter (15% of that time) (Gailiušis, 2000). The predominant flow of water is from the south to the north, discharging approximately 22 km^3 per year into the Baltic Sea. Brackish water intrusions are most common during August to October when 70% of the total annual input occurs (Pustelnikovas, 1994; MRC, 1996).

Nutrient-limitation experiments were conducted using 10 l transparent polyethylene enclosures (capped), which have been shown to be sufficient for phytoplankton investigations by Parsons (1982). Due to the small enclosure volume and to prevent problems such as lack of nutrients or sedimentation, experiments were run for 48 h (Bukaveckas & Shaw, 1998). A batch of enclosures covered by fishing net was placed in the upper water layer, 0.15 m below the surface, in the northern part of the Curonian lagoon in 1998, while during 2000–2002 enclosures were placed in the central part of the lagoon. A total 26 nutrient enrichment experiments were performed. The experiment schedule was selected taking into account the seasonal phytoplankton succession and water salinity fluctuations. Experiments were carried out only when salinity was below 1 PSU to avoid effects of Baltic Sea water intrusions into the northern part of the lagoon. Extraordinary weather conditions (heavy westerly winds) caused massive inflow of Baltic Sea water into the lagoon in August 1998. These conditions prevented us from continuing experiments in the late summer and autumn.

Due to the important role of large filamentous cyanobacteria in the Curonian lagoon phytoplankton, it was impossible to remove zooplankton from samples by filtering. Nitrogen, phosphorus, and silica were added in amounts sufficient to double natural Curonian lagoon concentrations (following the recommendations of Schelske (1984) that nutrient additions should be 'realistic' and no more than twice the maximum ambient concentrations): in the spring period 114 μmol N l^{-1} as KNO_3, 4.52 μmol P l^{-1} as K_2HPO_4, and 16.8 μmol Si l^{-1} as Na_2SiO_3, in the summer 28.6, 1.94 and 6.6 μmol l^{-1} N, P, and Si respectively. For each treatment, type-control conditions, enriched in phosphorus, enriched in nitrogen and enriched in both phosphorus and nitrogen, as well enriched in silica experiments (three replicates) were carried out. Enclosures were filled with lagoon water taken from the upper layer (0–0.3 m).

Chlorophyll *a* (Chl *a*), temperature and nutrient (NO_3-N, PO_4-P, SiO_2-Si, NH_4-N) concentrations were measured at the start and end of each experiment. Chl *a* samples were filtered through Gelman A/E glass fibre filters. Filters were extracted with 96% ethanol, and after 24 h in the dark, measured by spectrophotometry (HELCOM, 1988). Phytoplankton samples were fixed with acid Lugol solution. Rough taxonomic phytoplankton species composition (except picoplankton species) and phytoplankton cell density (counting units) were identified and counted using a MEIJI inverted microscope (300×, 600×) according to HELCOM recommendations. The nutrients were analysed according to standard

HACH methods (HACH, 1992). Additionally, more frequent (biweekly) nutrient concentration data collected by Marine Research Centre at the experiment location site were analysed. The effects of nutrient manipulations on Chl *a* concentration were evaluated statistically using analysis of variance (factorial ANOVA) using STATISTICA™ software (StatSoft Inc., 2001; version 6.0).

The long-term data (1984–2001) for the June–August period were analysed. The data originated from the national monitoring programme performed by MRC and Hydrometeorological centre of Lithuania. The effects of abiotic parameters on the most abundant microalgal species, *Aphanizomenon flos-aquae* (Linne) Ralfs ex Bornet et Flahautwere tested by stepwise multiple regression analysis using the STATISTICA™ software package. Additionally, to test the significance of the factors used in the multiple regression analysis, principle component analysis (PCA) was applied. The chosen significance level was $\alpha = 0.05$ for all analyses. Due to heterogeneity in the Curonian lagoon phytoplankton community the cross-station median values of *A. flos-aquae* abundance were used. Samples taken from brackish water (>0.5 PSU) were excluded. The concentration of phosphate, dissolved inorganic nitrogen and silica, used as abiotic factors, were measured at the same dates as the phytoplankton abundance. The effects of 5 and 7 days wind velocity maxima and 10 days average temperature before sampling were also included as independent variables.

Results

Seasonal dynamics of temperature, nutrients, phytoplankton and Chl a

The nutrient concentration dynamics patterns observed in this study were similar to the existing data on the annual nutrient cycle in the Curonian lagoon (MRC, 1996, 1999). During the study period the highest concentrations of nutrients (~120 μmol l^{-1} NO_3-N, and ~8.7 μmol l^{-1} PO_4-P), were observed in early spring (Fig. 2). The rapid decrease in the concentration of phosphate (lower than 0.3 μmol l^{-1} PO_4-P) occurred in mid-April. Nitrogen concentration decreased to a minimum (analytical zero) in May. Due to the fast

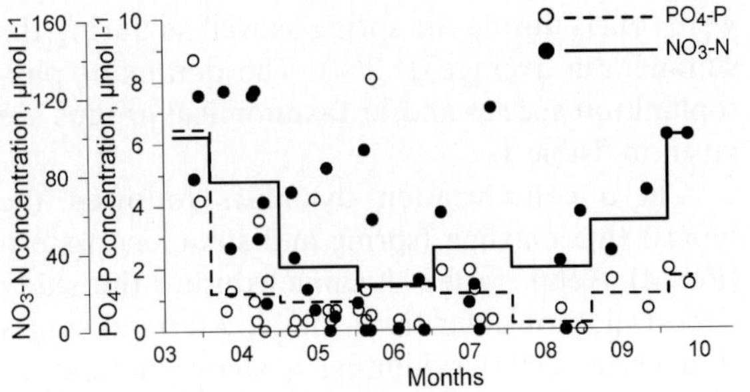

Figure 2. The changes of phosphate PO_4-P and nitrate NO_3-N concentrations over the study period. Lines indicate monthly averages. All values measured at the start of the experiments.

regeneration of phosphorus compounds, phosphate concentration began to increase in early summer. However, mean phosphate levels remained lower than early spring values. Nitrate concentration also tended to increase from midsummer. The ammonium concentrations did not show a pronounced variation during the studied season. Rather, it fluctuated randomly from analytical zero up to 24.3 μmol l^{-1} NH_4-N (Fig. 3). The silica concentration exhibited a clear seasonal pattern. The lowest concentrations were recorded during spring after the diatom bloom. Silica concentration remained low almost throughout every summer and, in early autumn, it started to increase again (Fig. 3).

Diatoms prevailed in the spring (March–May) and constituted, on average, 59.7% of total cell density. During the summer, their abundance and biomass decrease down to 22.8%. During spring, cyanobacteria were rare, reaching only 2.6%, on average, of total cell density. They predominated during the summer (starting from May) attaining, on average, 37% of total cell density. Green algae

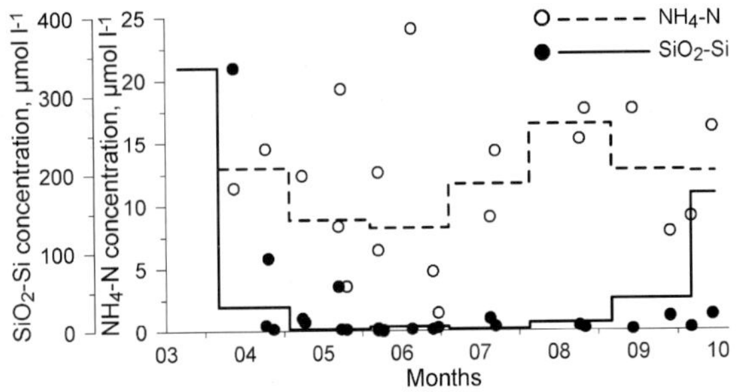

Figure 3. The changes of silica SiO_2-Si and ammonia NH_4-N concentrations over the study period. Lines indicate monthly averages. All values measured at the start of the experiments.

were found during the spring as well as during the summer (in average 21.2%). The dominant phytoplankton species and/or taxonomical groups are given in Table 1.

Chl *a* concentration dynamics followed the typical two maxima (spring and summer) pattern (Fig. 4). Temperature dynamics during the study period did not differ from the long-term averages (Pilkaitytė, 2003) exhibiting a sharp increase in April (from 6 to 17 °C) and maximum values (over 20 °C) during July–August (Fig. 4).

Nutrient enrichment effects

The difference between initial and final (after 48 h) Chl *a* concentration was used as a measure of nutrient enrichment effect. These effects were varied throughout the study period. In spring, the additions of phosphorus and silica were associated with the highest increase in chl *a* concentrations (Fig. 5). The relative increase from initial values in P manipulation was 16.5%, in the PN, –35.8%, and in the Si manipulation –20.4%. The phosphorus impact was statistically significant during the spring season ($p=0.049$, df = 109). Statistically significant differences were found only between PN manipulation and the control, as well as between PN and N manipulations. The silica enrichment did not differ significantly from the control ($p>0.05$). During summer, the addition of nitrogen benefited the development of phytoplankton ($p>0.001$, df = 162). The increase from initial values in N and PN manipulations were, on average, 13.1 and 19.6% respectively, while in the control, P and Si manipulations, a decrease of about 16% on average was observed.

Long-term data analysis

Multiple regression analysis was used to examine whether the development of *Aphanizomenon flos-*

Table 1. Dominant phytoplankton taxa during the study period

Spring	Summer
Small *Centrales*	*Aphanizomenon*
Aulacoseira spp.	*flos-aquae*
Diatomaceae	Small *Centrales*
Ankistrodesmaceae	*Planktonema lauterbornii*

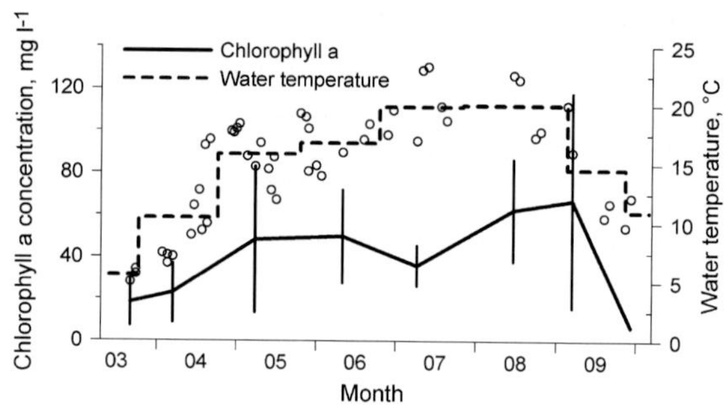

Figure 4. Seasonal development of Chl *a* (±SD) and temperature (line indicates monthly averages) over the study period (1998, 2000–2002).

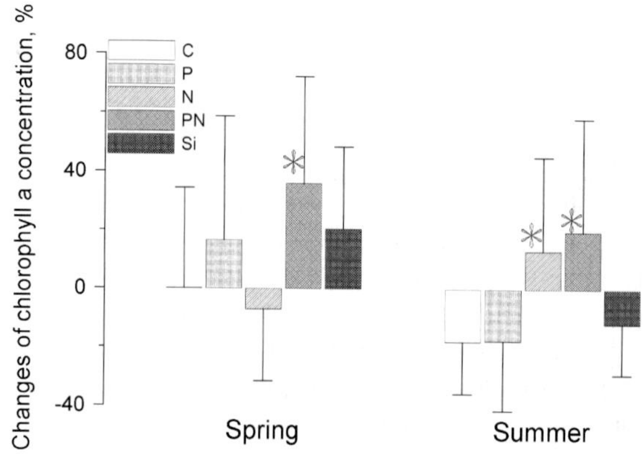

Figure 5. Relative changes in Chl *a* concentration (+SD) compared to initial values depending on different season and treatment. The stars indicate statistically significant increases. Manipulations: C – control, P – enriched with phosphorus, N – enriched with nitrogen, PN – enriched with phosphorus and nitrogen, Si – enriched with silica.

aquae in the Curonian lagoon is due to the differences in nutrient, wind velocity or temperature. The results indicated that the temperature, wind velocity (5 days maximum) and phosphate concentration could explain the variation in cyanobacteria biomass ($R^2=0.7, p=0.001$). Temperature had a positive effect on biomass, while concentration of PO_4-P and wind velocity had a negative one (Table 2). A similar approach was used to test wind speed as a proxy for the effect of resuspension on nutrient concentration. Only phosphate concentration was found to be statistically related to 7 days wind maximum ($\beta=-0.4$; $p=0.03$). To test the significance of the factors used in the multiple regression analysis a principle component analysis (PCA) was applied. The results revealed that the

Table 2. Forward stepwise multiple regression standardised coefficient beta and *p* of independent variables. Dependent variable: *Aphanizomenon flos-aquae*

Independent variable	Beta	*P*
Temperature	0.59	0.001
Wind velocity	−0.32	0.043
PO_4-P	−0.48	0.004
DIN	−0.24	>0.05

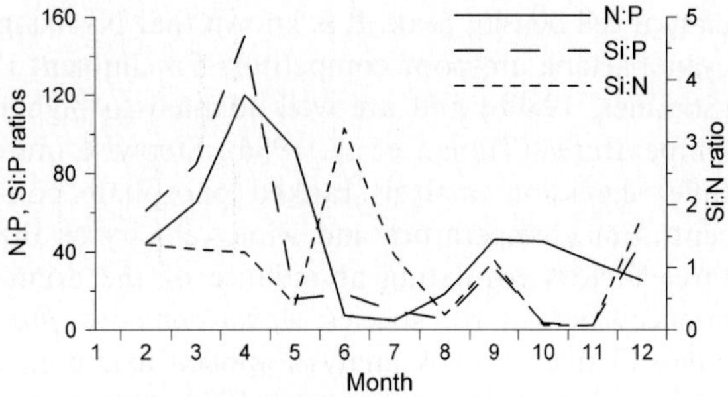

Figure 6. Ratios of inorganic nutrient concentration (after MRC, 1998, 2000–2001).

first two components were responsible for 65% of the total variability. The loadings of the first two components were similar to the results of the multiple regression analysis (Fig. 7).

Discussion

Long-term studies have provided valuable insight on phytoplankton ecology in the Curonian lagoon (Krylova, 1985; Olenina, 1998). However, these monitoring studies did not explain the relationship between phytoplankton and nutrients on a small temporal scale. More detailed studies on mesozooplankton–phytoplankton interactions (Gasiūnaitė & Olenina, 1998; Razinkovas & Gasiūnaitė, 1999) considered 'top-down' interactions as a possible mechanism of seasonal succession in the lagoon. The results of our short (48 h) enclosure experiments revealed another possible 'bottom-up' regulation mechanism, based on nutrient limitation. Nutrient concentrations observed during our experiments were similar to those reported earlier (MRC, 1999), with highest nutrient concentrations being observed in spring (Figs. 2 and 3). Generally, in early spring, phytoplankton can be regarded as limited only by ambient physical conditions as the temperatures are below 7 °C (Fig. 4). The phytoplankton community at this stage is dominated by diatoms (Table 1, see as well Krylova, 1985; Olenina, 1997).

The sharp silica depletion in April (Fig. 6), coinciding with a rapid increase in temperature (Fig. 4) points towards a possible forcing of diatom succession due to silica limitation. Observed values of Si:P and Si:N ratios in April and May (Fig. 6) fall in the diatom dominance threshold range (Sommer, 1999). This suggestion, however, is supported only partially by the experimental results since the observed positive effect of silica addition was not statistically significant (Fig. 5).

Phosphate addition during spring was found to have a positive effect on phytoplankton development. However, only combined P and N manipulations had a statistically significant influence on Chl *a* concentrations (Fig. 5). This can be explained as both P and N concentrations were quite low and, most probably, P is the major limiting nutrient while N is just modulating the effects. The low absolute phosphate concentrations (<0.03 μmol P l^{-1}), near the limitation threshold (Lindenschmidt & Chorus, 1998), that were observed at that time, further confirm the evidence of a P-limitation in spring.

Later, from late spring to early summer, diatoms in the Curonian lagoon phytoplankton community are replaced mainly by green algae and cyanobacteria (Table 1, see also Olenina, 1997). This short 'clear water' period is characterised by increased importance of the top-down interactions. Zooplankton grazing is assumed to be the driving factor for phytoplankton community succession at that moment (Gasiūnaitė & Razinkovas, 2000). The positive response of the phytoplankton community to nitrogen addition in summer (June–August; $p>0.001$, df$=162$), however, brings evidence of another mechanism – nitrogen limitation. In that case, contrary to the spring situation, N is supposed to be the major limiting nutrient while P is assumed to be modulative (Fig. 5). Observed low N:P ratios (below 8:1) in June–July support this suggestion (Fig. 6). Green algae and especially cyanobacteria are known to be good competitors at low N:P ratios (Tilman et al., 1986; Oliver & Ganf, 2000), starting to dominate after the spring diatom cell density peak. It is known that N_2-fixing cyanobacteria are poor competitors for ambient P (Sommer, 1989b) and are well adapted to higher temperatures (Tilman et al., 1986). Stepwise multiple regression analysis ranked phosphate concentration, temperature and wind velocity as the three factors regulating abundance of the dominant cyanobacteria species *Aphanizomenon flos-aquae* (Table 2). PCA analysis applied to a multivariate long-term data set (1984–2001) also grouped the abundance of *A. flos-aquae* along with average temperature and DIN against the maximum wind speed (Fig. 7). The dominance of cyanobacteria in the Curonian lagoon phytoplankton community could also be explained by other factors. Grazing by mesozooplankton on small green algae (Gasiūnaitė & Olenina, 1998; Razinkovas & Gasiūnaitė, 1999), and N-limitation give the competitive advantage to nitrogen-fixing filamentous cyanobacteria.

Climatic and hydrodynamic conditions could also have affected the phytoplankton community in summer. Shallow lagoon ecosystems are characterised by frequent Langmuire spirals that, on average, can transport phytoplankton between high light intensities at the surface and nearly darkness at the bottom every 10 to 20 min (Schubert et al., 2001). Over short-term intervals of about 90 min, cyanobacteria are much better able to cope with such light stresses than are green algae (Schubert et al., 2003). Some of the cyanobacteria species, particularly *Aphanizomenon flos-aquae*, are characterised by diurnal vertical migrations (Kononen, 1992; Hägerhäll-Aniansson, 2001). This adaptation provides the possibility to use the high light intensity in the surface layer during the daytime while taking advantage of high nutrient concentrations near the bottom at night (Reynolds et al., 1987). Intensive water turbidity due to wind action could disturb such migrations (Ibelings & Maberly, 1998). The negative relation found between wind velocity and cyanobacteria abundance during summer over the long-term data series (Table 2) shows the negative effect of that factor on the dominant cyanobacteria species *A. flos-aquae*. The wind regime can also have important effects on nutrient recycling through the resuspension of recently-deposited sediments, including the fluffy layer formed during cyanobacteria blooms. Unfortunately, at present there are no experimental data related to resuspension effects. However, the negative relationship between phosphate concentration and wind speed (used as a proxy for resuspension) in July–August (long-term time series) points towards possible resuspension effect on phosphorus retention, which is important in freshwater systems (Conley, 1999). In our case, no clear decision about the importance of the different effects of wind-induced resuspension of the upper sediment layer to P concentration could be made. Studies in the Darss-Zingst estuary have shown that these processes are a complex of biological as well as chemico–physical interactions which must be investigated in detail to quantify their relative importance (Köster et al., 2000; Rieling et al., 2000). Nitrogen internal loads could also be modified by resuspension. Analysis of the long-term data, however, did not provide any evidence of such mechanism. Further succession is mostly determined by ambient physical factors including the light regime (transparency reaches minimum by the end of summer, mainly due to self shading; Razinkovas & Pilkaitytė, 2002) and the decline in temperature and photoperiod.

On the basis of our results, it can be concluded that regulatory patterns of phytoplankton

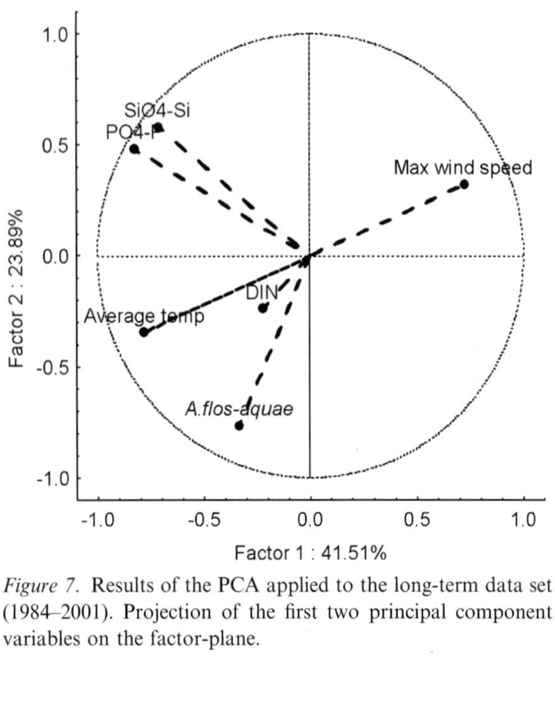

Figure 7. Results of the PCA applied to the long-term data set (1984–2001). Projection of the first two principal component variables on the factor-plane.

seasonal succession in the Curonian lagoon occur in three phases. The first phase, restricted to early spring, is characterised by phytoplankton growth controlled only by ambient physical conditions. The second phase is characterised by phosphorus-limited conditions. However, low Si:N and Si:P ratios during the period starting from the peak of the diatom bloom in late March to its decline in May point also to possible silica limitation. The third phase is nitrogen-limited and, possibly, light-limited, characterised by dominance of cyanobacteria. In this phase, phytoplankton biomass and density attain the highest values, and phytoplankton community structure is easily modified by climatic factors, particularly temperature and wind speed.

Although the Curonian lagoon is mostly composed of freshwater, its hydraulic regime is strongly different from most limnic systems (Gasiūnaitė & Razinkovas, 2004), therefore exhibiting complex phytoplankton growth limitation patterns. Similarly, in Chesapeake Bay (USA), as well as other estuarine systems, P-limitation of biomass accumulation was observed during winter and spring and N-limitation during summer (D'Elia et al., 1986; Webb, 1988; Fisher et al., 1992; Conley, 1999). Given the variability in the nutrient limitation in freshwater and seawater (being possible to observe P-limitation in seawater and N-limitation in freshwater), the hydraulic regime can be suggested as a key factor due to the significant external discharge sources (Aldridge et al., 1993; Kocum et al., 2002). Generally, estuaries may act like freshwater lakes under strong riverine influence, and like marine ecosystems in summer and autumn (Fisher et al., 1995). In the case of the nearly freshwater Curonian lagoon, variations in external nutrient supply are mainly responsible for the seasonal pattern of nutrient limitation. The influence of internal nutrient loads and possible wind induced resuspension still require further experimental evaluation.

Acknowledgements

We would to thank Dr Paul Bukaveckas of the University of Louisville for inspiring that study as well as for practical help with experiment setup and Dr Rosemary Russo of the EPA Athens laboratory and Prof Hendrik Schubert, of the University of Rostock for editorial and language comments on the manuscript. Thanks go as well to the Marine Research Centre and Hydrometeorological Centre of Lithuania for kindly provided long term data. Two anonymous referees are acknowledged for constructive comments on the manuscript. This work was also partially supported by the EU FP5 COLAR project (Contract No.EVK3–2002–00057).

References

Aldridge, F. J., C. L. Scheske & H. J. Carrick, 1993. Nutrient limitation in a hypereutrophic Florida lake. Archiv für Hydrobiologie 127: 21–37.

Bukaveckas, P. & W. Shaw, 1998. Phytoplankton responses to nutrient and grazer manipulations among northeastern lakes of varying pH. Canadian Journal of Fisheries and Aquatic Sciences 55: 958–966.

Conley, D. J., 1999. Biogeochemical nutrient cycles and nutrient management strategies. Hydrobiologia 410: 87–96.

D'Elia, C. F., J. G. Sanders & W. R. Boynton, 1986. Nutrient enrichment studies in a coastal plain estuary: phytoplankton growth in large-scale, continuous cultures. Canadian Journal of Fisheries and Aquatic Sciences 43: 397–406.

Fisher, T. R., E. R. Peele, J. W. Ammerman & L. W. Harding, 1992. Nutrient limitation of phytoplankton in Chesapeake bay. Marine Ecology Progress Series. 82: 51–63.

Fisher, T. R., J. M. Melack, J. U. Grobbelaar & R. W. Howarth, 1995. Nutrient limitation of phytoplankton and eutrophication of inland, estuarine, and marine waters. In Tiessen H. (ed), Phosphorus in the Global Environment – Transfers, Cycles and Management, SCOPE 54: 301–323.

Gailiušis, B., 2000. The analysis of hydrometeorological regime and hydraulic environment. In Lazauskienė, L. L., G. Vaitonis, & A. Draugelis (eds) Klaipėda port. Economy and Ecology (in Lithuanian)., Baltic ECO: 37–48.

Gasiūnaitė, Z. & A. Razinkovas, 2000. The role of plankton and nektobenthos crustaceans in the fate of primary production in the Curonian lagoon (in Lithuanian). Ekologija 3: 22–27.

Gasiūnaitė, Z. R. & A. Razinkovas, 2004. Temporal and spatial patterns of the crustacean zooplankton dynamics in transitional lagoon ecosystem. Hydrobiologia 514: 139–149.

Gasiūnaitė, Z. R. & I. Olenina, 1998. Zooplankton-phytoplankton interactions: a possible explanation of the seasonal succession in the Kursiu Marios lagoon. Hydrobiologia. 363: 333–339.

HACH, 1992. Water Analysis Handbook. Loveland, Colorado, USA, 400–407, 537–543 pp.

Hägerhäll-Aniansson, B. (ed), 2001. Nitrogen fixation in the Baltic Sea. Brief guide for environmental managers, University of Kalmar, Vol. 1.

Hecky, R. E., P. Campbell & L. L. Hendzel, 1993. The stoichiometry of carbon, nitrogen, and phosphorus in particulate matter of lakes and oceans. Limnology and Oceanography 38: 709–724.

HELCOM, 1988. Biological determinants. Guidelines for the Baltic monitoring programme for the third stage. Part D. 27D, 164.

Hutchins, D. A., 1995. Iron and the marine phytoplankton community. In Round F. E., D. J. Chapman (eds), Progress in Phycological Research, Biopress Ltd. 11: 1–49.

Ibelings, B. W. & S. C. Maberly, 1998. Photoinhibition and the availability of inorganic carbon restrict photosynthesis by surface blooms of cyanobacteria. Limnology and Oceanography 43: 408–419.

Kawaguchi, T., A. J. Lewitus, C. M. Aelion & H. N. McKellar, 1997. Can urbanization limit iron availability to estuarine algae? Journal of Experimental Marine Biology and Ecology 213: 53–69.

Kocum, E., D. B. Nedwell & G. J. C. Underwood, 2002. Regulation of phytoplankton primary production along a hypernutrified estuary. Marine Ecology Progress Series 231: 13–22.

Kononen, K., 1992. Dynamics of the toxic cyanobacterial blooms in the Baltic Sea. Finnish Marine Research 261: 3–36.

Köster, M., H.-D. Babenzien, H. J. Black, S. Dahlke, S. Gerbersdorf, J. Meyercordt, L.-A. Meyer-Reil, T. Rieling, I. Stodian & A. Voigt, 2000. Significance of aerobic and anaerobic mineralization processes of organic carbon in sediments of a shallow coastal inlet in the southern Baltic Sea. In Flemming, B. W., M. T. Delafontaine, & G. Liebezeit (eds) Muddy coast dynamics and resource management. Elsevier Science B. V., Amsterdam: 185–194.

Krylova, O. I., 1985. Functioning Ecological Differences of Plankton and Benthos in the Kursiu and Vistula Lagoons of the Baltic Sea. (in Russian) Kaliningrad, AtlantNIRO, 224 pp.

Lindenschmidt, K.-E. & I. Chorus, 1998. The effect of water column mixing on phytoplankton succession, diversity and similarity. Journal of Plankton Research 20: 1927–1951.

MRC, 1996. Annual report of scientific work. Center of Marine research, Environmental Protection Ministry. 94 pp.

MRC, 1999. Results of hydrologic, hydrochemic and hydrobiologic investigations in Curonian lagoon during 1998. Centre of Marine research, Environmental Protection Ministry. 120 pp.

Olenina, I., 1997. Phytoplankton and its developpment in the South-Eastern coastal Baltic and the Curonian lagoon, Doctoral dissertation thesis (in Russian). Vilnius, 161 pp.

Olenina, I., 1998. Long-term changes in the Kursiu Marios lagoon: eutrophication and phytoplankton response. Ekologija 1: 56–65.

Oliver, R. L. & G. G. Ganf, 2000. Freshwater blooms. In B. A. Whitton & M. Potts (eds), The Ecology of Cyanobacteria. Their Diversity in Time and Space, Kluwer Academic Publishers: 149–194.

Parsons, T. R., 1982. The future of controlled ecosystem enclosure experiments. In Grice, G. D. & M. R. Reeve, (eds) Marine Mesocosms. Biological and Chemical Research in Experimental Ecosystems. Springer-Verlag., New York: 411–418.

Pilkaitytė, R., 2003. Phytoplankton seasonal succession and abundance in the eutrophic estuarine lagoons, Doctoral dissertation thesis, Klaipėda, 97 pp.

Pustelnikovas, O., 1994. Transport and accumulation of sediment and contaminants in the lagoon of Kuršių marios (Lithuania) and Baltic sea. Netherlands Journal of Aquatic Ecology 28: 405–411.

Razinkovas, A. & Z. Gasiūnaitė, 1999. The potential grazing of zooplankton and nektobenthic species on phytoplankton in an estuarine system: The Kursiu marios (Curonian) lagoon. The Finnish Environment 263: 44–51.

Razinkovas, A. & R. Pilkaitytė, 2002. Factors limiting phytoplankton development in the Curonian lagoon (in Lithuanian). Jūra ir aplinka 1: 39–46.

Reynolds, C. S., R. L. Oliver & A. E. Walsby, 1987. Cyanobacterial dominance: the role of buovancy regulation in dynamic lake environments. New Zealannd Journal of Marine and Freshwater Research 21: 379–390.

Rieling, T., S. Gerbersdorf, I. Stodian, H. J. Black, S. Dahlke, M. Köster, J. Meyercordt & L.-A. Meyer-Reil, 2000. Benthic microbial decomposition of organic matter and nutrient fluxes at the sediment-water interface in a shallow coastal inlet of the southern Baltic Sea (Nordrügensche Bodden). In Flemming, B. W., M. T. Delafontaine, & G. Liebezeit (eds) Muddy Coast Dynamics and Resource Management. Elsevier Science B.V., Amsterdam: 175–184.

Schelske, C. L., 1984. In situ and natural phytoplankton assemblage bioassays. In Schubert, L. E. (ed.), Algae as Ecological Indicators. Academic Press., London: 15–47.

Schubert, H., S. Sagert & R. M. Forster, 2001. Evaluation of the different levels of variability in the underwater light field of a shallow estuary. Helgoland Marine Research 55: 12–22.

Schubert, H., L. Schlüter & P. Feuerpfeil, 2003. The ecophysiological consequences of the underwater light climate in a shallow Baltic estuary. In Dahlin, H., B. Dybern, & S. Petersson (eds) Proceedings of the Baltic Marine Science Conference, Rønne, Denmark, 22–26 October 1996. ICES Cooperative Research Report, 257. ICES: Copenhagen., Denmark: 29–37.

Sommer, U., 1989a. Nutrient status and nutrient competition of phytoplankton in a shallow, hypertrophic lake. Limnology and Oceanography 34: 1162–1173.

Sommer, U., 1989b. The Role of Competition in Phytoplankton Succession. In Sommer U. (ed.), Plankton Ecology: Succession in Plankton Communities, Springer-Verlag: 57–106.

Sommer, U., 1999. A comment on the proper use of nutrient ratios in microalgal ecology. Forum, Archiv fur Hydrobiology 146: 55–64.

Tilman, D., R. Kiesling, R. Sterner, S. S. Kilham & F. A. Johnson, 1986. Green, bluegreen and diatom algae: taxonomic differences in competitive ability for phosphorus, silicon and nitrogen. Archiv fur Hydrobiology 106: 473–485.

Webb, K. L., 1988. Comment on "Nutrient limitation of phytoplankton growth in brackish coastal ponds" by Caraco, Tamse, Boutros, and Valiela (1987). Canadian Journal of Fisheries and Aquatic Sciences 45: 381–382.

Wetzel, R. G., 2001. Limnology. Lake and river ecosystems. Academic Press, London 1006 pp.

Hydrobiologia (2006) 555:49–58
H. Queiroga, M.R. Cunha, A. Cunha, M.H. Moreira, V. Quintino, A.M. Rodrigues, J. Serôdio & R.M. Warwick (eds),
Marine Biodiversity: Patterns and Processes, Assessment, Threats, Management and Conservation
DOI 10.1007/s10750-005-1105-5

Opinion paper

The near-bottom layer as an ecological boundary in marine ecosystems: diversity, taxonomic composition and community definitions

Jean-Claude Dauvin[1,*] & Carole Vallet[2]
[1]*Station Marine de Wimereux, UMR CNRS 8013 ELICO, Université des Sciences et Technologies de Lille, B.P. 80, 28 Avenue Foch, F62930, Wimereux, France*
[2]*Maison de la Recherche en Environnement Naturel, UMR CNRS 8013 ELICO, Université du Littoral Côte d'Opale, 32 Avenue Foch, F62930, Wimereux, France*
*(*Author for correspondence: E-mail: Jean-Claude.Dauvin@univ-lille1.fr)*

Key words: benthic boundary layer, near-bottom community, ecological boundary, ecocline

Abstract

The near-bottom layer of the ocean represents a boundary between two oceanic biotopes (pelagial and benthal), and as a result, the animal populations living in this habitat belong to various diverse ecological groups. There is a profusion of terms to designate the organisms which live near the sea bottom, both in relation to their behaviour and to boundary-layer hydrodynamics. Do the fauna living above the sea bottom form a true community? Should the fauna in this habitat be considered as a true community or a mixed assemblage comprised of benthic and pelagic organisms? Between 1988 and 1996, more than 500 suprabenthic hauls were taken with a modified Macer-GIROQ sledge at 15 sites in the English Channel and the Seine Estuary (5–70 m), at 13 sites on the southern edge of the Cap Ferret Canyon (Bay of Biscay, 350–1100 m), and at 8 sites on the Atlantic seamounts south of the Azores (260–2235 m). This intensive sampling permitted the collection of more than several hundred species and will serve to facilitate discussion concerning the biodiversity of the fauna collected near the sea bottom. This paper proposes that in the estuary, the near-bottom layer is colonized by a mixed assemblage of both pelagic and benthic organisms, while in the coastal and in the bathyal zones, the response to the gradual extinction of light and the decreasing benthic resuspension at near-bottom depths leads to an ecocline.

Introduction

The Benthic Boundary Layer (BBL) exhibits an accumulation and diversity of organisms swimming near the sea bottom in all marine environments (Mees & Jones, 1997). Because the near-bottom layer represents a boundary between two oceanic biotopes (pelagial and benthal), the animal populations living in this habitat belong to various diverse ecological groups. However, there is a profusion of terms to define the fauna living in the BBL, such as hyperbenthos, suprabenthos, holohyperbenthos, merohyperbenthos, nektobenthos, hypoplankton, BBL macrofauna, demersal zooplankton, bottom plankton, near-bottom plankton, and semiplankton. According to Mees & Jones (1997), these terms should be considered as synonyms in the literature. However, the absence of a clear definition for this habitat, as well as the disagreement concerning the biological components living in this habitat at the boundary between the pelagial and benthal biotopes, would seem to preclude such a perception. The presence of organisms in this layer is determined by two general factors: organism behaviour and boundary layer hydrodynamics. It has been well established that the turbulent BBL region contains a higher concentration of particulate matter in the water immediately overlying the bottom (Wishner & Gowing, 1987). This layer is maintained by suspension of superficial sediments at the sediment-water interface and by material sinking from above that is temporarily carried along in the near-bottom turbulence (Lampitt, 1985). Two main 'transitional area' categories have been defined for such boundary: ecotone and ecocline (see Attrill & Rundle, 2002). An ecotone is an area of relatively rapid change, producing a narrow ecological zone between two different and relatively homogeneous communities. An ecocline represents a boundary of more gradual, progressive change between two systems and is a response to a gradual transformation of at least one major environmental factor. The question to be answered, then, is whether the BBL fauna should be considered as a true community (a group of interdependent organisms living in the same habitat and interacting with each other), an ecotone, an ecocline or a mixed assemblage (a collection of individuals accumulated together in the same place) constituted of benthic and pelagic organisms.

Based on a data set of more than 500 suprabenthic hauls taken with a modified Macer-Giroq sledge in three types of biotopes (estuary, coastal zone and deep water) in the Northeastern Atlantic, this note seeks to respond to three questions: (1) What kinds of organisms live near the sea bottom? (2) Why is there such a high diversity and subsequently high abundance and biomass of organisms near the sea bottom? and (3) Do the fauna living above the sea bottom form a true community, a mixture of benthic and planktonic faunas, an ecotone, an ecocline or a mixed assemblage?

Materials and methods

The BBL was sampled in three main regions between 1988 and 1996:

- In the English Channel – a total of 460 samples were taken from 15 sites exhibiting a variety of sediment types ranging from muddy fine sand to coarse sand, located at depths between 0 and 75 m, in the area extending from the Seine estuary to the circalittoral, in both the eastern and western parts of the Channel (Dauvin et al., 2000; Mouny et al., 2000). The Channel is an epicontinental shallow sea with high turbulence due to the tide, which induces an absence of vertical stratification in the water column, with the exception of the area south of Plymouth where a summer thermocline occurs from May to October. The Channel is also a bio-geographical crossroad for temperate species and shows a high diversity of bottom habitats and a high diversity of benthic organisms (Dauvin et al., 2000).
- In the Bay of Biscay – 13 samples were taken from13 sites along a transect of bathyal sand and mud from the continental slope (350–1100 m) in the Cap Ferret Canyon, which is affected both by the transfer of organic matter from the Gironde estuary and the Aquitanian shelf to the nearby deep-sea environment, and a seasonal vertical flux from the euphotic zone (Elizalde et al., 1991, 1993, 1999; Dauvin & Sorbe, 1995; Sorbe & Weber, 1995; Marquiegui & Sorbe, 1999; Sorbe, 1999).
- On the isolated Atlantic seamounts in the southern part of the Azores (Meteor, Atlantis, Cruiser, Irving) – 8 samples were taken at 8 sites (260–2235 m). The sediment at the top of the seamount is organic sand, and mud at bathyal depths. This oceanic area has both a low primary production of organic matter, and a very low flux of such matter from the water column (Zouhiri, 1999).

All hauls were collected with a new version of the Macer-Giroq sledge (Dauvin et al., 1995). This sledge allows the simultaneous sampling of the fauna at four levels between 0.10 and 1.45 m above the bottom, using four WP2 plankton nets (0.5 mm mesh size). A Tsurimi-Seiki-Kosakusho (TSK) flowmeter in each net measures the volume of the water filtered. The mean speed, during a 10–15 min haul against the tide, was 1.5 knots. Organisms were fixed with 10% neutralized formaldehyde, rinsed and transferred into 70% ethanol. All the organisms were sorted under a dissection microscope, counted, and identified to species level.

In this paper, the number of species corresponds to the total number of species found in each haul (sum of all four nets).

In the Bay of Saint Brieuc (western part of the English Channel) zooplankton was also sampled with a WP2 net (500 μm mesh size, vertical hauls) and the macrobenthic fauna with a 0.25 m^2 Hamon grab (1 mm sieve mesh).

Results and discussion

What kinds of organisms live near the sea bottom?

The total sampling volume of filtered water was generally important and varied between 200 and 800 m^3. The fauna was quite diverse because the sledge concentrates all the organisms swimming just above the sea bottom as well as those having the ability to leave the bottom due to various morphological or behavioural mechanisms.

The organisms can be classified in eight groups according to their swimming ability:

1. *Permanent swimmers present throughout the water column.* Including organisms, such as planktonic organisms, copepods, euphausiids, which are concentrated near the sea bottom only during the day in the coastal zone of the English Channel (Vallet & Dauvin, 2001).
2. *Temporary swimmers present throughout the water column.* Including organisms such as meroplankton (crustacean and polychaete larvae etc.), which can be collected daily, year-round in the BBL (Vallet & Dauvin, 2001), and ichthyoplankton, which can be collected in great quantity only at sunset and sunrise in the coastal zone of the English Channel and in the Region Of Frehwater Influence (ROFI) zone of the Seine estuary (Wang & Dauvin, 1994; Zouhiri & Dauvin, 1995).
3. *Permanent swimmers present only in the BBL.* This group corresponds to a limited number of deep-water species specialized for living exclusively in the BBL. For example, several chaetognaths, such as *Hemispadella dauvini* (Casanova, 1996), were collected in the bathyal depth of the Azores seamounts (Casanova, 1996). Copepods (harpacticoids cerviniid; calanoid misophrioid) and ostracods (halocyprid), which were not found in our samples, could be also classified in this category (see Ohtsuka et al., 1993; Jaume & Boxshall, 1997, for example).
4. *Burrowers/swimmers.* This group corresponds to the hyperbenthos, including benthic organisms living close to, but not on or in the sea bottom (Beyer, 1958). Later Brunel et al. (1978) named suprabenthos 'all bottom-dependent animals, mainly crustaceans which perform – with varying amplitude, intensity and regularity – seasonal or daily vertical migrations above the sea-floor', including decapods and peracarids particularly. Suprabenthos was collected mostly at night, with two peaks at sunset and sunrise, and maximum during mating (ampeliscids, cumaceans) in the Channel (Dauvin et al., 2000).
5. *Burrowers and temporary nuptial swimmers.* Including organisms, such as polychaet syllids and nereids, which emerged during the nuptial period (Zouhiri, 1999).
6. *Benthic mobile species.* Including organisms, such as the queen scallops *Aequipecten opercularis* (L., 1758), which were common on the coarse sand offshore near Roscoff in the Channel (Zouhiri, 1999). The swimming behaviour of these species, which evade dredge and sledge capture by swimming up into the water column, has been well documented (see Jenkins et al., 2003).
7. *Benthic drift species.* Comprising numerous mucus-secreting invertebrates (including bivalves, such as *Abra alba* (Wood, 1802) and *Macoma balthica* (L., 1758), and polychaets, such as *Pectinaria koreni* Malmgren, 1866, which were common in the Channel, especially in the ROFI zone of the Seine bay as a result of tidal hydrodynamism (Olivier et al., 1996).
8. *Fish.* Including demersal fish, e.g., flat fish, but also others, such as clupeids and gadiids, which were collected accidentally by the sledge (Zouhiri, 1999).

We also found epibenthic and endobenthic organisms in the nets, due to contamination resulting from the removal of sediment by the sledge. The Macer-Giroq sledge, it should be noted, caused very little contamination regardless of the type of sediment, including soft bathyal mud (Dauvin et al., 1995; Sorbe, 1999).

Table 1. Number of species for the eight groups of organisms according to their swimming ability in the ROFI zone of the Seine estuary (see text for the definition of the groups)

Group	Number of species	Frequency (%)
1	16	11.2
2	23	16.1
3	0	0
4	48	33.6
5	3	2.0
6	1	0.7
7	41	28.7
8	11	7.7
Total	143	

As an illustration of the number of taxa caught in each of the eight groups, the number of species of each group collected by the sledge is given for the ROFI zone of the Seine estuary (Table 1). It appears that the group 4 (burrowers and swimmers) and group 7 (benthic drift species) dominate in this turbulent coastal zone.

Why is there such a high diversity of organisms near the sea bottom?

Figure 1a shows the total number of species in each group of swimming organisms collected in the three biotopes for a seasonal sampling over 1 year in the Bay of Saint Brieuc (BBL: 20,000 m^{-2}, water column 30 m^{-2}, and benthos: 8 m^{-2}) (Vallet & Dauvin, 1998, 1999, 2001). The total number of species reached 203. All of the organisms sampled throughout the year can be divided into two main categories: the organisms that colonized the three biotopes, such as decapods, peracarids, and pycnogonids; and the organisms that are only found in the water column and near the sea bottom, such as copepods, crustacean larvae, and other planktonic organisms. The number of species from the first category was higher in the BBL than in the two other biotopes, due to a high concentration of organisms swimming near the sea bottom, especially amphipods and mysids. The number of species from the second category was similar for both the BBL and the water column: all the planktonic species could be collected near the sea bottom at any time of the day or night.

As the total surface of sampling varied strongly between the three compartments (BBL, water column and benthos), the taxa diversity was compared using Hurlbert's (1971) rarefaction methodology. The normalized expected number of taxa $E(S_{61})$ (Fig. 1b) confirmed the highest diversity in the BBL.

Because the near-bottom layer of the sea lies between two principal oceanic biotopes, the pelagial and the benthal, it would be reasonable to expect the animal population here to be the most diverse, consisting of a variety of ecological groups (Vereshchaka, 1995). In fact, the sledge sampled representative species from the eight groups described above. Nevertheless, species richness in the BBL was positively correlated with depth (Fig. 2) and with the material sinking from the water column (Fig. 3). In the Channel, species richness was also correlated with salinity; accordingly, the Seine estuary sites showed lower diversity than the circa-littoral offshore sites (Dauvin et al., 2000). Conversely, abundance (number of individuals. 100 m^3) was higher in shallow water and in the Seine estuary than at bathyal depths (Zouhiri, 1999; Dauvin et al., 2000). For the bathyal depths, abundance was also positively correlated with the input of organic matter sinking from the water column (Zouhiri, 1999).

Do the fauna living above the sea bottom form a true community, a mixture of benthic and planktonic fauna, an ecotone, an ecocline or a mixed assemblage?

Vereshchaka (1995) defined the benthopelagial layer as a specific biotope, which extends along the water-sediment interface bordering the pelagial layer from above and the benthal biotope from below. Thus, the benthopelagial layer would appear to be the same as the BBL. Two principal planktonic groups can dwell in this benthopelagial layer: pelagic and benthopelagic. The pelagic group is comprised of animals living in the water column, regardless of their exact location in the column. Their existence in the benthopelagial layer is accidental and episodic: they may approach the sea bottom at moments of diel migration, water mass advection near sea-floor rises, or passive sinking. Benthopelagic organisms, on the other hand, are obligatorily connected to the benthopelagial layer and spend at least part of their life there. Since

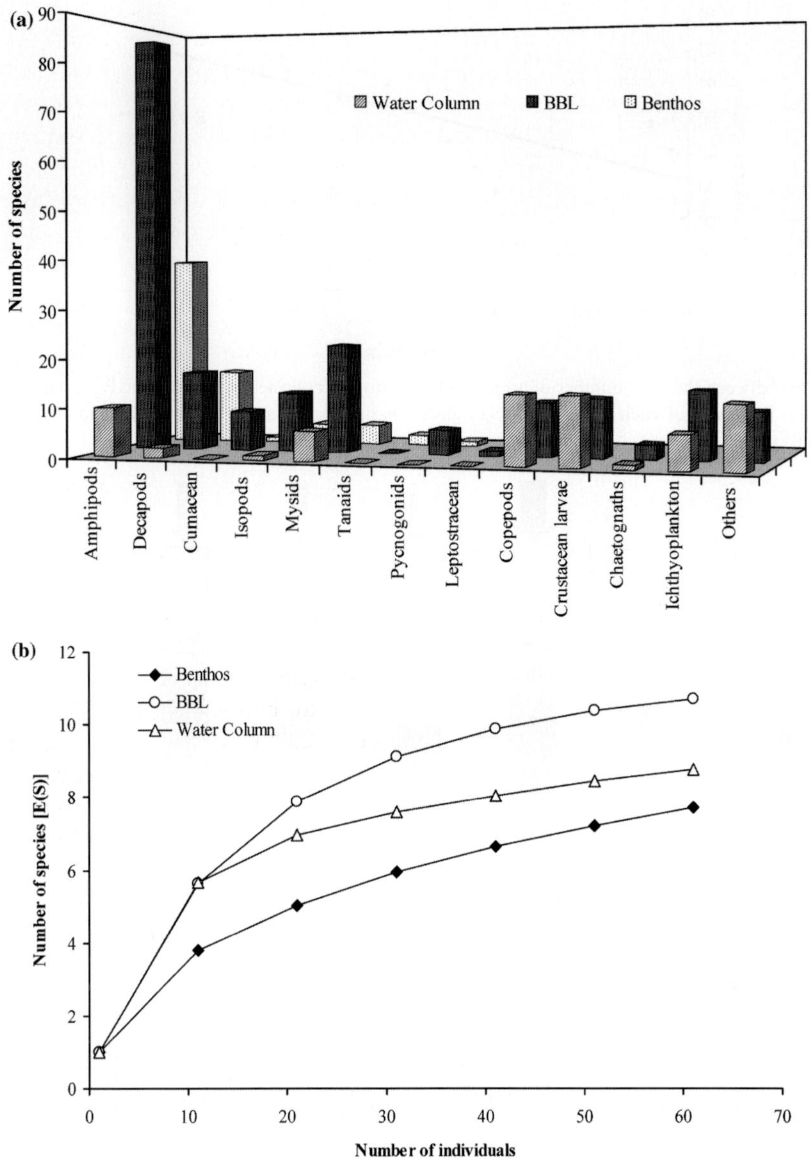

Figure 1. Number of swimming species recorded (a) and Hurlbert's rarefaction curves (b), for the three habitats (benthos, Benthic Boundary Layer (BBL), water Column) in the Bay of Saint Brieuc (western part of the Channel) based on seasonal sampling over a period of 1 year.

benthopelagic animals migrate, they can be linked to either of the neighbouring pelagial or benthal biotopes. Vereshchaka (1995) defined three ecological subgroups among benthopelagic organisms:

1. Hypobenthopelagic animals, which spend part of their life in the benthopelagial layer and the rest further below, in the benthal biotope; they correspond to the hyperbenthos described by Beyer (1958) or the suprabenthos described by Brunel et al. (1978).
2. Epibenthopelagic animals, which spend part of their life in the benthopelagial layer and the rest further above, in the pelagial biotope; they belong, in part, to the category of zooplankton.
3. Amphibenthopelagic animals, which periodically live in all three biotopes – pelagial, benthopelagial and benthal. The planktonic

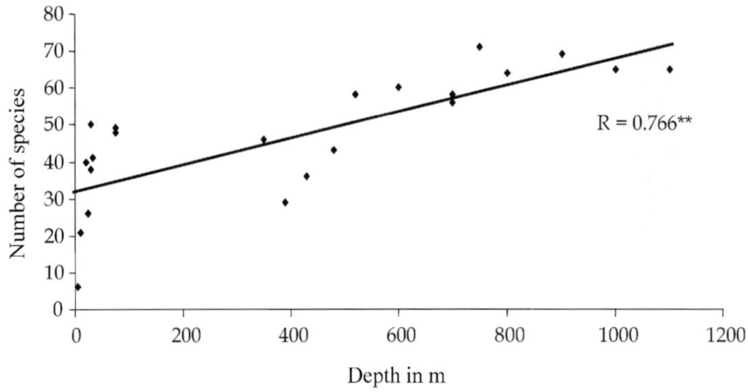

Figure 2. Relationships between the total number of amphipods and mysids collected in the Benthic Boundary Layer (total number of species collected in the four nets of each haul) for the English Channel and Bay of Biscay sites and the depth in m (*r*: Pearson coefficient, ** $p < 0.01$).

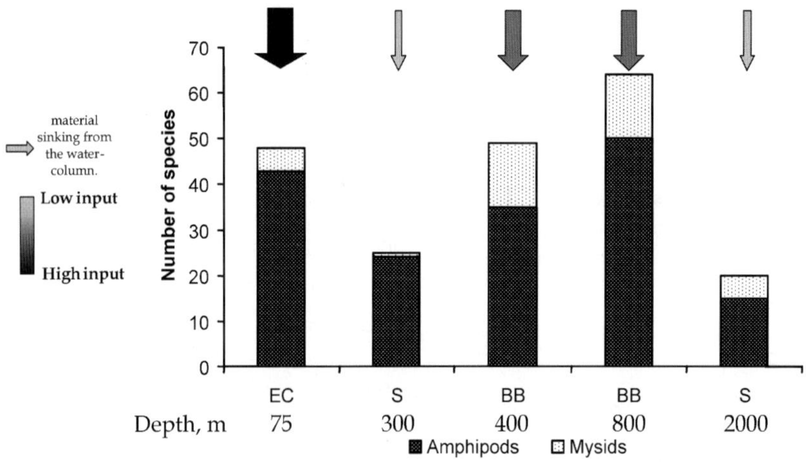

Figure 3. Relationships between the total number of amphipods and mysids for the English Channel (EC), Bay of Biscay (BB) and Seamount (S) sites, the depth in m, and the hypothetical input of material sinking from the water column in the Benthic Boundary Layer.

larvae of bottom-dwelling animals should also be included; larvae in the process of metamorphosis descend to the benthopelagial layer, while adults live in the benthal.

In their review, Mees & Jones (1997) defined the hyperbenthos as the association of all the animals living in the water layer close to the seabed named hyperbenthon with holohyperbenthos for permanent members and merohyperbenthos for temporary members as larvae. Large fish and epibenthic crustaceans, though they feed in the hyperbenthal (the zone occupied by the hyperbenthos) and are thus an integral part of the community, were considered as accidental catches.

It is clear that the organisms living in the BBL cannot be considered a true community compared to those defined communities on the sea bottom and in the water column, because the species living in the benthopelagial layer, with the exception of some specialized deep chaetognaths and copepods, belong both to pelagial and benthal biotopes. Instead, the benthopelagial layer should be considered an ecological boundary between two biotopes.

Figure 4 summarizes the main abiotic factors and faunal characteristics found in the BBL in three distinct habitats: estuarine, coastal and bathyal zones. The abiotic conditions and the

Figure 4. Main abiotic factors and faunal characteristics in estuarine (a), coastal (b) and bathyal (c) zones, including a schematic diagram of the swimming activities of organisms in the Benthic Boundary Layer and in the water column.

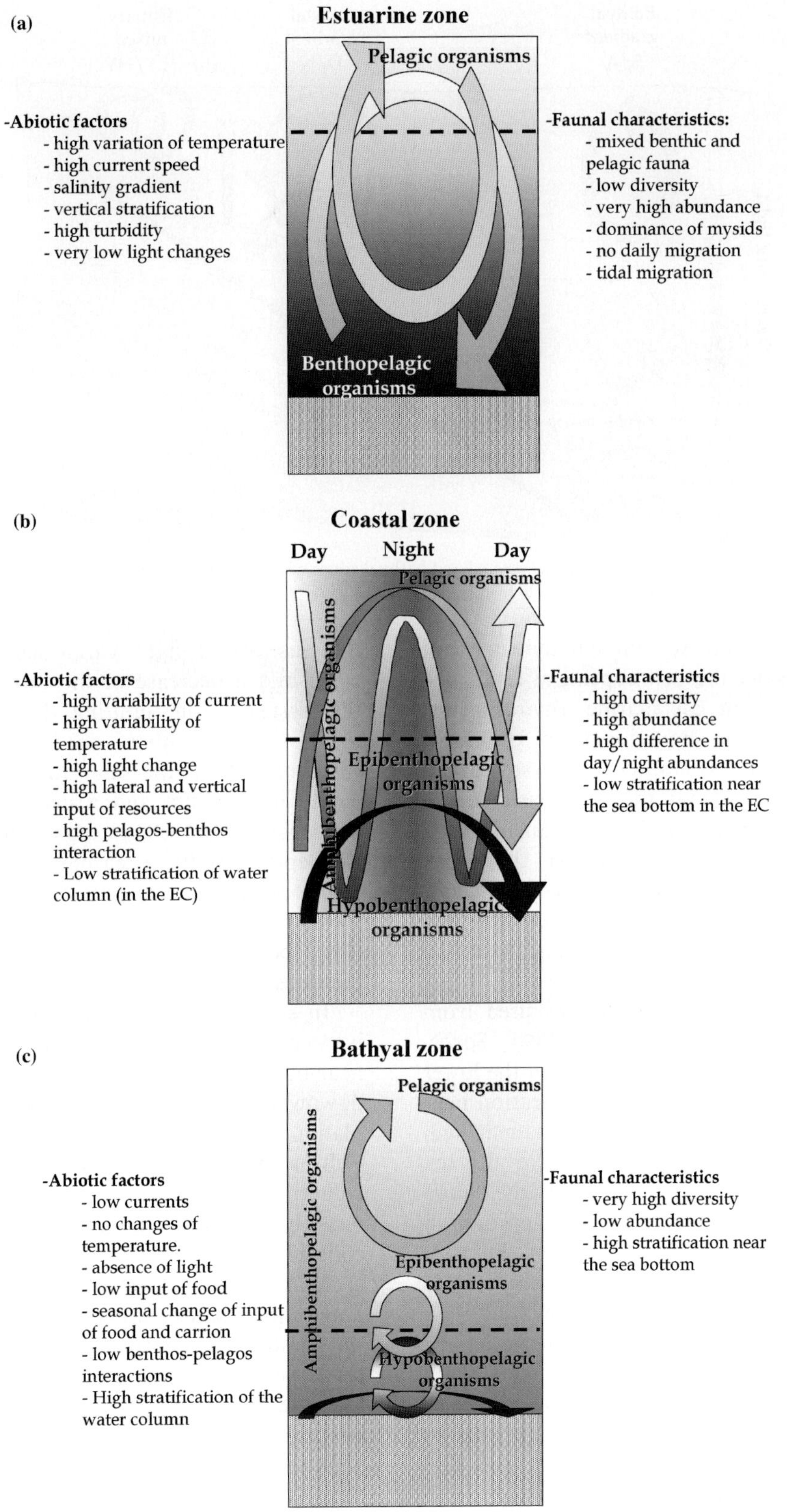
(a)
Estuarine zone
Pelagic organisms
Benthopelagic organisms
-Abiotic factors
- high variation of temperature
- high current speed
- salinity gradient
- vertical stratification
- high turbidity
- very low light changes
-Faunal characteristics:
- mixed benthic and pelagic fauna
- low diversity
- very high abundance
- dominance of mysids
- no daily migration
- tidal migration
(b)
Coastal zone
Day
Night
Day
Pelagic organisms
Amphibenthopelagic organisms
Epibenthopelagic organisms
Hypobenthopelagic organisms
-Abiotic factors
- high variability of current
- high variability of temperature
- high light change
- high lateral and vertical input of resources
- high pelagos-benthos interaction
- Low stratification of water column (in the EC)
-Faunal characteristics
- high diversity
- high abundance
- high difference in day/night abundances
- low stratification near the sea bottom in the EC
(c)
Bathyal zone
Pelagic organisms
Amphibenthopelagic organisms
Epibenthopelagic organisms
Hypobenthopelagic organisms
-Abiotic factors
- low currents
- no changes of temperature.
- absence of light
- low input of food
- seasonal change of input of food and carrion
- low benthos-pelagos interactions
- High stratification of the water column
-Faunal characteristics
- very high diversity
- low abundance
- high stratification near the sea bottom

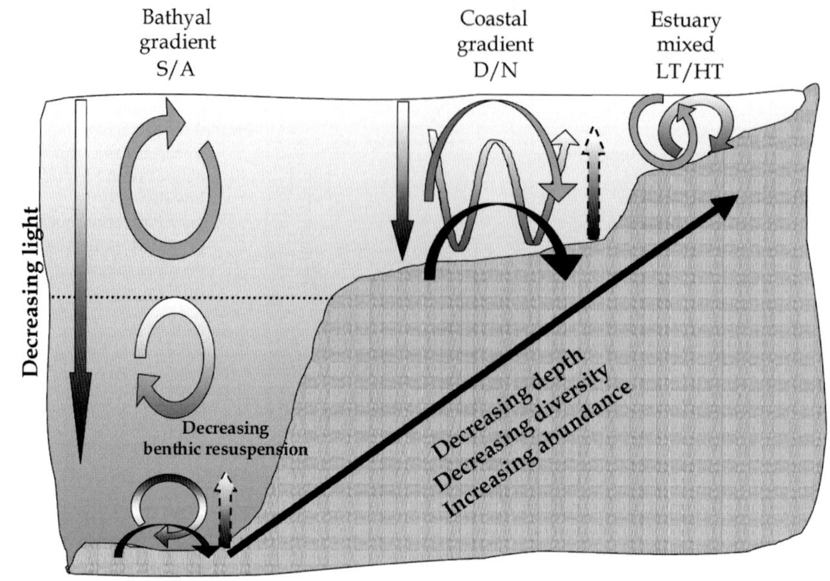

Figure 5. Diagram of the boundaries in estuarine, coastal and bathyal zones with indications of the main temporal changes. S: spring; A: autumn; D: day; N: night; LT: low tide; HT: high tide.

faunal composition are so different that it is not possible to consider the benthopelagial layer as a unique boundary in all marine environments. Figure 5 depicts the boundaries in estuarine, coastal and bathyal zones. Given these boundaries, it would seem that the turbulence in the estuary is so high that there is a permanent mixed assemblage of organisms living in the BBL, including pelagic and benthopelagic organisms, and with strong changes of diversity between high tide (sampling of marine species) and low tide (sampling of freshwater species). However, in the Bay of Biscay, where the sledge was used from 400 to 3000 m (Dauvin & Sorbe, 1995; Sorbe, 1999), the fauna was collected mainly in the lower net of the sledge, indicating a concentration near the sea bottom as the result of a permanent high stratification of the water column near the sea bottom. Conversely, in the Channel, the fauna sampled was of the same order of magnitude in all four nets, probably as a result of the high turbulence of this megatidal sea affected by tide changes (alternation of high and low tide, ebb and flow) and monthly changes (alternation of spring tide and neap tide) (Dauvin et al., 2000). For the latter two zones, the benthopelagial layer is affected by two main abiotic factors related to depth: a decrease in light, reducing the diel changes of the faunal composition in reaction to the absence of light variations in this aphotic zone; and a decrease of the turbulence and benthic resuspension, limiting these factors to the first few meters above the bottom in the deep environment. At the same time, there are an increase in diversity and a decrease in the abundance of benthopelagic organisms in relation to the depth. For these reasons, we suggest that the benthopelagial layer showing a gradient of progressive changes could be considered as an ecocline between the pelagial and the benthal biotopes.

In summary, the fauna collected near the sea bottom contain endemic species residing in the near-bottom environment, species derived from downward extensions of pelagic planktonic populations, which are often seasonal in nature, as well as infaunal species emerging into the water column, often on diel cycles. Several papers have recently discussed the definitions of the fauna collected in the BBL (see Vereshchaka, 1995; Mees & Jones, 1997; Dauvin et al., 2000). Unfortunately, the authors studying the BBL neither use the same terms consistently, nor do they consistently select the same categories of organisms in their studies of the benthopelagial layer or hypobenthal. Some of them have even included endobenthic resuspended organisms as a part of the 'Hyperbenthos'. These semantic problems lead to a lack of coherence. As suggested by Mees & Jones (1997), semantic agreement is needed to facilitate progress towards a better understanding of the role played by this group of BBL animals in a functioning ecosystem.

Acknowledgements

This study was carried out as a part of the 'Programme National d'Environnement Côtier' (PNEC) co-ordinated by IFREMER and CNRS and the framework of the Seine-Aval program coordinated by the Haute Normandie Region. The authors thank all the participants from these both programs, Lisa Ellen Spencer Services for help in English and two anonymous referees for their very useful comments on the first version of the typescript.

References

Attrill, M. J. & S. D. Rundle, 2002. Ecotone or ecocline: ecological boundaries in estuaries. Estuarine Coastal Shelf Science 55: 929–936.

Beyer, F., 1958. A new, bottom-living Trachymedusa from the Oslofjord. Description of the species, and a general discussion of the life conditions and fauna of the fjord deeps. Nytt Magasin for Zoologi 6: 121–143.

Brunel, P., M. Besner, D. Messier, L. Poirier, D. Granger & M. Weinstein, 1978. Le traîneau Macer-Giroq: appareil amélioré pour l'échantillonnage quantitatif de la petite faune nageuse au voisinage du fond. Internationale Revue der Gesamten Hydrobiologie 63: 815–829.

Casanova, J. P., 1996. A new genus and species of deep-sea chaetognath from the Atlantic:a probable link between the families Heterkrohniidae and Spadellidae. Journal of Natural History 30: 1239–1245.

Dauvin, J. C. & J. C. Sorbe, 1995. Suprabenthic amphipods from the southern margin of the Cap-Ferret Canyon (Bay of Biscay, Northeastern Atlantic Ocean): abundance and bathymetric distribution. Polskie Archiwum Hydrobiologii 42: 441–460.

Dauvin, J. C., J. C. Sorbe & J. C. Lorgeré, 1995. The Benthic Boundary Layer macrofauna from the upper continental slope and the Cap-Ferret Canyon (Bay of Biscay). Oceanologica Acta 18: 113–122.

Dauvin, J. C., C. Vallet, P. Mouny & S. Zouhiri, 2000. Main characteristics of the boundary layer macrofauna in the English Channel. Hydrobiologia 426: 139–156.

Elizalde, M., J. C. Dauvin & J. C. Sorbe, 1991. Les mysidacés suprabenthiques de la marge sud du canyon du Cap-Ferret (Golfe de Gascogne): répartition bathymétrique et activité natatoire. Annales de l'Institut Océanographique Paris 67: 129–144.

Elizalde, M., J. C. Sorbe & J. C. Dauvin, 1993. Las comunidades suprabentonicas batiales del golfo Vizcaya (margen sur del canon de Cap-Ferret): composicion faunistica y estructura. Publicaciones Especiales del Instituto Espanol de Oceanografia 11: 247–258.

Elizalde, M., O. Weber, A. Pascual, J. C. Sorbe & H. Etcheber, 1999. Benthic response of *Munnopsurus atlanticus* (Crustacea Isopoda) to the carbon content of the near-bottom sedimentary environment on the southern margin of the Cap-Ferret canyon (Bay of Biscay; northeastern Atlantic Ocean). Deep-Sea Research II 46: 2331–2344.

Hurlbert, S. H., 1971. The non-concept of species diversity: a critique and alternative parameters. Ecology 52: 577–586.

Jaume, D. & G. A. Boxshall, 1997. A new genus and two new species of misophrioid copepods (Crustacea) from the North Atlantic hyperbenthos. Sarsia 82: 39–54.

Jenkins, S. R., W. Lart, B. J. Vause & A. R. Brand, 2003. Seasonal swimming behaviour in the queen scallop (*Aequipecten opercularis*) and its effect on dredge fisheries. Journal of Experimental Marine Biology and Ecology 289: 163–179.

Lampitt, R. S., 1985. Evidence for the seasonal deposition of detritus to the deep-sea floor and its subsequent resuspension. Deep-Sea Research I 32: 885–897.

Marquiegui, M. A. & J. C. Sorbe, 1999. Influence of near-bottom environmental conditions on the structure of bathyal macrobenthic crustacean assemblages from the Capbreton canyon (Bay of Biscay, NE Atlantic). Acta Oecologia 20: 353–362.

Mees, J. & M. Jones, 1997. The hyperbenthos. Oceanography and Marine Biology: an Annual Review 35: 221–255.

Mouny, P., J. C. Dauvin & S. Zouhiri, 2000. Benthic Boundary Layer fauna from the Seine estuary (eastern English Channel, France): spatial distribution and seasonal changes. Journal of the marine biological Association of the United Kingdom 80: 959–968.

Ohtsuka, S., H. S. J. Roe & G. A. Boxshall, 1993. A new family of calanoid copepods, the hyperbionycidae, collected from the deep-sea hyperbenthic community in the northeastern Atlantic. Sarsia 78: 69–82.

Olivier, F., C. Vallet, J. C. Dauvin & C. Retière, 1996. Deposition-resuspension mechanisms involved in post-larval and juvenile populations dynamics in the *Abra alba* community of the eastern part of the Bay of Seine (English Channel). Journal of Experimental Marine Biology and Ecology 199: 89–109.

Sorbe, J. C., 1999. Deep-sea macrofaunal assemblages within the Benthic Boundary Layer of the Cap-Ferret canyon (Bay of Biscay, NE Atlantic). Deep-Sea Research II 46: 2309–2329.

Vallet, C. & J. C. Dauvin, 1998. Composition and diversity of the Benthic Boundary Layer macrofauna from the English Channel. Journal of the marine biological Association of the United Kingdom 78: 387–409.

Vallet, C. & J. C. Dauvin, 1999. Seasonal changes of macrozooplankton and Benthic Boundary Layer macrofauna from the Bay of Saint Brieuc (western English Channel). Journal of Plankton Research 21: 35–49.

Vallet, C. & J. C. Dauvin, 2001. Biomass changes and bentho-pelagic transfers throughout the Benthic Boundary Layer in the English Channel. Journal of Plankton Research 23: 903–922.

Vereshchaka, A. L., 1995. Macrozooplankton in the near-bottom layer of continental slopes and seamounts. Deep-Sea Research I 42: 1639–1668.

Wang, X. & J. C. Dauvin, 1994. The suprabenthic crustacean fauna of the infralittoral fine sand community in the Bay of seine (eastern English Channel): composition, swimming activity and diurnal variations. Cahiers de Biologie Marine 35: 135–155.

Wishner, K. F. & M. M. Gowing, 1987. *In situ* filtering and ingestion rates of deep-sea benthic boundary layer zooplankton in the Santa Catalina Basin. Marine Biology 94: 358–366.

Zouhiri, S., 1999. Composition qualitative de la faune démersale et structure trophique du suprabenthos en trois sites de l'Atlantique Nord-est (Manche, Golfe de Gascogne et seamounts du sud des Açores). Thèse de Doctorat du Muséum National d'Histoire Naturelle Paris, 109 pp.

Zouhiri, S. & J. C. Dauvin, 1995. Diel changes of the Benthic Boundary Layer macrofauna over coarse sand sediment in the western English Channel. Oceanologica Acta 19: 141–153.

Hydrobiologia (2006) 555:59–74
H. Queiroga, M.R. Cunha, A. Cunha, M.H. Moreira, V. Quintino, A.M. Rodrigues, J. Serôdio & R.M. Warwick (eds),
Marine Biodiversity: Patterns and Processes, Assessment, Threats, Management and Conservation
DOI 10.1007/s10750-005-1132-2

Seasonal and spatial patterns of distribution of subtidal benthic invertebrate communities in the Mondego River, Portugal – a poikilohaline estuary

P. Chainho[1,*], J.L. Costa[1], M.L. Chaves[1], M.F. Lane[2], D.M. Dauer[2] & M.J. Costa[1,3]
[1]*Instituto de Oceanografia, Faculdade de Ciências da Universidade de Lisboa, Campo Grande, 1749-016, Lisboa, Portugal*
[2]*Department of Biological Sciences, Old Dominion University, Norfolk, Virginia, 23529, USA*
[3]*Departamento de Biologia Animal, Faculdade de Ciências da Universidade de Lisboa, Campo Grande, 1749-016, Lisboa, Portugal*
*(*Author for correspondence: E-mail: pmchainho@fc.ul.pt)*

Key words: benthic condition, spatial patterns, seasonal variations, environmental gradients, Mondego River estuary, Venice salinity system

Abstract

The use of benthic assemblages to assess ecological quality of estuarine environments is a major tool for the implementation of the Water Framework Directive (2000/60/CE) for European aquatic ecosystems. Benthic communities show spatially heterogeneous distributions and experience seasonal variations due to both natural and anthropogenic stresses. The major goal of this study was to quantify the relationships between environmental gradients and the spatial and temporal patterns of the benthic communities along a Portuguese estuary. Seasonal and spatial variations relating macrobenthic communities and measures of water and sediment quality along the northern branch of the Mondego River estuary were examined at seven sampling stations from July 2000 to June 2001. Cluster analysis of biological data indicated three major groups of communities based on spatial distribution patterns: (1) a lower sector with stronger marine influence and dominated by *Streblospio shrubsolii* and *Cerastoderma glaucum*; (2) a middle sector with dominance of *S. shrubsolii* and *Corophium multisetosum*; (3) and an upper sector where *C. multisetosum* dominates a community characterized by a lower number of species. Canonical correspondence analyses of biological and environmental data determined a major salinity gradient influencing benthic communities. Seasonal changes of benthic communities were mainly determined by freshwater input and salinity changes that imposed a strong decrease in densities and number of species during winter, followed by a recovery during spring. Benthic ecological structure and contaminant levels indicated that the Mondego northern branch is moderately disturbed, although opportunistic species dominated the benthic community, suggesting that natural and anthopogenic sources of stress may be acting together. The Mondego River estuary, a poikilohaline-type estuary, characterized by strong seasonal changes in water flow and salinity, cannot be consistently stratified into salinity regions based upon the Venice classification system. Biotic communities, exemplified here by the benthic communities, are seasonally displaced, compared to a homiohaline-type estuary where the Venice system can be applied without modification. Future identification of reference conditions and design of monitoring programs cannot be accomplished without understanding how interactions between biotic and physico-chemical dynamics differ between homiohaline and poikilohaline estuaries. Results obtained in this study could be used to assist future assessments in other Portuguese estuaries.

Introduction

Benthic invertebrate communities have been widely used as indicators of the ecological status of coastal and estuarine ecosystems (e.g. Pearson & Rosenberg, 1978; Hily et al., 1986; Dauer, 1993; Weisberg et al., 1997; Borja et al., 2003) and were included in the biological elements indicated by the Water Framework Directive (WFD) (2000/60/EC) for use in environmental monitoring. Benthic infaunal species live in the sediments and show relatively low mobility, being exposed to stress due to contaminants, low dissolved oxygen, limiting nutrient levels and physical disturbances (Dauer et al., 1992; Weisberg et al., 1997; Cowie et al., 2000). Benthic communities include species with different life cycles and specific tolerances to stress events, which make them suitable to be classified into different functional groups that reflect the magnitude of disturbances (e.g. Bilyard, 1987). They also play an important role in the chemical fluxes of the water/sediment interface (Aller & Aller, 1998; Aller et al., 2001) and are one of the main compartments of aquatic food webs, being effective indicators of impacts at higher levels of biological organization (Bilyard, 1987; Alden et al., 1997).

In spite of all the advantages mentioned above some biological characteristics of benthic communities have to be taken into account when interpreting results of benthic condition assessment. Benthic communities show high spatial heterogeneity in estuaries that are related to the influence of natural gradients of different factors contributing to the overall distribution of species. Many benthic species occur along a wide spectrum of an estuarine environment while some others are confined to a narrower habitat, according to their tolerance to environmental variables such as salinity, sediment type, depth, etc. In addition to spatial patterns, temperate estuarine invertebrate communities also show important temporal variations related to seasonal and interannual changes. Seasonal fluctuations in abundance and composition can be due to recruitment pulses that occur during spring and autumn for most species, but also to the occurrence of extreme environmental conditions such as low temperatures, floods and droughts (Alden et al., 1997; Attrill & Power, 2000; Salen-Picard & Arlhac, 2002). Freshwater flow variability is one of the main factors influencing the high temporal and spatial changes in physical, chemical and biological conditions in estuaries, particularly in rivers that show strong seasonal changes (Kimmerer, 2002). These hydrodynamic fluctuations have an important effect on the erosion and depositional cycles, influencing the sediment composition and therefore the colonization by particular benthic communities. In addition, widely varying salinity patterns in an estuary will alter local benthic community composition due to seasonal flow patterns (Boesch, 1977b) or extreme episodic storm events (Boesch et al., 1976a, b).

The Mondego River estuary, located in the western Atlantic Portuguese coast is divided into two branches that diverge 7.5 km upstream from the river mouth and have different hydrographic characteristics (Fig. 1). In the southern branch the water circulation is mainly driven by tidal excursion and the only freshwater input comes from the Pranto River, a small tributary. The northern branch receives most of the freshwater input and is strongly influenced by seasonal water flow fluctuations (Flindt et al., 1997). These two branches were originally in contact through a small channel in the upstream area but the southern branch became gradually silted up and the connection occurs only during strong spring tides (Cunha & Dinis, 2002). The southern branch subsystem has been widely studied concerning physical and chemical variables that influence ecological processes. These studies identified a eutrophication gradient as one of the main factors influencing benthic communities (Flindt et al., 1997; Marques et al., 1997; Lillebø et al., 1999; Martins et al.,

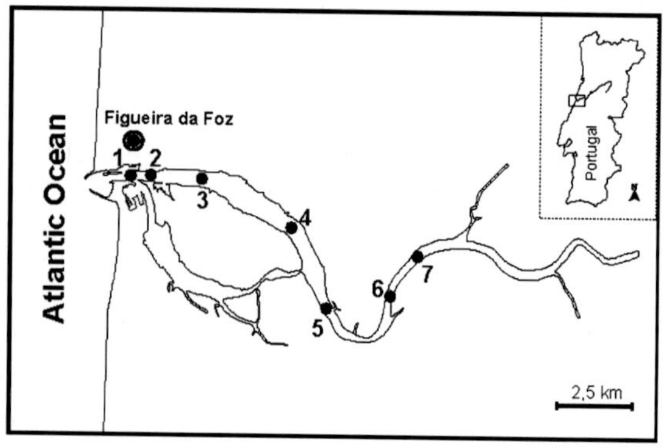

Figure 1. Location of sampling stations selected in the northern branch of the Mondego River estuary.

2001; Cardoso et al., 2002). In contrast few studies have been carried out in subtidal communities of the northern branch (Marques et al., 1993; Pardal et al., 1993) and they did not cover the region upstream of the link between the southern and northern branches.

Understanding the spatial and temporal variations of the benthic communities is a basic tool for discriminating between pollution induced changes and natural variations (Boesch, 1973; Holland et al., 1987) and the first step towards the development of environmental indicator tools for Portuguese estuaries. Assessment of the benthic condition of the northern branch of the Mondego River estuary can give a reasonable representation of both natural and man-made impacts in the entire river basin since it receives most of the freshwater generated upstream. This paper gives an overview of the spatial distribution of benthic communities in the northern branch of the Mondego River estuary, identifying the main environmental gradient generating the distribution of those communities. Seasonal changes were also analyzed by examining variations in the dominant species. These findings emphasize the ecological importance of the poorly understood biotic differences between homiohaline and poikilohaline estuaries.

Methods

Study area

The Mondego River estuary is located on the Portuguese Atlantic coast (40°08′ N; 8°50′ W), a temperate-warm region influenced both by Atlantic and Mediterranean climates (Fig. 1). This region is characterized by a rainfall period that extends from November to May and a drought period of very low water flow between June and October (Loureiro et al., 1986). River flow data for the period between 1987 and 1997 measured at the Coimbra dam indicated an annual mean flow of 812 m^3 s^{-1}. Maximum flows were measured from December to March and minimum flows occurred between June and October (Fig. 2). Average monthly flows varied between a maximum value of 167 m^3 s^{-1} in January and a minimum average flow of 15 m^3 s^{-1} in September. The Mondego River flow regime is very irregular and important daily changes can occur due to the action of dams controlling the discharges. This estuary is affected by a mesotidal semi-diurnal regime and is normally totally mixed, except for periods of extreme floods or droughts when it can be only partially mixed (Cunha & Dinis, 2002). During the last decades the Mondego River estuary was severely altered by the construction of dams located upstream, the drainage of some mudflat areas, embankment of the river margins, dredging activities to maintain a navigation channel and intensive agriculture use. Sediment grain size has a very heterogeneous spatial distribution along the estuary, much coarser in the upstream areas where it consists mainly of very coarse sand and some marginal locations covered with fine sand. The lower section has a dominant composition of medium to fine sand although there are small areas near the river banks were sandy mud sedimentation occurs (Cunha & Dinis, 2002).

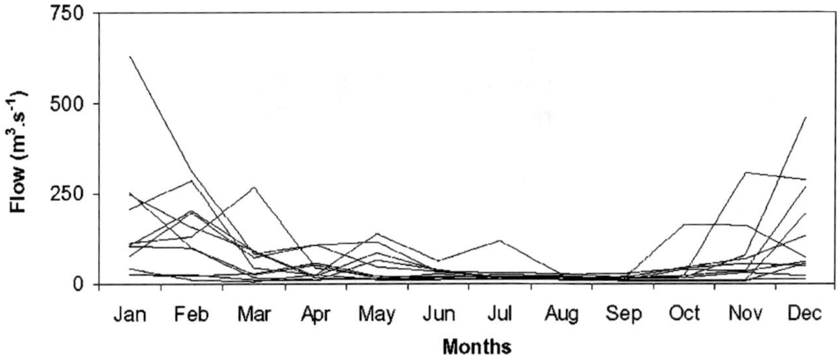

Figure 2. Average monthly flows measured at Coimbra dam between 1987 and 1997 (source: Portuguese Water Institute).

Sampling design

A total of seven sampling stations were selected along the longitudinal extension of the northern branch of the estuary (Fig. 1) in an attempt to cover the salinity range between the vicinity of the river mouth and the upper tidal reaches. Three benthic invertebrate samples were taken at each station using a modified van Veen grab (0.05 m^2) and their contents were fixed and preserved with 4% buffered formalin. Grab contents were sieved in the laboratory using a 500 μm mesh and preserved in 70% alcohol. All samples were sorted using a microscope and invertebrates were identified to the highest possible taxonomic separation. Bottom dissolved oxygen (DO) (mg l^{-1}), water temperature (°C) and salinity were measured *in situ* using a Data Sonde Surveyor 4 and a Sechii disc was used to measure transparency (m). Additional water samples were collected and frozen for subsequent analysis of nitrates (mg l^{-1}), nitrites (mg l^{-1}), phosphates (mg l^{-1}) and ammonia (mg l^{-1}). The analyses were made by the laboratory of the Portuguese Environmental Institute using methods certified by the Portuguese Quality Institute. At each station, another grab sample was taken and frozen for sediment grain size, total organic content (TOC) and heavy metals analysis. Sediment grain size composition was determined using an AFNOR type sieve battery (0.063 mm; 0.25 mm; 0.5 mm; 2 mm; 9.25 mm) after drying the sediment (60 °C) for a period of 48 h. Samples were classified using the Roux (1964) scale. TOC was obtained as the difference between dry weight, measured after drying the sample at 60 °C during 24 h and ash weight, obtained after ignition at 480 °C for a period of 12 h. Heavy metal concentrations (mg kg^{-1}) in the sediment (arsenic, chromium, lead, cooper and zinc) were also determined by the Portuguese Environmental Institute, using certified methods.

Seasonal variations were assessed by repeating the surveys every 3 months, namely in July (summer) and October 2000 (autumn) and in February/March (winter) and June (spring) 2001. All samples were collected on low tide, during spring tide periods. The winter cruise took place immediately after a flood event and Stations 1 and 7 could not be sampled due to the strong currents impeding the use of the collecting devices.

Data analysis

Spatial patterns

Stations were classified into spatial groups using the "mean variance per comparison" technique described by Williams & Stephenson (1973). By applying this technique it was possible to group stations into spatial groups independent of the effects associated with collections conducted during different seasons. The technique estimates the variance between stations and between sampling events by calculating the Euclidean distance between stations and sampling events (over all species) after stations and sampling events were centred to their respective means. The variance estimates were used as a measure of dissimilarity between stations for cluster analyses in order to assign the spatial groups. A flexible sorting strategy was used for the cluster analysis with an intensity coefficient or value of −0.25 (Boesch, 1977a). Dissimilarity coefficients were calculated using a program written in the SAS/IML® matrix programming language while the dendrograms for this analysis were produced using PROC CLUSTER of the SAS /Stat® software package. All species counts were standardized to have an overall mean value of zero and a standard deviation of one prior to conducting this analysis. The analysis was conducted using data collected during only three seasons (summer, autumn and spring) due to the missing observations in winter. An overall test for a significant difference in species composition between site groups was accomplished using a MANOVA (Wilk's Lambda) while pairwise comparisons between individual site groups were made using an *F*-test on pairwise squared Mahalanobis distances. Descriptive measures determined for the benthic communities of each group identified by the cluster analysis were the average density, the Shannon-Wiener diversity index (H′ log *e*), the number of species (considering each taxon as a species), the Simpson's dominance index (λ') and the top 10 dominant species. Similarity percentage breakdown procedure (SIMPER) (Clarke & Warwick, 1994), included in the PRIMER software package was used to determine the contribution of individual taxa towards the dissimilarity between and similarity within the groups identified by Cluster analysis.

Seasonal variation
Multivariate ordination techniques were applied to analyze spatio-temporal variation in the species density reduced data set and identify the relations between environmental and biological data, using CANOCO 4.5 software. Relations between environmental and biological data were analyzed by performing a Canonical Correspondence Analysis (CCA). CCA constrains the axes to be linear combinations of the environmental variables (ter Braak & Šmilauer, 2002).

The species data matrix was reduced by eliminating taxa that occurred in less than five samples and accounted for less than 0.1% of the total abundance, to obtain an interpretable ordination diagram. A matrix of explanatory variables was also constructed to determine the variation in the species data that was related to environmental factors. The reduced environmental matrix excluded variables that showed collinearity (silt and clay) and variables which registered values under the detection limit (nitrates, nitrites, phosphates and ammonia). Metals were also excluded because they never exceeded Long et al.'s (1995) effects range-median (ERM) or Crommentuijn et al.'s (2000) maximum permissible concentrations (MPC) and any differences found between stations would not have a biologically meaningful interpretation. All other variables measured were included, namely dissolved oxygen, water temperature, salinity, transparency, depth, sediment type (stones, coarse sand, medium sand, fine sand) and total organic content.

Selection of variables was based on stepwise multiple regressions and the statistical significance of the variables added to the analysis was tested using a Monte Carlo permutation test (499 unrestricted permutations) (ter Braak & Verdonschot, 1995). Decision about the ordination model to use in the constrained analysis (CCA) was based on length of gradient calculated by a previous Detrended Canonical Correspondence Analysis (DCCA). Since the longest gradient was 4.05 the CCA was based on a unimodal model. DCCA was also used to obtain the ordination of species and samples, since an "arch effect" was apparent in the initial canonical correspondence analysis (CCA) (Gauch, 1982).

Seasonal variations were analyzed by plotting major taxonomic groups and ecological groups for each station group-season combination based on the classification of Borja et al. (2000) (Appendix 1). These authors assigned taxa to one of five ecological groups, according to their sensitivity to an increasing stress gradient (I to V). Contributions from individual taxa to dissimilarities between seasons were estimated using the similarity percentage breakdown procedure (SIMPER). Densities of the dominant species were plotted for each season.

Results

Benthic macrofauna general characterization

A total of 38 394 invertebrate specimens were collected and 84 taxa identified (Appendix 1). The highest densities were registered during spring at Station 9 (56,613 ind m^{-2}) while the lowest values were observed during winter at Station 4 (160 ind m^{-2}). The 10 dominant species accounted for 93% of the total average density with the amphipod *Corophium multisetosum* Stock, 1952 (47%) and the polychaete *Streblospio shrubsolii* (Buchanan, 1890) (21%) being the two dominant species. The dominance of *C. multisetosum* was due to the high numbers collected during spring at the three upstream stations. *S. shrubsolii* was the most abundant species during all the other seasons. Dominant species found during summer and autumn were typical estuarine surface and subsurface deposit feeders (e.g. *S. shrubsolii, Spio martinensis* Mesnil, 1896, *Caulleriella caputesocis* (Saint-Joseph, 1894)) but some freshwater invertebrates such as insect larvae (Ephemeroptera and Diptera) were collected during winter and spring, mainly in the upstream stations.

Spatial patterns in benthic communities

Cluster analysis indicated three main groups of stations in the northern branch of the Mondego River estuary (Fig. 3). There was a significant difference in species composition across all site groups (Wilk's Lambda = 0.063; DF = 28:33; $p < 0.0001$) and between each of the individual site groups (Table 1). Spatial Group A consisted of Stations 1 and 2, which are located in the lower sector of the estuary. These stations are characterized by medium sand sediments, a stronger tidal influence and a salinity decrease associated with

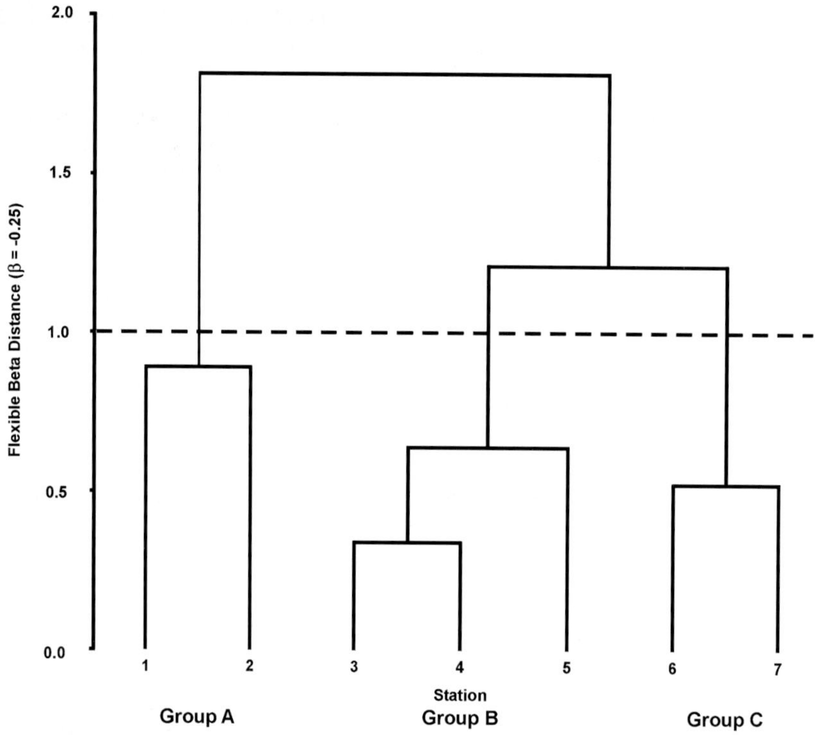

Figure 3. Cluster analysis of the density data set collected in the northern branch of the Mondego River estuary. Stations were grouped into spatial groups independent of the effects associated with collections conducted during different seasons using the "mean variance per comparison" method described by Williams and Stevenson (1973).

flood events that occur primarily during winter. This group had the lowest dissolved oxygen concentrations, as well as higher levels for TOC, ammonia, lead, copper and zinc concentrations (Table 2). This may be due to the spatial group's proximity to the urban area of Figueira da Foz that has harbour facilities, sewage outfalls and a large bridge with automobile traffic. Lead and zinc concentrations exceeded Long et al.'s (1995) effects range-low (ERL) at the station located closer to the bridge during winter. The benthic community identified for this group was dominated by some polychaete and bivalve species. The spionid polychaete *Streblospio shrubsolii* showed the higher average abundance (29%), followed by the suspension feeding bivalves *Cerastoderma glaucum* (Poiret, 1789) (25%) and *Scrobicularia plana* (da Costa, 1778) (10%) (Table 2). Group B clustered Stations 3, 4 and 5, located in the middle sector of the estuary. At this location salinity decreased during winter and spring but tidal influence maintained relatively high salinities during the

Table 1. Pairwise comparisons of species composition between site groups

Site Group	A	C
B	15.37	16.07
	3.26	3.41
	0.0007	0.0005
A		22.74
		4.02
		< 0.0001

Shown are the pairwise squared Mahalanobis distances between site groups, the F values and their associated probabilities.

drought period. It is a transition zone for sediment that changes gradually from medium to coarse sand moving upstream. The lower average percentages of TOC were found in this spatial group (Table 2). The benthic community was dominated by only two species, *S. shrubsolii* (48%) and the amphipod *Corophium multisetosum* (37%) (Table 2). The two uppermost stations formed Group C, characterized by lower salinities, higher

Table 2. Descriptive biological and environmental parameters of groups of stations identified in the Mondego River estuary (A, B and C)

	Group A (Stations 1 and 2)	Group B (Stations 3, 4 and 5)	Group C (Stations 6 and 7)
Biological parameters			
Density (ind m^{-2})	380 W – 9420 Sp	504 W – 9651 Sp	433 W – 39720 Sp
Diversity	1.00 W – 2.24 A	0.70 Sp – 1.52 W	0.63 Sp – 1.27 S/W
N Species	4 W – 40 S	14 W – 38 S	8 W – 14 Sp
Dominance	0.31 S/A – 0.04 W/Sp	0.24 W – 0.67 Sp	0.34 S – 0.72 Sp
Dominant Species	*Streblospio shrubsolii* (29%)	*Streblospio shrubsolii* (48%)	*Corophium multisetosum* (77%)
	Cerastoderma glaucum (25%)	*Corophium multisetosum* (37%)	Oligochaeta (10%)
	Scrobicularia plana (10%)	Oligochaeta (3%)	*Corbicula fulminea* (8%)
	Hydrobia ulvae (7%)	*Spio martinensis* (2%)	*Ephoron virgo* (2%)
	Spio martinensis (6%)	*Hydrobia ulvae* (2%)	Nemertea (1%)
	Caulleriella caputesocis (5%)	Nemertea (1%)	*Cyathura carinata* (<1%)
	Mediomastus fragilis (5%)	*Hediste diversicolor* (1%)	Chironomidae (<1%)
	Angulus tenuis (3%)	*Gammarus subtypicus* (1%)	*Gammarus subtypicus* (<1%)
	Heteromastus filiformis (2%)	*Capitella capitata* (1%)	*Boccardiella ligerica* (<1%)
	Oligochaeta (1%)	*Corbicula fulminea* (1%)	*Saduriella losadai* (<1%)
Environmental parameters			
Depth (m)	3.00–5.50	0.80–5.00	3.50–6.00
Salinity (‰)	7.00–40.20	7.00–31.60	2.00–14.20
DO (mg l^{-1})	3.40–9.00	6.40–9.50	6.40–9.60
NO_3 (mg l^{-1})	1.00–4.70	2.50–5.40	3.20–5.40
NO_2 (mg l^{-1})	< 0.05	0.05–0.10	0.05–0.07
NH_4^+ (mg l^{-1})	0.08–0.33	0.08–0.23	0.08–0.20
As (mg kg^{-1})	1.20–9.00	0.90–4.60	1.00–13.00
Pb (mg kg^{-1})	2.40–92.00	4.00–28.30	2.40–40.00
Cu (mg kg^{-1})	1.30–28.00	0.60–9.00	0.70–23.00
Cr (mg kg^{-1})	2.40–59.00	5.30–44.00	3.00–70.00
Zn (mg kg^{-1})	9.40–184.00	13.00–71.00	12.00–115.00
TOC (%)	0.40–9.90	0.20–4.00	1.20–10.10
Sediment type	Medium sand	Coarse-Medium sand	Coarse sand

Maximum and minimum average values seasonally determined (summer – S; autumn – A; winter – W; spring – Sp) are presented for each group, concerning density, Shannon-Wiener diversity, number of species and Simpson's dominance. Top 10 dominant species are listed with the respective contributions to total density

oxygen levels and coarser sand. Chromium and zinc measured at these sites exceeded Crommentuijn et al.'s (2000) Negligible Concentration (NC) values and Long et al.'s (1995) ERL values and higher nitrates and TOC levels were also measured, probably due to the intensive agriculture in this area. The benthic community was strongly dominated by *C. multisetosum* (77%), mainly because of high numbers collected during spring. Oligochaetes (10%) and an introduced bivalve species *Corbicula fulminea* (Müller, 1774) (8%) were also abundant. Although in low numbers, insects larvae (*Ephoron virgo* (Olivier, 1791) and Chironomidae) occurred in this area associated with higher winter and spring freshwater input (Table 2).

SIMPER analysis showed higher dissimilarities between Groups A and C (91%) and higher closeness of stations of Group B to A than to C. *Streblospio shrubsolii* gave the highest contribution to the dissimilarities between the lower and the middle sector of the estuary (16%), except for the

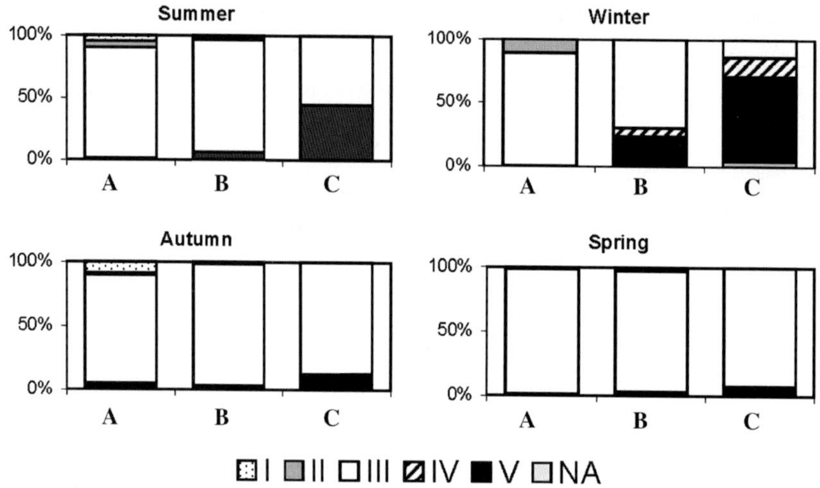

Figure 4. Ecological groups (Borja et al., 2000) identified seasonally for groups of stations defined in the Mondego River estuary.

spring assemblage. Although dominant in both assemblages this spionid occurs with lower density at Group A stations. On the other hand, *Cerastoderma glaucum*, the polychaete *Spio martinensis* and the gastropod *Hydrobia ulvae* (Pennant, 1777) were found with much higher densities in Group A and explained 14%, 7% and 6% of the dissimilarity between Groups A and B, respectively. The high density of *Corophium multisetosum* in the upstream stations accounted for 27% of the dissimilarity between Groups B and C. Oligochates, *S. shrubsolii* and *Corbicula fulminea* were also important in separating these two groups and explained 15%, 15% and 12% of the dissimilarities between Groups B and C, respectively.

Ecological groups obtained by classifying taxa according to their sensitivity to pollution (Borja et al., 2000) were used as a measure of the ecological structure of the assemblages showing an average density domination of 18 taxa belonging to Group III, which includes species tolerant to organic enrichment. Over 92% of the organisms collected during the study were classified in this group, while only two taxa, Oligochaeta and *Capitella capitata* (O. Fabricius, 1780), or about 5% of the total number of organisms were classified into Group V (Fig. 4), considered as first-order opportunistic species.

Seasonal variation in benthic communities

Major variations were found in winter assemblages that showed a reduced number of species and extremely low abundances (Table 2). In contrast, higher numbers and stronger dominance were registered during spring in all station groups with increased densities in Group C. As previously mentioned the ecological structure of the Mondego River estuary was numerically dominated by Group III species (Fig. 4). However, some variations of the specific composition occurred between seasons. Summer and autumn registered a higher heterogeneity of the ecological groups represented in Group A, mainly because of the occurrence of rare species of marine influence. Spatial Group C registered the highest numerical representation of Group V during summer and winter, due to the density of Oligochaetes (Fig. 4). Spring showed higher homogeneity due to the strong numerical abundance of Group III species that lowered the relative contribution of other groups.

SIMPER analysis showed which species contributed the most to the seasonal variations within each station group:

Group A – higher dissimilarities were found between winter assemblages and all other seasons. Only four taxa were collected with very low numbers in this period, namely *Streblospio shrubsolii*, Hesionidae, the isopod *Cyathura carinata* (Kroyer, 1847) and Turbellaria. *Cerastoderma glaucum*, *Scrobicularia plana*, *Spio martinensis*, *Mediomastus fragilis* (Rasmussen, 1973) and *Caulleriella caputesocis* were the species which best discriminated between summer and all other seasons, showing higher densities during summer, decreasing towards winter (Fig. 5). *S. shruboslii*

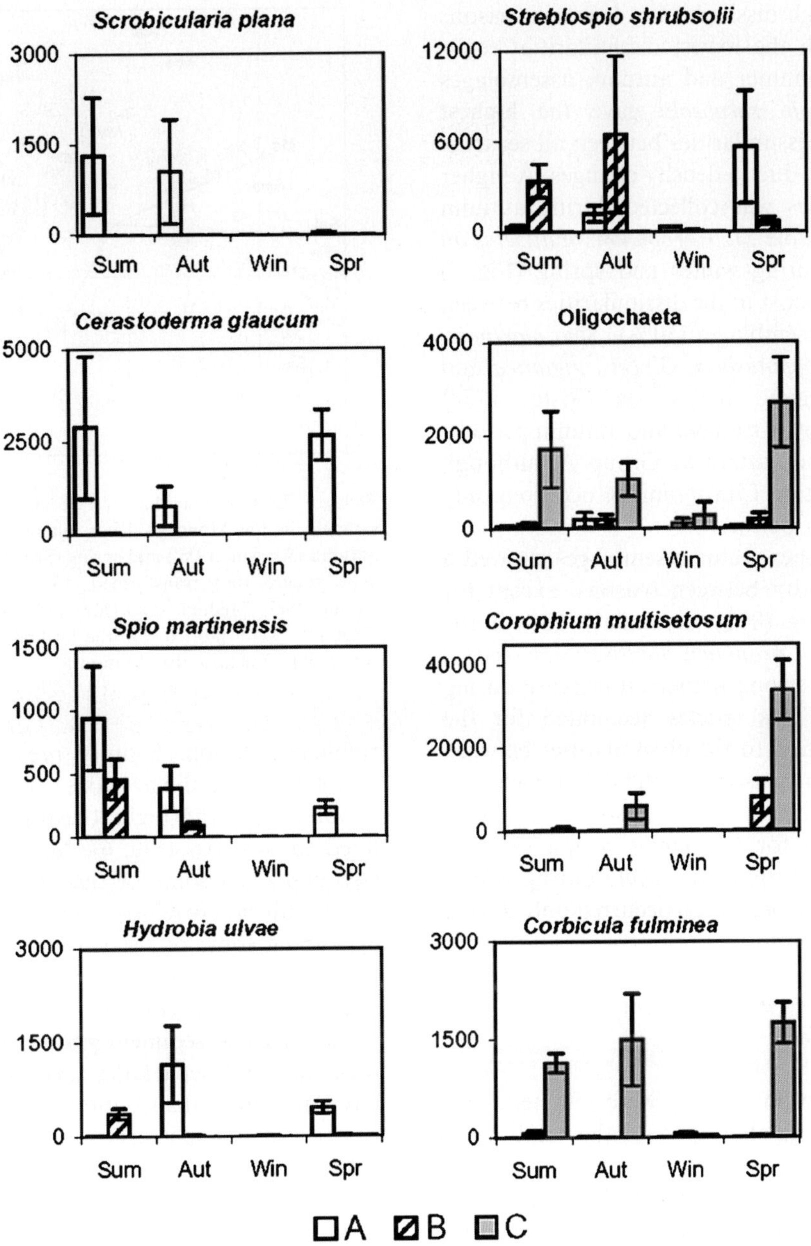

Figure 5. Seasonal variations on the densities of the eight dominant species in the benthic community of the northern branch of the Mondego River estuary. Standard error is indicated.

was the best discriminating species (>20%) between spring and other seasons due to the higher densities found in this period (Fig. 5). Summer was also characterized by the presence of rare species such as *Eumida sanguinea* (Örsted, 1843), *Glycera gigantea* de Quatrefages, 1866, *Nephtys cirrosa* Ehlers, 1868, *Nepthys hombergii* de Lamarck, 1818 and *Modiolus barbatus* (Linnaeus, 1758), that were absent during other seasons. Some other species revealed a relatively important presence during summer and autumn but were absent from the winter and spring assemblages such as *Heteromastus filiformis* (Claparède, 1864) and *C. caputesocis*.

Group B – high dissimilarities between seasons were found, with the lowest dissimilarities registered between summer and autumn assemblages (57%). *Streblospio shrubsolii* gave the highest contribution to dissimilarities between all seasons, except for spring, due to density changes. A higher number of species was collected during autumn and summer (Table 2). *Corophium multisetosum* occurred only during winter and spring (Fig. 5) and contributed most to the dissimilarities between these seasonal assemblages (30%). *Spio martinensis*, *Caulleriella caputesocis*, *Glycera gigantea* and the bivalve *Angulus tenuis* (da Costa, 1778) occurred only during summer and autumn periods, showing the same pattern as Group A, although with lower densities. Chironomidae occurred only during the winter period.

Group C – upper sector assemblages showed a higher homogeneity between seasons, except for winter and spring (80% dissimilar) due to the contribution of *Corophium multisetosum* (60%) that registered a strong numerical increase during spring (Fig. 5). This species accounted for the highest contribution to the dissimilarities between spring and other seasons (>50%). Oligochaetes and *Corbicula fulminea* were also abundant in Group C, except for the winter period (Fig. 5). Chironomidae occurred exclusively during winter and spring and *Ephoron virgo* occurred only during spring.

Relationships between environmental and biological variables

After data reduction 28 taxa were retained, that accounted for more than 99% of the total abundance.

The constrained ordination (CCA) of species density with stepwise forward selection of environmental variables retained only three environmental variables ($p < 0.05$): salinity, medium sand and TOC. The ordination plot showed an apparent arch effect on species and samples, suggesting the need for detrending. DCCA of species abundance data produced an ordination in which the first four axes were statistically significant ($p < 0.01$), with respective eigenvalues of 0.66, 0.14, 0.06 and 0.03. The first two axes explained 34.9% of the total variance in species data and 56.8% of the total variance on the species–environment relation. Salinity presented a negative correlation with the first axis (−0.90) and it was the variable with the highest explanatory power related to this axis (Fig. 6). Medium sand content also explained some of the variation on the first axis, with a correlation of −0.67. In spite of explaining some of the variance in species distribution, TOC showed low correlations with the first two axes (<0.30) (Fig. 6).

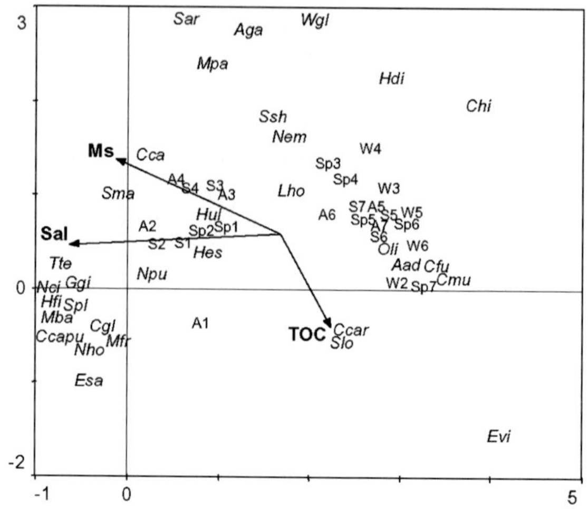

Figure 6. DCCA of benthic invertebrate density for the stations sampled in the Mondego River estuary during summer (S), autumn (A), winter (W) and spring (Sp). Taxa abundance data were square root transformed. Environmental variables – Salinity (Sal), Medium sand (Ms) and Total Organic Content (TOC) – were plotted on the ordination as arrows. See Appendix 1 for taxa abbreviations.

Salinity and sediment grain size established the main gradient separating species that apparently have an optimal distribution in lower salinities and coarser sands, such as *Corophium multisetosum, Corbicula fulminea, Ephoron virgo*, Chironomidae, *Amage adspersa* (Grube, 1863) and Oligochaeta (Fig. 5). These species were mainly identified in the Group C assemblage. The ordination indicates a medium location of *Streblospio shrubsolii, Cyathura carinata, Hediste diversicolor* (O.F. Müller, 1776), *Saduriella losadai* Holthuis, 1964, Nemertea, *Lekanesphaera hookerii* (Leach, 1814) *and Websterinereis glauca* (Claparède, 1870) along the saline gradient. Some of these species occurred with higher densities in Group B stations but all of them demonstrated tolerance to changes occurring along the estuary since they showed a wide distribution along the three different regions identified. In the DCCA plot the remaining species were located closer to the positive end of the salinity and sediment vectors, associated with spatial Group A stations. A small part of the species variation was still explained by the second axis (6.2%), separating species characteristic of Groups A and C assemblages from Group B. Although the correlation of TOC to this axis was not very high, this variable apparently explains part of the differences mentioned, since the Group B stations registered lower values. DCCA ordination of species and samples also revealed temporal differences in similarity within groups, separating the plots of the same stations in different seasons (Fig. 6).

Discussion

The identification of the spatial and seasonal patterns of change in the benthic communities of the Mondego River estuary may be an important contribution to the development of biological criteria to assess the environmental condition of Portuguese estuaries. The study design led to the identification of three spatial groups with distinct environmental and biological characteristics in the northern branch of this estuary, instead of the five classes established by the Venice system. That system, based on the identification of salinity classes, is widely accepted and indicated by the WFD to identify different water types. Salinity decreases moving upstream from spatial Group A near the mouth of the River to the spatial Group C upstream and is coupled with a transition from medium to coarse sands upstream. Several sources of anthropogenic disturbance were identified in Group A, such as a well developed urban area with harbour facilities that require periodic dredging of the access channel, dumping and a bridge with high traffic levels. This sector showed organic enrichment and higher levels of heavy metals were measured in the bridge vicinity. Lower organic enrichment was found in spatial Group B while the Group C stations showed higher levels of organic matter and nutrient concentrations, probably related to nutrient loads from intensive agriculture.

Benthic communities presented similar patterns of longitudinal change along the estuary. The lower sector assemblages were dominated by the surface deposit feeder *Streblospio shrubsolii* and some suspension feeding (*Cerastoderma glaucum*) and deposit-feeding (*Scrobicularia plana*) bivalve species. Group B showed a stronger dominance of *S. shrubsolii* populations, shared with the amphipod *Corophium multisetosum.* The last species mentioned accounted for almost 80% of the upper sector assemblages due to high densities found during spring. DCCA identified a saline gradient influencing species distribution. Seasonal differences found in community composition and structure, were apparently also related to salinity changes caused by high freshwater input from upstream. Although salinity values were obtained with a single measurement for each season, daily and monthly flows measured upstream strongly indicate that they were representative of seasonal changes (Fig. 2). Colonization of Group A stations by marine species was observed during the drought season while Group C stations were colonized by freshwater species during higher flow periods. Species with higher tolerances to salinity changes such as *S. shrubsolii*, *Hediste diversicolor*, *Cyathura carinata* and *Saduriella losadai* showed a wider distribution and persistence between seasons. These results are consistent with previous studies in the Mondego River estuary (Marques et al., 1993; Pardal et al., 1993), although the number of species and densities collected during this study were much higher. The higher densities in the present study are probably due to the use of a 1 mm mesh sieve during previous studies, compared to the use of a 0.5 mm mesh sieve in the present study.

Salinity effects may be acting together with physical disturbances caused by the strong currents occurring during the rainfall period and consequent alterations of the erosion–deposition cycles. Winter communities were very impoverished both numerically and in terms of the number of species and stronger changes occurred in the lower sector of the estuary, where most stenohaline species were found. In the middle sector lower salinities persisted for a longer period and the benthic community was apparently adapted to this saline regime showing lower changes between seasons. *Streblospio shrubsolii*, a typical estuarine species dominated the community over all seasons, except for spring when *Corophium multisetosum* increased its density. This polychaete is classified as an opportunistic species that colonises organically enriched sediments (Pearson & Rosenberg, 1978;

Sardá & Martin, 1993). This change in the dominance may be due to competition as both species occupy similar spatial niches (surface feeders dwelling in the upper 1 cm of the sediment) and occur in high densities. Previous studies carried on Portuguese estuaries showed the prevalence of this amphipod in salinities ranging from 2.5 to 10 and coarser sediment (Queiroga, 1990). Cunha et al. (2000) concluded that *C. multisetosum* abundance may be associated with increased freshwater inflow following rainy periods and subsequent decreasing abundance with the higher summer temperatures, since it is a cold-temperature species, with its southern limit in Portugal. During spring, *C. multisetosum* seems to replace the dominant spionid *S. shrubsolii* in Group B, since increasing densities occur in the lower sector for this species in the same period. The benthic community of the upper sector seems to be more stable with lower fluctuations in specific composition and densities, except for the spring boom of *C. multisetosum*.

The AZTI Marine Biotic Index (AMBI) approach of Borja et al. (2000) produces an overall evaluation of benthic community condition, relative to anthropogenic stress, by scoring benthic species based upon relative tolerance to pollution. Using this approach the Mondego River estuary would be considered moderately disturbed since Group III species are dominant. These species are tolerant of excess organic matter and occur in normal conditions but increase their numbers when stimulated by organic enrichment. Based on AMBI values, winter communities in the upper sector would be considered severely degraded due to the dominance of oligochaetes and summer and autumn communities in Group A would be less disturbed since a higher number of pollution sensitive species occur (e.g. *A. tenuis*, *Nephtys* spp., *Glycera* spp., *Owenia fusiformis* delle Chiaje, 1842, *Eteone picta* de Quatrefages, 1866, *Eumida sanguinea*, *Diopatra neapolitana* delle Chiaje, 1841). Pollution sensitive species did occur in all sectors but generally in low densities and with great variation between seasons. These results suggest caution in applying and interpreting the AMBI approach when strong seasonal patterns occur. Nevertheless, benthic communities are still under seasonal physical and physiological stress that favours the settlement of species with opportunistic life histories and/or wide tolerance to changes in environmental conditions, such as *Streblospio shrubsolii* (Sardá & Martin, 1993; Rossi & Lardicci, 2002). The present data demonstrate the ability of this species to colonize and produce high-density local populations in the lower sector of the Mondego River between winter and spring. Lardicci et al. (1997) indicated peaks of fecundity occurring every 2 months for this species when food availability is not a limiting factor.

Estuaries may be classified in many ways including geomorphology (e.g. drowned river valleys, fjords, etc.), degree of saline stratification (e.g. highly stratified, partially stratified, etc.), and tidal mixing (e.g., macrotidal, microtidal, etc.) (Lauff, 1967; Elliott & McLusky, 2002). Major variations in amount of freshwater flow may be due to strong seasonal differences in rainfall and such differences can be amplified by dam placement. Boesch (1977b) states that poikilohaline estuaries, where strong seasonal changes in salinity and water flow occur at specific locations, cannot be assessed using the Venice system. He also states that euryhaline marine species are displaced as an effect of poikilohalinity, allowing the domination of estuarine endemic species. The dominance of *Streblospio shrubsolii* in Group B during summer and autumn and in Group A during winter and spring indicates an agreement of the Mondego River estuary results with poikilohalinity effects theory. Apparently, stronger freshwater inputs displace euryhaline species found in the lower sector of the estuary during lower flow periods, allowing the dominance of the spionid *S. shrubsolii*. At the same time, this estuarine endemic species is displaced from the middle sector by freshwater species such as the amphipod *Corophium multisetosum*.

The Mondego River estuary, a poikilohaline-type estuary, characterized by strong seasonal changes in water flow and salinity, cannot be consistently stratified into salinity regions based upon the Venice classification system as indicated by the WFD. Biotic communities, exemplified here by the benthic communities, are seasonally displaced, compared to a homiohaline-type estuary where the Venice system can be applied without modification. Future identification of reference conditions and design of monitoring programs cannot be accomplished without understanding how interactions between biotic and physicochemical dynamics differ between homiohaline and poikilohaline estuaries (Elliott & McLusky, 2002). Our results indicate that seasonal and spatial stratification may be necessary to be able to separate natural and anthropogenic stresses. In addition, to the biotic effects further research is necessary in understanding the relative roles of river flow versus groundwater as sources of pollutants. For example, during low river flow seasons groundwater flow may become relatively much more important than during times of high river flow. Understanding ecological patterns, particularly where spatial and/or temporal variation, is great, also requires a comprehensive appreciation of the interactions of geomorphology, hydrology, and climatology.

Acknowledgments

This study was financially supported by two Ph.D. fellowships (SFRH/BD/5144/2001 and SFRH/BD/6365/2001) granted by FCT (Science and Technology Foundation) and ESF in the aim of the III European Community Support Framework. Financial support was also given to a project sponsored by Instituto de Ambiente. The SAS/IML program used here was written by Dr. Raymond W. Alden III of the University of Nevada, Las Vegas. We would also like to thank Professors Jean-Claude Dauvin and João Castro for the help with the identification of some invertebrate species. Finally, we thank Gilda Silva and Nuno Prista for their valuable comments and contributions to the final draft of this manuscript.

References

Alden, R. W., S. B. Weisberg, J. A. Ranasinghe & D. M. Dauer, 1997. Optimizing temporal sampling strategies for benthic environmental monitoring programs. Marine Pollution Bulletin 34: 913–922.

Aller, R. C. & J. Y. Aller, 1998. The effect of biogenic irrigation intensity and solute exchange on diagenetic reaction rates in marine sediments. Journal of Marine Research 56: 905–936.

Aller, R. C., J. Y. Aller & P. F. Kemp, 2001. Effects of particle and solute transport on rates and extent of remineralization in bioturbated sediment. In Aller, J. Y. S. A. Woodin, & R. C. Aller (eds) Organism–Ssediment Interactions. University of South Carolina Press, Columbia, USA: 402 pp.

Attrill, M. J. & M. Power, 2000. Effects on invertebrate populations of drought-induced changes in estuarine water quality. Marine Ecology Progress Series 203: 133–143.

Bilyard, G., 1987. The value of benthic infauna in marine pollution monitoring studies. Marine Pollution Bulletin 18: 581–585.

Boesch, D. F., 1973. Classification and community structure of macrobenthos in the Hampton Roads area, Virginia. Marine Biology 21: 226–244.

Boesch, D. F., 1977a. Application of Numerical Classification in Ecological Investigations of Water Pollution. EPA Report 600/3–77–033 U.S. EPA. Corvallis, oregan, USA: 115 pp.

Boesch, D. F., 1977b. A new look at the zonation of the benthos along the estuarine gradient. In Coull, B. (ed.) Ecology of Marine Benthos. University of South Carolina Press, Columbia, USA: 467 pp.

Boesch, D. F., R. J. Diaz & R. W. Virnstein, 1976a. Effects of tropical storm Agnes on soft-bottom communities of the James and York estuaries and the lower Chesapeake Bay. Chesapeake Science 17: 246–259.

Boesch, D. F., M. L. Wass & R. W. Virnstein, 1976b. The dynamics of estuarine benthic communities. In Wiley, M. (ed.) Estuarine Processes, Vol. I. Academic Press, New York: 541 pp.

Borja, A., J. Franco & V. Pérez, 2000. A Marine Biotic Index to establish the ecological quality of soft bottom benthos within European estuarine and coastal environments. Marine Pollution Bulletin 40: 1100–1114.

Borja, A., I. Muxika & J. Franco, 2003. The application of a marine biotic index to different impact sources affecting soft-bottom benthic communities along European coasts. Marine Pollution Bulletin 46: 835–845.

ter Braak, C. J. F. & P. Šmilauer, 2002. CANOCO Reference Manual and CanoDraw for Windows Useŕs Guide: Software for Canonical Community Ordination (version 4.5). Microcomputer Power, Ithaca, New York, USA: 500 pp.

ter Braak, C. J. F. & P. F. M. Verdonschot, 1995. Canonical correspondence analysis and related multivariate methods in aquatic ecology. Aquatic Sciences 57: 255–289.

Cardoso, P. G., A. I. Lillebø, M. A. Pardal, S. M. Ferreira & J. C. Marques, 2002. The effect of different primary producers on *Hydrobia ulvae* population dynamics: a case study in a temperate intertidal estuary. Journal of Experimental Marine Biology and Ecology 277: 173–195.

Clarke, K. R. & R. M. Warwick, 1994. Change in Marine Communities: an Approach to Statistical Analysis and Interpretation. Natural Environmental Research Council, Plymouth Marine Laboratory, United Kingdom: 144 pp.

Cowie, P. R., S. Widdicombe & M. C. Austen, 2000. Effects of physical disturbance on an estuarine intertidal community: field and mesocosm results compared. Marine Biology 136: 485–495.

Crommentuijn, T., D. Sijm, J. de Bruijn, M. Van den Hoop K. van Leeuwen & E. van de Plassche, 2000. Maximum permissible and negligible concentrations for metals and metalloids in the Netherlands, taking into account background concentrations. Journal of Environmental Management 60: 121–143.

Cunha, M. R., M. H. Moreira & J. C. Sorbe, 2000. The amphipod Corophium multisetosum (Corophiidae) in Ria de

Aveiro (NW Portugal). II. Abundance, biomass and production. Marine Biology 137: 651–660.

Cunha, P. P. & J. Dinis, 2002. Sedimentary dynamics of the Mondego estuary. In Pardal, M. A. J. C. Marques, & M. A. S. Graça (eds) Aquatic Ecology of the Mondego River Basin: Global Importance of Local Experience. Imprensa da Universidade de Coimbra, Portugal: 43–62.

Dauer, D. M., 1993. Biological criteria, environmental health and estuarine macrobenthic community structure. Marine Pollution Bulletin 26: 249–257.

Dauer, D. M., J. A. Ranasinghe & A. J. Rodi, 1992. Effects of low dissolved oxygen events on the macrobenthos of the lower Chesapeake Bay. Estuaries 15: 384–391.

Elliott, M. & McLusky, 2002. The need for definitions in understanding estuaries. Estuarine, Coastal and Shelf Science 55: 815–827.

Flindt, M. R., L. Kamp-Nielsen, J. C. Marques, M. A. Pardal, M. Bocci, G. Bendoricchio, J. Salomonsen, S. N. Nielsen & S. E. Jørgensen, 1997. Description of three shallow estuaries: Mondego river (Portugal), Roskilde Fjord (Denmark) and the lagoon of Venice (Italy). Ecological Modelling 102: 17–31.

Gauch, H. G., 1982. Multivariate Analysis in Community Ecology. Cambridge University Press, Cambridge, United Kingdom: 314 pp.

Hily, C., H. le Bris & M. Glémarec, 1986. Impacts biologiques des emissaires urbains sur les ecosystemes benthiques. Océanis 12: 419–426.

Holland, A. F., A. T. Shaughnessy & M. H. Hiegel, 1987. Long-term variation in mesohaline Chesapeake Bay macrobenthos: spatial and temporal patterns. Estuaries 10: 227–245.

Kimmerer, W. J., 2002. Effects of freshwater flow on abundance of estuarine organisms: physical effects or trophic linkages?. Marine Ecology Progress Series 243: 39–55.

Lardicci, C., G. Ceccherelli & F. Rossi, 1997. Streblospio shrubsolii (Polychaeta: Spionidae): temporal fluctuations in size and reproductive activity. Cahiers de Biologie Marine 38: 207–214.

Lillebø, A. I., M. A. Pardal & J. C. Marques, 1999. Population structure, dynamics and production of *Hydrobia ulvae* (Pennant) (Mollusca: Prosobranchia) along an eutrophication gradient in the Mondego estuary (Portugal). Acta Oecologica 28: 289–304.

Lauff, G. H., 1967. Estuaries. American Association for the Advancement of Science, Washington, D.C. 757 pp.

Long, E. R., D. D. McDonald, S. L. Smith & F. D. Calder, 1995. Incidence of adverse biological effects within ranges of chemical concentrations in marine and estuarine sediments. Environmental Management 19: 81–95.

Loureiro, J. J., M. C. Almeida, M. L. Machado & E. Teixeira, 1986. Bacia hidrográfica do rio Mondego. In D.G.R.A.H., (ed.) Monografias Hidrológicas dos principais cursos de água de Portugal continental. Lisboa, Portugal: 240–278.

Marques, J. C., P. Maranhão & M. A. Pardal, 1993. Human impact assessment on the subtidal macrobenthic community structure in the Mondego estuary (western Portugal). Estuarine, Coastal and Shelf Science 37: 403–419.

Marques, J. C., M. A. Pardal, S. N. Nielsen & S. E. Jorgensen, 1997. Analysis of the properties of exergy and biodiversity along an estuarine gradient of eutrophication. Ecological Modelling 102: 155–167.

Martins, I., M. A. Pardal, A. I. Lillebø, M. R. Flindt & J. C. Marques, 2001. Hydrodynamics as a major factor controlling the occurrence of green macroalgal blooms in a eutrophic estuary: a case study on the influence of precipitation and river management. Estuarine, Coastal and Shelf Science 52: 165–177.

Pardal, M. A., J. C. Marques & G. Bellan, 1993. Spatial distribution and seasonal variation of subtidal polychaete populations in the Mondego estuary (western Portugal). Cahiers de Biologie Marine 34: 497–512.

Pearson, T. H. & R. Rosenberg, 1978. Macrobenthic succession in relation to organic enrichment and pollution of the marine environment. Oceanography and Marine Biology. Annual Review 16: 229–311.

Queiroga, H., 1990. *Corophium multisetosum* (Amphipoda: Corophiidae) in Canal de Mira, Portugal: some factors that affect its distribution. Marine Biology 104: 397–402.

Rossi, F. & C. Lardicci, 2002. Role of the nutritive value of sediment in regulating population dynamics of the deposit-feeding polychaete *Streblospio shrubsolii*. Marine Biology 140: 1129–1138.

Roux, R. M., 1964. Les sediments de l'étange de Reine. Recueil des Travaux de la Station Maritime d'Endoume 35: 257–285.

Salen-Picard, C. & D. Arlhac, 2002. Long-term changes in a Mediterranean benthic community: relationships between the Polychaete assemblages and hydrological variations of the Rhône river. Estuaries 25: 1121–1130.

Sardá, R. & D. Martin, 1993. Populations of *Streblospio* Webster (Polychaeta: Spionidae) in temperate zones: demography and production. Journal of the Marine Biological Association of the United Kingdom 73: 769–784.

Weisberg, S. B., J. A. Ranasingue, C. Schaffner, R. J. Diaz, D. M. Dauer, L. Frithsen & J.B. , 1997. An estuarine benthic index of biotic integrity (B-IBI) for Chesapeake Bay. Estuaries 20: 149–158.

Williams, W. T. & W. Stephenson, 1973. The analysis of three-dimensional data (sites × times × species) in marine ecology. Journal of Experimental Marine Biology and Ecology 11: 207–227.

Appendix

List of taxa identified in the northern branch of the Mondego River estuary and respective ecological groups, assigned according to Borja et al. (2000) criteria. Abbreviations of taxa used in the DCCA are also shown.

Phyllum Plathyelmintha		Family Phyllodocidae		Family Gammaridae	
ClassTurbellaria		*Eteone picta* **(II)**		*Gammarus subtypicus* **(I)**	
Turbellaria n.i. **(NA)**		*Eumida sanguinea* **(II)**	*Esa*	Family Haustoriidae	
Phyllum Nemertea		*Ophiodromus flexuosus* **(II)**		*Haustorius arenarius* **(I)**	
Nemertea n.i. **(III)**	*Nem*	Family Pilargiidae		Family Hyalidae	
Phyllum Annelida		*Sigambra tentaculata* **(NA)**		*Hyale pontica* **(I)**	
Class Oligochaeta		Family Pisionidae		Family Melitidae	
Oligochaeta n.i. **(V)**	*Oli*	*Pisione remota* **(II)**		*Melita palmata* **(III)**	*Mpa*
Class Hirudinea		Family Polygordiidae		Order Decapoda	
Hirudinea n.i. **(NA)**		Polygordiidae n.i. **(NA)**		Family Crangonidae	
Class Polychaeta		Family Saccocirridae		*Crangon crangon* **(I)**	
Family Ampharetidae		*Saccocirrus papillocercus* **(NA)**		Family Portunidae	
Amage adspersa **(NA)**	*Aad*	Family Sabellariidae		*Carcinus maenas* **(III)**	
Amage gallasii **(NA)**	*Aga*	Sabellariidae n.i. **(I)**		Family Processidae	
Family Capitellidae		Family Spionidae		*Processa sp.* **(I)**	
Capitella capitata **(V)**	*Cca*	*Boccardiella ligerica* **(IV)**		Class Insecta	
Heteromastus filiformis **(III)**	*Hfi*	*Polydora* ciliata **(IV)**		Order Colembola	
Mediomastus fragilis **(III)**	*Mfr*	*Prionospio fallax* **(IV)**		Colembola n.i. **(NA)**	
Family Cirratulidae		*Prionospio cirrifera* **(IV)**		Order Ephemeroptera	
Caulleriella caputesocis **(III)**	*Ccapu*	*Pygospio elegans* **(III)**	*Sma*	Ephemeroptera n.i. **(NA)**	
Family Glyceridae		*Spio martinensis* **(III)**	*Ssh*	Family Polymitarcidae	
Glycera tridactyla **(II)**		*Streblospio shrubsolii* **(III)**		*Ephoron virgo* **(II)**	
Glycera gigantea **(II)**	*Ggi*	**Phyllum Arthropoda**		Family Caenidae	
Goniada emerita **(II)**		Class Crustacea		*Caenis* sp. **(NA)**	
Goniada norvegica **(II)**		Order Ostracoda		Order Plecoptera	
Family Hesionidae		Ostracoda n.i. **(NA)**		Family Leuctridae	
Hesionidae n.i. **(NA)**	*Hes*	Order Tanaidacea		Leuctridae n.i. **(NA)**	
Family Nereididae		Family Paratanaidae		Order Diptera	
Hediste diversicolor **(III)**	*Hdi*	*Leptochelia dubia* **(NA)**		Family Chironomidae	
Websterinereis glauca **(III)**	*Wgl*	*Heterotanais* sp. **(NA)**		Chironomidae n.i. **(IV)**	
Family Nephtyidae		Order Mysidacea		**Phyllum Mollusca**	
Nephtys caeca **(II)**		Family Mysidae		Class Gastropoda	
Nephtys cirrosa **(II)**		*Acanthomysis longicornis* **(NA)**		Family Hydrobiidae	
Nephtys hystricis **(II)**		*Gastrosaccus spinifer* **(NA)**		*Hydrobia ulvae* **(III)**	*Hul*
Nephtys hombergii **(II)**	*Nho*	*Leptomysis gracilis* **(NA)**		Class Bivalvia	
Nephtys longesetosa **(II)**		Order Isopoda		Family Corbicularidae	
Nephtys pulchra **(II)**	*Npu*	Family Anthuridae	*Ccar*	*Corbicula fulminea* **(III)**	*Cfu*
Family Opheliidae		*Cyathura carinata* **(III)**		Family Cardiidae	
Ophelia radiata **(I)**		Family Chaetiliidae	*Slo*	*Cerastoderma glaucum* **(III)**	*Cgl*
Family Onuphidae		*Saduriella losadai* **(NA)**		Family Mytilidae	
Diopatra neapolitana **(I)**		Family Gnathiidae		*Modiolus barbatus* **(I)**	*Mba*
Family Orbiniidae		*Paragnathia formica* **(III)**		Family Solenidae	
Scoloplos armiger **(I)**	*Sar*	Family Sphaeromatidae	*Lho*	*Solen marginatus* **(I)**	
Family Oweniidae		*Lekanesphaera hookeri* **(II)**		Family Scrobiculariidae	

Continued on p. 74

Appendix (Continued)

Owenia fusiformis **(I)**		Order Amphipoda		*Scrobicularia plana* **(III)**	*Spl*
Family Paraonidae		Amphipoda n.i. **(NA)**		Family Tellinidae	
Paraonidae n.i. **(NA)**		Family Ampeliscidae		*Angulus tenuis* **(I)**	
Family Pectinariidae		*Ampelisca lusitanica* **(I)**		Class Opisthobranchia	
Pectinaria koreni **(I)**		Family Amphithoidae		Opisthobranchia n.i. **(NA)**	
Family Pholoidae		*Amphithoe* sp. **(NA)**		**Phyllum Echinodermata**	
Pholoe minuta **(II)**		Family Corophiidae		Class Ophiuroidea	
		Corophium acherusicum **(III)**	*Cmu*	Ophiuroidea n.i. **(NA)**	
		Corophium multisetosum **(III)**			

Hydrobiologia (2006) 555:75–87
H. Queiroga, M.R. Cunha, A. Cunha, M.H. Moreira, V. Quintino, A.M. Rodrigues, J. Serôdio & R.M. Warwick (eds),
Marine Biodiversity: Patterns and Processes, Assessment, Threats, Management and Conservation
DOI 10.1007/s10750-005-1107-3

Benthic ecology of semi-natural coastal lagoons, in the Ria Formosa (Southern Portugal), exposed to different water renewal regimes

Sofia Gamito
IMAR, Faculdade de Ciências do Mar e do Ambiente, Universidade do Algarve, Campus de Gambelas, 8005-139, Faro, Portugal (E-mail: sgamito@ualg.pt)

Key words: lagoon ecology, benthic macrofauna, water renewal, environmental stress, taxonomic resolution

Abstract

Several studies in semi-natural coastal lagoons in the Ria Formosa lagoonal system have been carried out. These man-made water reservoirs behave as small lagoons with one opening to the tidal channels, which may be intermittent. Because of their size, these reservoirs are ideal sites for ecological studies. Water quality and macrobenthic fauna were analysed in five water reservoirs. All reservoirs received the same incoming water through a tidal channel, but they differed in water renewal regime. Multidimensional Scaling (MDS) and Discriminant Analysis were used to evaluate the similarity among sites, stations and sampling occasions. Different levels of taxonomic resolution (family, large taxonomic groups and phylum level) were also evaluated. The separation of sites and stations became unclear using high taxonomic levels. Results from the multivariate analyses suggest a slight differentiation of the stations according to sampling occasion but a clear differentiation of the several water reservoirs. Some of the lagoons studied with low water renewal rates showed strong environmental variations. They were characterised by low diversity indexes and abundance of small-sized organisms. Other lagoons, with high water renewal rates, showed low environmental variation and well diversified and structured benthic communities. The main environmental factor that seems to affect the benthic communities was the variation in salinity between neap and spring tides, which is related with the water renewal regime. Coastal lagoons offer a protected shallow habitat, which can be highly productive. Well structured communities, controlled by *k*-strategists, can develop and settle in leaky lagoons, that is, lagoons with wide entrance channels and tidal currents which guarantee a good water renewal. In these lagoons, biomass can accumulate in large organisms. In contrast, lagoons with a single narrow entrance, that may be closed for long periods, are characterised by persistent physical stress and are dominated by communities of small-sized *r*-strategists.

Introduction

Coastal lagoons are often highly productive but, at the same time, they can be highly stressed by anthropogenic inputs and human activities (Kjerfve, 1994). Depending on their hydrology these shallow coastal environments may be characterised by gradual or sharp daily and seasonal variations in physical–chemical water parameters. Assessing patterns in the structure of benthic communities has several advantages over experimental methods for the detection of anthropogenic disturbance (Warwick et al., 1990) or ecological conditions. Because of the sessile nature of most benthic organisms, these can integrate environmental conditions over time better than pelagic fauna. However, identification of benthic fauna to the species level are time consuming and analysis of benthic data at lower levels of taxonomic resolution has been proven to be an efficient method for detecting environmental change in strong environmental gradients (Warwick, 1988; Olsgard et al., 1998; Rumohr & Karakassis, 1999).

The Ria Formosa lagoonal system lies along the southern coast of Portugal, extending for about 55 km. It is a true barrier island system, comprising mainland, backbarrier lagoons, inlet deltas, barrier islands, barriers platforms and shoreface (Pilkey et al., 1989). Tidal amplitude varies from a maximum of 3.5 m at spring tide to 0.5 at neap tide, which causes an important semi-diurnal and fortnightly tidal amplitude fluctuation of the water volume inside the system (Falcão & Vale, 1990). The total area covered by water during spring tides varies between 14.1 and 63.1 km^2 and during each tidal cycle there is an exchange of almost all the water mass (Águas, 1986). There is no significant freshwater input to the system and salinity remains around 36 psu throughout the year, except during sporadic, short run-off periods (Falcão & Vale, 1990).

The Ria Formosa lagoonal system covers a total area of 163 km^2, of which 20 km^2 are occupied by salinas and aquaculture ponds (CCRA, 1984). The main water reservoirs of the salinas and the extensive aquaculture ponds behave like small lagoons where there are one or more openings to a tidal channel. The water comes into the reservoir through a tidal gate, which can be manual or automatic, and is transferred to other water reservoirs by gravity or pumping, according to the salt or fish production needs. In these reservoirs, the main factors that can influence the community are the quality of the water that comes in, the reservoir size and depth and the water renewal rate. When the renewal rate is low the quality of the water inside the reservoirs may decrease. Small sizes and shallow depths can also contribute to a decrease in water quality.

These reservoirs are ideal sites for ecological studies because of their small size. Four water reservoirs were studied during 2 years, and a fifth one some years later, with the aim of analysing the macrofauna and their relation to water quality and water renewal regimes. All reservoirs received the same incoming water through a tidal channel, but they differed in water renewal regime. Therefore, we expected that the physical and chemical characteristics of the water, and consequently the macrofauna communities would differ due to this main environmental factor.

In previous analyses of data of the first four reservoirs data (Gamito, 1989, 1997) a separation of the sites according to water renewal rate was evident. One of the sites (Site B) had similar characteristics to the shallow-water or coastal marine systems, whereas another site (Site A) had characteristics of an environment under great stress, due to restricted water renewal. The other two sites, C and D, showed intermediate stress characteristics, but were subjected to dystrophic crisis accompanied by degradation in the water quality. These data can serve as a basis for comparing other reservoirs from the Ria Formosa lagoonal system.

The fifth reservoir (site E), studied some years later, is located between reservoirs C and D. Due to the fact that the water is renewed every day, it is expected to be similar to site B in terms of ecological characteristics. A variety of multivariate analysis techniques were used to evaluate the similarity among sites, stations and also sampling occasions. Different levels of taxonomic resolution (family, large taxonomic groups and phylum level) were also evaluated.

Methods

A 2-year sampling programme was carried out at four sites subjected to different water renewal regimes, in 1985 and 1986, with the aim of determining their current ecological conditions and relating them to the dynamics of water circulation. Some years later, in 1996/1997, a fifth water reservoir, from an aquaculture facility, was also studied. All sites, situated near Olhão, on the Ria Formosa, received the same incoming water through the Marim Channel, but they differed in water renewal regime (Fig. 1 and Table 1). At site A, a water reservoir of an inactive salina, the water was only renewed fortnightly, during the spring tides. At site B, a tidal-mill water reservoir, the water was partially renewed every day. Similarly, at site E, a fish farm water reservoir, the water is pumped daily to the fish production reservoirs. At two other sites (sites C and D) the water was renewed according to salt production requirements, but more intensively during spring tides.

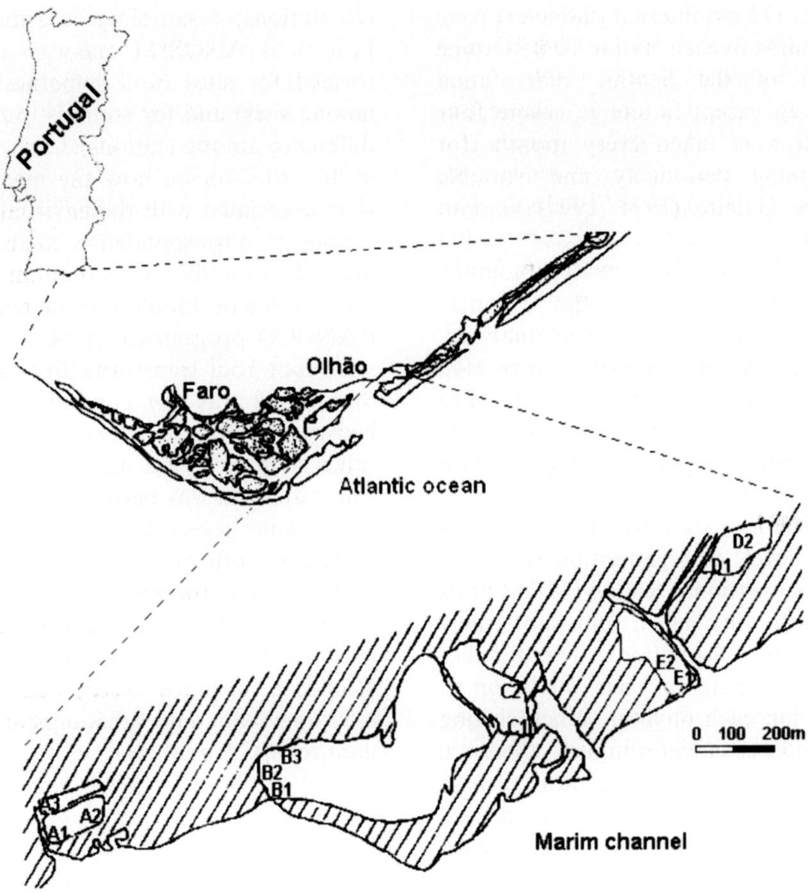

Figure 1. The study area and location of the sampling stations. Stations A1, B1, C1 and D1 are near the tidal gates. The water reservoir of site E had two tidal gates, one near station E1 and the other at some distance from station E2.

Table 1. Characteristics of the water reservoirs and number of stations in each site

Sites	Tide gate	Water renewal	Total area (ha)	Mean depth (m)	Number of Stations	Vegetation % cover
A – Old salinas	Manual	Every 14 days	1.5	1/0.4	3	< 10
B – Tidal mill	Automatic	Every day	9	1/0.6	3	100
C – Working salinas	Manual	Spring tides	2	1/0.6	2	50
D – Working salinas	Automatic	Spring tides	1	1/0.8	2	100
E – Aquaculture	Manual	Every day	1.4	0.8/2.0	2	< 10

The deeper stations were near the gates (stations A1, B1, C1 and D1), except in site E, where it was in the middle of the water reservoir (station E2).

Ten sampling stations were selected at the first four sites (A, B, C and D), and another two at the fifth (site E) (Fig. 1). Two water samples were taken every month at each station, one during the neap tides, when the water renewal is minimal, and the other a week later, during the spring tides. In site E water samples were taken from a site situated between the two stations. The following variables were determined: water temperature, salinity, dissolved oxygen, biochemical oxygen demand (BOD_5), chlorophyll *a* and phaeopigments. In the first four sites determination of nutrient concentration was also carried out (nitrates, nitrites, silicates and phosphates).

Two core samples (12 cm internal diameter) were taken every 2 months in each station to determine the composition of the benthic macrofauna (1 mm mesh sieve), except in site E, where four cores per station were taken every month (for details on sampling periodicity and variable determination see Gamito (1994, 1997) Gamito et al. (2003).

Total number of species, density, biomass, Shannon–Wiener diversity index and evenness (based on density data) in each site were analysed. Since the total number of species was very high and also because not all taxa could be identified to the species level, the data of all stations was grouped by families and explored with ordination techniques.

The differentiation of stations and sites at different taxonomic levels of organisation was evaluated with Multidimensional scaling (MDS) plots (Bray–Curtis dissimilarity index, root–root transformation or no transformation). The families were organised according to their phyla or to major groups within each phylum. When sorting out the animals after sediment sampling, these are usually placed in large taxonomic groups such as bivalves, gastropods, polychaetes, oligochaetes, amphipods, isopods, and so on, before the identification procedure.

MDS was also carried out on a reduced data set of the most abundant families in each station and sampling occasion. A mean density higher than 30 individuals m^{-2}, considering all the stations and sampling occasions was the criteria used to select the most abundant families. The main reason for this reduced data set is that the interpretation of ordination diagrams is clear with reduced data sets than with tens or even hundreds of taxa.

With this reduced data matrix, the similarities among sites, stations and sampling occasions were then investigated with ANOSIM and Discriminant analysis. The softwares used were PRIMER (v. 5.2) and Brodgar (v. 2.1). A ANOSIM two-way crossed analysis was performed, with sampling station as factor 1 and sampling month as factor 2. The analysis was repeated with sites as factor 1 (Null hypotheses: no site or station effect; no time effect). Only data from the first 2 years were used because the sampling periodicity at site E was different (every month, during 1 year, compared with every 2 months during 2 years, at sites A–D (10 stations×6 sampling occasions with two replicates). A ANOSIM one-way analysis was performed for sites (null hypothesis: no difference among sites) and for stations (null hypothesis: no difference among stations).

In order to see how the macrofauna families were associated with the environmental variables, canonical correspondence analysis (CCA) was applied to the data (mean annual densities of the most abundant families, in each station) using the CANOCO programme (v. 4.0). The family data were root–root transformed prior to the analysis. Station E2 was not considered in this analysis, because its depth was much greater than that of all other stations of the other sites and also because water analysis was performed at only one station in this water reservoir. The Bio-ENV routine from PRIMER software was used to select the more important environmental variables by comparing two triangular matrices, the Bray–Curtis dissimilarity matrix on root–root transformed data and a Euclidean distance matrix of the standardised environmental data (the same data matrices as used for CCA analysis).

Results

Salinity varied greatly in site A, as well as in in site C during the second year. This was due to reduced water renewal and high evaporation rates during summer months and rainfall during winter. Dissolved oxygen was always relatively high, which can be explained by the time of sampling (between 9 and 12 a.m). The occasional occurrence of high BOD values indicates some instability of the ecosystem (Table 2).

While sites B and E were characterised by higher faunal diversity, the biomass in site B was more than 5 times higher than in site E, mainly due to the presence of large specimens of molluscs and polychaete species (Table 3 and Fig. 2). Site A had the lowest diversity but a high density of small body sized organisms, with *r*-strategists characteristics.

A total of 96 families were identified in the five lagoons. 27 families had a mean density higher than 30 individuals m^{-2}, considering all the stations and sampling occasions, and represent 95.6% of all the data.

Table 2. Mean, maximal and minimal values of water temperature (Temp, °C), salinity (Sal, p.s.u.), dissolved oxygen (DO, mg l^{-1}), biochemical oxygen demand (BOD, mg l^{-1}), chlorophyll *a* (Chlo, mg m^{-3}) and phaeopigments (Phae, mg m^{-3}) in each site and year

		1985				1986				1996/97
		A	B	C	D	A	B	C	D	E
Temp	min	6.7	9.3	8.0	8.2	11.2	11.5	9.7	10.5	12.0
	mean	19.8	19.7	18.9	19.4	21.3	20.3	19.7	20.1	19.9
	max	28.7	27.4	28.1	28.8	29.8	27.4	26.7	26.9	25.0
Sal	min	13.3	31.3	26.6	29.9	29.2	34.5	30.5	34.9	27.2
	mean	37.6	36.7	39.8	40.5	47.4	37.2	43.1	43.1	35.2
	max	76.5	40.7	51.0	50.0	83.7	41.6	71.8	54.9	37.4
DO	min	3.3	2.0	2.8	3.5	4.2	5.0	2.7	3.3	4.2
	mean	8.3	7.4	6.7	6.9	8.4	7.6	6.8	6.1	7.0
	max	12.3	10.8	10.1	11.3	13.0	10.4	15.3	8.7	10.8
BOD	min	1.1	0.4	1.0	0.8	0.3	0.6	1.0	1.1	0.7
	mean	2.1	2.1	2.6	3.1	2.2	2.5	3.7	3.2	2.7
	max	7.2	7.4	7.3	9.2	6.4	8.0	8.4	8.1	6.2
Chlo	min	0.1	0.3	0.1	0.1	0.1	0.3	0.2	0.5	0.6
	mean	1.6	2.2	2.9	3.9	1.3	3.0	5.6	3.0	1.4
	max	7.7	8.2	11.5	15.7	7.7	11.1	23.3	11.9	2.9
Phae	min	0.1	0.2	0.2	0.2	0.1	0.4	0.5	0.0	0.2
	mean	0.8	1.5	1.3	1.4	0.6	1.6	2.0	1.3	0.7
	max	7.8	5.2	4.6	5.1	3.3	5.9	9.1	5.0	1.6

Table 3. Mean, maximal and minimal values of number of species (S), density (N, number of individuals m^{-2}), biomass (B, ash free dry weight g m^{-2}), production (P, ash free dry weight g m^{-2} y^{-1}), Shannon–Wiener diversity index (H′, bits) and evenness, in each site

Site		S	N	B	P	H	J
A	Min	6	6322	2.9	28.5	0.6	0.22
	Mean	9	11 400	9.5		1.8	0.58
	Max	12	21 148	15.9		2.9	0.80
B	Min	31	4848	18.9	67.5	3.5	0.66
	Mean	39	8967	45.3		3.7	0.71
	Max	47	12 364	70.0		4.1	0.78
C	Min	10	2365	6.1	27.2	2.1	0.56
	Mean	16	5624	22.7		2.8	0.71
	Max	27	8511	53.2		3.5	0.81
D	Min	17	2962	22.8	49.0	2.9	0.67
	Mean	21	6213	44.4		3.4	0.77
	Max	24	10 279	79.0		3.6	0.82
E	Min	31	3133	1.8	22.3	3.2	0.58
	Mean	48	8798	7.4		4.0	0.72
	Max	63	12 630	11.8		4.4	0.78

When the 96 families or the 27 most abundant families are considered, four distinct groups can be observed (Fig. 3), formed by the stations of sites A, B, D and E, regardless of the sampling date. Site C stations were projected in between the samples of sites A and D.

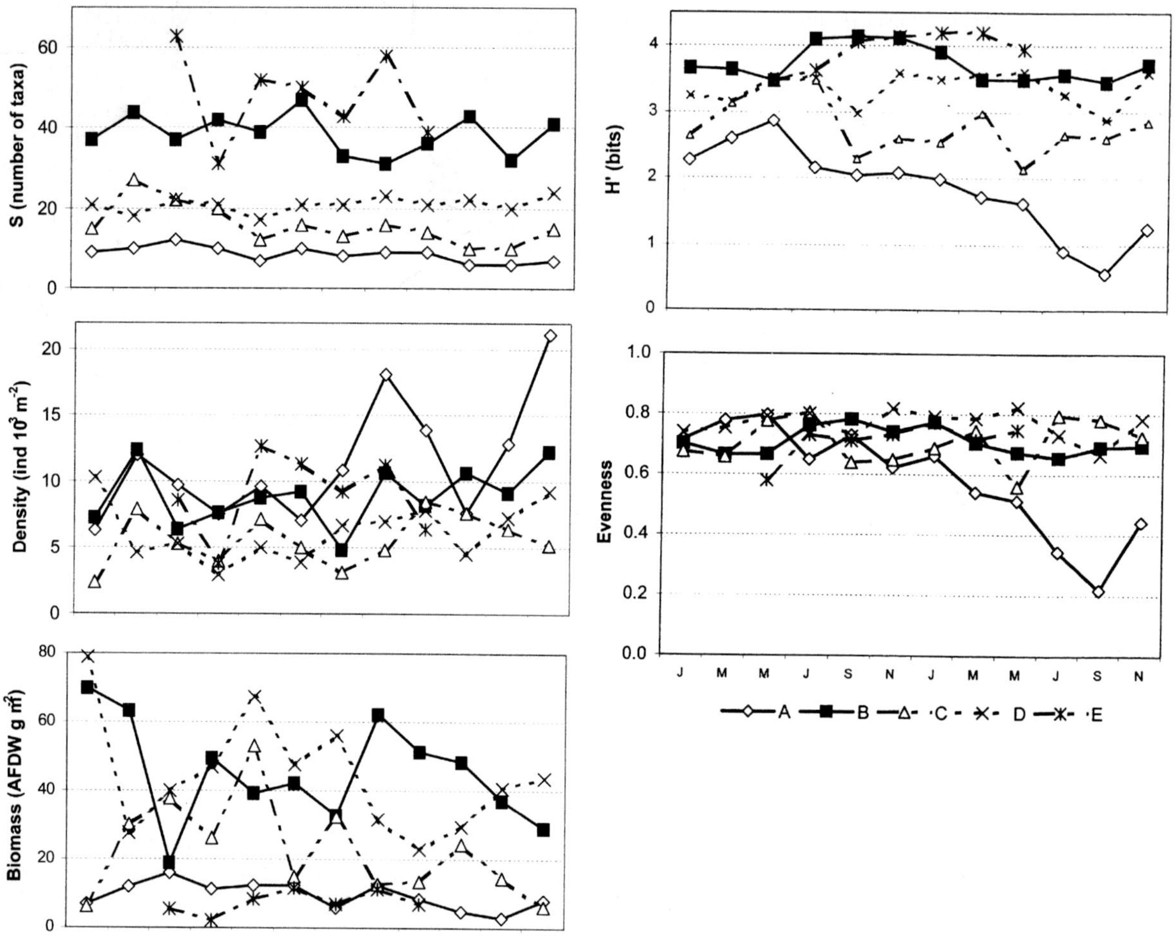

Figure 2. Variation of number of taxa, density, Shannon–Wiener diversity index, evenness and biomass per sampling occasion, in each site.

The families belonged to nine phyla (Annelida; Arthropoda; Coelenterata; Echinodermata; Mollusca; Nemertini; Phoronida, Sipuncula and Chordata). Within some of these phyla other large taxonomic groups were also considered: Olygochaeta and Polychaeta in the Annelida phylum; Amphipoda, Cirripedia, Cumacea, Decapoda, Isopoda, Insecta, Leptostraca and Tanaidacea in the Arthropoda phylum; Bivalvia, Gastropoda and Polyplacophora in the Mollusca phylum. Inspection of MDS plots resulting from these two data matrices indicates that the separation of the stations and sites are not clear (Fig. 3). Stations from sites B and E are mixed as well as with some stations from sites A, C and D. The differentiation of stations was even less evident for the analysis when no transformation of the data was previously performed.

Two-way crossed ANOSIM clearly separates sampling stations (Global R = 0.713, significance = 0.1%) but not the sampling occasion (Global R = 0.048, significance = 76.7%), The same analysis for sites and sampling occasion also separates sites (Global R = 0.808; significance = 0.1%) but not the sampling occasion (Global R = 0.031; significance = 17.7%). One-way ANOSIM confirms the separation of sites and stations (Global R = 0.843 and R = 0.809, respectively; significance = 0.1 %), although the stations within each site show some similarities (Pairwise comparisons: barely separable at all (R < 0.25): A1, A2, A3 and E1, E2; overlapping

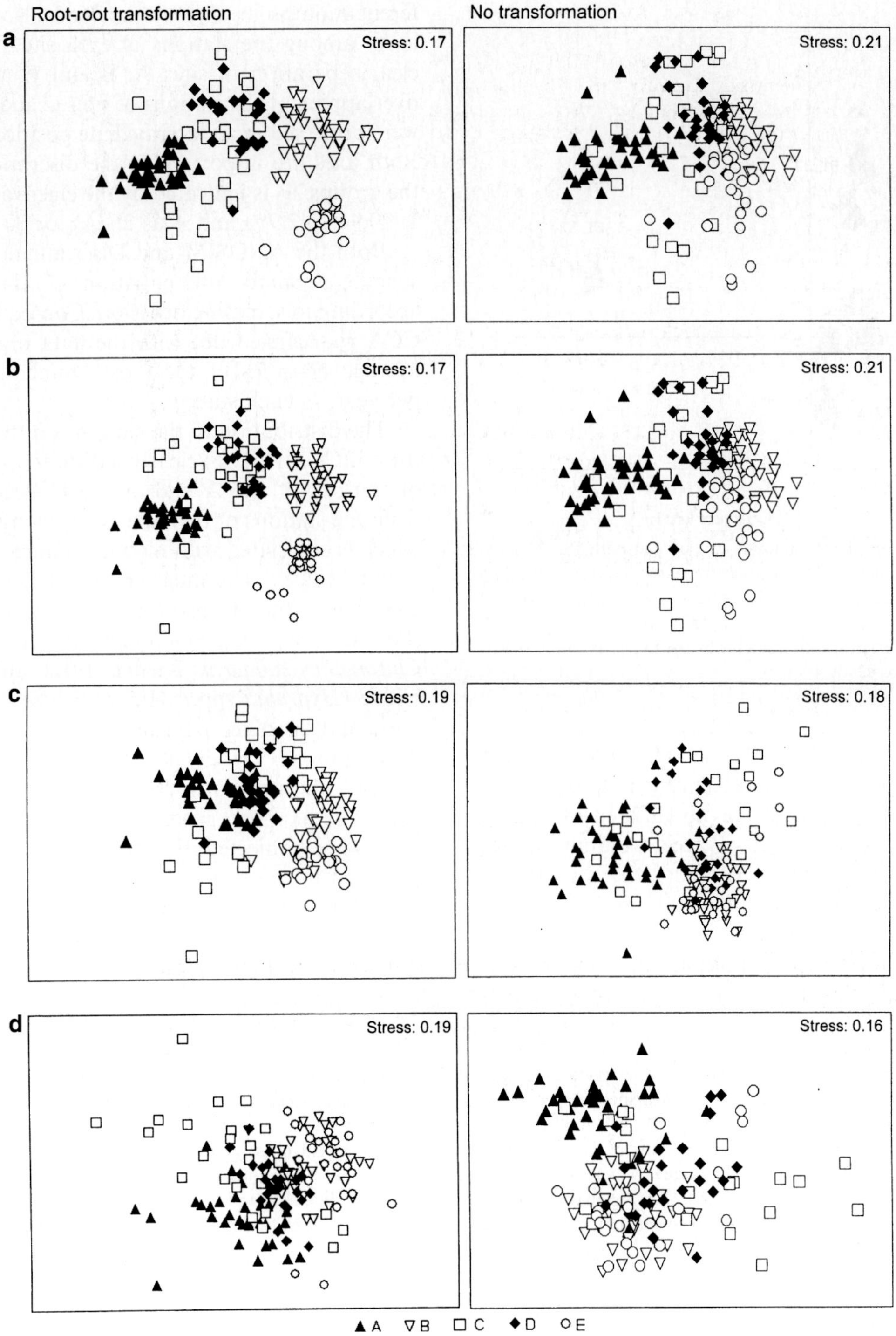

Figure 3. Multidimensional Scaling carried out with different data sets: (a) 96 families; (b) 27 most abundant families; (c) 19 high taxonomic groups (see text for details) or (d) nine Phyla; Bray–Curtis dissimilarity index, root–root transformation, or no transformation, 144 samples (12 stations, 12 sampling occasions in each station).

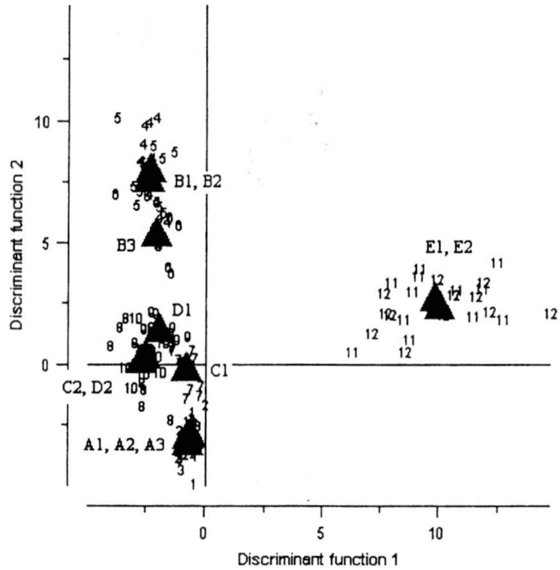

Figure 4. Discriminant analysis carried out with the 27 most abundant families. Stations codes (repeated numbers: sampling occasions in each station): A1 – 1; A2 – 2; A3 – 3; B1 – 4; B2 – 5; B3 – 6; C1 – 7; C2 – 8; D1 – 9; D2 – 10; E1 –11; E2 – 12. Percentage variance accounted for by the two axes: 73.3% (Axis 1: 42.9%; Axis 2: 30.4%).

($R < 0.50$): B1, B2; B2, B3 and C2, D2). From the three ANOSIM tests we accept the null hypothesis in the first test: there are no differences among sampling occasions but we accept the alternative hypothesis of the two other tests: there are differences within sites and within stations.

The discriminant analysis reveals the great similarity among the samples taken along the different months in each station and also the similarity among the stations of each site. There is a clear separation of sites A, B and E, with some overlapping of the stations of sites C and D, which were projected in an intermediate position (Fig. 4). Both axes are important in the discrimination of the groups, as is indicated by the eigenvalues (Axis 1: 20.9 or 42.9% and Axis 2: 14.8 or 30.4%).

Both the ANOSIM and Discriminant analyses suggest a small differentiation of the stations according to sampling occasion. Consequently, the CCA was carried out with the data organised as average densities of the most abundant families per year, in each station.

The distribution of the stations on the first two axes of CCA (Fig. 5) clearly indicates a separation of stations of site A and of site C (second year) from the stations of the other sites, along axis one, which is correlated with high maximum–minimum salinity values. The most abundant families were Cardiidae (the bivalve *Cerastoderma glaucum* (Poiret, 1789)); Chironomidae (the insect larvae *Chironomus salinarius* Kieffer, 1915*)* and Hydrobiidae (*Hydrobia* spp.). Higher values of temperature and dissolved oxygen were also observed in these stations. Depth, phaeopigment and chlorophyll a concentrations were higher in the negative section of axis one, where stations from the other sites were projected. Higher BOD values were observed in the stations projected in the positive part of axis two.

Station E1 was separated from all the other stations, with high densities of the families Capr-

Figure 5. Canonical correspondence analysis carried out with the 27 most abundant families (mean annual values per station). Percentage variance accounted for by axes: Species: I = 40.2%; I + II = 58.5%; species-environment relation: I = 55.7%; I + II = 81.0%. Monte Carlo test P = 0.005 (the relation between taxa and environmental variables is highly significant). Stations codes: the first character indicates the year of survey (I): 1985; (II): 1986; and subsequent ones the site and station. Environmental data: TEMP (water temperature; mean annual values in each station); SAL (salinity: difference between maximal and minimal value observed in each year, in each station); DO (dissolved oxygen: difference between maximal and minimal value observed in each year, in each station); BOD (biochemical oxygen demand; mean annual values in each station); CA and PA (chlorophyll a and phaeopigments; mean annual values in each station). Families codes:

F2	Ampeliscidae	F21	Chironomidae	F41	Idoteidae	F70	Paraonidae
F7	Aoridae	F22	Cirratulidae	F46	Leptocheliidae	F73	Phoronidae
F9	Apseudinae	F25	Corophiidae	F49	Lucinidae	F80	Semelidae
F15	Capitellidae	F27	Cumacea	F52	Maldanidae	F85	Spionidae
F16	Caprellidae	F30	Dorvilleidae	F58	Nassariidae	F94	Turritellidae
F17	Cardiidae	F34	Gammaridae	F62	Nereididae	F95	Veneridae
F19	Cerithiidae	F40	Hydrobiidae	F64	Oligochaeta		

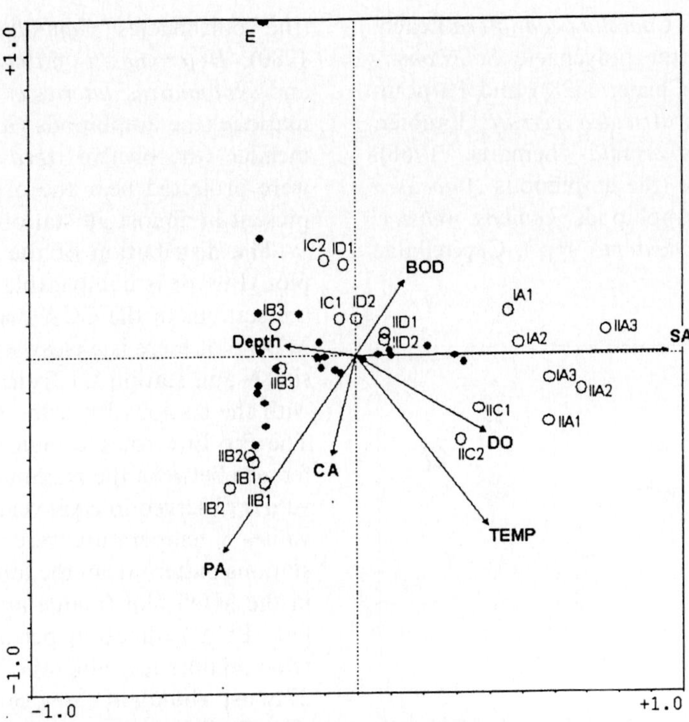
+1.0
E
IC2
ID1
BOD
IA1
IC1
ID2
IIB3
IID1
IIA3
Depth
IID2
IA2
SAL
IB3
IIB3
IA3
IIA2
IIC1
IIA1
DO
IIC2
IIB2
CA
IB1
IIB1
IB2
TEMP
PA
-1.0
-1.0
+1.0

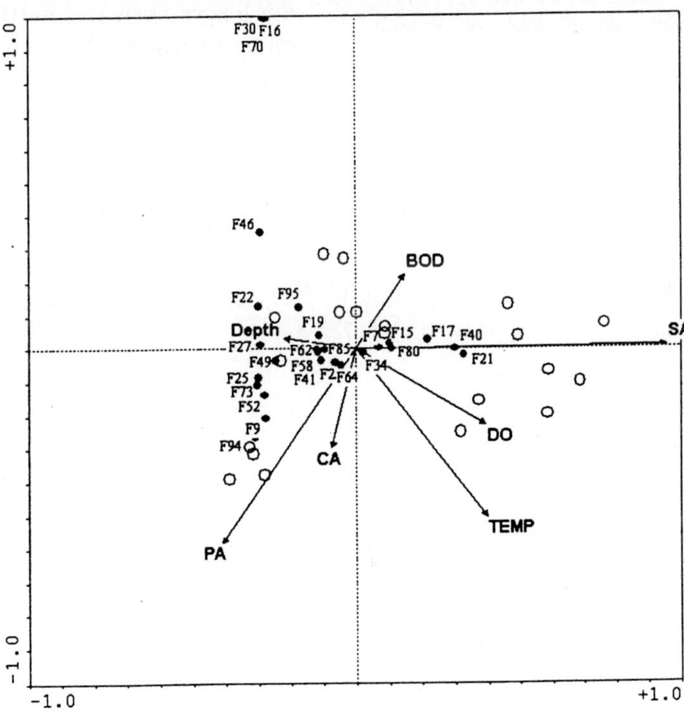
+1.0
F30 F16
F70
F46
BOD
F22
F95
F19
Depth
F7
F15
F17
F40
SAL
F27
F62
F85
F80
F21
F49
F58
F34
F41
F2
F64
F25
F73
F52
F9
F94
CA
DO
TEMP
PA
-1.0
-1.0
+1.0

ellidae (the amphipods *Caprella acanthifera* Leach, 1814), Dorveilleidae (the polychaete *Schistomeringos rudolphi* (Delle Chiaje, 1828)) and Paraonidae (the polychaetes *Aricidea cerrutii* Laubier, 1967 and *Paradoneis armata* Glémarec, 1966). Families Ampeliscidae (the amphipods *Ampelisca* spp.), Aoridae (the amphipods *Lembos websteri* Bate, 1857 and *Microdeutopus* spp.), Capitellidae

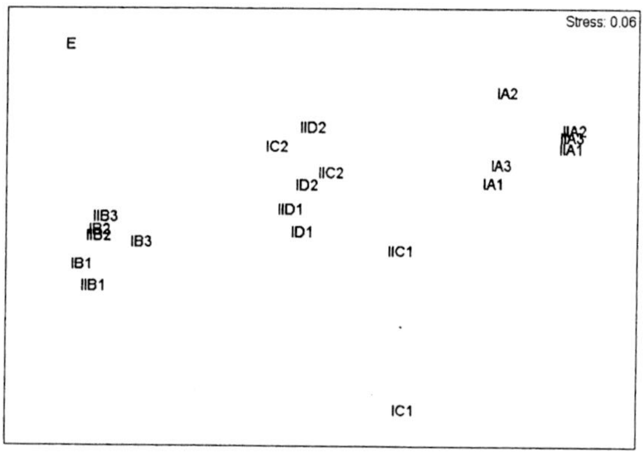

Figure 6. Multidimensional Scaling carried out with the 27 most abundant families (mean annual values per station). Stations codes as in Figure 5.

(the polychaetes *Capitella capitata* (Fabricius, 1780), *Heteromastus filiformis* (Claparède, 1864) and *Notomastus latericeus* M.Sars, 1851), Gammaridae (the amphipods *Gammarus* spp.) and Semelidae (the bivalve *Abra ovata* (Philippi, 1836) were projected near the origin of the axis, being present in almost all stations.

The distribution of the stations on the MDS plot (Fig. 6) is comparable to the distribution of the stations in the CCA diagram (Fig. 5a). In the MDS plot there is a clear separation of stations of site A and station C1 from the stations of site B, with the stations from the other sites in between. The Bio-Env routine indicated that only the difference between the maximal and minimal salinity values observed in each year and the mean annual values of temperature were required to capture the stations pattern from the families assemblages seen in the MDS plot (matching correlation 0.698). In fact, PCA ordination performed with only these two variables resembles the MDS plot based on the 27 most abundant families (mean annual values) (Fig. 7). Salinity, phaeopigments and temperature, with longer arrows, were the main environmental factors singled out by CANOCO analysis.

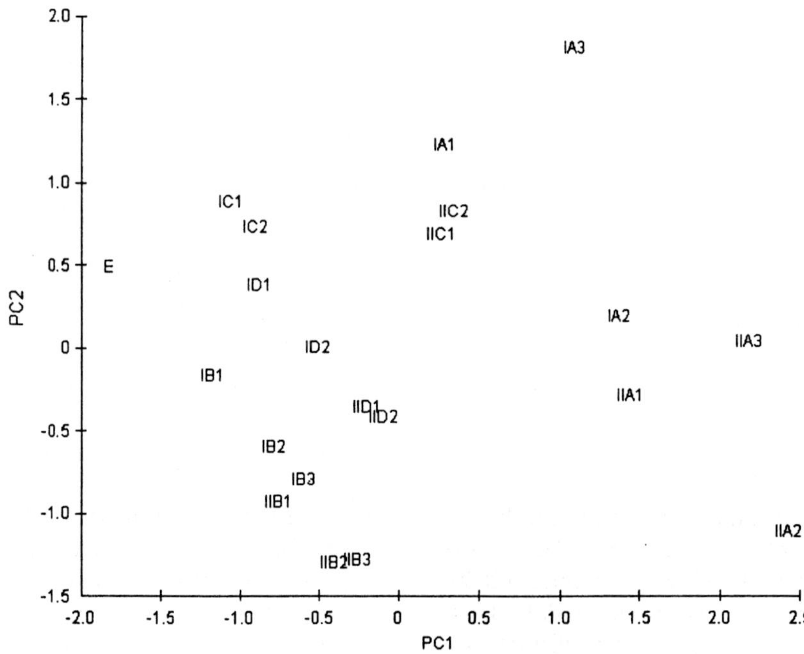

Figure 7. PCA ordination carried out with the two most important environmental variables, salinity and temperature, selected with the BIO-ENV Primer routine. (Axis 1: 63.9%; Axis 2: 36.1%). Stations codes as in Figure 5.

Discussion

The several ordination techniques clearly show an aggregation of stations corresponding with individual water reservoirs that is independent of the time of sampling. Sites B and E stations were separated from site A stations, while sites C and D remained in an intermediate position. However, when higher taxonomic levels were considered, such as large taxonomic groups, instead of families, the separation of stations among sites was not so clear. Stations from sites B and E are all mixed. The same happens with some stations from sites A, C and D. Differentiation of sites when Phyla were considered was even worse.

In previous analysis of the first four sites, where data was organised at the species level (full data set or reduced data set with only the most abundant species), or in 12 biomass categories, the differentiation of stations/sites was consistent whichever data set was used (Gamito, 1997). Apparently there was a degree of redundancy in the data, since a subset of taxa closely reproduced the patterns defined by the full data set (cf. Gray et al., 1988). This is probably because site differences were so great that any reasonable group of taxa would show the same pattern of variation. However, this consistency was not observed when large taxonomic groups or Phyla were considered. In an analysis of macrobenthic data, Gray et al. (1988) indicated that no information would be lost if the organisms had been identified to family level only. Warwick (1988), for the same macrobenthic data, pointed out that the multivariate analysis of Phyla results varied, depending on whether abundance or biomass was used and in the strength of data transformation, but generally the agreement between species or family and phyla configurations were good. This author recommended a weak or no data transformation to accentuate the differentiation between sites. In our case the differentiation between sites was even worse when no transformation was performed. It seems that the use of taxonomic resolution higher than the family level is not recommended for the present data set. Karakassis & Hatziyanni (2000) also concluded that for macrofauna data from sites subjected to a varying degree of disturbance due to sea bream and sea bass farming in the Mediterranean, the best balance between results and taxonomic resolution was obtained at the family level.

The observation of the CCA plot made it possible to detect a strong environmental gradient, related to increasingly stressful conditions due to limited water exchange, with large environmental variations occurring at some sites (A and C, second year). A second gradient with higher values of BOD might be related to increasing eutrophication and a consequent deterioration of water quality. However, this environmental factor, with a small arrow length, is apparently not so important for station differentiation, in contrast to the findings of a previous analysis of data from the first four sites (Gamito, 1997). In that study more environmental variables were included, and the first CCA axis was related, as here, with a gradient of increasing variability of the environmental conditions due to reduced water renewal rates. The organisation of the data matrix in families and not in species apparently did not modify the main results.

The water reservoirs studied here presented some differentiation that can be related to water renewal regimes. Site A, where the water was only renewed every fourteen days, can be characterised by large environmental variations between neap and spring tides. The salinity varied from 13 to 84 psu and primary production and macrofauna biomass was always low. The low diversity community was composed of abundant small sized organisms. Migration of juveniles into the reservoir apparently was not a limitation, as the reservoir was open twice a month, but the harsh environmental conditions might have caused low survival rates.

Two reservoirs, sites B and E, showed the highest diversity indexes. Here the water was renewed daily and the environmental variation between neap and spring tides was low. However, primary production and macrofauna biomass was higher in site B than in site E. In site E, the greater depth, along with the naturally high water turbidity of the Ria Formosa, might have prevented macroalgae and seagrass from rooting and developing on the bottom, in contrast to site B where the bottom was covered by phanerogams. In fact it is light availability that mainly governs seagrass growth in moderately nutrient enriched regions (Moore & Wetzel, 2000). The presence of vegetation allows small benthic animals to settle, develop and reproduce, protected from predation. Seagrasses increase the spatial complexity of the community structure by amplifying both surface and volume five fold, when compared with non-vegetated areas. The development of seagrass meadows stimulates the abundance and biomass of the indigenous macrobenthic assemblages, and it also increases diversity and specialisation of feeding gilds (Hily & Bouteille, 1999).

The other two water reservoirs displayed intermediate characteristics. However, mean BOD concentrations were higher than in the other water reservoirs, along with chlorophyll a concentration, especially in site C during the second year. In fact, in the beginning of the second year, there was an intense development of phytoplankton at this site. During that period, the water was green and very turbid, and some dead fish were found. Krom (1989) considered that in 'low flow rate fish ponds' the algal populations have the tendency to overshoot and then collapse.

Occasional environmental instability is usual in these water reservoirs where there is an excessive production of organic matter that can be followed, in extreme situations, by total consumption of oxygen during the night and massive mortalities (Krom et al., 1985, 1989; Sorokin et al., 1996; Lardicci et al., 1997). During the neap tides the water quality deteriorated, due to a decrease in the water renewal rate and also due to the lower quality of the incoming water. This deterioration was more pronounced in sites C and D and extreme in site A (Gamito, 1994, 1997).

The macrofauna observed in the five water reservoirs is characteristic of estuarine or shallow-water habitats (Gamito, 1997) and common in the Ria Formosa lagoonal system (Sprung, 1993, 1994). Apparently, when the environmental conditions are more stable, more species can settle and find the appropriate conditions for growth and reproduction. Organisms with larger body size and higher biomasses develop in these sites. The species present in large abundance in the more extreme conditions may remain, but at lower density, mainly due to the lower competitive capacity. The stressful environmental conditions encouraged high-density populations of small sized organisms, such as *Capitella capitata*, *Cerastoderma glaucum*, *Hydrobia* spp. and *Chironomius salinarius*. These findings agree with those of Hargrave & Thiel (1983) and Schwingammer (1981), who found that in the absence of exogenous disturbances, biomass may accumulate in large organisms, while environments characterised by variability or persistent physical stress appear to be dominated by communities of small-sized organisms.

Coastal lagoons offer a protected shallow habitat, which can be highly productive (Alongi, 1998). Well structured communities, controlled by *K*-strategists, can develop and settle in leaky lagoons, that is, lagoons with wide entrance channels and tidal currents which permit good water renewal. In these lagoons, biomass can accumulate in the form of large organisms, while lagoons with a single narrow entrance that can be closed for long periods are characterised by persistent physical stress, and are dominated by communities of small-sized r-strategists organisms (Gamito et al., 2005).

Acknowledgements

This work was partially supported by a INIC PhD scholarship and by the project PBIC/C/MAR/2226/95. Special thanks are due to several colleagues and students from the FCMA, University of Algarve, and from IPIMAR (Olhão), who helped in the field and laboratory work and macrofauna identification, and Karim Erzini for English revision of the text. The comments of two anonymous referees greatly improved the manuscript.

References

Águas, M. P., 1986. Simulação da circulação hidrodinâmica na Ria Formosa. in: Os sistemas lagunares do Algarve. Univ. do Algarve, Faro pp. 78–90.

Alongi, D. M., 1998. Coastal Ecosystem Processes. CRC Press, Boca Raton. Florida.

CCRA (Comissão de Coordenação da Região do Algarve), 1984. Programa de ordenamento e desenvolvimento da Ria Formosa. Vol. 1 - Caracterização, objectivos e acções, Faro.

Falcão, M. M. & C. Vale, 1990. Study of the Ria Formosa ecosystem: benthic nutrient remineralization and tidal variability of nutrients in the water. Hydrobiologia 207: 137–146.

Gamito, S., 1989. The benthic macrofauna of some water reservoirs of salt-pans from Ria Formosa (Portugal). Scientia Marina 53: 639–644.

Gamito, S., 1994. The benthic ecology of some Ria Formosa lagoons, with reference to the potential for production of the gilthead seabream (Sparus aurata L.). PhD Thesis, UCTRA, University of Algarve. 255 p.

Gamito, S., 1997. Application of canonical correspondence analysis to environmental and benthic macrofauna data of four sites in the Ria Formosa (Portugal). Publicaciones Especiales Instituto Español de Oceanografia 23: 41–52.

Gamito, S., A. Pires, C. Pita & K. Erzini, 2003. Food availability and the feeding ecology of ichthyofauna of a Ria Formosa (South Portugal) water reservoir. Estuaries 26: 938–948.

Gamito, S., J. Gilabert, C. Marcos Diego & A. Pérez-Ruzafa, 2005. Effects of changing environmental conditions on lagoon ecology. In Gönenç, I. E. & J. Wofflin (eds.), Coastal Lagoons: Ecosystem processes and modeling for sustainable use and development, CRC Press, Boca Raton. pp: 193–229.

Gray, J. S., M. Aschan, M. R. Carr, K. R. Clarke, R. H. Green, T. H. Pearson, R. Rosenberg & R. M. Warwick, 1988. Analysis of community attributes of the benthic macrofauna of Frierfjord/Langesundfjord and in a mesocosm experiment. Marine Ecology Progress Series 46: 151–165.

Hargrave, B. T. & H. Thiel, 1983. Assessment of pollution-induced changes in benthic community structure. Marine Pollution Bulletin 14: 41–46.

Hily, C. & M. Bouteille, 1999. Modifications of the specific diversity and feeding guilds in an intertidal sediment colonized by an eelgrass meadow (*Zostera marina*) (Brittany, France). Comptes Rendus de L'Academie des Sciences – Sciences de la Vie 322: 1121–1131.

Karakassis, I. & E. Hatziyanni, 2000. Benthic disturbance due to fish farming analysed under different levels of taxonomic resolution. Marine Ecology Progress Series 203: 247–253.

Kjerfve, B., 1994. Coastal lagoons. In Kjerfve, B. (ed.) Coastal lagoon processes. Elsevier, Amsterdam: 1–8.

Krom, M. D., C. Porter & H. Gordin, 1985. Description of the water quality conditions in a semi-intensively cultured marine fish pond in Eilat, Israel. Aquaculture 49: 141–157.

Krom, M. D., A. Neori & J. van Rijn, 1989. Importance of water flow rate in controlling water quality processes in marine and fresh water fish ponds. The Israelli Journal of Aquaculture 41: 23–33.

Lardicci, C., F. Rossi & A. Castelli, 1997. Analysis of macrozoobenthic community structure after severe dystrofic crises in a Mediterranean coastal lagoon (Orbetello, Italy). Marine Pollution Bulletin 34: 536–547.

Moore, K. A. & R. L. Wetzel, 2000. Seasonal variations in eelgrass (*Zostera marina* L.) responses to nutrient enrichment and reduced light availability in experimental ecosystems. Journal of Experimental Marine Biology and Ecology 244: 1–28.

Olsgard, F., P. J. Somerfied & M. R. Carr, 1998. Relationships between taxonomic resolution, macrobenthic community patterns and disturbance. Marine Ecology Progress Series 172: 25–36.

Pilkey, O. H., W. J. Neal, J. H. Monteiro & J. M. A. Dias, 1989. Algarve barrier islands: a noncoastal-plain system in Portugal. Journal of Coastal Research 5: 239–261.

Rumohr, H. & I. Karakassis, 1999. Comparison of multivariate patterns: different taxonomic levels in macrofaunal analysis versus sediment profiling imagery (SPI). Marine Ecology Progress Series 190: 125–132.

Schwinghamer, P., 1981. Characteristic size distribution of integral benthic communities. Canadian Journal of Fisheries and Aquatic Sciences 38: 1255–1263.

Sorokin, Y. I., P. Y. Sorokin & A. Gnes, 1996. Structure and functioning of the anthropogenically transformed Commacchio lagoonal ecosystem (Ferrara, Italy). Marine Ecology Progress Series 133: 57–71.

Sprung, M., 1993. Estimating macrobenthic secondary production from body weight and biomass: a field test in a non-boreal intertidal habitat. Marine Ecology Progress Series 100: 103–109.

Sprung, M., 1994. Macrobenthic secondary production in the intertidal zone of the Ria Formosa – a lagoon in Southern Portugal. Estuarine Coastal and Shelf Science 38: 539–558.

Warwick, R. M., 1988. Analysis of community attributes of the macrobenthos of Frierfjord/Langesundfjord at taxonomic levels higher than species. Marine Ecology Progress Series 46: 167–170.

Warwick, R. M., H. M. Platt, K. R. Clarke, J. Agard & J. Gobin, 1990. Analysis of macrobenthic and meiobenthic community structure in relation to pollution and disturbance in Hamilton Harbour, Bermuda. Journal of Experimental Marine Biology Ecology 138: 119–142.

Hydrobiologia (2006) 555:89–98
H. Queiroga, M.R. Cunha, A. Cunha, M.H. Moreira, V. Quintino, A.M. Rodrigues, J. Serôdio & R.M. Warwick (eds), Marine Biodiversity: Patterns and Processes, Assessment, Threats, Management and Conservation
DOI 10.1007/s10750-005-1108-2

Inter-annual variability of soft bottom macrofaunal communities in two Ionian Sea lagoons

Artemis Nicolaidou[1,*], Konstantinos Petrou[1], Konstantinos Ar. Kormas[1] & Sofia Reizopoulou[1,2]
[1]*Department of Zoology – Marine Biology, School of Biology, University of Athens, 157 84 Panepistimiopoli, Athens, Greece*
[2]*Hellenic Centre for Marine Research, Institute of Oceanography, PO Box 712, 190 13 Anavissos, Attiki, Greece*
(**Author for correspondence: Tel.: +30-210-727-4359; Fax: +30-210-727-4359; E-mail: anikol@biol.uoa.gr*)

Key words: Lagoons, benthos, Amvrakikos, Greece, temporal variations

Abstract

Inter-annual variation in the composition of the soft bottom macrobenthic communities of two undisturbed lagoons, in Amvrakikos Gulf, Ionian Sea, Greece, was investigated over three consecutive summers. The environmental parameters that showed the greatest variability were organic carbon of the sediment and salinity. The species found were typical of lagoonal systems, the most common and abundant of which were *Abra ovata*, *Mytilaster minimus* and, in the most enclosed areas, larvae of chironomids. Multivariate analysis registered community changes, which were mostly caused by changes in species dominance. Structural community characteristics such as number of species, number of individuals and diversity did not show significant differences among years except in the stations with least water exchange with the sea.

Introduction

Coastal lagoons in the Mediterranean Sea are usually shallow water bodies receiving variable amounts of fresh water. Due to their geomorphological and hydrological characteristics, environmental conditions in the lagoons undergo frequent fluctuations on a daily and seasonal basis. This instability causes changes in the distribution of benthic species and the structure of communities (Millet & Guelorget, 1994), which are often accentuated by anthropogenic influences. The most common and serious anthropogenic impact is nutrient enrichment, which often instigates a shift from sea-grasses to opportunistic green macroalgae (Valiela et al., 1997) and leads to oxygen depletion known as 'dystrophic crises' (Amanieu et al., 1977; Izzo & Hull, 1991; Sfriso et al., 1992; Viaroli et al., 1996). Temporal variations in benthic communities associated with such eutrophication phenomena have been the subject of numerous studies (e.g., Lardicci et al., 1997; Tagliapietra et al., 1998; Koutsoubas et al., 2000; Mistri et al., 2000, 2001a; Lardicci et al., 2001). Conversely, very little is published concerning yearly changes in undisturbed lagoons (Gravina et al., 1989; Bachelet et al., 2000). Yet, the information on natural temporal variability in lagoons is also very important. Establishing patterns of community change under relatively undisturbed conditions provides some understanding of the functioning of lagoonal communities and allows a better assessment of the extent of anthropogenic disturbances.

The present paper investigates the inter-annual changes in the benthic communities of the lagoonal system of Tsoukalio/Rodia in Amvrakikos Gulf, Ionian Sea over a period of 3 years.

The lagoons are far from any point source of pollution. Kormas et al. (2001) measured monthly the nutrient and chlorophyll concentrations in the lagoons over an annual cycle and attributed the temporal changes they observed to climatic factors and to the geomorphological characteristics of the lagoons.

Materials and methods

Tsoukalio and Rodia lagoons are formed in the common delta of the rivers Arachthos and Louros, on the north coast of Amvrakikos Gulf (Fig. 1). Tsoukalio is separated from the sea by a narrow barrier with eight narrow openings allowing limited water exchange with it. Rodia is connected to Tsoukalio through a very narrow opening approximately 10 m wide and less than 2 m deep. The depth in Tsoukalio ranges from 0.6 to 1.8 m and in Rodia from 1.0 to 3.5 m. The lagoons are used for the extensive culture of various species of mullet and eels, which are trapped as fry and allowed to grow naturally in the lagoon. Small numbers of gilthead (*Sparus aurata*, L.) and bass (*Dicentrarchus labrax* L.) are also caught.

Sampling was carried out once in the late summer of three consecutive years (1993–1995) in Tsoukalio and two in Rodia (1994–1995), at four and three stations, respectively.

Temperature, salinity and dissolved oxygen were measured just above the bottom using temperature/salinity and oxygen probes (Yellow Spring Inc.). Five replicate benthic samples were taken at each station with a Ponar grab sampling 0.05 m^2 of the bottom. Samples were sieved through a 1 mm mesh sieve, preserved in 4% formalin, and stained with Rose Bengal. A small amount of sediment was kept for granulometry and analysis of organic carbon.

In the laboratory, the macrofauna was sorted, identified to species level where possible, and counted. Grain size analysis of the sediment was carried out according to the methods described by Buchanan (1984). Samples with high content of sand were dry sieved, while the rest were analysed with the pipette method.

Univariate and multivariate methods were used to statistically analyse the data. Values of environmental parameters were log-transformed and normalized before being subjected to ordination by means of principal component analysis (PCA). The scores of the stations on the first two axes were correlated with environmental parameters using the Spearman Rank correlation. For the macrofauna, univariate measures included number of taxa, abundance, Shannon-Wiener diversity H' (log_2 basis) and Pielou's Evenness J. The significance of the differences of the above indices among the years in each station was tested with analysis of variance (ANOVA) and Tukey's least significant difference (LSD) test. Community changes were described by non-metric multidimentional scaling (MDS). This was based on a similarity matrix constructed using the Bray-Curtis similarity index. The data were first transformed by $Y = \log(x + 1)$. The PCA and MDS were performed using the PRIMER v5 software package, developed in the Plymouth Marine Laboratory. The routine BIO-ENV of the above programme was used to identify potential correlation of the environmental variables and the community pattern. For the rank correlation and the ANOVA the STATISTICA software was used.

Results

Environmental parameters

The values of environmental parameters measured are shown in Table 1.Temperatures measured at the times of sampling reached 29 °C with small variation between stations. Dissolved oxygen also showed some variation (4.8–6.6 mg l^{-1}) but never dropped to hypoxic levels. Salinity, on the contrary, was more variable both in space and time. Lower salinity was observed in Rodia, where it ranged from 17 to 19.5 psu. Salinity in Tsoukalio was highest in the summer of 1993 and ranged from 24 to 27 psu and lowest in 1994 when it ranged from 18 to 22 psu. Data from the National Meteorological Service showed that total rainfall was higher by approximately 20% in the year preceding the 1994 sampling.

The sediment granulometry did not change considerably in the 3 years, with the exception of stations T3 and T4. Following limited dredging to maintain the entrance to Rodia, which preceded the 1994 sampling, station T3 changed from muddy sand to sandy mud while the opposite happened in T4, which changed from sandy mud to muddy sand. Muddy sand was also found in station T2, while in the rest of the stations the sediment was sandy mud with varying amount of sand (8.3–45.3%).

The percentage of organic carbon was inversely related to the amount of sand in the sediment, as would be expected, but it varied more with time within each station. It ranged from 5.1% in T3 in 1993 to 1.21% in R3 in 1995.

The variations in the environmental parameters are registered in the PCA (Fig. 2), where high salinity stations in 1993 are located at the lower part of the graph, and low salinity at the higher. For example, salinity at station T2 in 1993 is 27 and at R2 in 1994 is 17. Sandy stations with low organic carbon are found at high scores on Axis 1 and muddy stations with high organic carbon are found at low scores. Correlation of the scores on Axis 1 with organic carbon was highly significant ($r = -0.942$, $p < 0.001$) while the scores on Axis 2 were significantly correlated with salinity ($r = -0.817$, $p < 0.001$).

Together the two axes accounted for 70% of the variability in the environmental data. Thus, variation in the environmental conditions concerned mostly sediment organic carbon and salinity.

The benthic fauna

The number of species, the number of individuals, the diversity and the evenness at each station and year are shown in Table 1, while the abundance of each species is found in the Appendix. A total of 48 species were found over the 3 years; however, a much smaller number was found at each station at any one time. Thus, larger number of species (29) was found in T2 in 1993 while only three species in Station R1 in 1995.

Some species were present in almost all stations and times such as *Abra ovata* (Philippi), *Cyclope neritea* (L.) and *Nephtys hombergi* (Savigny). Some species had a wide distribution and reached extremely high numbers on some occasions such as *Mytilaster minimus* (Poly) with density of 10 812 indiv m^{-2}, while others were characterised by occasional appearance at high densities such as *Erichthonius difformis* Milne-Edwards (622 indiv m^{-2}), *Idothea balthica* (Pallas)(1624 indiv m^{-2}) and Chironomidae (4839 and 1180 indiv m^{-2}). As a result, the dominant species differed with time at each station. For example, in station T2 *C. neritea*, *N. hombergi* and *M. minimus* were almost equally dominant except in 1993 where *M. minimus* numbering 3456 indiv m^{-2} accounted for 55.2% of the individuals. Similarly, in station T4 1993 *C. neritea* and *N. hombergi* had an important contribution but *E. difformis* dominated with 45% of the individuals.

The changes in community composition are reflected in the MDS plot of Figure 3. No clear groups or gradients are formed, however the stations of the more enclosed lagoon Rodia are situated at the left of the diagram. There is also a shift of most 1995 stations to the left of those of the previous low salinity year. Furthermore, there is a tendency of stations with high organic carbon to be located at the upper right part of the graph. The BIOENV results showed only a very weak correlation between the environmental parameters and the pattern produced by the MDS. The highest correlation obtained ($\rho_w = 0.317$) was for the combination of temperature, salinity and % organic carbon in the sediment, while the single most important environmental variable was % organic carbon ($\rho_w = 0.285$).

The high dominance of some species, together with the number of species present, had a pronounced effect on the diversity (Table 1). Thus, lowest diversity (0.79) was observed in station R1 in 1995 with only three species and evenness J' 0.50. In fact, high dominance was responsible for the low diversity in the rest of the Rodia stations in that year. Highest diversity (3.27) was observed in T3 in 1995 with 24 species and J' 0.71.

The availability of five samples per station enabled the testing of the significance of the differences in the diversity H, the number of species S and the number of individuals N with time. The diversity showed no significant changes with time in any of the stations. The number of species differed significantly only in three stations. Station T2 in 1993 had a higher number of species than in the other 2 years ($F = 8.452$, $p < 0.01$), station T6 had a lower number of species in 1993 than in 1994 ($F = 1.939$, $p < 0.05$) and R2 had more species in 1995 than in 1994 ($F = 6.255$, $p < 0.05$). Concerning individuals, station T2 had a higher number in 1993 ($F = 10.398$, $p < 0.01$) and stations R1 and R2

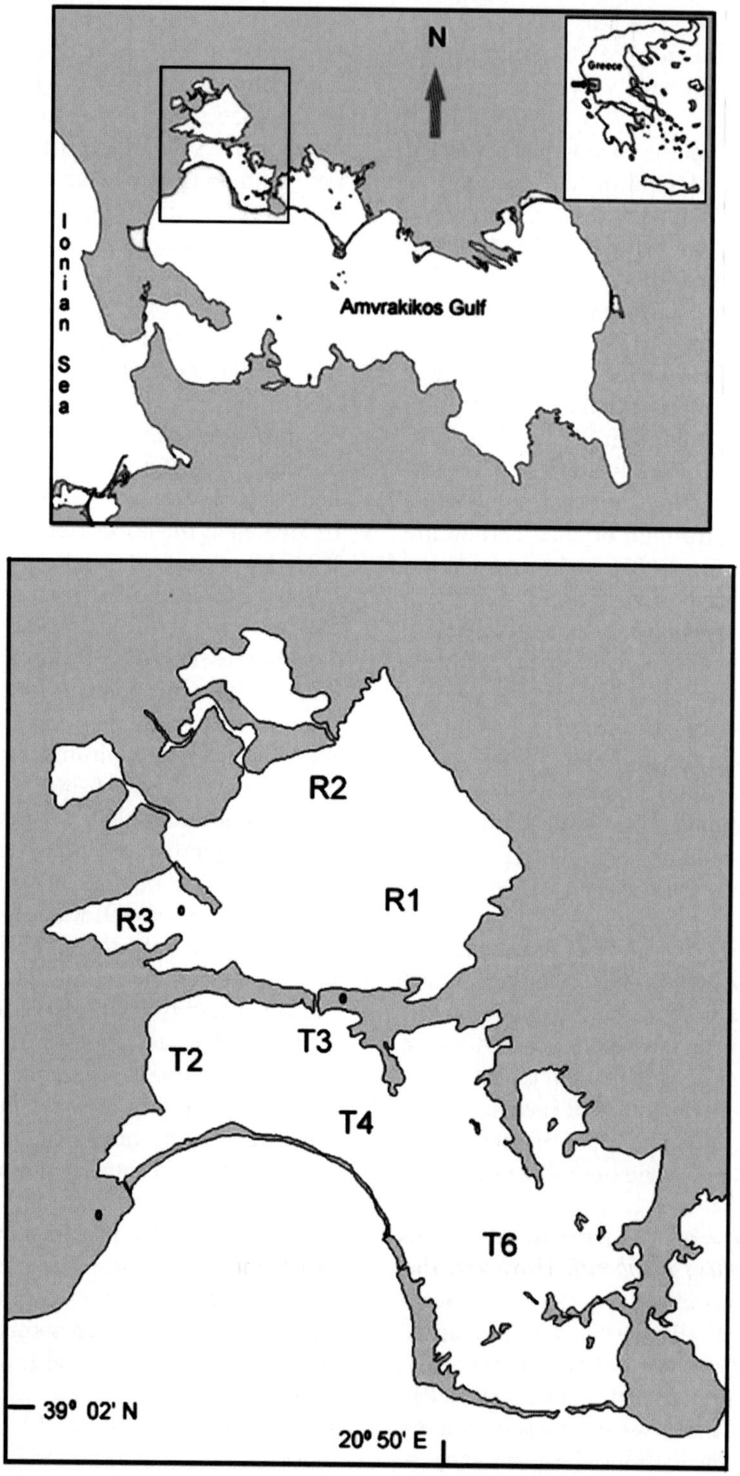

Figure 1. Location of sampling area. T: Tsoukalio, R: Rodia.

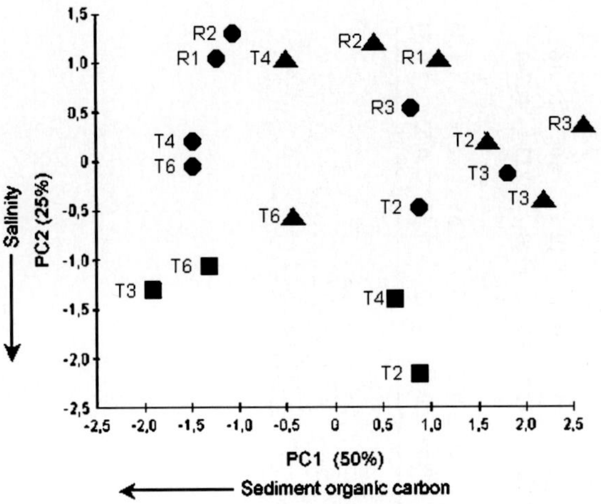

Figure 2. Plot of the first two axes of the PCA ordination on environmental parameters. ■, 1993; ●, 1994; ▲, 1995.

in 1995 ($F=93.453$, $p<0.001$ and $F=118.343$, $p<0.001$, respectively). In the rest of the stations there were no statistically significant changes in any of the ecological indices considered.

Discussion

The species composition in Tsoukalio and Rodia was similar to that of other Mediterranean lagoons (Reizopoulou et al., 1996; Mistri et al., 2001b). According to Lardicci et al. (1997) three main groups of benthic species are found in coastal brackish water lagoons, depending on their hydrologic and trophic status: euryhaline brackish water species, marine species preferring shallow sheltered areas and opportunistic species abundant in organically enriched areas. In the present case, opportunistic species were only occasionally present and then only in small numbers, confirming the undisturbed conditions of the lagoons. The second group was represented by species such as *Loripes lacteus* (L.) and *Nephtys hombergi*. However, the species with the highest abundance were those typical of eurythermal and euryhaline communities such as *Abra ovata*, and *Mytilaster minimus* and, in the most enclosed stations, larvae of Chironomidae. The dominance of these species changed from year to year and the differences in community were registered as shifts of stations in the MDS. Thus, the community pattern was regulated by certain species, as was also observed in other lagoons (Mistri et al., 2001a).

Of the environmental parameters studied, those, which showed the greatest variability, were sediment organic carbon and salinity. However, neither of them appeared directly related with the benthic community pattern of the lagoons. This is not surprising since the environmental measurements represent only an instance in time, while the benthic community reflects the conditions, which prevailed throughout the year prior to sampling. Thus, yearly ranges of environmental factors, were they available, might have shown a better relationship with the benthic fauna. In fact, there was a definite separation of the innermost Rodia stations, in which conditions are expected to be more variable in time due to their seclusion. Guelorget & Perthuisot (1983) introduced the concept of 'confinement' of coastal lagoons, to describe the degree of their isolation and the time required for the renewal of their marine elements. The effect of 'confinement' on benthic communities in coastal lagoons has been documented by many authors (Gravina et al., 1989; Guelorget et al., 1994; Reizopoulou et al., 1998; Koutsoubas et al., 2000). In addition, changes in the regime of freshwater input are known to influence drastically the macrobenthic communities (Stora et al., 1995). According to Mistri et al. (2000) the communities in inner parts of lagoons are physically regulated, while Bachelet et al. (2000) recognized that the higher the water exchange in a lagoon the more stable their benthic communities.

The most enclosed stations were also those, which showed some statistically significant differences in the number of species and individuals from year to year. On the other hand, diversity did not vary significantly with time in any of the stations. This confirms that species diversity is not a sensitive parameter in expressing variations in coastal lagoons. Reizopoulou et al. (1996) found that it was inefficient in distinguishing even between undisturbed and impacted lagoons. Due to their environmental instability and natural enrichment, coastal lagoons can be considered as naturally stressed environments where diversity is maintained at low levels. Nevertheless, a fact, which should not be overlooked, is the patchy distribution of some species, which, in the present case, would have increased the variability among replicate samples.

Table 1. Environmental parameters and ecological indices at each station and year

Station	T2			T3			T4			T6			R1		R2		R3	
Year	93	94	95	93	94	95	93	94	95	93	94	95	94	95	94	95	94	95
Temperature °C	26.5	28.0	21.0	26.5	27.0	19.0	27.0	29.0	21.0	27.0	28.0	20.0	26.5	19.0	27.0	19.0	25.0	18.5
Salinity psu	27.0	20.0	20.0	27.5	18.0	22.0	24.0	21.0	20.0	26.0	22.0	26.0	18.0	19.5	17.0	19.0	18.0	19.0
Dissolved oxygen mg/l	4.8	6.4	5.6	6.2	6.0	6.0	4.8	6.2	5.6	6.2	6.0	6.5	6.4	6.0	6.2	6.3	6.6	6.0
% Sand	66.7	60.3	56.6	18.0	79.0	76.0	60.1	11.7	15.4	13.1	8.3	28.0	23.3	14.2	12.0	25.1	45.3	77.5
% Organic carbon	1.7	2.0	1.6	5.1	1.5	1.6	2.0	4.2	4.0	3.7	4.2	3.8	5.0	1.2	4.1	3.1	2.3	1.2
Number of species (S)	29	6	7	16	22	24	23	13	12	11	9	12	7	3	10	12	16	16
Number of individuals/m2 (N)	6260	96	292	552	1682	1808	1404	380	248	1215	556	564	1900	312	184	1720	5232	16744
Diversity (H')	2.32	2.2	2.2	2.86	3.12	3.27	3.09	2.25	3.09	2.04	1.91	2.42	1.26	0.79	3.05	1.65	2.16	1.35
Evenness (J)	0.48	0.71	0.79	0.72	0.7	0.71	0.68	0.61	0.862	0.59	0.6	0.67	0.45	0.5	0.92	0.46	0.53	0.34

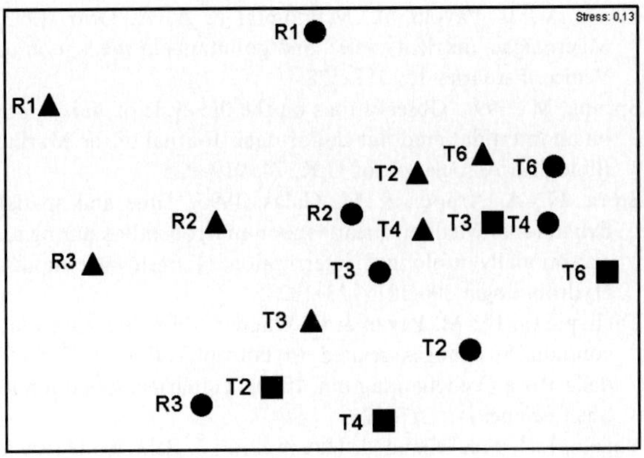

Figure 3. Plot of non-metric MDS ordination of stations based on species abundances. ■, 1993; ●, 1994; ▲, 1995.

Changes in community composition as shown by the MDS and at the same time constancy in ecological indices, suggest that most of the observed variability is caused by changes in the relative abundance of the species. Obviously, the increase in abundance of some species was compensated by the decrease in others. According to Mistri et al. (2001a) compensatory mechanisms in benthic communities in lagoons make them resistant to environmental changes, which generate fluctuations in species abundance.

Changes in the dominance of species could be caused not only by changes in the physical environment, but also by biological interactions such as competition and predation, or they may be related to intrinsic characteristics of the species. For example, interspecific competition between species of amphipods, has been observed by Nicolaidou & Karakiri (1989) in another lagoon of Amvrakikos Gulf. Furthermore, *Abra ovata* is thought to adopt different life strategies depending on environmental characteristics (Nicolaidou & Kostaki-Apostolopoulou, 1988; Sprung, 1994). Clarifying the mechanisms, which produce the changes in dominance of species in coastal lagoons, will contribute towards better understanding of the dynamics of the benthic communities.

Acknowledgements

The authors would like to thank the Tsoukalio–Rodia Lagoons Fishermen Cooperatives for their valuable assistance during sampling. The project was supported financially by the Greek Ministry of Agriculture.

References

Amanieu, M., O. Guerlorguet & P. Michel, 1977. Richesse et diversite de la macrofaune benthique d'une lagune littorale mediterraneenne. Vie Milieu 27: 85–109.

Bachelet, G., X. De Montaudouin, I. Auby & P. Labourg, 2000. Seasonal changes in macrophyte and macrozoobenthos assemblages in three coastal lagoons under varying degrees of eutrophication International Committee for Exploration of the Seas. Journal of Marine Sciences 57: 1495–1506.

Buchannan, J. B., 1984. Sediment analysis. In Holme, N. A. & A. D. McIntyre (eds.) Methods for the study of benthos. Blackwell, Oxford: 41–65.

Gravina, M. F., G. D. Ardizzone, F. Scaletta & C. Chimenz, 1989. Descriptive analysis and classification of benthic communities in some Mediterranean coastal lagoons (Central Italy). Marine Ecology 10: 141–166.

Guelorget, O. & J. P. Perthuisot, 1983. Le domain paralique Expression geologique, biologique du confinement. Travaux du Laboratoire de Geologie, Ecole Normale Superieure de Paris 16, Paris 136 pp.

Guelorget, O., J. P. Perthuisot, N. Lamy & A. Lefebvre, 1994. Structure et organisation de l' Etang de Thau d'apres la faune benthique (macrofaune-meiofaune) Relations avec le confinement. Oceanoligica Acta 17: 105–114.

Izzo, G. & V. Hull, 1991. The anoxic crises in dystrophic processes of coastal lagoons: an energetic explanation. In Rossi, C. & E. Tiezzi (eds.) Ecological physical chemistry. Elsevier, Amsterdam: 559–572.

Kormas, K. A., A. Nicolaidou & S. Reizopoulou, 2001. Temporal variations of nutrients, chlorophyll a and particulate organic matter in three coastal lagoons of Amvrakikos Gulf (Ionian Sea, Greece). Marine Ecology 22: 201–213.

Koutsoubas, D., C. Dounas, C. S. Arvanitidis, S. Kornilios, G. Petihakis, G. Triantafyllou & A. Eleftheriou, 2000. Macrobenthic community structure and disturbance assessment in Gialova Lagoon, Ionian Sea. International Committee for Exploration of the Seas. Journal of Marine Sciences 57: 1472–1480.

Lardicci, C., S. Como, S. Corti & F. Rossi, 2001. Recovery of the macrozoobenthic community after severe dystrophic crises in a Mediterranean coastal lagoon (Orbetello, Italy). Marine Pollution Bulletin 42: 202–214.

Lardicci, C., F. Rossi & A. Castelli, 1997. Analysis of macrozoobenthic community structure after severe dystrophic crises in a Mediterranean coastal lagoon. Marine Pollution Bulletin 34: 536–547.

Millet, B. & O. Guelorget, 1994. Spatial and seasonal variability in the relationships between benthic communities and physical environment in a lagoon ecosystem. Marine Ecology Progress Series 108: 161–174.

Mistri, M., E. A. Fano, G. Rossi, K. Caselli & R. Rossi, 2000. Variability in macrobenthos communities in the Valli di Comacchio, Northern Italy, a hypereutrophized lagoonal ecosystem. Estuarine, Coastal and Shelf Science 51: 599–611.

Mistri, M., E. A. Fano & R. Rossi, 2001a. Redundancy of macrobenthos from lagoonal habitats in the Adriatic Sea. Marine Ecology Progress Series 215: 289–296.

Mistri, M., R. Rossi & E. A. Fano, 2001b. Structure and secondary production of a soft bottom macrobenthic community in a brackish lagoon (Sacca di Goro, north-eastern Italy) Estuarine. Coastal and Shelf Science 52: 605–616.

Nicolaidou, A. & M. Karakiri, 1989. The distribution of Amphipoda in a brackish-water lagoon in Greece. Marine Ecology 10: 131–139.

Nicolaidou, A. & M. Kostaki-Apostolopoulou, 1988. The growth of *Abra ovata* in a brackish water lagoon. Vie Marine 9: 7–10.

Reizopoulou, S., K. A. Kormas & A. Nicolaidou, 1998. Benthic biodiversity in five coastal brackish water lagoons of Amvrakikos Gulf, Hellas. Rapports de Committee Internationale de Mer Mediterranee 35: 580–581.

Reizopoulou, S., M. Thessalou-Legaki & A. Nicolaidou, 1996. Assessment of disturbance in Mediterranean lagoons: An evaluation of methods. Marine Biology 125: 189–197.

Sfriso, A., B. Pavoni, A. Macromini & A. A. Orio, 1992. Macroalgae, nutrient cycles, and pollutants in the lagoon of Venice. Estuaries 15: 517–528.

Sprung, M., 1994. Observations on the life cycle of *Abra ovata* on an intertidal mud flat in Portugal. Journal of the Marine Biological Association of U.K. 74: 919–925.

Stora, G., A. Arnoux & M. Galas, 1995. Time and spatial dynamics of Mediterranean lagoon macrobenthos during an exceptionally prolonged interruption of freshwater inputs. Hydrobiologia 300/301: 123–132.

Tagliapietra, D., M. Pavan & C. Wagner, 1998. Macrobenthic community changes related to eutrophication in Pallude della Rosa (Venetian lagoon, Italy). Estuarine, Coastal and Shelf Science 47: 217–226.

Valiela, I., J. McClelland, J. Hauxwell, P. J. Behr, D. Hersh & K. Foreman, 1997. Macroalgal blooms in shallow estuaries: controls and ecophysiological and ecosystem consequences Limnology and Oceanography 42: 1105–1118.

Viaroli, P., M. Naldi, C. Bondavalli & S. Bencivelli, 1996. Growth of the seaweed *Ulva rigida* C Agardh in relation to biomass densities, internal nutrient pools and external nutrient supply in the Sacca di Goro lagoon (Northern Italy). Hydrobiologia 329: 93–103.

Appendix

Table A1. Species abundance (indiv/m^2), in each station and year

Stations	T2			T3			T4			T6			R1		R2		R3	
Years	93	94	95	93	94	95	93	94	95	93	94	95	94	95	94	95	94	95
Species																		
Polychaeta																		
Capitella capitata (Fabricius, 1780)	68	0	0	16	0	0	0	0	4	0	0	4	0	0	0	0	0	0
Harmothoe spinifera (Ehlers, 1864)	132	4	0	24	64	20	52	4	8	15	0	4	0	0	0	0	8	0
Microspio mecznikovianus (Claparede, 1868)	12	0	0	20	12	0	12	0	0	55	4	0	0	0	0	0	0	0
Naineris laevigata Grube, 1855	1336	0	0	0	8	52	12	0	0	5	0	0	0	0	28	0	8	0
Nephtys hombergi (Savigny, 1820)	140	16	56	164	140	64	84	116	48	270	152	76	40	0	8	0	60	48
Nereiphylla rubiginosa (Saint-Joseph, 1888)	4	0	0	0	0	0	0	0	0	0	0	0	0	0	0	0	0	0
Ophiodromus pallidus (Claparede, 1868)	16	0	0	0	0	0	4	0	0	0	0	0	0	0	0	0	0	0
Pectinaria koreni (Malmgren, 1865)	0	0	0	0	8	0	0	0	0	0	0	0	0	0	0	0	0	0
Phyllodoce sp.	48	0	16	8	0	32	4	0	0	0	0	4	0	0	0	20	0	260
Platynereis dumerilii	396	0	0	0	24	328	72	0	0	0	0	0	0	0	0	12	0	0

Continued on p. 97

Table A1. (Continued)

Stations	T2			T3			T4			T6			R1		R2		R3	
Years	93	94	95	93	94	95	93	94	95	93	94	95	94	95	94	95	94	95
Aoudouin & Milne-Edwards, 1834																		
Prionospio malmgreni	0	0	0	0	0	0	0	0	0	0	8	0	0	0	0	0	8	0
Claparede, 1870																		
Protoaricia oerstedi	0	0	0	0	12	0	0	0	0	0	0	0	0	0	12	0	0	0
(Claparede, 1864)																		
Pseudopolydora antennata	0	0	0	4	0	0	0	4	0	25	0	0	0	0	0	0	0	0
(Claparede, 1870)																		
Schistomeringos rudolphi	0	0	0	0	4	0	0	0	0	0	0	0	0	0	0	0	0	0
(Delle Chiaje, 1828)																		
Syllis gracilis	32	0	0	0	0	4	36	0	0	0	0	0	0	0	0	0	0	0
Grube, 1840																		
Mollusca																		
Abra ovata	56	0	12	16	308	20	12	16	44	10	20	236	924	0	36	40	600	32
(Philippi, 1836)																		
Bittium reticulatum	0	4	0	0	0	20	0	0	4	0	0	0	0	0	0	8	0	68
(Da Costa, 1788)																		
Cerastoderma glaucum	44	0	16	80	4	12	8	40	12	55	268	84	4	0	4	8	4	0
(Poiret, 1789)																		
Cyclope neritea	36	16	144	160	100	64	40	168	44	90	92	104	8	0	16	68	28	28
(Linnaeus, 1758)																		
Gastropoda juv.	0	0	0	0	0	0	0	0	0	0	0	4	0	0	0	0	0	0
Gibbula adansonii	40	0	0	0	0	0	4	0	0	0	0	0	0	0	0	0	0	4
(Pyraudeau, 1826)																		
Haminea hydatis	0	0	0	0	4	0	0	4	0	0	0	0	0	0	0	0	0	0
(Linnaeus, 1758)																		
Hinia reticulata	0	0	0	0	0	4	0	0	0	0	0	0	0	0	0	0	0	0
(Risso, 1826)																		
Hydrobia sp.	4	0	0	0	0	0	0	0	0	0	0	8	0	0	0	0	0	0
Loripes lacteus	156	0	28	24	164	160	20	0	20	0	0	8	904	48	24	40	0	316
(Linnaeus, 1758)																		
Mytilaster minimus	3456	16	20	0	188	576	52	0	12	0	4	0	0	8	36	288	2040	10812
(Poli, 1795)																		
Philine aperta	12	0	0	0	12	0	0	4	0	0	0	0	0	0	0	0	0	4
(Linnaeus, 1767)																		
Retusa truncatula	0	0	0	0	0	4	0	0	8	0	0	0	0	0	0	8	0	0
(Bruguiere, 1792)																		
Rissoa venusta	8	0	0	0	0	20	0	0	0	0	0	4	0	0	0	0	0	8
(Philippi, 1844)																		
Venerupis aurea	12	0	0	4	24	0	0	0	0	0	4	0	0	0	0	0	0	0
(Gmelin, 1791)																		

Continued on p. 98

Table A1. (Continued)

Stations	T2			T3			T4			T6			R1		R2		R3	
Years	93	94	95	93	94	95	93	94	95	93	94	95	94	95	94	95	94	95
Amphipoda																		
Corophium acherusicum (A. Costa, 1853)	4	0	0	0	0	16	8	0	0	0	0	0	0	0	0	0	0	4
Dexamine spinosa Montagu, 1813	28	0	0	0	0	4	0	0	0	0	0	0	0	0	0	0	0	0
Ericthonius difformis Milne-Edwards, 1830	0	0	0	0	0	52	632	0	0	10	0	0	0	0	0	0	0	0
Microdeutopus sp.	16	0	0	4	40	12	120	4	0	0	4	0	0	0	12	0	76	4
Oedicerotidae	0	0	0	0	0	0	0	0	0	0	0	0	0	0	0	0	8	0
Isopoda																		
Idotea balthica (Pallas)	0	0	0	0	0	20	100	0	4	0	0	0	0	0	0	0	1624	0
Sphaeroma ghigii (Arcangeli, 1941)	0	0	0	0	12	12	0	0	0	0	0	0	0	0	0	0	712	4
Tanaidacea																		
Leptochelia savignyi (Kroyer, 1842)	76	0	0	0	0	0	68	0	0	0	0	0	0	0	0	0	0	0
Cumacea																		
Iphinoe serrata (Norman, 1867)	0	0	0	4	0	0	4	4	0	670	0	0	0	0	0	0	4	0
Decapoda																		
Hippolyte sp.	12	0	0	0	0	0	0	0	0	0	0	0	0	0	0	0	0	0
Paguridae	0	0	0	0	4	0	0	0	0	0	0	0	0	0	0	0	0	0
Miscellanea																		
Actiniaria	8	40	0	12	536	208	36	8	40	0	0	28	12	0	8	24	8	4
Chironomidae	8	0	0	0	0	4	0	0	0	0	0	0	0	256	0	1180	8	4836
Copepoda	4	0	0	8	0	0	0	0	0	0	0	0	0	0	0	0	0	0
Echinodermata	0	0	0	0	0	0	4	0	0	0	0	0	0	0	0	0	0	0
Nemertea	96	0	0	4	10	100	20	4	0	10	0	0	8	0	0	24	36	312
Phoronis sp.	0	0	0	0	4	0	0	0	0	0	0	0	0	0	0	0	0	0
Turbellaria	0	0	0	0	0	0	0	4	0	0	0	0	0	0	0	0	0	0

Hydrobiologia (2006) 555:99–113
H. Queiroga, M.R. Cunha, A. Cunha, M.H. Moreira, V. Quintino, A.M. Rodrigues, J. Serôdio & R.M. Warwick (eds),
Marine Biodiversity: Patterns and Processes, Assessment, Threats, Management and Conservation
DOI 10.1007/s10750-005-1109-1

Spatial patterns of benthic macroinvertebrates in intertidal areas of a Southern European estuary: the Tagus, Portugal

Ana Maria Rodrigues*, Sónia Meireles, Teresa Pereira, Alice Gama & Victor Quintino
Departamento de Biologia, Centro de Estudos do Ambiente e Mar – CESAM, Universidade de Aveiro, Campus Universitário de Santiago, 3810-193 Aveiro, Portugal
*(*Author for correspondence: Tel.: +351-234370769; Fax: +351-234426408; E-mail: anarod@bio.ua.pt)*

Key words: spatial patterns, benthic macroinvertebrates, intertidal, Tagus estuary

Abstract

This study characterizes the composition and spatial distribution patterns of the benthic macrofauna in the intertidal mudflats of the Tagus estuary, western Portugal. A total of 68 species, more than 226,000 specimens with a total wet weight biomass of approximately 1170 g were identified in 380 sites. The species *Streblospio shrubsolii*, *Cyathura carinata*, *Tharyx* sp., *Hydrobia ulvae* and Tubificids were the most common and abundant. *Scrobicularia plana* strongly dominated the biomass. The invertebrate macrofauna of the Tagus estuary shows similarities to what is known from other temperate mudflats. The diversity of species, their overall abundance and the ratio of Molluscs plus Crustaceans to Polychaete species corroborate the distinctiveness between temperate and tropical mudflats and sandflats. The spatial distribution of the fauna reflects the sediment characteristics but the relationship between the environmental and the biological data is not as strong as obtained for sublittoral areas. This relationship diminishes from the sublittoral shelf to sublittoral estuarine areas, showing minimum values in this study, suggesting that such a relationship is less straightforward as natural disturbance increases. Nevertheless, a mixture of grain-size, elevation (inundation time) and particular habitats (relic oyster beds) form the best explanatory factors for the spatial distribution patterns of the intertidal benthic macrofauna of the Tagus estuary.

Introduction

Intertidal invertebrate communities show vertical zonation patterns, which are more evident on rocky shores but can also be seen in soft sediments (Petterson, 1991; Raffaelli & Hawkins, 1996). These patterns are the result of the differential ability of species to cope with the changes in physical and biological factors associated with the major environmental gradients – changes in salinity, sediment particle size, exposure, shore level – and the way these gradients interact (Beukema, 1976; Whitlatch, 1977; Raffaelli & Hawkins, 1996; Dittmann, 2000; Little, 2000).

Eltringham (1971) considered a three zone pattern for British muddy shores consisting of an upper zone characterized by the absence of fauna or the presence of *Nereis diversicolor* Müller, 1776, a mid-zone with the bivalve molluscs *Cerastoderma edule* (Linnaeus, 1758), *Scrobicularia plana* (da Costa, 1778) and *Macoma balthica* Linnaeus, 1758 and a low-shore zone with the bivalves *Mya arenaria* Linnaeus, 1758 or *Tellina fabula* Gmelin, 1791, depending on the sediment particle size, together with some polychaete species. Mudflats are not flat; they may have gentle slopes or form steep banks. Beukema (1976; 1988) proposed a model for the intertidal flats of the Wadden Sea in which the crease from the lower to the upper level of the mudflat. This gradient acts exactly in the opposite direction on the Alcochete beaches, where fines increase towards the lower shore (Fig. 2). The fact that these beaches are less sheltered and often exposed to wind and tidal driven waves could explain the predominant sandy sediment with low fines content in the upper shore area.

Mud and mud with shells comprise almost 90% of the sites. The remainder include mainly silty and very silty fine and very fine sand, located at the edge of the intertidal banks closer to the navigation channels. The exception corresponds to the

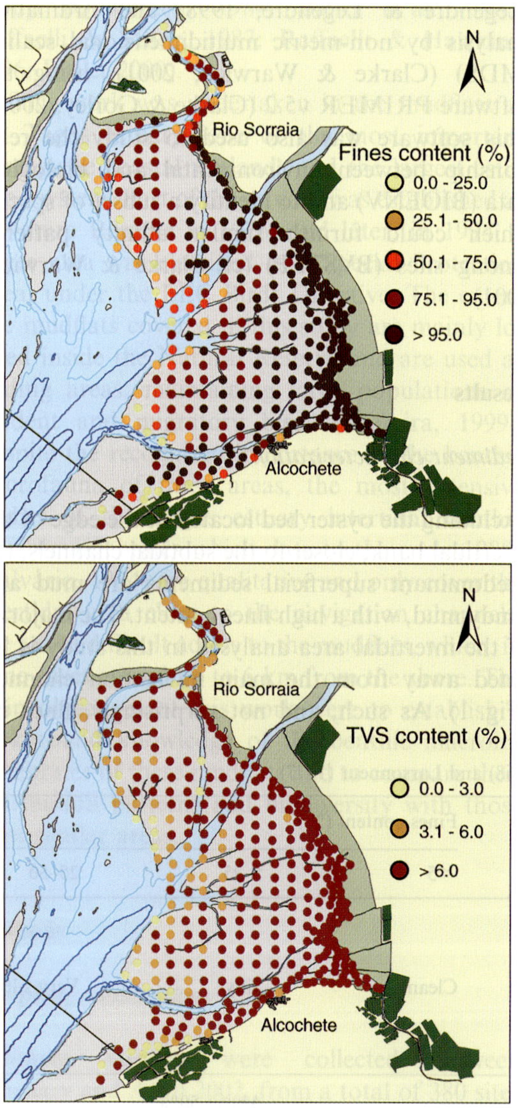

Figure 2. Spatial patterns of sedimentary descriptors.

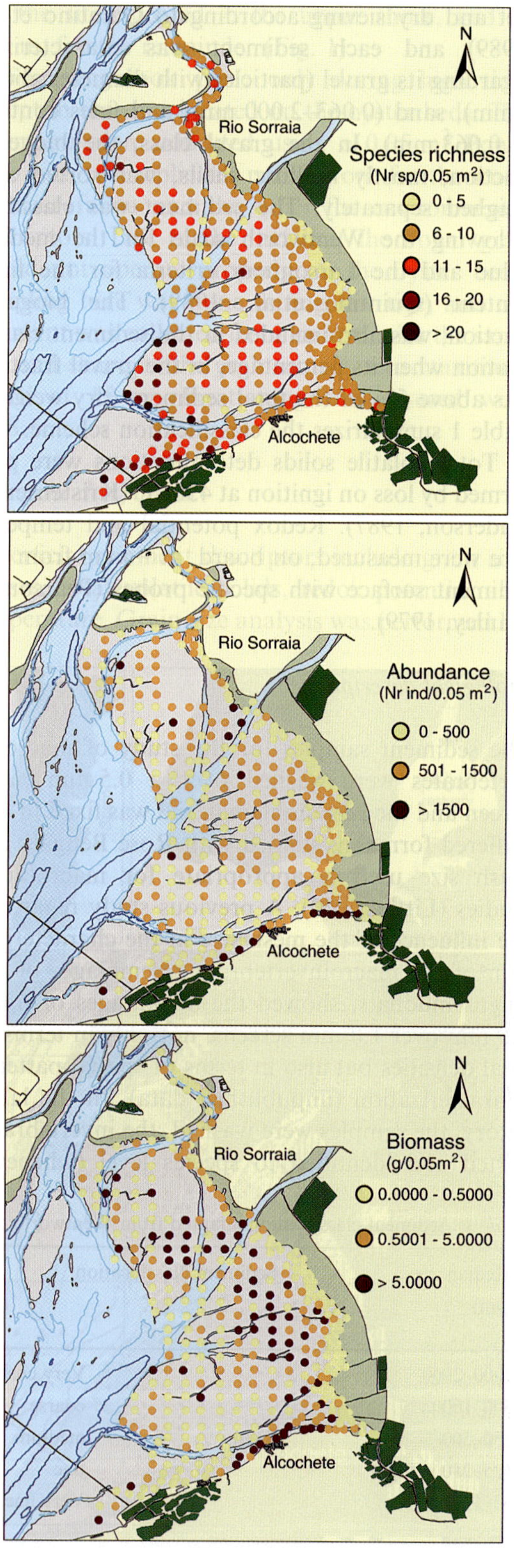

Figure 3. Spatial distribution of benthic macrofauna descriptors.

Table 2. Total number of individuals (A), total biomass, in g (B) and total number of presences for each species (P) in all the sampling sites (380 unit samples of 0.05 m^2 each)

TAXA	Total A	A (%)	Total B	B (%)	Total P	P (%)
Abra alba (Wood, 1802)	2	0.00	0.0002	0.00	1	0.26
Abra prismatica (Montagu, 1808)	2	0.00	0.0537	0.00	2	0.53
Abra tenuis (Montagu, 1803)	152	0.07	0.1250	0.01	26	6.84
Alkmaria romijni Horst, 1919	1291	0.57	0.2339	0.02	110	28.95
Ampithoe valida Smith, 1873	3	0.00	0.0068	0.00	3	0.79
Anthozoa ind.	60	0.03	0.1940	0.02	4	1.05
Autolytus cf. langerhansi Gidholm, 1967	4	0.00	0.0006	0.00	2	0.53
Balanus sp.	234	0.10	**25.1464**	2.15	19	5.00
Boccardia polybranchia (Haswell, 1885)	9	0.00	0.0052	0.00	5	1.32
Capitella spp.	1282	0.57	0.2136	0.02	50	13.16
Carcinus maenas (Linnaeus, 1758)	36	0.02	6.7073	0.57	21	5.53
Cerastoderma edule (Linnaeus, 1758)	144	0.06	7.6852	0.66	60	15.79
Copepode ind.	11	0.00	0.0009	0.00	9	2.37
Corbula gibba (Olivi, 1792)	5	0.00	0.0089	0.00	5	1.32
Corophium volutator (Pallas, 1766)	2995	1.32	3.4470	0.29	57	15.00
Cossura sp.	103	0.05	0.0032	0.00	29	7.63
Crangon crangon (Linnaeus, 1758)	45	0.02	1.6480	0.14	27	7.11
Crustacea ind.	10	0.00	0.0267	0.00	9	2.37
Cyathura carinata (Krøyer, 1847)	**3248**	1.43	**13.0873**	1.12	**265**	69.74
Fabricinae ind.	2075	0.92	0.0363	0.00	**201**	52.89
Foraminifera ind.	37	0.02	0.0005	0.00	5	1.32
Glycera tridactyla Schmarda, 1861	10	0.00	0.3672	0.03	9	2.37
Heteromastus filiformis (Claparède, 1864)	35	0.02	0.2830	0.02	20	5.26
Hydrobia ulvae (Pennant, 1777)	**4942**	2.18	**13.5653**	1.16	**249**	65.53
Idotea chelipes (Pallas, 1766)	3	0.00	0.0033	0.00	3	0.79
Idotea emarginata (Fabricius, 1793)	3	0.00	0.0412	0.00	3	0.79
Insecta ind.	78	0.03	0.1561	0.01	38	10.00
Lanice conchilega (Pallas, 1766)	1	0.00	0.1348	0.01	1	0.26
Lekanesphaera hookeri (Leach, 1814)	2	0.00	0.0250	0.00	2	0.53
Lekanesphaera levii (Argano & Ponticelli, 1981)	29	0.01	0.0591	0.01	18	4.74
Leptocheirus pilosus Zaddach, 1844	155	0.07	0.0648	0.01	17	4.47
Lysidice ninetta Audouin & Milne-Edwards, 1833	7	0.00	0.0301	0.00	5	1.32
Mediomastus fragilis Rasmussen, 1973	1	0.00	0.0032	0.00	1	0.26
Melita palmata (Montagu, 1804)	314	0.14	0.2625	0.02	41	10.79
Melinna palmata Grube, 1870	1	0.00	0.0001	0.00	1	0.26
Merceriella enigmatica Fauvel, 1923	23	0.01	0.2347	0.02	9	2.37
Mesopodopsis slabberi (van Beneden, 1861)	47	0.02	0.3206	0.03	35	9.21
Micronephtys minuta (Theél, 1879)	20	0.01	0.0145	0.00	16	4.21
Mysella sp.	4	0.00	0.0015	0.00	3	0.79
Mytilus galloprovincialis Lamarck, 1819	1	0.00	17.1322	1.46	1	0.26
Nematoda ind.	2460	1.09	0.0440	0.00	**229**	60.26
Nematonereis unicornis (Grube, 1840)	14	0.01	0.0049	0.00	4	1.05
Nemertea ind.	96	0.04	0.0521	0.00	49	12.89
Neomysis integer (Leach, 1814)	2	0.00	0.0022	0.00	2	0.53
Nephtys hombergii Savigny, 1818	6	0.00	0.5814	0.05	5	1.32

Continued on p. 104

Table 2. (Continued)

TAXA	Total A	A (%)	Total B	B (%)	Total *P*	*P* (%)
Nereis diversicolor Müller, 1776	1794	0.79	**26.3224**	2.25	**273**	71.84
Nereis virens (M. Sars, 1835)	4	0.00	0.5107	0.04	4	1.05
Notomastus latericeus M. Sars, 1851	2	0.00	0.0091	0.00	1	0.26
Ostracoda ind.	**5559**	2.46	0.7802	0.07	105	27.63
Palaemon longirostris H. Milne-Edwards, 1837	1	0.00	0.2875	0.02	1	0.26
Paragnathia formica (Hesse, 1864)	27	0.01	0.0161	0.00	25	6.58
Pholoe synophthalmica Claparède, 1868	1	0.00	0.0004	0.00	1	0.26
Podarkeopsis cf. arenicolus (La Greca, 1947)	1	0.00	0.0015	0.00	1	0.26
Polydora caeca (Oersted, 1843)	943	0.42	0.4418	0.04	27	7.11
Polydora cf. socialis (Schmarda, 1861)	7	0.00	0.0034	0.00	2	0.53
Polydora ciliata (Johnston, 1838)	12	0.01	0.0050	0.00	3	0.79
Polydora ligni Webster, 1879	2730	1.21	0.8756	0.07	75	19.74
Protodorvillea kefersteini (McIntosh, 1869)	1	0.00	0.0001	0.00	1	0.26
Pycnogonida ind.	12	0.01	0.0023	0.00	4	1.05
Pygospio elegans Claparède, 1863	1194	0.53	0.4047	0.03	45	11.84
Saduriella losadai Holthuis, 1964	1	0.00	0.0006	0.00	1	0.26
Scrobicularia plana (da Costa, 1778)	2050	0.91	**1005.4482**	**85.91**	**258**	67.89
Spionidae ind.	11	0.00	0.0003	0.00	3	0.79
Streblospio shrubsolii (Buchanan, 1890)	**80413**	**35.52**	**16.8156**	1.44	**378**	99.47
Tanaidacea ind.	18	0.01	0.0007	0.00	6	1.58
Tharyx ind.	**14376**	6.35	3.9881	0.34	**190**	50.00
Tubificidae ind.	**97224**	**42.95**	**21.9224**	1.87	**360**	94.74
Tubificoides benedeni (Udekem, 1855)	1	0.00	0.0001	0.00	1	0.26

Alcochete beaches where sandy sediments sometimes with very low fines content occur on the upper shore.

Total volatile solids range from 0.6 to 11.6% of the total sediment and increase gradually in the same direction as increasing fines content (Fig. 2). An inverse pattern was observed for redox potential. The lowest values were obtained in sites with high fines content, and the highest values in sand and oyster bed sites, with low fines content.

Macrofauna patterns

Table 2 presents the total abundance, presences and biomass (wet weight) for each species in the study area. Sixty-eight species/taxa, in total, were recorded. Annelids represented 90% of the total abundance (mean = 536 individuals. 0.05 m^{-2}), arthropods 6% (mean = 34 individuals. 0.05 m^{-2}) and molluscs 3% (mean = 19 individuals. 0.05 m^{-2}). Annelids (polychaetes) also dominated the species richness with a total of 30 species. Arthropods (crustaceans) and molluscs (bivalves) had 20 and 8 species, respectively.

Species richness ranged from 0 species (one site) to 23 spp. 0.05 m^{-2}. Its general pattern of distribution (Fig. 3) shows the lowest values in the upper mudflat near the salt-marsh, and the highest values in the oyster beds and the Alcochete flat (with sandier sediments). The species *Streblospio shrubsolii* (Buchanan, 1890), the Tubificids, *Nereis diversicolor*, *Cyathura carinata* (Krøyer, 1847), *Scrobicularia plana*, and *Hydrobia ulvae* (Pennant, 1777) were the most common species, present in more than 65% of the sites (cf. Table 2).

Abundance ranged from 0 individuals to 2338 ind. 0.05 m^{-2}. The overall abundance pattern was not as clear as the species richness pattern (Fig. 3). The lowest values were found in the upper mudflat near the salt-marsh and the highest values either in scattered sites or in the upper sandy shore near Alcochete (Fig. 3). The most abundant species were the oligochaetes of the Tubificidae family and the polychaetes *Streblos-*

pio shrubsolii and *Tharyx* sp. representing, respectively, 76% and 14% of the total number of polychaetes. Tubificidae spp. and *S. shrubsolii* represented 78% of the total macrofauna abundance. The most abundant crustaceans were ostracods (44% of the total number of crustaceans) followed by the isopod *Cyathura carinata* (26%) and the amphipod *Corophium volutator* (Pallas, 1766) (24%). The species *Scrobicularia plana* dominated among the bivalves (87% of the total number of bivalves) and the species *Hydrobia ulvae* was the only gastropod found in the study area.

The biomass pattern reflects either the distribution of the bivalve *Scrobicularia plana*, which comprises almost 86% of the total biomass in the study area, or the high diversity and abundance of some sites. The species *Nereis diversicolor*, *Balanus* sp., Tubificidae ind., *Streblospio shrubsolii*, *Hydrobia ulvae* and *Cyathura carinata* were also important contributors to the total biomass (Table 2). The biomass pattern is shown in Figure 3. The sites located in the upper mudflat, at the edge of salt-marsh, presented the lowest biomass (<1 g. 0.05 m^{-2}), the sites located in the middle of the flats presented the highest values, ranging between 10 and 55 g. 0.05 m^{-2} (Fig. 3).

Multivariate analysis

The results of the multivariate analysis of the abundance data, and the spatial patterns emphasized, are presented in Figure 4. Eight sites were excluded from this analysis due to low species richness and abundance (in one of the sites no macrofauna was found). Seven affinity groups, named A to G, were considered (Fig. 4). The most important species for each group are shown in Table 3 and the global characterization of the affinity groups is presented in Table 4. Among the highlighted species in Table 3, the list of the characteristic species included in Table 4 are those whose mean abundance is equal or above 1.0 ind. 0.05 m^{-2}.

Group A comprises 24 sampling sites located at the edge of the salt marsh (Fig. 4), in an area of mud (95% fines) with high total volatile solids content (7.6%) and the lowest redox potential (−12.4 mV) (Table 4). Mean species richness and abundance present here the lowest values of the whole study area (6 spp. 0.05 m^{-2}; 107.5 ind. 0.05 m^{-2}). The mysid *Mesopodopsis slabberi* (van Beneden, 1861) had its highest abundance here (Table 3).

Group B comprises 217 sites (Fig. 4) and corresponds to a wide area of muds with the highest organic content (fines: 93.7%; TVS: 8.1%). In general, the area presents low values of species richness and abundance (9 spp. 0.05 m^{-2}; 751.6 ind. 0.05 m^{-2}), except for the Tubificids (oligochaetes) which characterize this group. In fact, group B includes the most abundant sites chiefly due to Tubificids. Also, this group includes 2 unique species, *Lekanesphaera hookeri* (Leach, 1814) and *Notomastus latericeus* M. Sars, 1851 (Table 3).

Group C is composed of 21 sites, located mainly in the northern part of the study area in muddy sand and has the most irregular distribution pattern of all groups (Fig. 4). The polychaete *Streblospio shrubsolii*, widely distributed in the study area, had its highest mean abundance here (Table 3).

Group D is defined by 9 sites, all located close to the upper shore in the Alcochete flat (Fig. 4). This area is characterized by medium sand (sand: 74.6%; fines 25%) with the lowest organic content (2.4%), and high species richness (15.1 spp. 0.05 m^{-2}). The gastropod *Hydrobia ulvae* is the most abundant species and the polychaetes *Capitella* spp. and *Alkmaria romijni* Horst, 1919 are also important, having their highest abundances here (Table 3).

Group E comprises 68 sites, located at the outward edge of the intertidal area (Fig. 4), consisting of mud and muddy sand. This group also includes most of the subtidal sites sampled in this study. It has low mean species richness and abundance (7.9 spp. 0.05 m^{-2}; 276.7 ind. 0.05 m^{-2}). The polychaete *Tharyx* sp. had high abundance in this area (Table 3).

Group F is composed of 19 sites in the northern navigation channel (Fig. 4), essentially with mud and muddy sand sediments. It shows intermediate values of species richness and abundance (9.3 spp. 0.05 m^{-2}; 485.5 ind. 0.05 m^{-2}). The polychaete *Polydora caeca* (Oersted, 1843) and the amphipod *Corophium volutator* had their highest mean abundance here (Table 3).

Finally, group G comprises 14 sites located on the oyster beds (Fig. 4), and shows the highest mean species richness (19.1 spp. 0.05 m^{-2}); mean

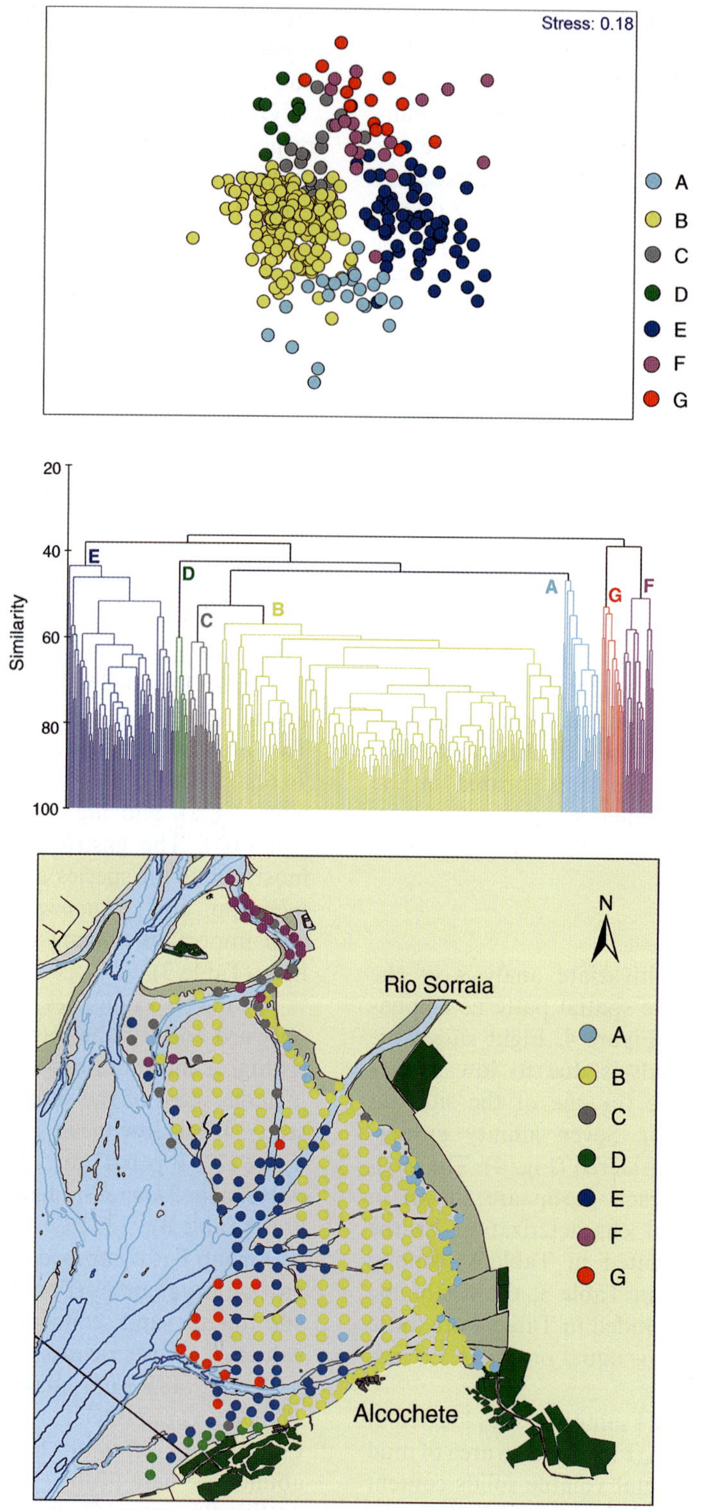

Figure 4. Multivariate analyses of the macrofauna abundance data showing the ordination and classification diagrams and the spatial distribution of affinity groups.

Table 3. Characteristic species for each affinity group identified by multivariate analysis. The taxa are represented by their mean abundance per unit sample (0.05 m^2). The darker highlight indicates the group where each species presents the highest abundance and the brighter highlight indicates the exclusive species in each group

	A	B	C	D	E	F	G
Mesopodopsis slabberi	0.83	0.37	0.39		0.32	0.26	0.14
Lekanesphaera hookeri		0.93					
Notomastus latericeus		0.93					
Idotea chelipes		0.47					0.14
Glycera tridactyla		0.93		0.11	0.44		0.29
Nereis diversicolor	4.46	6.29	3.52	5.22	1.41	3.22	4.00
Ostracoda ind.	9.50	24.77			0.13		
Tubificidae ind.	38.17	436.78	51.52	46.56	9.13	8.79	5.43
Palaemon longirostris			0.48				
Abra prismatica			0.48		0.15		
Neomysis integer		0.47	0.48				
Tanaidacea ind.			0.52		0.15	0.32	
Streblospio shrubsolii	47.83	216.82	546.24	515.56	12.76	249.63	339.00
Foraminifera ind.				4.11			
Copepoda ind.				0.33	0.29	0.15	0.29
Crangon crangon	0.42	0.33	0.24	0.67	0.35		0.14
Abra tenuis		0.32		9.22	0.59		
Scrobicularia plana	0.92	7.78	2.14	13.78	2.32		1.43
Fabricinae ind.		8.17	4.19	21.78	0.56	0.15	0.57
Alkmaria romijni		2.20	8.95	65.56	0.34	0.74	0.29
Capitella sp.	0.58	0.82	0.95	116.11	0.39	0.15	0.21
Hydrobia ulvae	3.83	14.24	1.43	163.78	4.26	0.22	0.71
Abra alba					0.29		
Nephtys hombergii					0.88		
Crustacea ind.		0.14			0.74		0.14
Paragnathia formica	0.42	0.70	0.48	0.11	0.74	0.16	
Cossura sp.		0.93			1.25		1.14
Cerastoderma edule		0.26	0.14	0.56	1.44		0.71
Tharyx sp.	0.33	16.18	1.67	22.00	143.93	1.53	59.93
Saduriella losadai						0.53	
Nereis virens		0.14				0.53	
Idotea emarginata	0.42				0.15	0.53	
Leptocheirus pilosus			0.95		0.15	7.53	1.29
Corophium volutator		0.42	0.14	0.11	0.66	15.16	6.00
Polydora caeca		0.47	1.95			46.42	0.57
Spionidae ind.		0.47			0.44		0.50
Boccardia polybranchia		0.47					0.57
Mysella sp.				0.33			0.71
Autolytus cf. langerhansi					0.44		0.71
Polydora ciliata		0.47					0.79
Corbula gibba		0.47	0.48		0.29		0.71
Micronephtys minuta		0.23	0.48	0.11	0.18		0.71
Insecta ind.	0.17	0.25	0.29			0.16	0.79
Heteromastus filiformis		0.23			0.24		1.00

Continued on p. 108

Table 3. (Continued)

	A	B	C	D	E	F	G
Carcinus maenas		0.33	0.95	0.89	0.29		1.21
Lekanesphaera levii		0.28		0.22	0.74	0.53	1.71
Nemertea ind.	0.13	0.22	0.14	0.11	0.29		2.86
Balanus sp.	0.83	0.14	1.48			0.84	13.64
Melita palmata	0.83	0.47	1.43	0.11	0.32	2.53	15.00
Pygospio elegans	0.42	0.15	0.48	5.33	3.19		34.36
Cyathura carinata	0.38	9.55	15.00	19.22	0.96	5.53	37.64
Nematoda ind.	2.29	5.87	2.76	16.11	2.88	1.15	55.36
Polydora ligni	0.42	0.63	5.95	2.44	0.26	6.79	165.43
Ampithoe valida							0.21
Lysidice ninetta							0.50
Polydora cf. socialis							0.50
Lanice conchilega							0.71
Mediomastus fragilis							0.71
Melinna palmata							0.71
Mytilus galloprovincialis							0.71
Pholoe synophthalmica							0.71
Podarkeopsis cf. areniculos							0.71
Protodorvillea kefersteini							0.71
Tubificoides benedeni							0.71
Pycnogonida ind.							0.86
Nematonereis unicornis							1.00
Merceriella enigmatica							1.64
Anthozoa ind.							4.29

abundance is also high (485.5 ind. 0.05 m^{-2}) and the group has the highest number of exclusive species (Table 3). The sediment presents high values of the biogenic fraction (mostly oyster and *Scrobicularia* shells), sand content and redox potential (Table 4). The polychaete *Polydora ligni* Webster, 1879 is the most characteristic species, but the isopod *Cyathura carinata*, the amphipod *Melita palmata* Grube, 1870 and the polychaete *Pygospio elegans* Claparède, 1863 are also important species for this group (Table 3).

A similar overall pattern of macrobenthic assemblages was obtained with a simplified data matrix including a subset of only seven species: *Cyathura carinata*, *Corophium volutator*, *Hydrobia ulvae*, *Polydora ligni*, *Streblospio shrubsolii*, *Tharyx* sp. and Tubificidae ind.; this simplified subset of species was obtained with the BVSTEP routine in PRIMER and the rank correlation between its Bray–Curtis similarity matrix and the inter-site similarity matrix including the total fauna was higher than 0.95 (rho = 0.954; p = 0.001). These seven species are the most abundant in the affinity groups (Table 4).

Discussion

When compared to tropical mudflats, species richness in temperate mudflats tends to be lower and abundances higher (Macnae & Kalk, 1962; Reise, 1991; Dittmann, 1995, 2000). The species richness of benthic fauna in the Tagus mudflats reported in this study is lower than that found in tropical intertidal areas and the value is in agreement to what has been reported for temperate areas and particularly Portugal (Table 5). The exception seems to be the Sado estuary (Portugal) and the Schelde (The Netherlands) which have species richness values closer to those reported for tropical areas (Table 5).

Table 4. Affinity groups characterization. Mean values are reported to the unit sampling area (0.05 m^2)

Affinity Groups	A	B	C	D	E	F	G
Number of sampling sites	24	217	21	9	68	19	14
Gravel content (mean)	0.0	0.0	0.0	0.3	0.0	0.2	0.0
Sand content (mean)	4.9	6.3	43.5	74.6	23.7	43.9	46.8
Fines content (mean)	95.0	93.7	56.4	25.0	76.3	55.9	53.2
Biogenic fraction (mean)	0.1	1.5	1.9	2.6	1.7	2.4	16.7
Total Volatile Solids (mean)	7.6	8.1	4.7	2.4	6.3	4.2	4.5
Redox Potential (mean)	−12.4	19.6	99.0	92.0	42.2	4.1	126.5
Mean Abundance	107.5	751.6	650.6	1075.4	276.7	485.5	761.9
Mean Biomass	0.0056	0.0223	0.0248	0.0965	0.0082	0.0261	0.2531
Mean Species richness	6.0	9.0	8.6	15.1	7.9	9.3	19.1
Total Species richness	20	42	31	27	39	25	53
Number of exclusive species	0	2	1	1	2	1	15
Characteristic Species		Tubificidae ind.	*S. shrubsolii*	*H. ulvae*	*Tharyx* sp.	*P. caeca*	*P. ligni*
		N. diversicolor		*Capitella* spp.	*C. edule*	*C. volutator*	Nematoda ind.
		Ostracoda ind.		*A. romijni*	*Cossura* sp.	*L. pilosus*	*C. carinata*
				Fabricinae ind.			*P. elegans*
				S. plana			*M. palmata*
				A. tenuis			*Balanus* sp.
				Foraminifera ind.			*M. enigmatica*
							Nemertea ind.
							L. levii
							C. maenas
							H. filiformis
							Anthozoa sp.
							N. unicornis

Abundances are more difficult to compare due to different sampling devices and mesh sizes, the number of sites sampled in each area and the number of replicates per site. The total abundance reported in this study (approximately 12,000 ind. m^{-2}) is, however, comparable to what has been reported for temperate mudflats and thus higher than that found in tropical intertidal areas.

In temperate intertidal areas, Polychaets are the most diverse group, followed by Molluscs and Crustaceans (Reise, 1991). In tropical areas Mollusc and Crustacean species richness tends to increase (Reise, 1991; Dittmann, 1995, 2000) and may dominate the invertebrate fauna (Reise, 1991). In this study Polychaetes dominate both the species richness and the abundance. The ratio of Molluscs plus Crustaceans to Polychaete species is 1.03, corroborating the general trend for temperate areas for which this ratio is close to 1.0 (Reise, 1991). This author found ratios of 1.06 and 0.75 for Bahia Quillaipe (Chile) and Konigshafen (North Sea) respectively, and 1.44 for Ao Nam Bor (Thailand). Calculating this ratio for the examples reported in his paper, the ratio for seasonal climate areas varies from 1.18 to 1.29 and for tropical areas ranges from 3.5 to 8.2. Using Dittmann's (1995) data for the macrofauna, the values are 5.39 (Hinchinbrook channel, Australia) and 3.8 (Haughton estuary, Australia) and the data from Whitlatch (1977) for Barnstable Harbour (Massachusetts) is 0.5. Reise (1991) points out that the use of a 0.25 mm mesh screen in his study could have increased the Polychaete fraction. Dittmann (1995), however, used 0.5 and 1 mm mesh size screens and in the present study 0.5 mm was used and the trend seems to be maintained.

The species *Streblospio shrubsolii*, the Tubificids, *Cyathura carinata*, *Tharyx* sp. and *Hydrobia ulvae*

Table 5. Species richness in temperate and tropical mud and sandflats

Locality	Species richness	Reference
Tagus estuary	68	This study
Tagus estuary (right and left margins)	86	Calvário (1984)
Tagus estuary (right margin)	44	Silva et al. (this volume)
Tagus estuary (right margin)	33	Desgarrado Pereira (1997)
Ria de Aveiro (Portugal)	47	Moreira (1988)
Ria Formosa (Portugal)	53	Sprung (1994)
Mondego estuary (Portugal)	24	Marques et al. (1993)
Sado estuary (Portugal)	150	Costa et al. (1990)
Sado estuary (Portugal)	90	Rodrigues et al. (1994)
Bahia Quillaipe (Chile)	43	Reise (1991)
Konigshafen (North sea)	42	Reise (1991)
The Oosterschelde (Netherlands)	54	Meire & Dereu (1990)
Westerscheld (Belgium) – polluted site	9	Meire & Dereu (1990)
Berg river estuary (South Africa)	25	Kalejta & Hockey (1991)
Barnstable harbour (Massachussets, USA)	47	Whitlatch (1977)
Waden Sea (The Netherlands)	42	Beukema et al. (1978)
Schelde estuary (The Netherlands)	106	Ysebaert and Herman (2002)
South East England – 14 areas	13 to 53	Kay and Knights (1975)
Haughton estuary (North-East Australia)	77 to 96	Dittmann (2000)
Inhaca Island (Mozambique – East Africa)	465	Macnae and Kalk (1962)
Ao Nam Bor (Thailand)	118	Reise (1991)
Mangrove Swamps (Northwestern Australia)	112	Wells (1983)
Northern Great Barrier Reef	150	Gibbs (1978)

were the most common and abundant species found in this study. They are also common and abundant in other intertidal areas (Beukema, 1976; Kay & Knights, 1975; McLusky, 1987; Moreira, 1988; Marques et al., 1993; Desgarrado Pereira et al., 1997; Ysebaert & Herman, 2002). The species *S. shrubsolii, Tharyx* sp. and Tubificids are also frequent and abundant for the Tagus subtidal fauna (Rodrigues & Quintino, 1994) and *H. ulvae* is even referred to as an abundant species in South African intertidal areas (Kalejta & Hockey, 1991). Total abundance was in fact dominated by only a few species and if the species that contribute less than 3% to the total abundance of each site are excluded, the resulting matrix contains a subset of only 10 species; nevertheless, its Bray–Curtis similarity matrix presents a high rank correlation with the global fauna (rho = 0.971; $p = 0.001$). Although the overall macrobenthic assemblages pattern may be obtained with such a simplified data matrix and although this is a common situation in other intertidal areas, namely a strong dominance by very few species (Kay & Knights, 1975; Beukema, 1976; Kalejta & Hockey, 1991; Ysebaert & Herman, 2002), this should not hide the biological significance of rare species, namely in terms of their response to disturbance (Gray et al., 1990). The finding of one specimen of the isopod *Saduriella losadai* Holthuis, 1964 can be reported collected in a site of group *F* with muddy fine sand sediment. This is a new genus described by Holthuis (1964) from material collected in the river Ulla, in Ria de Arosa (Spain). After its description the species was found in the Minho river (Portugal–Spain Northern border) (Dexter, 1992), in the upper reaches of Ria de Aveiro (Cunha et al., 1999), in the Mondego estuary (Chainho et al., this volume) and in the Guadalquivir (Baldó et al., 2000). The finding of a specimen in the present study, confirms the distribution of the species further south along the Portuguese coast as a specimen had already been refereed in the stomach content of eels caught in the upper reaches of the Tagus estuary (Costa et al., 1992).

Eltringham (1971) considered a three zone pattern for British muddy shores and Beukema (1976; 1988) proposed a model for the intertidal flats of the Wadden Sea where maximal values for both species number and biomass were found at intermediate values for intertidal height and sediment parameters, such as silt content. In this study, although the characterising species were not the same as those referred to by Eltringham (1971), three main zones from the upper to the lower levels of the Tagus mudflat were also identified: the upper mudflat (Group A), with the lowest species richness, abundance and biomass of all the areas identified in this study; the lower mudflat (Group E), characterized by *Tharyx* sp., *Cerastoderma edule* and *Cossura* sp. and, between these, a mid-zone (Group B) with higher values for species richness, abundance and biomass than the two former. This mid-area is mainly characterized by *Nereis diversicolor*, Tubificids and Ostracoda spp. Apart from this succession, three other main areas were identified, corresponding to particular habitats. They have the highest species richness and biomass and are located in sediments with intermediate values for silt and sand content (Table 3): Group F located on the flats surrounding the northern navigation channel characterized by the amphipods *Corophium volutator* and *Leptocheirus pilosus* Zaddach, 1844 and the polychaete *Polydora caeca*; Group D, corresponding to the sandier shores near Alcochete characterized by *Hydrobia ulvae, Scrobicularia plana, Abra tenuis* (Montagu, 1803), Foraminifera, *Alkmaria romijni*, a Fabricidae species and *Capitella* spp. The former, a well known organic enrichment indicator (Pearson & Rosenberg, 1978), appeared almost exclusively in this group and reflects the anthropogenic organic inputs which can be seen along this shore. Finally Group G, with the highest species richness, biomass and number of exclusive species (Table 4), corresponds to the oyster beds. The sand and fines fractions in this group are almost the same as in group F but the biogenic fraction is much higher, due mainly to oyster shell deposits. Biogenic structures provided by macrobenthic species, such as shells, diversify the habitat and could accommodate other species. Kay & Knights (1975) in a study in intertidal areas of south east England, also found the highest species diversity and abundance in a mussel bed community.

Sediment characteristics are important in determining the distribution and abundance of the macrobenthic fauna (for example, Pearson & Rosenberg, 1978; Beukema, 1988; Rodrigues & Quintino, 1993; Raffaelli & Hawkins, 1996; Snelgrove, 1999; Dittmann, 2000; Little, 2000; Quintino et al., 2001; Ysebaert & Herman, 2002) and in this study the different macroinvertebrate affinity areas are well characterized by sediment descriptors (Table 3) and showed clear spatial patterns (Fig. 4). These descriptors cannot however explain the global pattern. This is emphasized by the low Spearman rank correlation coefficient between the species abundance data and the subset of environmental variables which best fit the biological data using the BioEnv routine in PRIMER (Clarke & Gorley, 2001). The best fit was obtained with a single environmental variable, fines, showing a correlation of 0.317.

Besides the environmental data analysed in this study, values of mudflat elevation (inundation time) were calculated from satellite images of the study area (Santos, unpublished data). Although values could only be obtained for about 2/3 of the benthic sites, when these values are included in the environmental matrix the best fit is obtained with the variables fines content (or sand content), elevation and biogenic fraction, and the correlation coefficient increased to 0.518. Even so the value is lower than the ones usually found in studies conducted on subtidal areas. In a series of such studies, also conducted during 2002 on several locations of the Portuguese coast, the best relationship between environmental and benthic macrofauna data sets ranged from a rank correlation of 0.851 on the shelf off the Tagus estuary, to 0.605 on the lagoon of Óbidos, with a value of 0.717 for a study conducted on the shelf off Aveiro (unpublished data). This suggests that the relationship between environmental and benthic macrofauna data could be less straightforward as natural disturbance increases. In this study, a mixture of grain-size, elevation (inundation time) and particular habitat (shell debris) form the best possible explanatory combination. Arguably, inundation time could account for the lower-upper flat succession, grain-size for the marked difference between muddy and sandy shores, and biogenic fraction for the particular biotope represented by the relic oyster banks. Other unmeasured factors, namely current velocities, could eventually account for the specific assemblage bordering the navigation channels.

Acknowledgements

This study was included in the project SAT-TAGIS – Modelling the habitats, invertebrates and birds in intertidal flats of the Tagus estuary using satellite images and GIS (PDCTM/P/MAR/15256/1999), financed by the Portuguese FCT (Fundação para a Ciência e a Tecnologia). Sónia Meireles, Teresa Pereira and Alice Gama benefited from research initiation funding. Our thanks to Rui Marques, Leandro Sampaio and Joana Oliveira for assisting in sediment sampling and to Carlos David Santos who kindly furnished the elevation data.

References

Baird, D. P., R. Evans, H. Milne & M. W. Pienkowski, 1985. Utilization by shorebirds of benthic invertebrate production in intertidal areas. Oceanography and Marine Biological: An Annual Review 23: 573–597.

Baldó, F., C. Megina & J. A. Cuesta, 2000. *Saduriella losadai*, 1964 (Isopoda, Valvifera) in the Gadalquivir estuary (S.W. Spain). Crustaceana 73: 1015–1017.

Beukema, J. J., W. De Bruin & J. J. M. Jansen, 1978. Biomass and species richness of the macrobenthic animals living on the tidal flats of the Dutch Wadden Sea: Long-term changes during a period with mild winters. Netherlands Journal of Sea Research 12(1): 187–206.

Chainho, P., J. L. Costa, M. L. Chaves, M. F. Lane, D. M. Dauer & M. J. Costa. Seasonal and spatial patterns of distribution of subtidal benthic invertebrate communities in the Mondego River, Portugal – a poikilohaline estuary. Hydrobiology 555: 59–73.

Clarke, K. R. & R. N. Gorley, 2001. Primer v5: User Manual/ Tutorial. Plymouth, PRIMER-E 91 pp.

Clarke, K. R. & R. M. Warwick, 2001. Change in Marine Communities: An Approach to Statistical Analysis and Interpretation (2nd ed.). Plymouth, PRIMER-E.

Costa, J. L., C. A. Assis, P. R. Almeida, F. M. Moreira & M. J. Costa, 1992. On the food of the European eel, *Anguilla anguilla* (L.), in the upper zone of the Tagus estuary, Portugal. Journal of Fish Biology 41: 841–850.

Costa, M. H., J. F. S. Oliveira & G. Bellan, 1990. Ecologia Bêntica em substratos móveis intertidais do Estuário do Sado. In Santana, F., M. C. R. Santos, M. H. Costa & D. Pereira (eds), Proceedings 2ª Conferência Nacional sobre a Qualidade do Ambiente, UNL 4–6 Abril 1990, I: A225–A239.

Cunha, M. R., J. C. Sorbe & M. H. Moreira, 1999. Spatial and seasonal changes of brackish peracaridan assemblages and their relation to some environmental variables in two tidal channels of Ria de Aveiro (N.W. Portugal). Marine Ecology Progress Series 190: 69–87.

Desgarrado Pereira, C., M. J. Gaudêncio, M. T. Guerra & M. T. Lopes, 1997. Intertidal macrozoobenthos of the Tagus estuary (Portugal): The Expo'98 area. Publicaciones Especiales Instituto Español de Oceanografia 23: 107–120.

Dexter, D. M., 1992. Soft bottom invertebrates of the Portuguese benthos. Boletim Instituto Nacional Investigação das Pescas 17: 61–88.

Dittmann, S., 1995. Benthos structure on tropical tidal flats of Australia. Helgoländer Meeresunters 49: 539–551.

Dittmann, S., 2000. Zonation of benthic communities in a tropical tidal flat of north-east Australia. Journal of Sea Research 43: 33–51.

Doeglas, D. J., 1968. Grain-size indices, classification and environment. Sedimentology 10: 83–100.

Eltringham, S. K., 1971. Life in Mud and Sand. English University Press, London 218 pp.

Gibbs, P. E., 1978. Macrofauna of the intertidal sand flats on low wooded islands, northern Great Barrier Reef. Philosophical Transactions of the Society London, B 284: 81–97.

Gray, J. S., K. R. Clarke, R. M. Warwick & G. Hobbs, 1990. Detection of initial effects of pollution on marine benthos: an example from the Ekofisk and Eldfisk oilfields, North Sea. Marine Ecology Progress Series 66: 285–299.

Holthuis, L. B., 1964. *Saduriella*, a new genus of Isopoda Valvifera from northwestern Spain. Zoologische Mededelingen, Leiden 40: 29–35.

Kalejta, B. & P. A. R. Hockey, 1991. Distribution, abundance and productivity of benthic invertebrates at the Berg River Estuary, South Africa. Estuarine, Coastal and Shelf Science 33: 175–191.

Kay, D. G. & R. D. Knights, 1975. The macro-invertebrate fauna of the intertidal soft sediments of south east England. Journal Marine Biological Association of the United Kingdom 55: 811–832.

Kristensen, E. & F. Ø. Anderson, 1987. Determination of organic carbon in marine sediments: a comparison of two CHN-analyser methods. Journal of Experimental Marine Biology and Ecology 109: 15–23.

Larsonneur, C., 1977. La cartographie des depots meubles sur le plateau continental français: méthode mise au point et utilisée en Manche. Journal Recherche Océanographic 2(2): 33–39.

Legendre, P. & L. Legendre, 1998. Numerical Ecology (2nd ed.). Elsevier Science B.V, Amsterdam 853 pp.

Little, C., 2000. The Biology of Soft Shores and Estuaries. Oxford University press, 252 pp.

Macnae, W. & M. Kalk, 1962. The fauna and flora of sand flats at Inhaca Island, Moçambique. Journal Animal Ecology 31: 93–128.

Marques, J. C., L. B. Rodrigues & A. J. A. Nogueira, 1993. Intertidal macrobenthic communities structure in the Mondego estuary (western Portugal): reference situation. Vie Milieu 43(2–3): 177–187.

Meire, P. M. & J. Dereu, 1990. Use of the abundance/biomass comparison method for detecting environmental stress; some considerations based on intertidal macrozoobenthos and bird communities. Journal of Applied Ecology 27: 210–223.

McLusky, D. S., 1987. Intertidal habitats and benthic macrofauna of the Forth estuary, Scotland. Proceedings of the Royal Society of Edinburgh 93B: 389–399.

Moreira, F., 1999. On the use by birds on intertidal areas of the Tagus estuary: implications for management. Aquatic Ecology 33: 301–309.

Moreira, M. H., 1988. Estudo da comunidade bêntica num banco de lodo intertidal da Ria de Aveiro, com especial incidência no crescimento, biomassa e produção do berbigão, *Cardium edule* (L.). Ciênia. Biologica Ecology and Systematics. (Portugal) 8: 47–75.

Pearson, T. H. & R. Rosenberg, 1978. Macrobenthic succession in relation to organic enrichment and pollution of the marine environment. Oceanography and Marine Biology Annual Review 16: 229–311.

Pearson, T. H. & S. O. Stanley, 1979. Comparative measurement of the redox potential of marine sediments as a rapid means of assessing the effect of organic pollution. Marine Biology 53: 371–379.

Petterson, C. H., 1991. Intertidal zonation of marine invertebrates in sand and mud. American Scientist 79: 236–249.

Quammen, M. L., 1984. Predation by shorebirds, fish and crabs on invertebrates in intertidal mudflats: an experimental test. Ecology 65: 529–537.

Quintino, V., A. M. Rodrigues & F. Gentil, 1989. Assessment of macrozoobenthic communities in the lagoon of Obidos, western coast of Portugal. Scientia Marina 53: 645–654.

Quintino, V., A. M. Rodrigues, A. Ré, M. P. Pestana, S. Silva & H. Castro, 2001. Sediment alterations in response to marine outfall operation off Lisbon, Portugal: a sediment quality triad study. Journal of Coastal Research Special issue 34: 535–549.

Raffaelli, D. & H. Milne, 1987. An experimental investigation of the effects of shorebird and flatfish predation on estuarine invertebrates. Estuarine, Coastal and Shelf Science 24: 1–13.

Raffaelli, D. & S. Hawkins, 1996. Intertidal Ecology. Chapman and Hall, London 356 pp.

Rees, H. L., D. C. Moore, T. H. Pearson, M. Elliott, M. Servile, M. Pomfret & D. Johnson, 1990. Procedures for the monitoring of marine benthic communities at UK sewage sludge disposal sites. Department of Agriculture and Fisheries for Scotland. Scottish Fisheries Information Pamphlet 18: 1–79.

Reise, K., 1991. Macrofauna in mud and sand of tropical and temperate tidal flats . In Elliott, M. & J.-P. Ducrotory (eds.). Olsen and Olsen, Fredenborg: 211–216.

Rodrigues, A. M. & V. Quintino, 1993. Horizontal biosedimentary gradients across the Sado estuary, W Portugal. Netherlands Journal of Aquatic Ecology 27(2–4): 465–482.

Rodrigues, A. M. & V. Quintino, 1994. Diagnóstico do estado de Qualidade da Região Subtidal do Barreiro-Seixal, estuário do Rio Tejo, através da análise do Ambiente Biossedimentar. In A. Rosa Pires, C. Pio, C. Boia & T. Nogueira (eds), Proceedings of 3ª Conf. Nacional sobre a Qualidade do Ambiente, UA, 5–7 Fevereiro 1992, III: 967–999.

Rodrigues, A. M., A. M. Picado, V. Quintino, E. Mendonça, F. Duarte, J. Blackstock & T. H. Pearson, 1994. Análise ao nível do indivíduo e da comunidade dos efeitos biológicos associados à descarga de efluentes industriais: um caso de estudo. In A. Rosa Pires, C. Pio, C. Boia & T. Nogueira (eds), Proceedings of 3ª Conf. Nacional sobre a Qualidade do Ambiente, UA, 5–7 Fevereiro 1992, III: 991–1010.

Silva, G., J. L. Costa, P. R. Almeida & M. J. Costa. Structure and dynamics of a benthic invertebrate community in an intertidal area of the Tagus estuary, western Portugal: a six year data series. Hydrobiology 555: 115–128.

Snelgrove, P. V. R., 1999. Getting to the bottom of marine biodiversity: sedimentary habitats. BioScience 49: 129–138.

Sprung, M., 1994. Macrobenthic secondary production in the intertidal zone of the Ria Formosa – a lagoon in Southern Portugal. Estuarine, Coastal and Shelf Science 38: 539–558.

Wells, F. E., 1983. An analysis of marine invertebrate distributions in a mangrove swamp in nortwestern Australia. Bulletin Marine Science 33: 736–744.

Whitlatch, R. B., 1977. Seasonal changes in the community structure of the macrobenthos inhabiting the intertidal sand and mud flats of Barnstable Harbour, Massachusetts. Biological Bulletin 152: 275–294.

Ysebaert, T. & P. M. J. Herman, 2002. Spatial and temporal variation in benthic macrofauna and relationships with environmental variables in an estuarine, intertidal soft-sediment environment. Marine Ecology Progress Series 244: 105–124.

Hydrobiologia (2006) 555:115–128
H. Queiroga, M.R. Cunha, A. Cunha, M.H. Moreira, V. Quintino, A.M. Rodrigues, J. Serôdio & R.M. Warwick (eds),
Marine Biodiversity: Patterns and Processes, Assessment, Threats, Management and Conservation
DOI 10.1007/s10750-005-1110-8

Structure and dynamics of a benthic invertebrate community in an intertidal area of the Tagus estuary, western Portugal: a six year data series

Gilda Silva[1,*], José Lino Costa[1], Pedro Raposo de Almeida[1,2] & Maria José Costa[1,3]
[1]*Instituto de Oceanografia, Faculdade de Ciências da Universidade de Lisboa, Campo Grande, 1749-016 Lisboa, Portugal*
[2]*Departamento de Biologia, Unidade de Biologia da Conservação, Universidade de Évora, 7002-554 Évora, Portugal*
[3]*Departamento de Biologia Animal, Faculdade de Ciências da Universidade de Lisboa, Campo Grande, 1749-016 Lisboa, Portugal*
*(*Author for correspondence: E-mail: gcsilva@fc.ul.pt)*

Key words: macrofauna, community structure, intertidal habitat, soft-sediments, environmental factors, Portugal

Abstract

The intertidal benthic invertebrate community adjacent to “Parque das Nações”, Tagus estuary, western Portugal, was studied at seven sites between October of 1996 and 2002. Grain size analysis revealed that the area was essentially muddy with total organic matter ranging from 5.4 to 11.4%. Annelida represented more than 85% of the total abundance. The dominant taxa were Oligochaeta, *Streblospio shrubsolii*, *Scrobicularia plana*, *Hediste diversicolor*, *Hydrobia ulvae* and Cirratulidae. Analysis of abundance revealed high seasonal and interannual heterogeneity, although clear aggregations between winter/spring and summer/autumn sampling periods were detected by multivariate techniques. *Paragnathia formica* and Dolichopodidae were closely related to winter/spring cluster while *S. shrubsolii*, *Cossura* sp., *S. plana* and *Corbula gibba* were associated to the summer/autumn aggregation. Although other taxa were abundant in several seasons, constancy in their numbers throughout the year could be observed. Temperature, rainfall and daylight hours were the environmental variables best related to the biological data. Temperature and daylight hours were important for the establishment of the two seasonal aggregations. Rainfall has also shown to be an essential factor on the structuring of this intertidal community, closely related to the abundance increase of Cirratulidade and decrease of *Polydora* sp., *Nephtys* spp., *Cerastoderma glaucum*, *Corophium* spp. and other Gammaridea. Sediment composition seems to play an important role in changing the community characteristics during the year since some species abundance appear to be favoured by coarser particles and many by fine sediments.

Introduction

Benthic invertebrates are relatively sedentary and play an important role in cycling nutrients and inorganic compounds between sediments and water column. These facts, associated with their importance as food sources for economically or recreationally important fish species and the ability to react to natural or human induced disturbances allow these communities to exhibit great potential for integrating long-term environmental conditions (Bilyard, 1987; Herman et al., 1999).

The study of temporal and spatial variability of benthic communities is essential to understand their structure and dynamics. Some studies have been done on the relations between the temporal and spatial variation of several environmental variables and the biological patterns and processes in macrobenthic assemblages of soft-sediments (e.g. Ysebaert & Herman, 2002; Bazaïri et al., 2003; Ysebaert et al., 2003). In muddy sediments of temperate intertidal habitats, such as lagoons and salt marshes, marked seasonal fluctuation in density of macrofaunal assemblages in which deposit feeders dominate has been observed (Marsh & Tenore, 1990). Physical variables fluctuate with greater amplitude in intertidal areas, thus subjecting organisms inhabiting this environment to great physical stress (Woodin, 1974).

Long-term data studies consist on the assessment of temporal changes in macrofaunal composition based on a regular sampling period or on the comparison at selected sites after a long time interval (Grémare et al., 1998). Long-term data series are common in many marine areas: Wadden Sea (e.g. Reise, 1982), western English Channel (e.g. Ibanez & Dauvin, 1988), French Mediterranean coast (e.g. Salen-Piccard & Arlhac, 2002), Chesapeake Bay (e.g. Dauer & Alden, 1995) and southern California (e.g. Desmond et al., 2002). However, few long-term studies have been published from the Iberian Peninsula (e.g. López-Jamar et al., 1995).

In Portugal, studies involving soft-sediments macrofaunal assemblages have taken place in several estuaries (e.g. Calvário, 1982; Quintino & Rodrigues, 1989; Rodrigues & Quintino, 1993; Mucha & Costa, 1999; Marques et al., 2002; Mucha et al., 2003) and have also investigated the ecology of several species (e.g. Guerreiro, 1998; Abrantes et al., 1999; Lillebø et al., 1999; Cunha et al., 2000). In the Tagus estuary the studies on the benthic communities were either dedicated to the study of spatial patterns (Calvário, 1982; Rodrigues et al., this volume) or involved monitoring assessment works (Pereira et al., 1997; Costa et al., 1999). Long-term series of benthic macrofauna communities of the Tagus estuary exist but were not published. Therefore, the present study, which began in October 1996, becomes an important data asset.

The aim of this research was to assess the structure of the benthic invertebrate community of an intertidal soft-sediment area in the Tagus estuary and evaluate the relationship between several environmental factors and the dynamics of this community.

Materials and methods

Sampling

The study area was located in the right bank of the Tagus estuary, in a 5 km long intertidal zone in front of "Parque das Nações", Lisbon (Fig. 1), being all sampling sites located at the same relative shore height. Salinity values ranged usually from 15 to 20 (Cabral, 1998). Until 1995 this area was polluted due to the nearby existence of several industries. The closing of these units and the rehabilitation of the area for the EXPO'98 World Exhibition and consequent implementation of the "Parque das Nações" changed this situation. Between 1995 and 1998 several dredging operations associated to the construction works of the "Parque das Nações" and the Vasco da Gama Bridge took place in this area. Benthic macrofauna samples were taken seasonally (although not always done in the same months for the different years), during high tide from a boat, in seven sites using a 0.05 m^2 Van Veen grab Sousa-Reis/LMG model, from autumn of 1996 until autumn of 2002. However, during this period, four seasons were not sampled (autumn of 1997, winter of 1998, spring and summer of 1999). At each site of the 21 sampling moments, six replicates were collected for macrofauna analysis (total area sampled = 0.30 m^2 per site) and another grab was taken for sediment analysis, which included grain size and total organic matter (TOM) determination.

Sediment analysis

Grain size analysis was carried out according to Gaudêncio et al. (1991). Sediments were classified as gravel (particles diameter above ≥2.000 mm), coarse sand (0.500–2.000 mm), medium sand (0.250–0.500 mm), fine sand (0.250–0.063 mm) and mud (<0.063 mm). TOM was determined by loss of weight on ignition at 500 °C (adapted from Pereira et al., 1997).

Other environmental variables

In order to evaluate the relationship between other environmental variables and the benthic invertebrate community, daylight hours (*D*) of each sampling date (Anonymous, 1992), mean monthly air

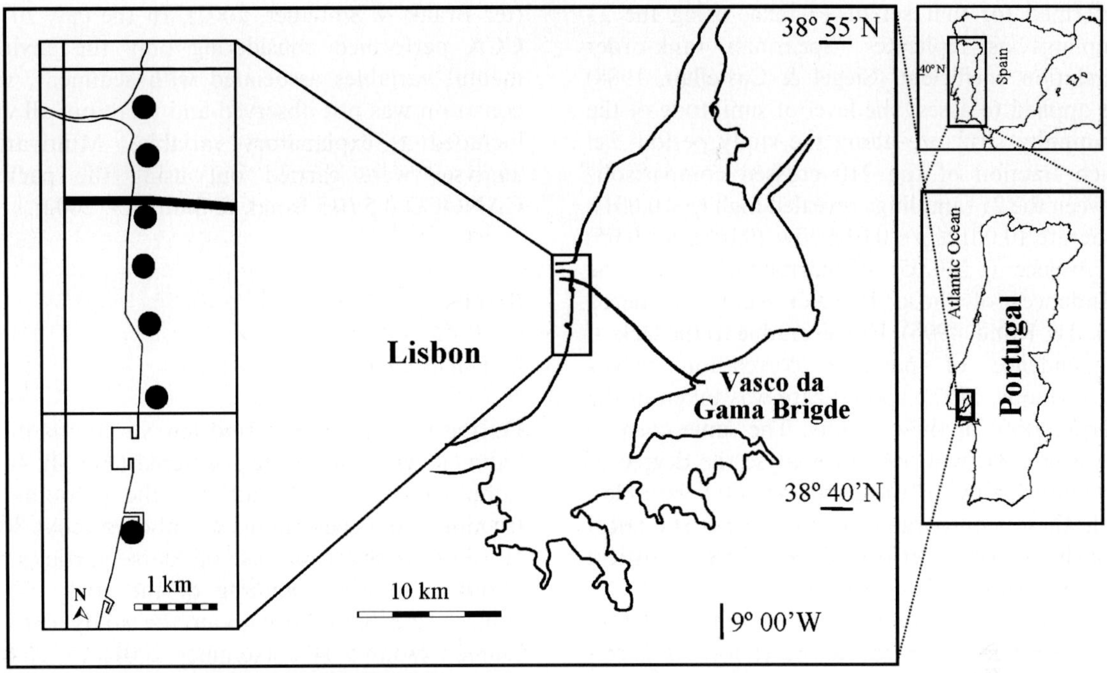

Figure 1. Map showing the location of the study area in the Tagus estuary and the sampling sites (●)

temperature (*T*) in °C, total monthly rainfall (*R*) in mm and river flow (*F*) data, presented as monthly mean discharge in $m^3\ s^{-1}$, for the River Tagus were analysed. These three last sets of data were obtained from the Ministry of Environment (DSM-DRAOT/LVT). Besides using the values of the sampling month for the air temperature, rainfall and river flow, community lagged response to these variables was also assessed when considering the values for the previous month (1 month) and for the cumulative previous 3 months since community responses may have lagged behind environmental conditions (Desmond et al., 2002). Due to the significant variations caused by the tides during the day, salinity was not considered in this study.

Macrofauna analysis

Benthic invertebrate samples were fixed in 5% neutralized formalin, washed through a 0.5 mm mesh sieve and preserved in 70% ethyl alcohol. Samples were sorted and organisms identified to the lowest taxonomic level possible. Each identified taxon was counted and density (number of individuals per 0.05 m^2) was determined. To evaluate mean individual biomass (milligrams per individual), dry weight was determined according to Rumohr (1999).

Statistical analysis

In total 41 invertebrate taxa were identified in 21 sampling occasions, but in subsequent statistical analyses only the 17 taxa corresponding to more than 0.05% of the total individuals captured were included in order to improve the robustness of the non-parametric tests applied to the data (avoiding the uncertainty of the reliability of the relative abundance of rare species) (Sokal & Rohlf, 1995) and the readability of the diagrams of the mutivariate technique used (ter Braak & Šmilauer, 2002). Altogether, these taxa accounted for 99.77% of total abundance. For these analyses several organisms had to be associated in higher taxonomic groups, either because specific identification was difficult, or the organisms were damaged to an extent that the identification of those individuals could not be possible.

To determine if the differences observed throughout the study period between the densities of the most abundant taxa were significant, Friedman test (Siegel & Castellan, 1988) was performed for that set of 17 taxa, using the 21 samplings as replicates. Spearman rank-order correlation coefficient (Siegel & Castellan, 1988) was applied to assess the level of similitude of the community structure along the study period, i.e. which fraction of the 210 crossed comparisons between the 21 samplings revealed high ($p < 0.001$), moderate ($0.001 \leq p < 0.01$), low ($0.01 \leq p < 0.05$) or absence ($p \geq 0.05$) of correlation when the abundances of those 17 taxa were considered (Sokal & Rohlf, 1995). However, due to the lack of independence of pairwise correlation values (Anonymous, 1997), special care was taken in the interpretation of these results. The same correlation coefficient was used to analyse the degree of correlation between the environmental variables. Both these non-parametric univariate statistical procedures were performed in SPSS statistical package (Anonymous, 1997).

Patterns of variation in community structure were investigated by means of canonical correspondence analysis (CCA) (ter Braak & Šmilauer, 2002): (i) performed considering all studied environmental variables; and also (ii) only those associated with sediment characterization, since the high correlation of these variables with the climatic variables might have obscured in the first CCA any specific relationship between the sediment variables and the biological data (ter Braak & Šmilauer, 2002). These analyses allowed the evaluation of the relationship between different abiotic factors and the structure of this soft-sediment intertidal community. For both situations a global Monte Carlo permutation test was used to evaluate the significance ($p < 0.05$) of the first ordination axis and the sum of all canonical axes (ter Braak & Šmilauer, 2002). In the first CCA the number of environmental variables had to be reduced by means of manual forward selection using a Monte Carlo permutation test ($p < 0.05$) (ter Braak & Šmilauer, 2002), since the proximity in the number of samples and environmental variables originated a non-significant relationship between species and environmental variables when the whole set of those environmental variables was considered (ter Braak & Verdonschot, 1995). However, since all non-selected environmental variables in that CCA were correlated to one of the explanatory variables, its addition as supplementary variables was performed and represented in the ordination diagram (ter Braak & Šmilauer, 2002). In the case of the CCA performed considering only the environmental variables associated with sediment, such constrain was not observed and therefore all were included as explanatory variables. Multivariate analyses were carried out using the package CANOCO 4.5 (ter Braak & Šmilauer, 2002).

Results

Sediment analysis

During the sampling period few variations in the sediment grain-size were registered (Fig. 2). In all sampling occasions, mud was the predominant fraction and its percentage was always above 80%. However, a slight increase of coarse particles was observed in the beginning of the study. TOM values were high throughout the study (Fig. 2), ranging from 5.04% (summer 2001) to 11.40% (winter 1999).

Other environmental variables

The mean monthly air temperature and total monthly rainfall values (Fig. 3) were characteristic of temperate zones. Mean air temperature values ranged between 7.50 and 24.30 °C and the lowest values were always registered in the winter months. Total monthly rainfall values ranged between 0.00 and 358.40 mm. The highest values of this parameter were registered in the winter months of 1996, 1997, 1998 and 2001 and were considerably lower in the remaining winters. Total monthly rainfall showed a variable pattern from year to year while mean air temperature revealed a more constant pattern.

Macrofauna analysis

A total of 18,140 individuals belonging to 41 taxa were collected between autumn of 1996 and 2002 in the study area (Table 1). Polychaeta accounted for 45% of the total abundance and Oligochaeta for 42%. Friedman test showed significant differences between the densities of the 17 considered taxa throughout the study ($\chi^2 = 220.95$; $df = 16$; $p < 0.001$). The benthic community of the study area was characterized by high abundances of six

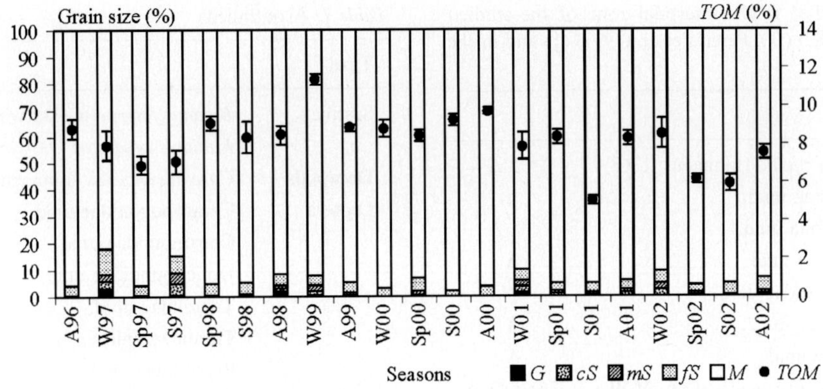

Figure 2. Variation of grain size defined as gravel (*G*), coarse sand (*cS*), medium sand (*mS*), fine sand (*fS*) and mud (*M*), and total organic matter (TOM - mean ± SE) in the sampled seasons (W – winter; Sp – spring; S – summer; A – autumn).

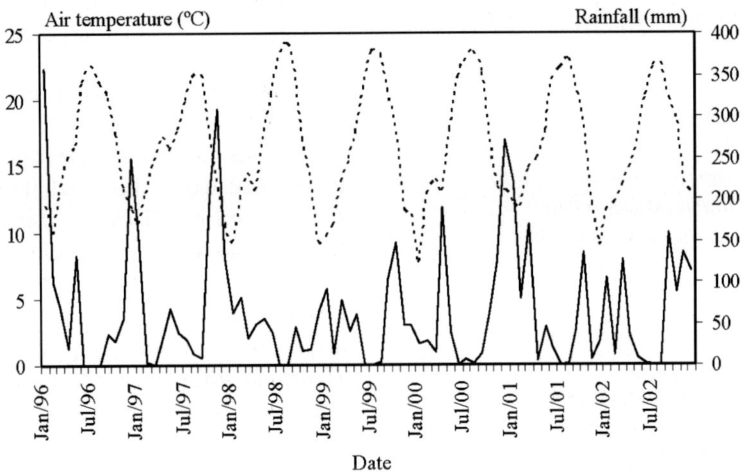

Figure 3. Monthly mean air temperature (discontinuous line) and total rainfall (continuous line) between January of 1996 and December 2002.

taxa. Oligochaeta, *Streblospio shrubsolii* (Buchanan, 1890) and *Scrobicularia plana* (da Costa, 1778) dominated the community and *Hediste diversicolor* (O.F. Müller, 1776), *Hydrobia ulvae* (Pennant, 1777) and Cirratulidae were also abundant in this area. All other taxa were less represented (Table 2).

Considering the abundance in each sampled season, Oligochaeta showed two density peaks (spring of 1998 and winter of 2001) with more than 50 ind. 0.05 m^{-2}. In the remaining seasons values of this parameter for this taxon were considerably lower (<15 ind. 0.05 m^{-2}). Two other species, *Streblospio shrubsolii* and *Scrobicularia plana*, although always present, never revealed such high abundances, except in the autumn of 2000, when *S. shrubsolii* registered 47.5 ind. 0.05 m^{-2}. In general, densities obtained for these two species were always below 10 ind. 0.05 m^{-2} presenting less variability than Oligochaeta.

In spite of the dominant taxa remaining the same throughout the sampled period, high degree of variation was observed in the community structure. In fact, using the Spearman correlation coefficient to compare the numerical relative importance of the most abundant taxa on different sampling moments (210 comparisons) only 26% of the comparisons revealed a high degree of association ($p < 0.001$; $N = 17$) and 15% of the correlations were found not significant ($p > 0.05$; $N = 17$).

To assess the influence of recruitment on the community structure changes, density and mean

Table 1. Taxa identified in the intertidal zone of the studied area by Calvário (1982) (◇), Pereira et al. (1997) (‡) and in the present study (Δ)

Taxa		
Anthozoa	*Actinia equina* (Linnaeus, 1768)	‡ Δ
Hydrozoa	Hydrozoa unid.	Δ
Nematoda	Nematoda unid.	‡ Δ
Nemertea	Nemertea unid.	Δ
Polychaeta	Aphroditidae unid.	‡
	Autolytus sp.	Δ
	Syllidae unid.	Δ
	Hediste diversicolor (O.F. Müller, 1776)	◇ ‡ Δ
	Nephtys hombergii (Savigni, 1818)	‡ Δ
	Nephtys cirrosa Ehlers, 1868	Δ
	Marphysa sanguinea (Montagu, 1815)	Δ
	Ophryotrocha puerilis (McIntosh, 1885)	Δ
	Cossura sp.	Δ
	Polydora sp.	‡ Δ
	Pygospio elegans Claparède, 1863	‡ Δ
	Prionospio sp.	‡
	Streblospio shrubsolii (Buchanan, 1890)	‡ Δ
	Cirratulus cirratus (Müller, 1776)	Δ
	Tharyx marioni (Saint-Joseph, 1894)	‡ Δ
	Caulleriella sp.	‡
	Capitellidae unid.	‡ Δ
	Sabellaria spinulosa Gravier, 1906	‡
Oligochaeta	Oligochaeta unid.	◇ ‡ Δ
Gastropoda	*Hinia reticulata* (Linnaeus, 1758)	‡
	Hydrobia ulvae (Pennant, 1777)	◇ ‡ Δ
Bivalvia	*Anomia ephippium* (Linnaeus, 1758)	‡
	Cerastoderma edule (Linnaeus, 1758)	‡
	Cerastoderma glaucum (Poiret, 1789)	Δ
	Abra alba (Wood, 1802)	Δ
	Scrobicularia plana (da Costa, 1778)	◇ ‡ Δ
	Corbula gibba (Olivi, 1792)	Δ
Isopoda	*Paragnathia formica* (Hesse, 1864)	◇ Δ
	Cyathura carinata (Krøyer, 1847)	Δ
	Idotea negleta Sars, 1899	Δ
	Lekanesphaera monodi (Arcangeli, 1934)	Δ
	Lekanesphaera rugicauda (Leach, 1814)	◇
	Porcellio pruinosus Brandt, 1833	◇
Amphipoda	*Corophium multisetosum* Stock, 1952	Δ
	Corophium sextonae Crawford, 1937	‡
	Jassa pusilla (O. Sars, 1894)	‡
	Melita palmata (Montagu, 1804)	Δ
	Orchestia kosswigi Ruffo, 1949	◇
	Orchestia mediterranea Costa, 1857	◇
	Orchestia stephenseni Cecchini, 1928	◇
	Stenothoe marina (Bate, 1856)	‡

Table 1. (Continued)

Taxa		
Sessilia	*Balanus improvisus* Darwin, 1854	‡
	Elminius modestus Darwin, 1854	‡
Decapoda	*Carcinus maenas* (Linnaeus, 1758)	◇ ‡ Δ
Diptera	Ceratopogonidae unid.	Δ
	Chironomidae unid.	Δ
	Dolichopodidae unid.	‡ Δ
	Tabanidae unid.	Δ
	Tipulidae unid.	Δ
	Psycodidae unid.	Δ
	Diptera unid.	◇ Δ
Lepidoptera	*Nymphula* sp.	Δ
	Bryozoa unid.	Δ
Ascidiacea	Molgulidae unid.	‡

individual biomass for every bimonthly period were computed for some selected taxa (Fig. 4). In all seasons recruitment peaks could be observed in these taxa: winter (Oligochaeta); spring (*Paragnathia formica* (Hesse, 1864)); late summer (*Streblospio shrubsolii* and *Cossura* sp.); and late summer/autumn (*Scrobicularia plana*). Three taxa revealed separated recruitment peaks: *Hediste diversicolor* in early spring and late summer; *Hydrobia ulvae* in early spring and late autumn and Dolichopodidae in winter and spring.

Relationships between environmental and biological variables

CCA revealed that the 3-month lagged rainfall, 3-month lagged air temperature and daylight hours were the environmental variables most related to the benthic community (Fig. 5). The first two CCA ordination axes explained 40.40% of taxa temporal variability and 96.30% of the relationship between abundance and the three selected environmental variables. The relative high correlation between taxa and environmental data for the first two axes (Table 3) suggests that the environmental variables explain the variability associated with taxa. The global permutation test showed that, for the first canonical axis (F-ratio = 7.07) as well as for the sum of all canonical axes (F-ratio = 4.10), relations between taxa abundance and those environmental variables were statistically significant ($p < 0.01$).

Table 2. Density (ind. 0.05 m^2) (median and interquartile range) and frequency of occurrence (FO) of the 17 most important taxa in the studied area and respective taxa codes

Taxa	Codes	Median	Interquartile range	FO (%)
Oligochaeta	Oli	2.40	1.43–1.43	100.00
Streblospio shrubsolii	Ssr	1.48	1.08–4.05	100.00
Scrobicularia plana	Spl	1.00	0.35–1.73	100.00
Hediste diversicolor	Hdv	0.34	0.25–0.85	95.24
Hydrobia ulvae	Hul	0.30	0.03–1.45	80.95
Cirratulidae	Crr	0.28	0.05–1.15	95.24
Dolichopodidae	Dch	0.05	0.05–0.13	71.43
Polydora sp.	Ply	0.03	0.00–0.10	57.14
Corophium spp.	Crp	0.03	0.00–0.08	61.90
Cerastoderma glaucum	Cgl	0.03	0.00–0.03	52.38
Gammaridea	Gmm	0.00	0.00–0.03	42.86
Nephtys spp.	Npt	0.00	0.00–0.05	38.10
Cyathura carinata	Cct	0.00	0.00–0.03	38.10
Cossura sp.	Css	0.00	0.00–0.00	23.80
Paragnathia formica	Pfm	0.00	0.00–0.00	14.29
Capitellidae	Cap	0.00	0.00–0.00	9.52
Corbula gibba	Cgb	0.00	0.00–0.00	9.52

CCA shows that the biological community mainly aggregated in four different taxa groups and two seasonal clusters (Fig. 5a). Group I (GI) is composed by the isopod *Paragnathia formica* and larval forms of the Dolichopodidae insect family. Group II (GII) comprises the polychaetes *Streblospio shrubsolii* and *Cossura* sp. and the bivalves *Scrobicularia plana* and *Corbula gibba* (Olivi, 1792). Group III (GIII) assembles the annelids *Hediste diversicolor* and Oligochaeta, the gastropod *Hydrobia ulvae* and the isopod *Cyathura carinata* (Krøyer, 1847). Group IV (GIV) gathers the polychaetes *Polydora* sp. and *Nephtys* spp., the bivalve *Cerastoderma glaucum* (Poiret, 1789) and the amphipods *Corophium* spp. and other Gammaridea.

The two main seasonal aggregations, winter/spring and summer/autumn, were associated to two of the taxa groups. Species group GI was related to winter/spring and GII with summer/autumn. These two main seasonal trends were linked with the increase of some of the aforementioned environmental variables, daylight hours and 3-month lagged air temperature, respectively (Fig. 5). GIII positioning indicated different taxa behaviour. In spite of being abundant all year, Oligochaeta showed an affinity towards winter/spring cluster and *Hydrobia ulvae* to summer/autumn grouping. With relatively constant abundances throughout the year *Hediste diversicolor* and *Cyathura carinata* appeared in different to any seasonal association. Almost all taxa included in these three groups did not seem particularly affected by a rainfall gradient. Conversely, taxa from GIV appeared in all seasons and seem inversely related to a precipitation gradient, such as *H. diversicolor* to a less extent. Whereas for Cirratulidae, this was the most important and directly related factor (Fig. 5). Therefore, this community presented two distinct assemblages associated to winter/spring and summer/autumn periods. The first one was composed by taxa of GI and GIII and the second by taxa of GII and GIII. However, *Corbula gibba* and *Cossura* sp. were only present in summer/autumn of warmer years as it can be inferred from the CCA diagram (Fig. 5). Cirratulidae and taxa from GIV can also be present in both assemblages, by association to heavy rainfalls for the first one and inversely for the latter group. Capitellidade showed the same trend as taxa of GIV, but was only present in those years of lesser rainfall (Fig. 5).

The other environmental variables were added as supplementary variables to the CCA (Fig. 5b)

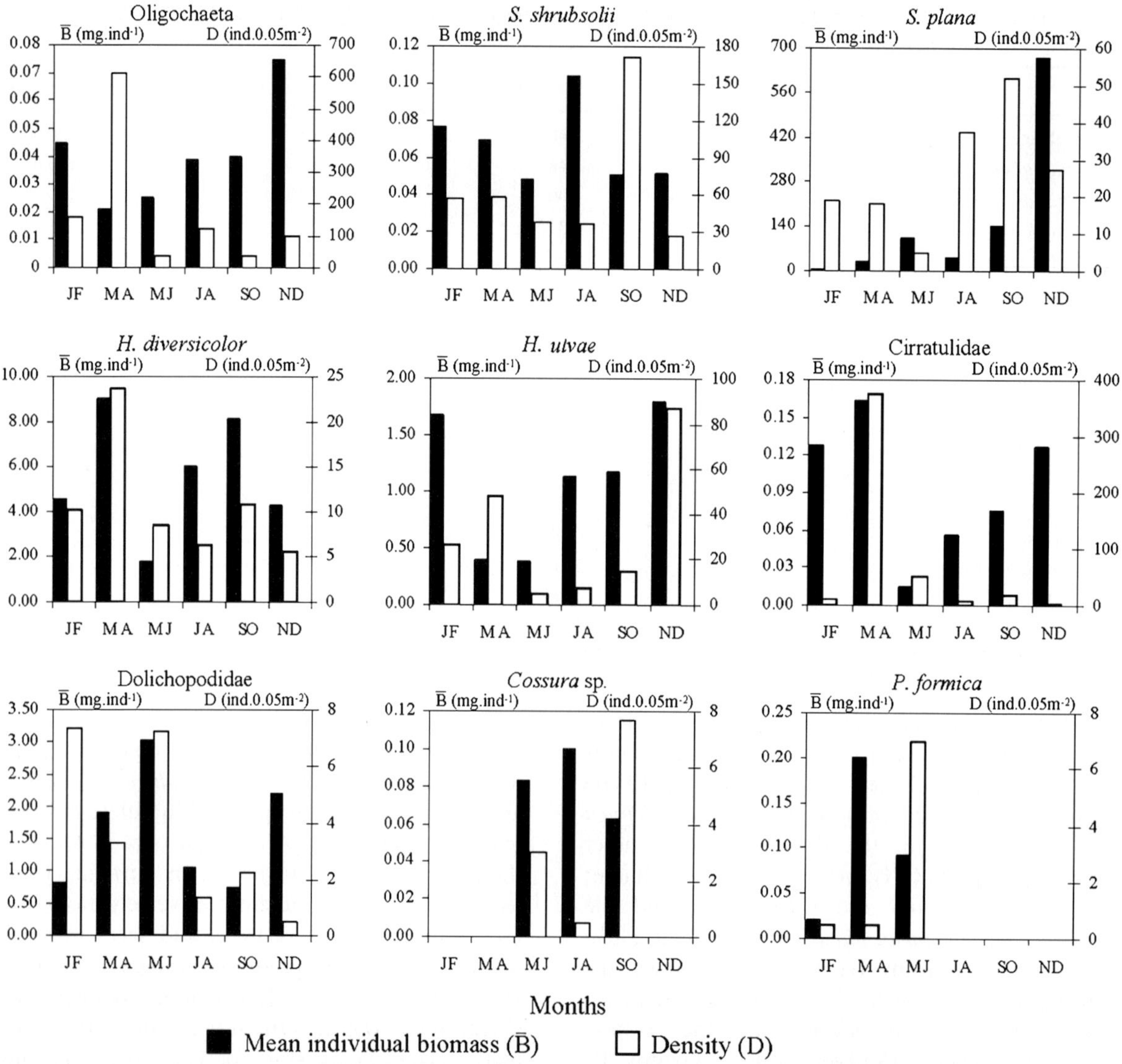

Figure 4. Mean individual biomass and density of several taxa during the studied period on a bimonthly periodicity (JF – January and February; MA – March and April; MJ – May and June; JA –July and August; SO – September and October; ND – November and December). The remaining taxa were not represented due to the reduced number of individuals in most seasons.

and revealed that river flow, rainfall and the larger grain size particles (gravel, coarse and medium sand) showed the same trend as 3-month lagged rainfall. Also, all temperature variables demonstrated the same tendency. Regarding muddy sediments and TOM it was obvious the association with the decrease of rainfall, since larger particles tend to appear when higher values of this parameter were registered. All these trends were confirmed statistically by means of Spearman rank-order correlation test ($p < 0.05$). Finally, fine sand increases with moderate rainfall levels, as it can be observed by the relationship between that sediment fraction and winter/spring samplings collected in lesser precipitation years (Fig. 5).

Figure 6 shows the CCA performed to analyse the 17 most abundant taxa distribution in function of sediment variables only, in order to reveal any existing relationships between them that might have been overlooked in the previous analysis. The first two ordination axes explained 6.50% of taxa abundance variability and 72.70% of the relation-

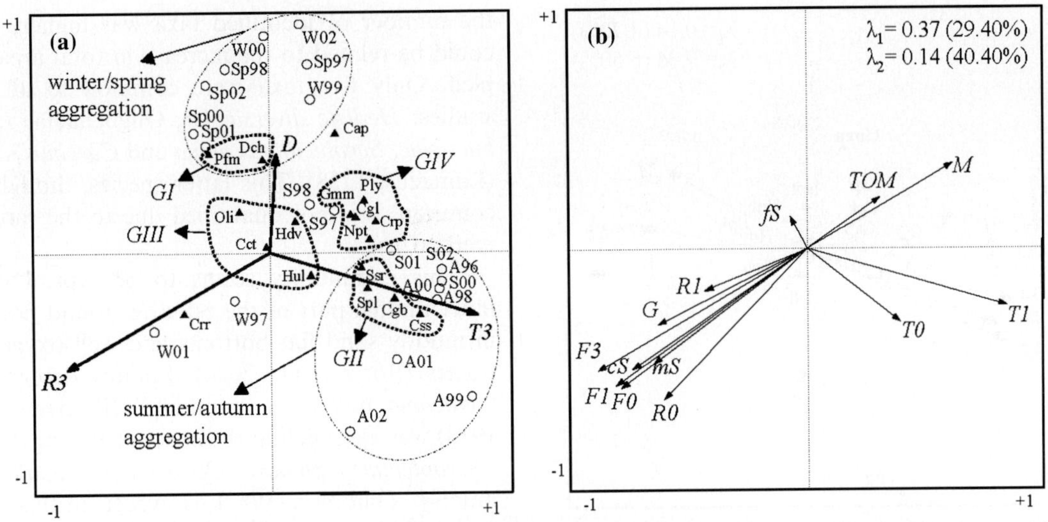

Figure 5. CCA ordination diagram with the 17 most abundant taxa and all environmental variables. Explanatory variables are represented apart (a) from the supplementary variables (b). Eigenvalues (λ_i) and cumulative percentage explained (between brakets) are also given. Taxa represented by triangles (for codes see Table 2), sampled seasons by circles (for codes see Fig. 2 caption) and environmental variables indicated by arrows (*D* – daylight hours, *F0* – sampling month river flow, *F1* – 1-month lagged river flow, *F3* – 3-month lagged river flow, *T0* – sampling month air temperature, *T1* – 1-month lagged air temperature, *T3* – 3-month lagged air temperature, *R0* – sampling month rainfall, *R1* – 1-month lagged rainfall, *R3* – 3-month lagged rainfall, *G* – gravel, *Cs* – coarse sand, *mS* – medium sand, *fS* – fine sand, *M* – mud, TOM-total organic matter).

Table 3. Results of the ordination by CCA performed for the 17 most abundant taxa considering all environmental variables

	Axis I	Axis II
Eigenvalues	0.37	0.14
Taxa/environment correlations	0.92	0.84
Intraset correlations of variables		
3-mo lagged rainfall	−0.79	−0.42
3-mo lagged air temperature	0.79	−0.22
Daylight hours	0.03	0.34

ship between abundance and sediment variables. The correlation between taxa and environmental variance obtained for the first two axes suggests that these variables still explain an important part of the variability associated to taxa abundance (Table 4). The global permutation test showed that the relations between taxa abundance and sediment variables were statistically significant ($p < 0.05$) for the first canonical axis (F-ratio = 6.34) as well as for the sum of all canonical axes (F-ratio = 2.54).

As previously mentioned the study area was typically muddy, however this analysis revealed that slight variations on sediment composition also played an important role on this biological community structure. Cirratulidae, *Cyathura carinata* and *Paragnathia formica* showed preference for coarser particles. *P. formica* simultaneously tended to be greatly related to fine sand sediments. Oligochaeta, *Cerastoderma glaucum* and Dolichopodidae also seemed slightly associated with fine sand grains and in contrast, abundance of Gammaridea and *Cossura* sp. were chiefly related to the decrease of this particular sediment. The remaining taxa (*Hediste diversicolor*, *Nephtys* spp., *Polydora* sp., *Streblospio shrubsolii*, Capitellidae, *Hydrobia ulvae*, *Corbula gibba*, *Scrobicularia plana*, *Corophium* spp.) seem closely associated with muddy sediments and the consequent increase of the TOM.

Discussion

In the study area sediment composition revealed that substrate was essentially muddy as previously observed in other studies (Calvário, 1982; Gaudêncio et al., 1991; Pereira et al., 1997). TOM contents were high and very similar to those mentioned by Pereira et al. (1997) for the same area. The registered values of this parameter reflect

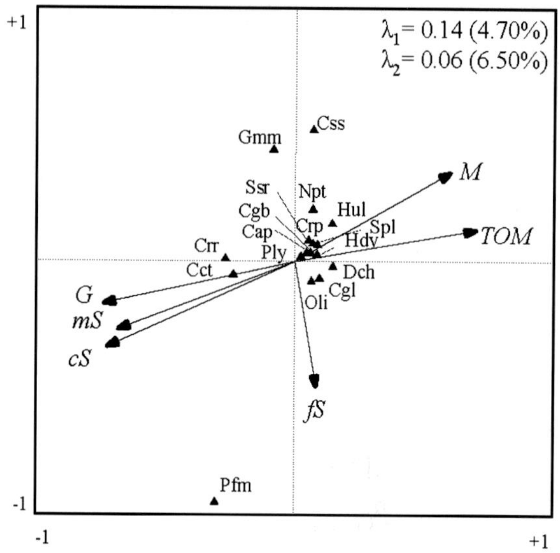

Figure 6. Ordination diagram of the CCA performed for the 17 most abundant taxa and sediment variables. Eigenvalues (λ_i) and cumulative percentage explained (between brakets) are also given. Taxa represented by triangles (for codes see Table 2) and sediment variables indicated by arrows (for codes see Fig. 5 caption).

Table 4. Results of the ordination by CCA performed for the 17 most abundant taxa considering only the sediments variables

	Axis I	Axis II
Eigenvalues	0.14	0.06
Taxa/sediment correlations	0.53	0.36
Intraset correlations of variables		
TOM	0.37	0.05
Mud	0.32	0.13
Fine sand	0.05	−0.18
Medium sand	−0.36	−0.10
Coarse sand	−0.38	−0.12
Gravel	−0.39	−0.06

that higher mud contents are directly related to the increase of organic matter (Creutzberg et al., 1984) since sediments with smaller grain size have reduced interstitial space thus meaning a decrease in permeability (Wolff, 1973).

Twenty-four new taxa were identified for this area and in the case of *Cossura* sp. it is the first reference for the Tagus estuary, together with Rodrigues et al. (this volume). In comparison to other studies done in the same area (Calvário, 1982; Pereira et al., 1997) it can be observed that the number of identified taxa was higher, which could be related to the increase in total area sampled. Only five taxa were common to all three studies: *Hediste diversicolor*, Oligochaeta, *Hydrobia ulvae*, *Scrobicularia plana* and *Carcinus maenas* (Linnaeus, 1758). This latter species, though very common, was not quantified due to the sampling method used.

This community seems to be typical of the inner middle part of the estuaries found normally in muddy sand flat bottoms and well oxygenated waters (Borja et al., 2004). Initially described as "*Macoma balthica* community" (Petersen, 1913, 1918) was classified in the coast of Portugal as the "*Scrobicularia plana – Cerastoderma edule* community" (Thorson, 1957). However, in this case it seems that *Cerastoderma edule* (Linnaeus, 1758) was replaced by *Cerastoderma glaucum*, a species with Mediterranean affinities and preferences for higher temperatures (Russell, 1971). As described for the Basque Country (Borja et al., 2004) the main taxa associated with this community are the Oligochaeta group, the polychaetes *Streblospio shrubsolii* and *Hediste diversicolor*, the gastropod *Hydrobia ulvae* and the crustaceans *Corophium* spp., *Cyathura carinata* and *Carcinus maenas*. The family Cirratulidae was one of the most important taxa in this intertidal community of the Tagus estuary but was not mentioned for other places (Junoy & Viétiez, 1990; Bazaïri et al., 2003; Borja et al., 2004). On the other hand, as for the Basque Country the spionid *Pygospio elegans* Claparède, 1863 was not found to be an important species, although characteristic of this type of community in other estuaries of the Cantabrian Sea (Junoy & Viétiez, 1990). This community also includes other, more or less abundant, polychaetes (*Polydora* sp.) or insect larvae, due to salinity decreases (e.g. Dolichopodidae) (Borja et al., 2004).

Polychaeta were the most important group in this intertidal area, which concurs with Fauchald's (1977) statement that polychaetes dominate macrofauna communities within fine sediments. The benthic community of this intertidal area was characterized by high number of individuals belonging to only six taxa: Oligochaeta, *Streblospio shrubsolii*, *Scrobicularia plana*, *Hydrobia ulvae*, *Hediste diversicolor* and Cirratulidae. These are often found in intertidal areas, particularly in muddy sediments (Dales, 1951; Wolff, 1973; Robineau, 1987; Guerreiro, 1998). Oligochaeta abundance showed higher variability in comparison to the others, probably related to rapid local increases in that population size after disturbances (Cowie et al., 2000).

In spite of the constant dominance of few abundant taxa, high variability of the community structure throughout the studied period was observed. Although two major human interventions took place in the area in the beginning of the study, it seems that the effects caused on community structure were considerably lesser than those due to natural events, otherwise aggregations of sampling seasons closer in time to the interventions (1996–1998) would be expected. Some of the existing variations may be a consequence of cyclic seasonal patterns (originated by air temperature and daylight hours variations) while others were related to isolated events (heavy rainfall). Irregular disturbances in some invertebrate assemblages appear to be more important than predictable seasonal cues, a conclusion supported by the density of more opportunistic species (Desmond et al., 2002). Several taxa in this study, including some of the most abundant, are considered to be opportunistic, such as Oligochaeta, *Streblospio shrubsolii*, Cirratulidae, Capitellidae and others tolerant to excess of organic matter, namely *Hediste diversicolor*, *Corophium* spp. and *Cyathura carinata* (Pearson & Rosenberg, 1978; Hily et al., 1986; Grall & Glémarec, 1997; Borja et al., 2000). Even without the occurrence of human induced disturbances these tolerant and opportunistic taxa are usually abundant in this area (Silva et al., 1999) and in many other intertidal areas of the estuary (Calvário, 1982; Rodrigues et al., this volume). Most likely natural instability of estuaries, due to the transition from marine to freshwater environment, as well as being an interface between aquatic and terrestrial habitats causes the fluctuations in this intertidal area.

CCA showed the existence of distinct assemblages associated to winter/spring and summer/autumn periods as a result of the increase of daylight hours and air temperature, respectively. These factors influencing the structure and dynamics of this community may be related to recruitment and mortality particularly of the most relevant taxa. In temperate latitudes, most coastal soft-bottom benthic communities exhibit variability in intrannual population density due to seasonal patterns of reproduction (López-Jamar et al., 1986; Desmond et al., 2002; Ysebaert et al., 2003).

The association between *Paragnathia formica* and Dolichopodidae and winter/spring aggregation was essentially a result of high abundances of those taxa between the months of January and June. Furthermore, for this isopod there were clear indications that in late spring post-feeding stages of larvae (praniza) occurred here, which like the adults live burrowed in the sediment (Upton, 1987). For the Dolichopodidae, all individuals were larval forms, and although present throughout the year seemed to register two different recruitment peaks, probably due to different species.

Corbula gibba and *Cossura* sp. were only present in summer/autumn of warmer years revealing the existence of certain thermal constraint for their settlement in the area. In the case of *C. gibba* this might be related to the fact that reproduction and settlement is thought to occur in that period (Jensen, 1990). *Scrobicularia plana* is a very common species in the Tagus estuary. Recruitment peak of this species took place in late summer/early autumn as also noted by Guerreiro (1998) in a previous study of this species in the Tagus estuary. *Streblospio shrubsolii* recruitment seemed to occur in early autumn as recorded for the western Mediterranean by Lardicci et al. (1997). However, this spionid might have constant recruitment throughout the year as detected in Alfacs Bay by Sardá & Martin (1993), a fact that should not be excluded in the Tagus estuary.

Oligochaeta, the most abundant taxon, tends to be related to winter/spring aggregation mainly because of the high number of individuals observed in late winter and early spring. This also appeared to be a recruitment period, because the highest abundance and the lowest mean individual biomass were registered in this season. *Hydrobia ulvae* associated to summer/autumn aggregation, chiefly because of the high abundances observed in autumn, although a recruitment peak could also be observed in the spring. Planas & Mora (1987) also observed high abundances in those periods at sites with considerable organic content. Lillebø et al. (1999) observed high percentage of individuals carrying egg masses in the spring and summer, and

September was one of the settlement periods detected.

Rainfall also seemed to be an important environmental factor affecting the structure and dynamics of this community. It was clear that some taxonomic entities, more than being associated to any seasonal aggregations observed in the CCA, were related to the rainfall gradient. This was the case of Amphipoda, *Polydora* sp., *Nephtys* sp. and *Cerastoderma glaucum* that even without revealing high abundances throughout the year were relatively constant showing significant decreases after heavy rainfalls. However, Cirratulidae abundance benefit from elevated values of this parameter showing high number of individuals towards the end of winter, especially in those years of intense precipitation. As in other studies (Desmond et al., 2002; Salen-Picard & Arlhac, 2002) freshwater inputs revealed to be important factors in structuring the macrobenthic communities, especially when lagged effects were considered.

Nevertheless, besides the direct impact of rainfall in the organisms' metabolism there should be also an indirect effect on the benthic community structure by promoting changes in sediments. In fact, after intense precipitation the fraction of coarser grains increased and therefore the fraction of smaller particles and TOM diminished. These seemed to be a consequence of coarse terrestrial sediments being transported to this intertidal area and/or of muddy and organic particles being washed away to adjacent subtidal zones. In years of lesser precipitation there was only a slight increase of the fine sand fraction in winter and spring.

Several species of the Capitellidae family are typical of muddy areas with high organic content, and usually are dominant taxa in polluted environments (Pearson & Rosenberg, 1978; Hily et al., 1986; Grall & Glémarec, 1997). Therefore, the appearance of this taxon in years of lesser rainfall was probably related to the increase of TOM and mud fraction in those situations.

Several studies have considered sediment variations as an important factor in the distribution of benthic assemblages in intertidal soft-sediments (Junoy & Viétiez, 1990; Ysebaert & Herman, 2002; Bazaïri et al., 2003; Ysebaert et al., 2003). However, in our study sediment variations were small and the majority of taxa seem to have a preference for sediments common to the seasons in which they were most abundant. For this reason it was impossible to isolate the cause for that preference and an association between those factors seems to be the probable explanation for the observed patterns. The only exceptions were *Cerastoderma glaucum*, Gammaridea and *Cyathura carinata* that seemed to have their sediment preference less influenced by temporal variations. In the latter species abundance was favoured by the increase of coarse sediments while Gammaridea preferred fine grain sizes avoiding even sediments with small fraction of fine sand.

Conclusion

This estuarine intertidal community revealed a high degree of natural instability through major structural temporal variations and the presence of opportunistic and tolerant taxa. Nevertheless, it showed a tendency for the occurrence of annual cycles, with typical assemblages of winter/spring and summer/autumn periods that were closely related to temperature and daylight hours. On the other hand, the main source of non-cyclic variations was rainfall, either by directly influencing the biological community or by inducing changes in sediments. In fact, sediment characteristics appear to play an important role in changing the community throughout the year, since in this predominantly muddy area some species abundance seemed favoured by coarser particles and many by fine sediments. This study proved that long-term data series are an essential tool to assess the structure and dynamics of benthic communities allowing the detection of inter- and intrannual fluctuations and their relationship with environmental factors.

Acknowledgements

We would like to thank our colleagues C. Pestana, C. Pereira, C. Teixeira, F. Jorge, J. Salgado, M. J. Correia, P. Chainho, T. Pereira and V. Amaral for the help given during field and laboratory work. We are thankful to M. Pardal and M. Cunha for the help with identification of some taxa. We thank Parque Expo, S.A. for the financial and logistic support given during the study.

References

Abrantes, A., F. Pinto & M. H. Moreira, 1999. Ecology of the polychaete *Nereis diversicolor* in the Canal de Mira (Ria de Aveiro, Portugal): populations dynamics, production and oogenic cycle. Acta Oecologica 20: 267–283.

Anonymous, 1992. Dados astronómicos para os almanaques de 1993 para Portugal. Observatório Astronómico de Lisboa, Lisboa, Portugal 73 pp.

Anonymous, 1997. SPSS Base 7.5 Applications Guide. SPSS, Chicago, USA 339 pp.

Bazaïri, H., A. Bayed, M. Glémarec & C. Hily, 2003. Spatial organisation of macrozoobenthic communities in response to environmental factors in a coastal lagoon of the NW African coast (Meja Zerga, Morocco). Oceanologica Acta 26: 457–471.

Bilyard, G. R., 1987. The value of benthic infauna in marine pollution monitoring studies. Marine Pollution Bulletin 18: 581–585.

Borja, A., F. Aguirrezabalaga, J. Martinez, C. Sola, L. Garcia-Arberas & M. Gorostiaga, 2004. Benthic communities, biogeography and resources management. In Borja, A. & M. Collins (eds), Oceanography and Marine Environment of the Basque Country. Elsevier Oceanography. (Series 70). Elsevier, Amsterdam, The Netherlands: 455–492.

Borja, A., J. Franco & V. Pérez, 2000. A marine biotic index to establish the ecological quality of soft bottom benthos within European estuarine and coastal environments. Marine Pollution Bulletin 40: 1100–1114.

ter Braak, C. J. F. & P. Šmilauer, 2002. CANOCO Reference Manual and CanoDraw for Windows User's Guide: Software for Canonical Community Ordination.Version 4.5. Microcomputer Power, Ithaca, New York, USA 500 pp.

ter Braak, C. J. & P. F. Verdonschot, 1995. Canonical correspondence analysis and related multivariate methods in aquatic ecology. Aquatic Sciences 57: 255–289.

Cabral, H. N. R., 1998. Utilização do estuário do Tejo como área de viveiro pelos linguados, *Solea solea* (L. 1758) e *Solea senegalensis* Kaup, 1858, e o robalo, *Dicentrarchus labrax* (L., 1758). Ph.D. Thesis, Universidade de Lisboa, Lisboa, Portugal, 396 pp.

Calvário, J., 1982. Estudo ambiental do estuário do Tejo: povoamentos bentónicos intertidais (substratos móveis). CNA TEJO 19, Rel 18, Comissão Nacional do Ambiente, Lisboa, Portugal, 53 pp.

Costa, M. J., P. R. Almeida, H. N. Cabral, J. L. Costa, M. J. Correia, L. S. Gordo, C. D. Pereira, J. F. Salgado & C. M. Teixeira, 1999. Monitorização da comunidade de invertebrados bentónicos durante a construção da Ponte Vasco da Gama. In Santana, F., L. Vasconcelos, M. R. Partidário, M. J. Seixas, & M. P. Sobral (eds), Actas da 6ª Conferência Nacional sobre a Qualidade do Ambiente. DCEA/FCT/UNL, Monte da Caparica, Portugal: 161–170.

Cowie, P. R., S. Widdicombe & M. C. Austen, 2000. Effects of physical disturbance on an estuarine intertidal community: field and mesocosm results compared. Marine Biology 136: 485–495.

Creutzberg, F., P. Wapenaar, G. Duineveld & N. Lopez Lopez, 1984. Distribution and density of the benthic fauna in the Southern North Sea in relation to bottom characteristics and hydrographic conditions. Rapports et Procès-Verbaux des Réunions du Conseil International pour l'Exploration de la Mer 183: 101–110.

Cunha, M. R., J. C. Sorbe & M. H. Moreira, 2000. The amphipod *Corophium multisetosum* (Corophiidae) in Ria de Aveiro (NW Portugal). I. Life hystory and aspects of reproductive biology. Marine Biology 137: 637–650.

Dales, R. P., 1951. An annual history of a population of *Nereis diversicolor* O.F. Müller. The Biological Bulletin 101: 131–137.

Dauer, D. M. & R. W. Alden III, 1995. Long-term trends in the macrobenthos and water quality of the lower Chesapeake Bay (1985–1991). Marine Pollution Bulletin 30: 840–850.

Desmond, J. S., D. H. Deutschman & J. B. Zedler, 2002. Spatial and temporal variation in estuarine fish and invertebrate assemblages: analysis of an 11-year data set. Estuaries 25: 552–569.

Fauchald K. 1977. The Polychaete worms: definitions and keys to the Orders, Families and Genera. Natural History Museum of Los Angeles County, Scientia (Series 28), Los Angeles, USA, 188.

Gaudêncio, M. J., M. T. Guerra & M. Glémarec, 1991. Recherches biosédimentaires sur la zone maritime de l'estuaire du Tage, Portugal: données sédimentaires préliminaires. In Elliot, M. & J. -C. Ducrotoy (eds), Estuaries and Coasts: Spatial and Temporal Intercomparisons. Olsen and Olsen, Fredensborg, Denmark: 11–16.

Grall, J. & M. Glémarec, 1997. Using biotic indices to estimate macrobenthic community perturbations in the Bay of Brest. Estuarine and Coastal Shelf Science 44: 43–53.

Grémare, A., J. M. Amouroux & G. Vétion, 1998. Long-term comparison of macrobenthos within the soft bottoms of the Banylus-sur-mer (northwestern Mediterranean Sea). Journal of Sea Research 40: 281–302.

Guerreiro, J., 1998. Growth and production of the bivalve *Scrobicularia plana* in two southern European estuaries. Vie et Milieu 48: 121–131.

Herman, P. M. J., J. J. Middelburg, J. van de Koppel & C. H. R. Heip, 1999. Ecology of estuarine macrobenthos. Advances in Ecological Research 29: 195–240.

Hily, C., H. Le Bris & M. Glémarec, 1986. Impacts biologiques des emissaries urbains sur les ecosystems benthiques. Océanis 12: 419–426.

Ibanez, F. & J. -C. Dauvin, 1988. Long-term changes (1977 to 1987) in a muddy fine sand *Abra alba* – *Melinna palmata* community from the Western English Channel: multivariate time-series analysis. Marine Ecology Progress Series 49: 65–81.

Jensen, J. N., 1990. Increased abundance and growth of the suspension-feeding bivalve *Corbula gibba* in a shallow part of the eutrophic Limfjord, Denmark. Netherlands Journal of Sea Research 27: 101–108.

Junoy, J. & J. M. Viétiez, 1990. Macrozoobenthic community structure in the Ria de Foz, an intertidal estuary (Galicia, Northwest Spain). Marine Biology 107: 329–339.

Lardicci, C., G. Ceccherelli & F. Rossi, 1997. *Streblospio shrubsolii* (Polychaeta: Spionidae): temporal fluctuations in size and reproductive activity. Cahiers de Biologie Marine 38: 207–214.

Lillebø, A. I., M. A. Pardal & J. C. Marques, 1999. Population structure, dynamics and production of *Hydrobia ulvae* (Pennant) (Mollusca: Prosobranchia) along an eutrophication gradient in the Mondego estuary (Portugal). Acta Oecologica 20: 289–304.

López-Jamar, E., O. Francesch, A. V. Dorrio & S. Parra, 1995. Long term variation of the infaunal benthos of La Coruna Bay (NW Spain): results from a 12-year study (1982–1993). Scientia Marina 59: 49–61.

López-Jamar, E., G. González & J. Mejuto, 1986. Temporal changes of community structure and biomass in subtidal macroinfaunal assemblages in La Coruña bay, NW Spain. Hydrobiologia 142: 137–150.

Marques, J. C., M. A. Pardal & P. Maranhão, 2002. Characterisation of the macroinvertebrate benthic communities in the Mondego estuary. In Pardal, M. A., J. C. Marques, & M. A. S. Graça (eds), Aquatic Ecology of the Mondego River Estuary Basin: Global Importance of Local Experience. Imprensa da Universidade de Coimbra, Coimbra, Portugal: 125–134.

Marsh, A. G. & K. R. Tenore, 1990. The role of nutrition in regulating the population dynamics of opportunistic, surface deposit feeders in a mesohaline community. Limnology and Oceanography 35: 710–724.

Mucha, A. P. & M. H. Costa, 1999. Macrozoobenthic community structure in two Portuguese estuaries: relationship with organic enrichment and nutrient gradients. Acta Oecologica 20: 363–376.

Mucha, A. P., M. T. S. D. Vasconcelos & A. Bordalo, 2003. Macrobenthic community in the Douro estuary: relations with trace metals and natural sediment characteristics. Environmental Pollution 121: 169–180.

Pearson, T. H. & R. Rosenberg, 1978. Macrobenthic succession in relation to organic enrichment and pollution of the marine environment. Oceanography and Marine Biology Annual Review 16: 229–311.

Pereira, C. D., M. J. Gaudêncio, M. T. Guerra & M. T. Lopes, 1997. Intertidal macrozoobenthos of the Tagus estuary (Portugal): the Expo'98 area. Publicaciones Especiales del Instituto Español de Oceanografía. 23: 107–120.

Petersen, C. G. J., 1913. Valuation of the sea. II. The animal communities of the sea-bottom and their importance for marine zoogeography. Report of the Danish Biological Station 21: 1–42.

Petersen, C. G. J., 1918. The sea bottom and the production of fish-food. Report of the Danish Biological Station 25: 1–62.

Planas, M. & J. Mora, 1987. Dinámica poblacional de *Hydrobia ulvae* (Pennant) en la ria de Pontevedra (NO España). Cahiers de Biologie Marine 28: 447–464.

Quintino, V. & A. M. Rodrigues, 1989. Environment gradients and distribution of macrozoobenthos in three Portuguese coastal systems: Obidos, Albufeira and Alvor. In Ryland, J. S. & P. A. Tyler (eds), Reproduction, Genetics and Distributions of Marine Organisms. Olsen and Olsen, Fredensborg, Denmark: 441 –450.

Reise, K., 1982. Long-term changes in the macrobenthic invertebrate fauna of the Wadden Sea: are polychaetes about to take over?. Netherlands Journal of Sea Research 16: 29–36.

Robineau, B., 1987. Caractérisation des peuplements macrozoobenthiques de l'estuaire de la Loire. Vie et Milieu 87: 67–76.

Rodrigues, A. M. & V. Quintino, 1993. Horizontal biosedimentary gradients across the Sado estuary, W Portugal. Netherlands Journal of Aquatic Ecology 27: 465–482.

Rodrigues A. M., S. Meireles, T. Pereira, A. Gama, V. Quintino. Spatial patterns of benthic invertebrates in intertidal areas of a southern European estuary: the Tagus, Portugal. Hydrobiologia 555: 99–113.

Rumohr, H., 1999. Soft bottom macrofauna: collection, treatment, and quality assurance of samples (Revision of No. 8). ICES Techiques in Marine Environmental Sciences 27: 1–19.

Russell, P. J. C., 1971. A reappraisal of the geographical distributions of the cockles, *Cardium edule* L. and *C. glaucum* Bruguièrre. Journal of Conchology 27: 225–234.

Salen-Picard, C. & D. Arlhac, 2002. Long-term changes in a Mediterranen benthic community: relationships between the polychaete assemblages and hydrological variations of the Rhône River. Estuaries 25: 1121–1130.

Sardá, R. & D. Martin, 1993. Populations of *Streblospio* (Polychaeta: Spionidae) in temperate zones: demography and production. Journal of the Marine Biological Association of the U.K. 73: 769–784.

Siegel, S. & N. J. Castellan, 1988. Nonparametric Statistics for the Behavioral Sciences (2nd ed.). McGraw-Hill International Editions, New York, USA 399 pp.

Silva, G., J. L. Costa, P. R. Almeida, L. S. Gordo & M. J. Costa, 1999. A comunidade de macroinvertebrados na monitorização da EXPO'98. In Santana, F., L. Vasconcelos, M. R. Partidário, M. J. Seixas, & M. P. Sobral (eds), Actas da 6ª Conferência Nacional sobre a Qualidade do Ambiente. DCEA/FCT/UNL, Monte da Caparica, Portugal: 271–276.

Sokal, R. R. & F. J. Rohlf, 1995. Biometry: The Principles and Practice of Statistics in Biological Research (3rd ed.). W.H. Freeman, New York, USA 887 pp.

Thorson, G., 1957. Bottom communities (sublittoral or shallow shelf). Memoirs of the Geological Society of America 67: 461–534.

Upton, N. P. D., 1987. Gregarious larval settlement within a restricted intertidal zone and sex differences in subsequent mortality in the polygynous saltmarsh isopod *Paragnathia formica* (Crustacea: Isopoda). Journal of the Marine Biological Association of the U.K. 67: 663–678.

Wolff, W. J., 1973. The estuary as a habitat. An analysis of data on the soft-bottom macrofuna of the estuarine area of the rivers Rhine, Meuse, and Scheldt. Zoologische Verhandelingen 126: 1–242.

Woodin, S. A., 1974. Polychaete abundance patterns in a marine soft sediment environment: the importance of biological interactions. Ecological Monographs 44: 171–187.

Ysebaert, T. & P. M. J. Herman, 2002. Spatial and temporal variation in benthic macrofauna and relationships with environmental variables in an estuarine, intertidal soft-sediment environment. Marine Ecology Progress Series 244: 105–124.

Ysebaert, T., P. M. J. Herman, P. Meire, J. Craeymeersch, H. Verbeek & C. H. R. Heip, 2003. Large-scale spatial patterns in estuaries: estuarine macrobenthic communities in the Schelde estuary, NW Europe. Estuarine Coastal Shelf Science 57: 335–355.

Hydrobiologia (2006) 555:129–141
H. Queiroga, M.R. Cunha, A. Cunha, M.H. Moreira, V. Quintino, A.M. Rodrigues, J. Serôdio & R.M. Warwick (eds),
Marine Biodiversity: Patterns and Processes, Assessment, Threats, Management and Conservation
DOI 10.1007/s10750-005-1111-7

Spatial and environmental factors affecting the distribution of the main decapod crustacean prey species in the NW Mediterranean

Marta M. Rufino*, Francesc Maynou, Pere Abelló & Francesc Sardá
Institut de Ciències del Mar – Centre Mediterrani d'Investigacions Marines i Ambientals (CSIC), Pg. Marítim de la Barceloneta, 37-49, 08003, Barcelona, Spain
*(*Author for correspondence: E-mail: mrufino@cmima.csic.es)*

Key words: decapod crustaceans, *Calocaris macandreae*, *Solenocera membranacea*, *Processa canaliculata*, *Alpheus glaber*, non-linear kriging, spatial distribution, environmental factors, sediment variables

Abstract

The decapod crustaceans *Alpheus glaber*, *Calocaris macandreae*, *Processa canaliculata* and *Solenocera membranacea* are some of the most common prey items for both fish and other decapod crustaceans in the muddy bottom communities of the lower continental shelf and upper slope in the western Mediterranean Sea. Despite their high densities and key ecological role, these species are often neglected in scientific studies due to their relatively small size and the associated difficulties in adequately sampling them, so little information is available on their biological and ecological characteristics. We studied the spatial distribution of these four main crustacean prey species in an area off the Ebro Delta (W Mediterranean), using non-linear geostatistics. We performed log-normal kriging, considering depth as a covariate in the spatial model (universal kriging with external trend). All semivariograms were fitted to a spherical model, but with different relationships to depth. All species were more abundant in a sub-area characterised by a gentler slope and the presence of submarine canyons. The spatial correlation between densities of all species and sediment variables were analysed. All species except *Solenocera membranacea* were negatively correlated with sorting coefficient (IGSD), showing a preference for well sorted, homogeneous sediments corresponding to low energy areas. Further, *Calocaris macandreae* higher densities were associated with areas of coarser grain size (φ) and *Processa canaliculata* density was negatively correlated with carbonate content of the sediment (CO_3).

Introduction

The analysis of species assemblages is often based on taxonomic groupings or, sometimes, feeding guilds. Rarely is the selection of species for ecological studies based on their ecological importance as prey species. Here we focus on the spatial analysis of four decapod crustaceans that are ecologically important as prey species of many components of the benthic assemblage, such as fish and larger decapods, which are the dominant taxa of the continental shelf and slope bottom communities (Cartes, 1998). An exhaustive analysis of the available literature on diet composition of the main fish and decapod crustacean species showed that *Alpheus glaber* (Olivi), *Calocaris macandreae* (Bell), *Processa canaliculata* (Leach) and *Solenocera membranacea* (Risso) constitute an important fraction of the dominant megafauna (fish and larger decapod crustaceans) of the benthic communities in the western Mediterranean (Table 1, see also Carrassón et al., 1992; Cartes et al., 1994a; Cartes, 1998). Due to their habits and small size, the actual abundance of these prey species is often underestimated in many fisheries trawl surveys using a standard gear, but when a smaller mesh size is used, these species dominate in the western Mediterranean crustacean community

Table 1. List of predators of the decapod species analysed in the current work

Prey	Predator	Source
Alpheus glaber	*Antonogadus megalokynodon*	Macpherson (1981)
	Coelorhyncus coelorhyncus	Macpherson (1981)
	Conger conger	Macpherson (1981)
	Etmopterus spinax	Macpherson (1981)
	Gadiculus argenteus	Macpherson (1981); Cartes et al. (1994b)
	Galeus melastomus	Macpherson (1981)
	Helicolenus dactylopterus	Froglia (1976); Macpherson (1981)
	large *Trisopterus minutus*	Gramitto (1999)
	Lepidohombus boscii	Sartor & De Ranieri (1996)
	Lepidotrigla cavillone	Moreno & Matallanas (1983)
	Lophius budegassa	Macpherson (1981)
	Merluccius merluccius	Cartes et al. (1994b); Bozzano et al. (1997)
	Ophichthus rufus	Casadevall et al. (1994)
	Ophidion barbatum	Matallanas (1980)
	Phycis blennoides	Macpherson (1979, 1981)
	Scyliorhinus canicula	Macpherson (1981)
	Simphurus nigrescens	Macpherson (1981)
	Trachynus draco	Morte & Sanz-Brau (1994a)
	Trisopterus minutus	Morte & Sanz-Brau (1994b); Gramitto (1999)
	Uranoscopus scaber	Sanz (1985)
Calocaris macadreae	Bathyal sharks	Carrasson et al. (1992)
	Coelorhyncus coelorhyncus	Macpherson (1981)
	Eledone cirrhosa	Sanchez (1981); Quetgas et al. (1999)
	Etmopterus spinax	Macpherson (1981)
	Galeus melanostomus	Macpherson (1981)
	Macrourids	Macpherson (1979, 1981)
Processa sp.	*Lepidotrigla cavillone*	Moreno & Matallanas (1983)
	Ophichthus rufus	Casadevall et al. (1994)
	Ophidion barbatum	Matallanas (1980)
	Phycis blennoides	Macpherson (1981)
	Serranus hepatus	Labropoulou & Eleftheriou (1997)
	Small *Trisopterus minutus*	Gramitto (1999)
Solenocera membranacea	*Galeus melastomus*	Macpherson (1981)
	Gnathophis mystax	Macpherson (1981)
	Hymenocephalus italicus	Macpherson (1981)
	Lepidorhombus boscii	Sartor & De Ranieri (1996)
	Lepidotrigla cavillone	Moreno & Matallanas (1983)
	Merluccius merluccius	Cartes et al. (1994b); Bozzano et al. (1997)
	Phycis blennoides	Macpherson (1981)
	Uranoscopus scaber	Sanz (1985)

Data from the Mediterranean only.

(Abelló et al., 1988; Cartes et al., 1994a). Despite their essential ecological role, other studies on the ecology and biology of these small decapods are scarce and mostly limited to community analysis (e.g. Minervini et al., 1984; Abelló et al., 1988; Fariña et al., 1997; Atkinson et al., 2003). The study of predator and prey characteristics and interrelationships in marine communities is not only important *per se* but also useful because it may allow energy transfer through the trophic web to be investigated.

Experimental surveys designed to assess the abundance of benthic organisms often do not take into account the spatial structure of these organisms or try to obliterate the autocorrelation arising from the existence of spatial structure by specific sampling designs, such as stratified random sampling (Rivoirard et al., 2000). There is, however, empirical evidence that the existence of structure at different spatial scales does induce autocorrelation between pairs of samples, making abundance estimates biased. Further, there is interest in ecology to quantify the spatial dependence among samples in order to gain knowledge on the spatial structure of benthic organisms and the factors determining this structure (Rossi et al., 1992).

The species selected constitute an important by-catch of the demersal trawl fishery in the Catalan sea, targeting Norway lobster and deep-water shrimps. It is essential to try to understand the complex distribution patterns of the species affected by deep-water trawling, especially those constituting a key link between the lower and the upper trophic levels. The aim of this work is to study the spatial distribution of four species of small crustaceans (*A. glaber*, *P. canaliculata*, *S. membranacea* and *C. macandreae*) using geostatistical techniques, and to analyse some of the environmental factors that determine this distribution.

Materials and methods

Survey

A trawl survey was conducted over muddy bottoms of the continental shelf and slope off Tarragona and the Ebro delta (40°20′ N–41°00′ N, NE Spain, Fig. 1) using the R/V 'García del Cid'. A regular grid of 1×2 nautical mile quadrats was established parallel to the coast. Each trawl was randomly ascribed a starting location within each quadrat. A few southerly quadrats were not sampled because of the presence of oil production platforms (Fig. 1). Sampling was conducted between 7 and 15 April 1994 at a total of 72 stations. The water depth varied from 83 to 713 m. The northern part of the study area was characterised by a narrow shelf (8–10 km wide) having a gentle slope seaward, indented by deep submarine canyons. South of ~40°50′ N, the shelf becomes progressively wider (38–40 km) and the slope becomes steeper with fewer submarine canyons. Samples were taken with a modified otter trawl drawn by a single warp ('Maireta System Trawl', Sardà et al., 1994). The codend stretched mesh width was 12 mm in order to retain small individuals not normally caught by commercial fishing gear. The working dimensions of the trawl mouth were measured acoustically (SCANMAR) and averaged 13.4 m across by 2.0 m high. The effective trawling time was 15 min and trawls were always made parallel to the depth contours. Towing speed varied between 2.3 and 2.7 knots (mean 2.5 knots). The surface area trawled in each tow was estimated from the GPS and the SCANMAR readings, averaging 3 ha (30 000 m^2). The total catch of each species was counted and standardised per swept area (individuals · ha^{-1}).

Sediment samples were collected using a Van Veen grab (Gray, 1981) on every second trawl, except on the continental shelf at depths shallower than 200 m, where sediment heterogeneity was presumed higher and the collection of one sediment sample per trawl was attempted (Fig. 1). The Van Veen grab employed collected samples of 0.012 m^2 and penetrated 10 cm deep into the sediment. Sediment samples were analysed following the methodology presented by Maynou & Sardá (1997) and Alonso et al. (1999). The surface sediment temperature (T, °C) was measured on-board using a double-joint, multimeter electrode (Orion Research model 250A) immediately after the sample collection. Sediment samples were stored at −20 °C for later laboratory analyses. Grain size analysis was carried out on the sediment samples using a SEDIGRAPH 5000D for the fine-grained fraction (<50 μm) and sedimentation tube analysis for the coarse-grained fraction. For each sample, the proportions of sand (62.5–1000 μm), silt (2–62.5 μm) and clay (0.06–2 μm) were determined. The grain size distribution of each sample was summarised by its median, $\varphi = -\log_2$ (median grain size in mm) (Gray, 1981), and its sorting

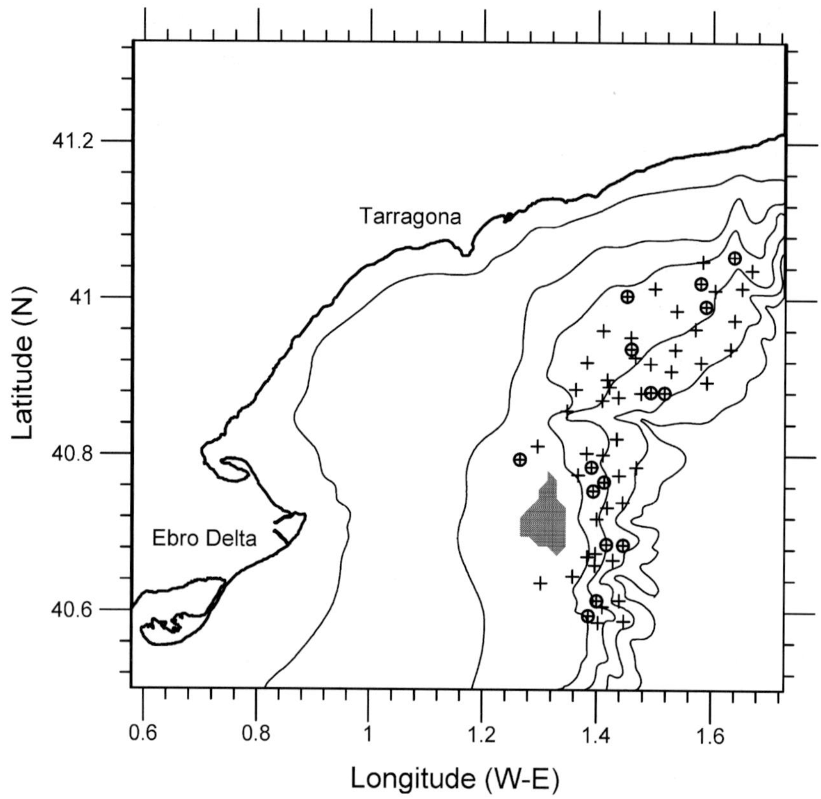

Figure 1. Map of the study area with the location of the sampling stations (crosses). The sites where sediment samples were also taken are marked by circles. Contour lines represent depth isolines, from left to right respectively: coast line, 50 m, 100 m, 200 m, 400 m, 600 m and 800 m depth. Grey area is an oil production area where trawling is not permitted.

coefficient IGSD (Inclusive Standard Deviation: Gray, 1981). High φ values indicate a greater percentage of small fraction contents (i.e. clay or silt), whereas lower values indicate coarser fractions (i.e. sand). High values of IGSD indicate more heterogeneous sediments, usually associated with high-energy environments (Gray, 1981), i.e. bottoms with higher hydrodynamism. The carbonate content (% of CO_3 by volume, CO_3) for each sediment sample was determined using a Bernard calcimeter, following the methodology presented by Vatan (1967). The organic matter contents (OM) were determined by ignition loss on three replicates of a dried sample (ignition at 500 °C for 6 h).

Higher CO_3 environments are indicative of more oceanic (as opposed to neritic) overlying production regimes (Gray, 1981). Sediments with high OM are representative of organic-rich sediments, and are used as an indirect indicator of the presence of endo- or epi-benthic smaller prey (annelids and other marine worms) which form the main prey of the decapod species analysed in this study (Maynou & Sardà, 1997).

Statistical methods

Geostatistics

The fundamentals of geostatistics, with emphasis on the non-linear methods employed here, are explained in several publications, so only a short summary is given (e.g. Cressie, 1991; Diggle et al., 2003). Linear geostatistics were not used in this work because of 'non-stationarity' in the data (see e.g. Cressie, 1991, on the assumptions and limitations of linear geostatistics).

Animals are not generally randomly distributed in space (Koenig, 1999). Their distribution in space may depend on different factors, such as the heterogeneity of the environment and the organism's ecology and life history. The spatial distribution of organisms can be described by spatial statistical tools. In this work we employ a spatial correlation function called 'semivariogram' (Cressie, 1991). The semivariogram is estimated from the data, taking into account the spatial position of the samples by the following equation:

$$\hat{\gamma}(h) = \frac{1}{2m(h)} \sum_{i=1}^{m(h)} \{Z(x_i) - Z(x_i + h)\}^2,$$

where $Z(x_i)$ is the abundance of a species at location x_i, h is a lag distance over which the local average is taken and $m(h)$ is the number of pairs of samples at the distance (h). The semivariogram is the average of the Euclidean distance between pairs of samples (h) plotted against average variance at distance h. Once the semivariogram is calculated, a function (the theoretical spatial covariance function or semivariogram model) is fitted to the empirical semivariogram through an automated fitting procedure. In this study we used a weighted least squares algorithm. We used the 'spherical' semivariogram model to fit the empirical semivariograms for all species, given by the following equation:

$$\gamma(h) = \begin{cases} c_0 + c_s\left(\frac{3}{2}\frac{h}{a} - \frac{1}{2}\left(\frac{h}{a}\right)^3\right), & h < a \\ c_0 + c_s, & h \geq a \end{cases}$$

This spatial covariance model is defined by three parameters: the nugget c_0 (indicating the variance not explained by the spatial model), the sill c_s (indicating the variance explained by the spatial model) and the range a (distance beyond which the importance of the spatial structure is negligible). The range has been interpreted as the average patch size in ecological studies by Sokal & Oden, (1978) and we follow this interpretation.

After calculating the parameters of the spherical semivariogram model, we used this spatial model to estimate the values of abundance of each species over a grid covering the study area in order to produce a high resolution map of each species, using the kriging algorithm (Cressie, 1991).

In the present study, the variables analysed (e.g. species abundance) were not stationary throughout the sampled area. Hence, a relationship with depth was examined to help fit the spatial model, through the semivariogram (which was produced with the residuals of the relationship between density and depth) and in the kriging procedure. When variables other than location are used in the kriging algorithm, the technique is termed Universal Kriging with External Trend (Cressie, 1991). The density and depth relationships examined were linear and quadratic. A linear trend of species abundance with depth shows that the abundance varies linearly with depth, while a quadratic trend shows that abundance peaks at a certain depth and then decreases.

The general Box–Cox set of transformations ($y = x^{\lambda}$, if $\lambda = 0$, then $y = \ln(x)$) (Sokal & Rohlf, 1994), was used to determine the coefficient λ that normalises the data. The analyses were conducted using the library 'geoR' (Ribeiro & Diggle, 1999) of the software package 'R' (Ihaka & Gentleman, 1996).

Mantel test

To analyse the degree of correlation between the sediment variables and species density, we used the Mantel test (Mantel, 1967), with the non-parametric Kendall's correlation coefficient r (Sokal & Rohlf, 1994). The Manhattan distance was used to produce a distance matrix of the variables. The Partial Mantel test was used to extract the effect of depth on the relationship. Hence, it was possible to determine whether the correlation between the sediment variable and species density was spurious because both were related to depth (which is known to be a major forcing factor in the marine environment). The significance of the Mantel and the partial Mantel tests was assessed by a permutation procedure using 10 000 permutations, as outlined in (Manly, 1985; Legendre, 2000). The Mantel test was performed using the library 'vegan' of the software package 'R' and the partial Mantel test was programmed by one of the authors (FM) in 'R', following the methods outlined in Legendre (2000).

Results

The Box–Cox analysis indicated that the appropriate transformation to normalise the density variables for the four species was $\lambda = 0$, hence the variables were log-transformed. Figure 2 shows the depth distribution of *A. glaber*, *C. macandreae*, *P. canaliculata* and *S. membranacea* $\log_e$ density.

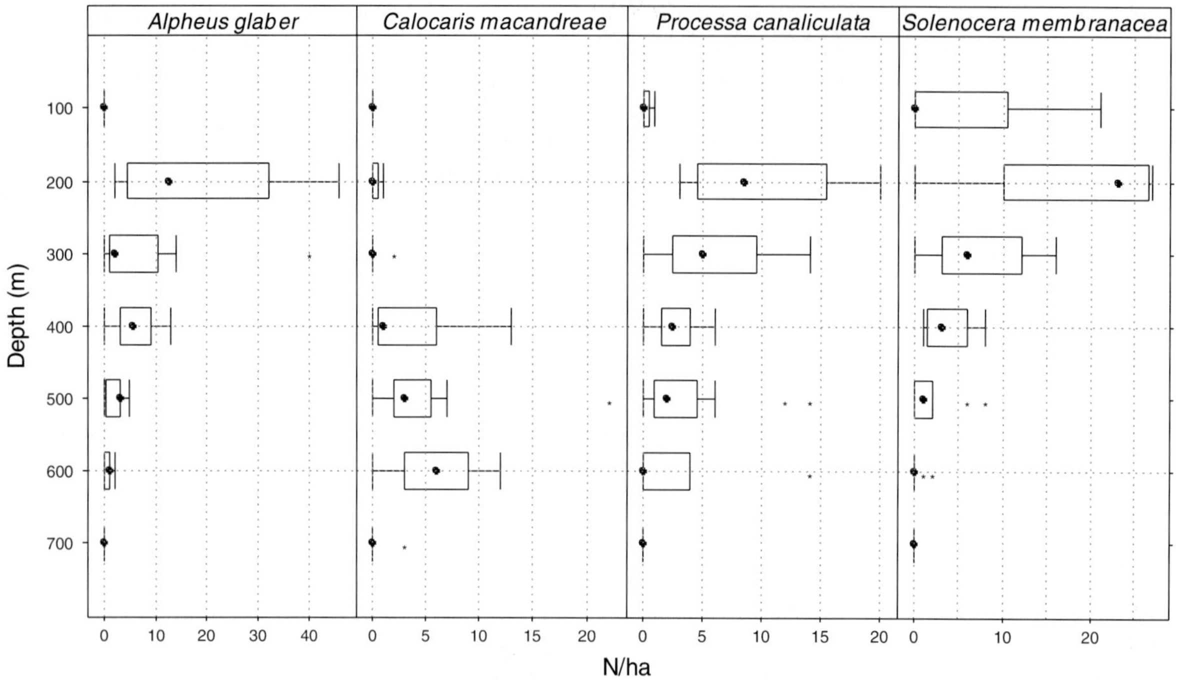

Figure 2. Depth distribution of species density at 100 m depth intervals indicating 5, 25 and 50 (median) percentiles. * denote outliers. An outlier of 313 ind.ha^{-1}, in *Calocaris macandreae* distribution was excluded from the graph.

All species showed maximum densities in the 200–300 m depth range, with the exception of *C. macandreae* that reached maximum numbers at 600 m. However, the sampling only encompassed the shallowest *C. macandreae* depth range.

Table 2 shows the summary statistics of the sediment variables analysed. The sediment temperature was around 13 °C, in agreement with water temperature reported by other authors at the depths sampled (Hopkins, 1985). CO_3 and IGSD showed lower values in the north and higher values in the south (CO_3 range: 23.0–41.1% and IGSD range: 1.7–3.5). Organic matter of the sediment showed a positive trend with depth, and was particularly high around the large submarine canyon present in the study area.

Table 3 shows the summary of the spatial covariance models of each studied species and respective ecological information, whereas Figure. 3 shows the semivariogram plots with the fitted models. The spherical model was the most statistically appropriate for an automatic model selection, since the minimising function always

Table 2. Summary statistics of sediment variables

	Min	Mean	SE	Max
Temperature (°C)	12.40	12.93	0.05	13.60
CO_3	23.00	28.86	0.96	41.10
ϑ	2.21	7.56	0.31	8.69
IGSD	1.66	2.19	0.10	3.39
OM	3.82	8.24	0.34	12.52

Min: minimum; SE: standard error of the mean; Max: maximum; CO_3: % of CO_3, ϑ: grain size; IGSD: sorting coefficient, OM: Organic Matter.

Table 3. Semivariogram models fitted by weighted least squares to the experimental semivariograms, produced for the density (ind · ha^{-1}) of each species

	Trend	Nugget	Sill	% SpD	Range
A. glaber	2nd depth	0.30	0.22	42	15.19
C. macandreae	1st depth	0.00	1.06	100	4.01
P. canaliculata	1st depth	0.37	0.39	51	24.24
S. membranacea	2nd depth	0.19	0.42	69	17.24

SpD indicates the % of the variance explained by the model: $(c_s - c_0/c_s)\times 100$; 1st depth indicates a linear trend with depth, 2nd depth a quadratic trend. All species semivariogram were fitted to a spherical model.

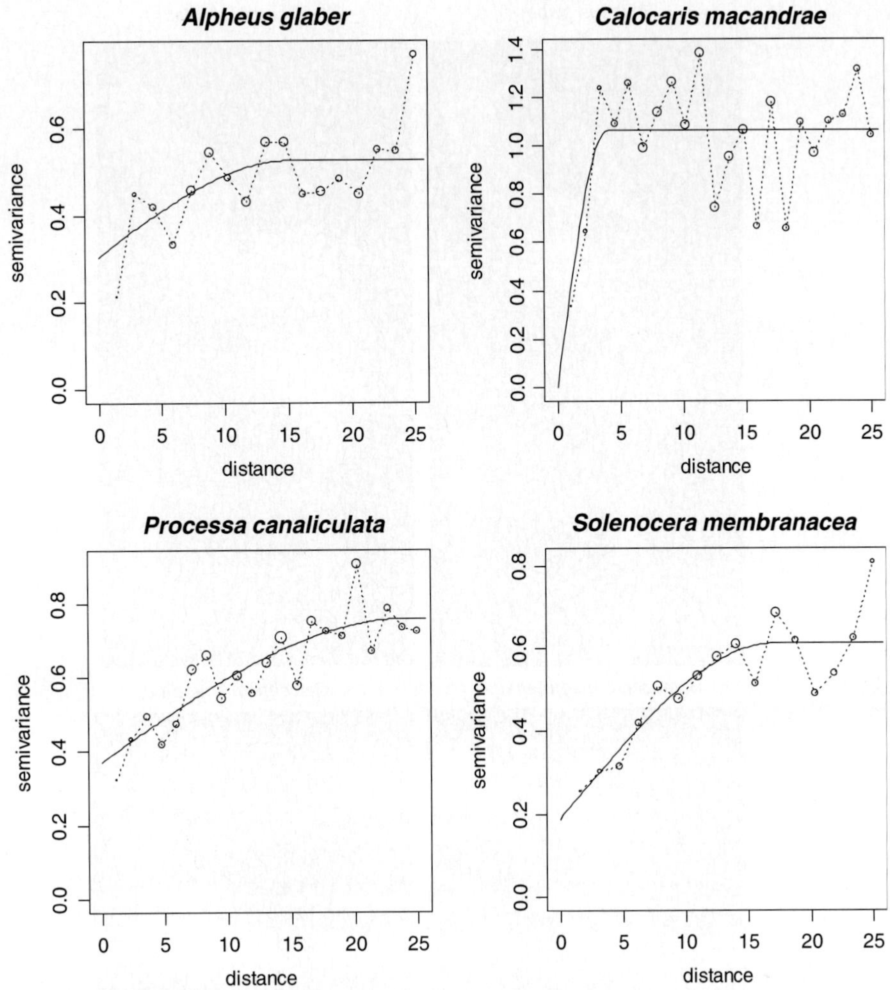

Figure 3. Experimental semi variogram (dashed line) and respective spherical model (solid line) fitted by weighted least squares (size of the circles is proportional to the number of pairs divided by the median number of pairs). Note that all spherical models are fitted to $\log(Z+1)$. See Table 3 for details of the models.

showed the lowest values. Further, the relationship with depth (quadratic or linear) varied with the species, according to the criterion of the minimising function. A quadratic relationship with depth presented a better fit for *A. glaber* and *S. membranacea*, while a linear depth trend was identified in *P. canaliculata* and *C. macandreae* (Table 3 and Fig. 3). Species presenting a linear relationship with depth showed the smallest and the largest ranges (patch size): 4 km in *C. macandreae*, which was also the deepest occurring species (Fig. 2), denoting a high degree of patchiness, and 24 km in *P. canaliculata* (with the broadest depth distribution; Fig. 2 and Table 2). The degree of spatial dependence (Robertson & Freckmann, 1995), i.e., the proportion of variance accounted for by the spatial model, varied between 42 and 100% (Fig. 3, Table 3).

Figure 4 shows the maps of the species density, produced by log-normal universal kriging using depth as an external variable. All species were more abundant in the northern part of the study area, showing a clear preference for a narrower continental shelf and a gentler continental slope. *Calocaris macandreae* showed a more clustered, deeper, distribution with smaller patches (4 km). The spatial distribution of the remaining species was more homogeneous, showing large high-density areas (15–24 km). *Alpheus glaber* and *S. membranacea* were concentrated in the extreme

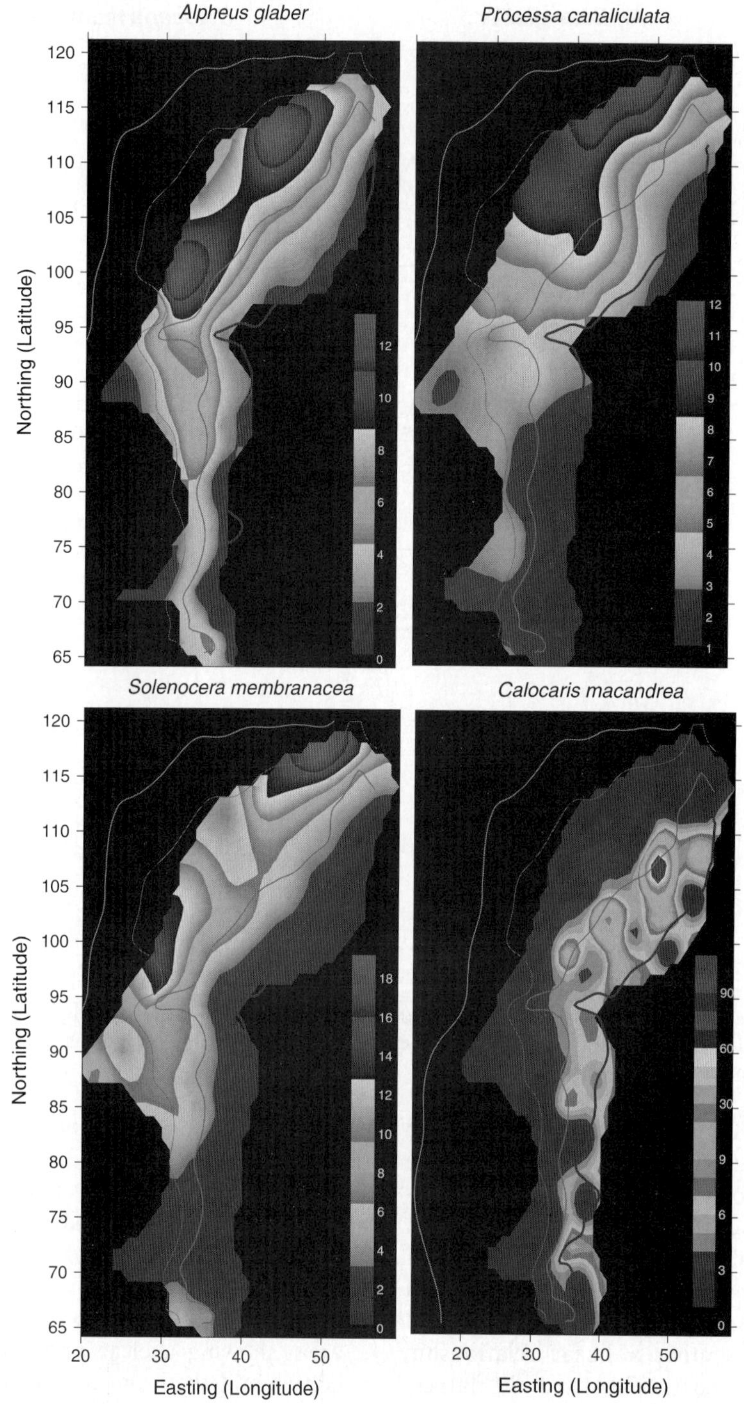

Figure 4. Maps produced by log-normal kriging with depth as external trend (see Table 4 for the variogram model results) of the species density (N · ha^{-1}). Contour lines, from left to right: 100 m, 200 m, 400 mand 600 m.

north and close to the central canyon, while *P. canaliculata* was spread throughout the north area from 200 to 400 mdepth (Figs. 3 and 4). *Processa canaliculata* and *S. membranacea* showed smaller densities than the remaining species (Figs. 3 and 4).

Table 4 shows the results of the Mantel test, indicating the degree of spatial overlap between species. Only *C. macandreae* did not overlap significantly with the remaining species. All the other species significantly overlapped in space with each other. No differences in species spatial correlation were found between the Mantel test and the partial Mantel test extracting the effect of depth. The correlations reported are therefore genuine and not spurious correlations with depth. The greatest spatial overlap was found between *S. membranacea* and *A. glaber* ($r = 0.3591$).

The Mantel and partial Mantel tests applied to species densities and sediment variables showed very similar results (Table 5). Sediment temperature was not correlated with population densities of any species, even when the effect of depth was extracted. The remaining sediment variables were correlated with population densities for at least one species, but the degree of overlap (r) was always low. All species except *S. membranacea* were significantly negatively correlated with IGSD (partial Mantel $-0.14 < r < -0.10$), showing preference for low energy, well-sorted sediments,

Table 4. Mantel test and partial Mantel test (extracting the effect of depth) using Kendall's correlation coefficient (10 000 permutations), between species distribution

	C. macandreae		*P. canaliculata*		*S. membranacea*	
	r	*p*	*r*	*p*	*r*	*p*
Mantel						
A. glaber	−0.0028	0.4867	**0.1912**	**0.0006**	**0.3591**	**<0.001**
C. macandreae			−0.0117	0.5548	−0.0712	0.9174
P. canaliculata					**0.2352**	**<0.001**
Partial Mantel						
A. glaber	−0.0030	0.4819	**0.1906**	**0.0011**	**0.3574**	**<0.001**
C. macandreae			−0.0118	0.5538	−0.0727	0.9194
P. canaliculata					**0.2351**	**<0.001**

Significant correlations are indicated in bold ($r > 0$ and $p < 0.05$ indicates spatial overlap).

Table 5. Mantel test and partial Mantel test (extracting the effect of depth) using Kendall's correlation coefficient (10 000 permutations), between species density and sediment variables

	A. glaber		*C. macandreae*		*P. canaliculata*		*S. membranacea*	
	r	*p*	*r*	*p*	*r*	*p*	*r*	*p*
Mantel								
Temp	−0.0584	0.8070	0.0746	0.1450	−0.0206	0.6080	0.0616	0.1580
CO_3	−0.0731	0.8840	−0.0771	0.8830	**−0.1012**	**0.9830**	−0.0559	0.7820
ϑ	−0.0406	0.6530	**−0.1493**	**0.9940**	−0.0706	0.8240	0.1292	0.0560
IGSD	**−0.1023**	**0.9590**	**−0.1324**	**0.9900**	**−0.1034**	**0.9770**	−0.0218	0.5900
OM	−0.0592	0.8290	0.0382	0.2650	−0.0091	0.4960	−0.0291	0.6430
Partial Mantel								
Temp	−0.0600	0.8578	0.0689	0.1564	−0.0141	0.5558	0.0306	0.2752
CO_3	−0.0735	0.9042	−0.0835	0.8894	**−0.0985**	**0.9913**	−0.0768	0.9125
ϑ	−0.0426	0.6889	**−0.1753**	**0.9994**	−0.0663	0.8174	0.0929	0.1166
IGSD	**−0.1029**	**0.9852**	**−0.1381**	**0.9959**	**−0.1010**	**0.9835**	−0.0385	0.6868
OM	−0.0441	0.7488	0.0384	0.2682	−0.0047	0.4767	−0.0384	0.7010

Significant correlations are indicated in bold ($r > 0$ and $p < 0.05$ indicates spatial overlap, while $r < 0$ and $p > 0.95$ indicates spatial exclusion). Temp: temperature (°C); CO_3: % of carbonate contents; φ: grain size; IGSD: sorting coefficient; OM: organic matter.

hence with low hydrodynamism. *S. membranacea* showed no significant correlation with any of the environmental variables, with a relative independence of sediment characteristics. *Processa canaliculata* was the only species showing a negative correlation with carbonates ($r = -0.1$, $p = 0.99$) and *C. macandreae* with φ ($r = -0.18$, $p = 1.0$). This last species was also strongly correlated with the sediment granulometry, as it was negatively correlated with both φ ($r = -0.164$, $p = 0.997$) and IGSD ($r = -0.134$, $p = 0.996$) (Table 3).

Discussion

Depth is known to play a key role in structuring demersal communities (Wenner & Boesch, 1979; Abelló et al., 1988; Macpherson, 1989; Wenner & Wenner, 1989; Setubal Pires, 1992; González-Gurriarán et al., 1993; Vassilopoulou et al., 1998; Williams et al., 2001) and has been often used as a covariate in geostatistical studies applied to fisheries (Sullivan, 1991; Simard et al., 1992). Using both visual and automatic model selection when fitting a semivariogram model to the empirical semivariogram, all species studied herein followed a spherical model, as in other crustaceans (Freire et al., 1991a; González-Gurriarán et al., 1993; Maynou & Sardà, 1997), showing that they are structured in high density patches, with the average patch size given by the range of the semivariogram. Using non-linear geostatistical techniques, we estimated a patch size of 4 km in *C. macandreae*, a smaller value than the 6 km observed by Maynou et al. (1996) using linear geostatistics. In the current study, performed in spring, *S. membranacea* showed a range of 17.24 km, which is similar to the 16 km (spring) and 12 km (autumn) patch size observed by Maynou et al. (1996) in the Catalan Sea. The same species, showed a range of 25 km off the Galician coast, which is larger than that found in the Mediterranean Sea (Freire et al., 1991b).

The four species studied showed greater densities in the northern half of the study area, which is characterised by a narrow shelf (8–10 km width) and a gentle slope cut by deep submarine canyons. Smaller densities were thus observed in the southern part, characterised by a wider shelf (38–40 km) and a steeper slope with a lower number of submarine canyons (Maynou & Sardà, 1997). High macrofaunal and meiofaunal biomass has been reported for some canyons (Cartes et al., 1994b; Stefanescu et al., 1994; Vetter & Dayton, 1998), probably related to higher organic matter and sediment advection from the shelf (Rowe et al., 1982). Both *A. glaber* and *S. membranacea* showed greater densities close to the main central canyon. In other areas of the Catalan Sea, *A. glaber* was more abundant near submarine canyons in autumn and in the middle slope in spring (Cartes et al., 1994a).

Spatial correlation between pairs of species (overlap) using the Mantel test was previously employed by Maynou et al. (1996), but using the Pearson coefficient which is able to detect linear relationships only, unlike the Kendall coefficient used in this study. Still, the only two species common to both studies were *C. macandreae* and *S. membranacea*, which were not significantly spatially correlated with each other in either study. Present data emphasise a different spatial behaviour of *C. macandreae* from all the other species, probably in agreement with its deep-sea habitat and burrowing habits (Anderson et al., 1991; Froglia et al., 1997; Rosenberg et al., 2000).

Sediment characteristics have been found to strongly influence the occurrence and distribution of many benthic decapod crustaceans, especially those with burrowing or burying habits (e.g. Atkinson & Taylor, 1988; Pinheiro et al., 1996; Comeau et al., 1998). Two species that occupy the same habitat and show significant spatial overlap may enter into direct competition for trophic resources. Most species in the current study showed significant spatial overlap between each other. However, as each species was significantly correlated with different sediment variables, there might be a niche partitioning in order to avoid direct trophic competition for resources linked to the sediment variables.

Both *A. glaber* and *C. macandreae* were associated with low energy and well-sorted sediments, and hence low hydrodynamism (Froglia et al., 1997; Atkinson et al., 2003). Low hydrodynamic homogeneous sediments appear to be important for constructing the burrows characteristic of those species. Furthermore, high densities of *C. macandreae* were observed in sediments with greater grain size, i.e. lower values of ϑ. *Processa canaliculata* was negatively correlated with IGSD and carbonates. No direct evidence for either burrowing or burying habits exists for this species, but other shallower species of the genus are known to bury in the substratum, mainly in sandy and sandy-muddy bottoms (Guerao & Abelló, 1996), so this species also probably shows similar habits. *Solenocera membranacea* did not show any significant correlation with any of the measured sediment variables within the sampled ranges. Previous studies found that *S. membranacea* shows burying habits (Heegaard, 1967; Somers, 1987; Demestre & Abelló, 1993), and high affinity for muddy and sandy bottoms (Zariquiey-Álvarez, 1968; Lagardère, 1973; González-Gurriarán & Olaso, 1987; Abelló et al., 1988; Abelló, 1993). However, its larger size and stronger swimming ability probably account for the lack of significant correlations with the sediment characteristics within the ranges sampled. Still, the influence of the substrate on this species may depend on other variables not measured in the current study.

Acknowledgements

We wish to thank all participants in the GEODELTA cruise on board R/V 'García del Cid', for all help and facilities provided. We would also like to thank Marta Coll for the data on crustacean species in fish stomach contents and Joan Cartes for reviewing the manuscript.

References

Abelló, P., 1993. Pautes de distribució de les espècies de la família Portunidae (Crustacea: Brachyura) als fons de substrat tou de la Mediterrània nord-occidental. Bulletí de la Institució Catalana d'Història Natural 61: 59–68.

Abelló, P., F. J. Valladares & A. Castellón, 1988. Analysis of the structure of decapod crustacean assemblages off the Catalan coast (North-West Mediterranean). Marine Biology 98(1): 39–49.

Alonso, B., G. Ercilla, F. Martínez-Ruiz, J. Baraza & A. Galimont, 1999. Pliocene-Pleistocene sedimentary facies at Site 976: Depositional history in the northwestern Alboran sea. Proceedings of the Ocean Drilling Program. Scientific Results 16: 157–168.

Anderson, S. J., R. J. A. Atkinson & A. C. Taylor, 1991. Behavioural and respiratory adaptations of the mud-burrowing shrimp *Calocaris macandreae* Bell (Thalassinidea: Crustacea) to the burrow environment. Ophelia 34(2): 143–156.

Atkinson, R. J. & A. A. C. Taylor, 1988. Physiological ecology of burrowing decapods. In Fincham, A. A. & P. S. Rainbow (eds), Aspects of Decapod Crustacean Biology. 59. Oxford Science Publications, London: 201–226.

Atkinson, R. J. A. & M. E. C. Gramitto Froglia, 2003. Aspects of the biology of the burrowing shrimp *Alpheus glaber* (Olivi) (Decapoda: Caridea: Alpehidae) from the central Adriatic. Ophelia 57(1): 27–42.

Bozzano, A., L. Recasens & P. Sartor, 1997. Diet of the European hake *Merluccius merluccius* (Pisces: Merluciidae) in the Western Mediterranean (Gulf of Lions). Scientia Marina 61(1): 1–8.

Carrassón, M., C. Stefanescu & J. E. Cartes, 1992. Diets and bathymetric distributions of two bathyal sharks of the Catalan deep sea (western Mediterranean). Marine Ecology Progress Series 82(1): 21–30.

Cartes, J. E., 1998. Dynamics of the bathyal Benthic Boundary Layer in the northwestern Mediterranean: depth and temporal variations in the macrofaunal–megafaunal communities and their possible connections within deep-sea trophic webs. Progress in Oceanography 41(1998): 111–139.

Cartes, J. E., J. B. Company & F. Maynou, 1994a. Deep-water decapod crustacean communities in the northwestern Mediterranean: Influence of submarine canyons and season. Marine Biology 120(2): 221–229.

Cartes, J. E., J. C. Sorbe & F. Sardà, 1994b. Spatial distribution of deep-sea decapods and euphausiids near the bottom in the northwestern Mediterranean. Journal of Experimental Marine Biology and Ecology 179(1): 131–144.

Casadevall, M. & J. T. Matallanas Bartoli, 1994. Feeding habits of *Ophichthus rufus* (Anguilliformes, Ophichthidae) in the western Mediterranean. Cybium 18(4): 431–440.

Comeau, M., G. Y. Conan, F. Maynou, G. Robichaud, J.-C. Therriault & M. Starr, 1998. Growth, spatial distribution, and abundance of benthic stages of the snow crab (*Chionoecetes opilio*) in Bonne Bay, Newfoundland, Canada. Canadian Journal of Fisheries and Aquatic Sciences 55: 262–279.

Cressie, N. A. C., 1991. Statistics for Spatial Data. John Wiley & Sons, Inc, New York, 928 pp.

Demestre, M. & P. Abelló, 1993. Growth and distribution of *Solenocera membranacea* (Risso, 1816) (Decapoda, Dendrobranchiata) in the northwestern Mediterranean Sea. Scientia Marina 57(2–3): 161–166.

Diggle, P. J., J. R. Ribeiro & O. F. Christensen, 2003. An introduction to model based geostatistics. Lecture notes in statistics. In J. Möller (ed.), Spatial Statistics and Computational Methods. Springer, Heidelberg.

Fariña, A. C., J. Freire & E. González-Gurriarán, 1997. Megabenthic decapod crustacean assemblages on the Galician continental shelf and upper slope (North–west Spain). Marine Biology 127(3): 419–434.

Freire, J., L. Fernandez & E. González-Gurriarán, 1991a. Spatial distribution of decapod crustaceans in the Galician continental shelf (NW Spain) using geostatistical analysis. Ices Council Meeting Papers 12.

Freire, J., E. González-Gurriarán, I. Olaso & L. Fernandez, 1991b. Análisis geoestadístico de la distribucíon espacial de *Plesionika heterocarpus* y *Solenocera membranacea* (Crustacea: Decapoda) en la plataforma continental de Galicia (NW España). Boletin del Instituto Español de Oceanografia 7(2): 79–88.

Froglia, C., 1976. Observations of the feeding of *Helicolenus dactylopterus* (Delaroche) (Pisces, Scorpaenidae) in the Mediterranean Sea. Rapp. P.: 47–48.

Froglia, C., R. J. A. Atkinson, I. Tuck & E. Arneri, 1997. Underwater television survey. A tool to estimate *Nephrops* stock biomass on the Adriatic trawling grounds (ed.), Tisucu godina prvoga spomena ribarstva u hrvata. Zagreb : 657–667.

González-Gurriarán, E. & I. Olaso, 1987. Cambios espaciales y temporales de los Crustáceos Decápodos de la plataforma continental de Galicia (NW España). Investigación Pesquera 51(suppl 1): 323–341.

González-Gurriarán, E., J. Freire & L. Fernández, 1993. Geostatistical analysis of spatial distribution of *Liocarcinus depurator*, *Macropipus tuberculatus* and *Polybius henslowii* (Crustacea: Brachyura) over the Galician continental shelf (NW Spain). Marine Biology 115(3): 453–461.

Gramitto, M. E., 1999. Feeding habits and estimation of daily ration of poor cod *Trisopterus minutus capelanus* (Gadidae) in the Adriatic Sea. Cybium 23(2): 115–130.

Gray, J. S., 1981. The ecology of marine sediments. Cambridge University Press, 200 pp.

Guerao, G. & P. Abelló, 1996. Patterns of activity in the sympatric prawns *Palaemon adspersus* and *Processa edulis* (Decapoda, Caridea) from a shallow Mediterranean bay. Scientia Marina 60: 319–324.

Heegaard, P., 1967. On behaviour, sex-ratio and growth of *Solenocera membranacea* (Risso) (Decapoda, Penaeidae). Crustaceana 13: 227–237.

Hopkins, T. S., 1985. Physics of the sea. In Margalef, R. (ed.), Key Environments: Western Mediterranean. Pergamon Press, New York: 100–125.

Ihaka, R. & R. Gentleman, 1996. R: A language for data analysis and graphics. Journal of Computational and Graphical Statistics 5(3): 299–314.

Koenig, W. D., 1999. Spatial autocorrelation of ecological phenomena. Trends in Ecology and Evolution 14(1): 22–26.

Labropoulou, M. & A. Eleftheriou, 1997. The foraging ecology of two pairs of congeneric demersal fish species: importance of morphological characteristics in prey selection. Journal of Fish Biology 50: 324–340.

Lagardère, J. P., 1973. Distribution des décapodes dans le Sud du Golfe de Gascogne. Revue des Travaux de l'Institut des Pêches Maritimes XXXVII(1): 77–95.

Legendre, P., 2000. Comparison of permutation methods for the partial correlation and partial Mantel tests. Journal of Statistical Computation and Simulation 67: 37–73.

Macpherson, E., 1979. Ecological overlap between macrourids in the western Mediterranean Sea. Marine Biology 53: 149–159.

Macpherson, E., 1981. Resource partitioning in a Mediterranean demersal fish community. Marine Ecology Progress Series 4: 183–193.

Macpherson, E., 1989. Influence of geographical distribution, body size and diet on population density of benthic fishes off Namibia (South West Africa). Marine Ecology Progress Series 50: 295–299.

Manly, B. F. J., 1985. The Statistics of Natural Selection. Chapman and Hall, New York, 484 pp.

Mantel, N., 1967. The detection of disease clustering and a generalised regression approach. Cancer Research 17(2): 209–220.

Matallanas, J., 1980. Feeding of *Ophidion barbatum* (Pisces, Ophidiidae) in the Mediterranean Sea. Cybium 10: 81–89.

Maynou, F. & F. Sardà, 1997. *Nephrops norvergicus* population and morphometrical characteristics in relation to substrate heterogeinity. Fisheries Research 30: 139–149.

Maynou, F., G. Y. Conan, J. E. Cartes & J. B. F. Company Sardà, 1996. Spatial structure and seasonality of decapod crustacean populations on the northwestern Mediterranean slope. Limnology and Oceanography 41(1): 113–125.

Minervini, R. & E. R. Fresi Manconi, 1984. Soft bottom decapod crustaceans distribution in the Gulf of Salerno. Nova Thalassia 6(Suppl.): 539–545.

Moreno, R. & J. Matallanas, 1983. Étude du régime alimentaire de *Lepidotrigla cavillone* (Lacepède, 1801) (Pisces, Triglidae) dans la Mer Catalane. Cybium 7(3): 93–103.

Morte, M. S. & A. Sanz-Brau, 1994a. A preliminary investigation on the diet of *Trachynus draco* (Pisces: Trachynidae) in the Gulf of Valencia, Spain. Journal of Fish Biology 45: 251.

Morte, M. S. & A. Sanz-Brau, 1994b. Observations on the food and feeding of the poor-cod, Trisopterus minutus (Linnaeus, 1758) in the Gulf of Valencia (Western Mediterranean). Journal of Fish Biology 45.

Pinheiro, M. A. A., A. Fransozo & M. L. Negreiros-Fransozo, 1996. Distribution patterns of *Areneus cribarius* (Lamark, 1818) (Crustacea, Portunidae) in Fortaleza Bay, Ubatuba (SP), Brazil. Revista Brasileira de Biologia 56(4): 705–716.

Quetgas, A., F. Alemany, A. Carbonell, P. Merella & P. Sánchez, 1999. Diet of the European flying squid *Todarodes sagittatus* (Cephalopoda: Ommastrephidae) in the Balearic Sea (western Mediterranean). Journal of the Marine Biological Association of the United Kingdom 79: 479–486.

Ribeiro, J. R. & P. J. Diggle, 1999. geoS: A geostatistical library for S-PLUS. Lancaster: Dept of Maths and Stats, Lancaster University.

Rivoirard, J., E. J. Simmonds, K. Foote, P. Fernandes & N. Bez, 2000. Geostatistics for estimating fish abundance. Blackwell Science, Oxford.

Robertson, G. P. & D. W. Freckmann, 1995. The spatial distribution of nematode trophic groups across cultivated ecosystem. Ecology 765: 1425–1432.

Rosenberg, R., H. C. Nilsson, B. Hellman & S. Agrenius, 2000. Depth correlated benthic faunal quantity and infaunal burrow structures on the slopes of a marine depression. Estuarine, Coastal and Shelf Science 50: 843–853.

Rossi, R. E., D. J. Mulla, A. G. Journel & E. H. Franz, 1992. Geostatistical tools for modelling and interpreting spatial dependence. Ecological Monographs 62: 277–314.

Rowe, G. T., P. T. Polloni & R. L. Haedrich, 1982. The deep-sea macrobenthos on the continental margin of the north-west Atlantic Ocean. Deep-sea Research 29: 257–278.

Sanchez, P., 1981. Regime alimentaire d'*Eledone cirrosa* (Lamarck, 1798) (Mollusca, cephalopoda) dans la mer catalane. Rapport Commission International Mer Médit 27(5): 209–212.

Sanz, A., 1985. Contribution to the study of the biology of *Uranoscopus scaber* Linnaeus, 1758 (Osteichthyes, Uranoscopidae) of the Western Mediterranean. Investigación Pesquera 49(1): 35–46.

Sardà, F., J. E. Cartes & J. B. Company, 1994. Spatio-temporal variations in megabenthos abundance in three different habitats of the Catalan deep-sea (Western Mediterranean). Marine Biology 120: 211–219.

Sartor, P. & S. De Ranieri, 1996. Food and feeding habits of *Lepidorhombus boscii* (Pisces, Scophtalmidae) in the southern Tuscan Archipelago, Tyrrhenian Sea. Vie et Milieu 46(1): 57–64.

Setubal Pires, A. M., 1992. Structure and dynamics of benthic megafauna on the continental shelf offshore of Ubatuba, southeastern Brazil. Marine Ecology Progress Series 86: 63–76.

Simard, Y., P. Legendre, G. Lavoie & D. Marcotte, 1992. Mapping, estimating biomass, and optimizing sampling programs for spatially autocorrelated data: Case study of the northern shrimp (*Pandalus borealis*). Canadian Journal of Fisheries and Aquatic Sciences 49(1): 32–45.

Sokal, R. R. & F. J. Rohlf, 1994. Biometry. Freeman, New York, 88 pp.

Sokal, R. S. & N. L. Oden, 1978. Spatial autocorrelation in biology. 1. Methodology. Biological Journal of the Linnean Society 10: 199–228.

Somers, I. F., 1987. Sediment type as a factor in the distribution of commercial prawn species in the western Gulf of Carpentaria, Australia. Australian Journal of Marine and Freshwater Research 38: 133–149.

Stefanescu, C., B. Morales-Nin & E. Massuti, 1994. Fish assemblages on the slope in the Catalan Sea (western Mediterranean): Influence of a submarine canyon. Journal of the Marine Biological Association of the United Kingdom 74(3): 499–512.

Sullivan, P., 1991. Stock abundance estimation using depth-dependent trends and spatially correlated variation. Canadian Journal of Fisheries and Aquatic Sciences 48: 1691–1703.

Vassilopoulou, V. & C. P. Papaconstantinou Bekas, 1998. Demersal fish community patterns in two areas of the Aegean Sea, Greece. Italian Journal of Zoology 65(Suppl): 293–297.

Vatan, A., 1967. Manuel de sédimentologie. Paris.

Vetter, E. W. & P. K. Dayton, 1998. Macrofaunal communities within and adjacent to a detritus-rich submarine canyon system. Deep-sea Research 45(1–3): 25–54.

Wenner, E. L. & D. F. Boesch, 1979. Distribution patterns of epibenthic decapod crustacea along the shelf-slope coenocline, middle Atlantic Bight, U.S.A. Bulletin of the Biological Society of Washington 3: 106–133.

Wenner, E. L. & C. A. Wenner, 1989. Seasonal composition and abundance of decapod and stomatopod crustaceans from coastal habitats, southeastern United States. Fishery Bulletin 87: 155–176.

Williams, A., J. A. Koslow & P. R. Last, 2001. Diversity, density and community structure of the demersal fish fauna of the continental slope off western Australia (20 to 35°S). Marine Ecology Progress Series 212: 247–263.

Zariquiey-Álvarez, R., 1968. Crustáceos decápodos ibéricos. Investigación Pesquera 32: 1–510.

Hydrobiologia (2006) 555:143–157
H. Queiroga, M.R. Cunha, A. Cunha, M.H. Moreira, V. Quintino, A.M. Rodrigues, J. Serôdio & R.M. Warwick (eds),
Marine Biodiversity: Patterns and Processes, Assessment, Threats, Management and Conservation
DOI 10.1007/s10750-005-1112-6

Probing diversity in the plankton: using patterns in Tintinnids (planktonic marine ciliates) to identify mechanisms

John R. Dolan*, Rodolphe Lemée, Stéphane Gasparini, Laure Mousseau & Céline Heyndrickx
Marine Microbial Ecology Group, Laboratoire d'Océanographie de Villefranche, CNRS UMR 7093, Station Zoologique, B. P. 28F-06230, Villefranche-Sur-Mer, France
*(*Author for correspondence: E-mail: dolan@obs-vlfr.fr)*

Key words: biogeography, biodiversity, zooplankton, Mediterranean, latitudinal gradient, phytoplankton, foraminifera

Abstract

In diversity research, the use of survey data appears to have declined in favour of experimental or modeling approaches because direct relationships are difficult to demonstrate. Here we show that use of field data can yield information concerning the mechanisms governing diversity. First, we establish that tintinnids display a global latitudinal pattern of diversity similar to other pelagic organisms; species numbers appear to peak between 20° and 30° north or south. This common large scale spatial trend has been attributed to the gradient in water column structure across the global ocean. We then examine the generality of a relationship between planktonic diversity and water column structure by considering data from the Mediterranean Sea, in which water column structure changes seasonally. Among populations of foraminifera, tintinnids, and the dinoflagellates of the genus *Ceratium*, we compare data from trans-Mediterranean sampling conducted at different times and monthly changes in species richness at single sites. We find that water column structure alone appears to be a poor predictor of temporal changes in diversity. Lastly, we present an example of temporal changes in tintinnid diversity based on data from an oceanographic sampling station in the N. W. Mediterranean where resources, as chlorophyll, appear distinctly unrelated to changes in water column structure. We show that short-tem temporal changes in diversity (week to week shifts) can be related to changes in chlorophyll concentration. We conclude that in tintinnids diversity can be directly linked to characteristics of food resources.

Introduction

Three major approaches are commonly employed in diversity research: experimentation, modeling, and survey, each of which possesses particular weaknesses. Here, after briefly reviewing the drawbacks of recent experimental and modeling efforts, we will illustrate the major problem involved in using survey data – that of distinguishing direct effects on diversity from co-varying or indirect factors. To show the utility of survey data we will begin by establishing the validity of using tintinnid ciliates as an example of a typical planktonic group. Then, we will use survey data to explore factors co-varying with tintinnid diversity to determine if a plausible controlling mechanism influencing diversity can be identified.

The experimental approach has been employed most commonly to investigate the relationship between a single ecosystem function (production, nutrient regeneration, etc.) and diversity. Typically experimentation with planktonic communities has involved very simple petri dish or test tube communities of a few microbial species (e.g., Naeem & Li, 1998; Naeem et al., 2000) or even strains of the same species (e.g. Buckling et al., 2000; Kassen et al., 2000). Thus, conclusions have been drawn from short-term results with very rudimentary communities. Extrapolation to the natural world has proven problematic especially as the little work which has been conducted by altering large, natural systems (such as whole lakes – e.g., Dodson et al., 2000), suggests that idiosyncratic, in other words unpredictable, results may be common. It is perhaps worth noting that with regard to marine benthic communities, experimentation with very simple communities has also shown 'idiosyncratic' effects of biodiversity on ecosystem function (Emmerson et al., 2001) or the existence of 'synergistic interactions' (Worm et al., 2002).

Modeling is an approach most often employed as a method to identify the factors governing diversity. The theoretical framework of modeling plankton communities can be traced back to Hutchinson's "Paradox of the Plankton" (1961) in which three major mechanisms were proposed: (1) symbiosis or commensalism, (2) predation, and (3) non-equilibrium conditions. In recent years, theoretical studies have suggested the primacy of each of these mechanisms. Thus, species oscillations can explain diversity as non-equilibrium conditions are likely the rule (Huisman & Weissing, 1999) or specialization among predators or consumers is important (Hulot et al., 2000) as well as facilitation or commensalism (Loreau & Hector, 2001).

The survey approach-determining, and then attempting to explain, natural patterns of species abundance is the most time-honored as it dates back at least to Darwin (1859). The obvious drawback is that mechanisms governing biodiversity can only be inferred by co-variation and co-variation need not indicate causation. An excellent example of this has been reported with regard to planktonic foraminifera.

From the point of view of past and present species distributions, foraminifera are no doubt the best-studied marine planktonic group (Caron & Swanberg, 1990). Paleotemperatures can be estimated from oxygen isotopes in fossil tests (e.g., Wilf et al., 2003) and at least in fossil assemblages, the co-variation of diversity and annual average sea surface temperature permit the estimation of past climatic change (e.g., Williams & Johnson, 1975). Diversity appears fairly predictable from the type of environment (Ottens & Nederbragt, 1992). Among living foraminifera communities in the Atlantic, species diversity was recently shown to be tightly correlated with annual sea-surface temperature, in a curvi-linear manner yielding maxima at 20–30° rather than a peak at the equator (Rutherford et al., 1999). The correlation was explained as due to the relation between annual sea surface temperature and both the depth and seasonality of the thermocline in the water column. Annual sea surface temperature appeared to be a proxy measure of the depth of the surface layer and therefore habitat volume for planktonic foraminifera somehow reflecting quantities of niches available for different species (Rutherford et al., 1999). However, the parameters which define or differentiate the niches are unknown.

Interestingly, molecular work has shown the existence of 'cryptic species' of foraminifera (de Vargas et al., 1999), that is morphologically very similar but genetically distinct populations. The genetically distinct populations appear to inhabit regions differing in productivity (de Vargas et al., 1999) or water column stability (de Vargas et al., 2002). Careful examination of the different 'strains' has also revealed apparently subtle morphological differences in, for example, shell porosity. However, once again what ecologically differentiates the populations is unknown.

In copepods, among the mechanisms thought to maintain diversity, specialization in feeding may be much less important than the effects of predators (McGowan & Walker, 1980). In an environment characterized by a structured water column, i.e., with a well-defined thermocline, predation pressure may be stronger yielding greater diversity than environments with mixed water columns (McGowan & Walker, 1980, 1985; Longhurst, 1985). However, with regard to planktonic foraminifera, we are in a poor position to begin to identify what mechanism (food, predation, etc.) may be important in influencing diversity. We know relatively little about foraminiferan nutrition and growth (Caron et al., 1995) or trophic specialization (Swanberg & Caron, 1991) or their predators (Caron & Swanberg, 1990). In contrast to foraminifera, a fair amount is known about tintinnid ciliates.

Tintininnids are ciliates of the microzooplankton; they are characterized by the possession of a species-specific shell or lorica, shaped like a bowl or vase or tube, within which the ciliate cell can withdraw (Fig. 1). While generally a small part of the ciliate community, they are much more abundant than foraminifera or radiolarians (Thompson et al., 1999) and there is a wealth of data on their ecology (see Dolan, 2000; Dolan et al., 2002).

As most planktonic ciliates, tintinnids are found largely in the surface layer of the sea where they feed on algae ranging in size from 2 to 20 microns. The generation times of tintinnids are similar to other planktonic ciliates and their algal prey (potentially a few hours). The dimensions of their lorica, specifically the diameter of the oral (or mouth) end, are related to their food. The oral diameter of the lorica is about four times the diameter of the prey ingested most efficiently (Dolan et al., 2002). Tintinnids are fed upon by a large variety of taxa. The stomach contents of larval fish contain tintinnids almost as frequently as copepods (Turner, 1984). Different metazoan zooplankters are also known to feed on tintinnids ranging from copepods to gelatinous zooplankton (Stoecker & Capuzzo, 1990). However, given their numerical dominance among the zooplankton, copepods are generally thought to be the major predators of tintinnids. Interestingly, there appears to be no clear relationship between the size of a tintinnid's lorica and its suceptability to predation by copepods (Stoecker & Capuzzo, 1990; Dolan & Gallegos, 2001), although most experimental work has been conducted with coastal species of copepods.

Like foraminifera and radiolarians, species identifications can be made using characteristics of gross morphology, with some caveats (for recent discussions see Cariou et al., 1999; Dolan, 2000; Dolan &

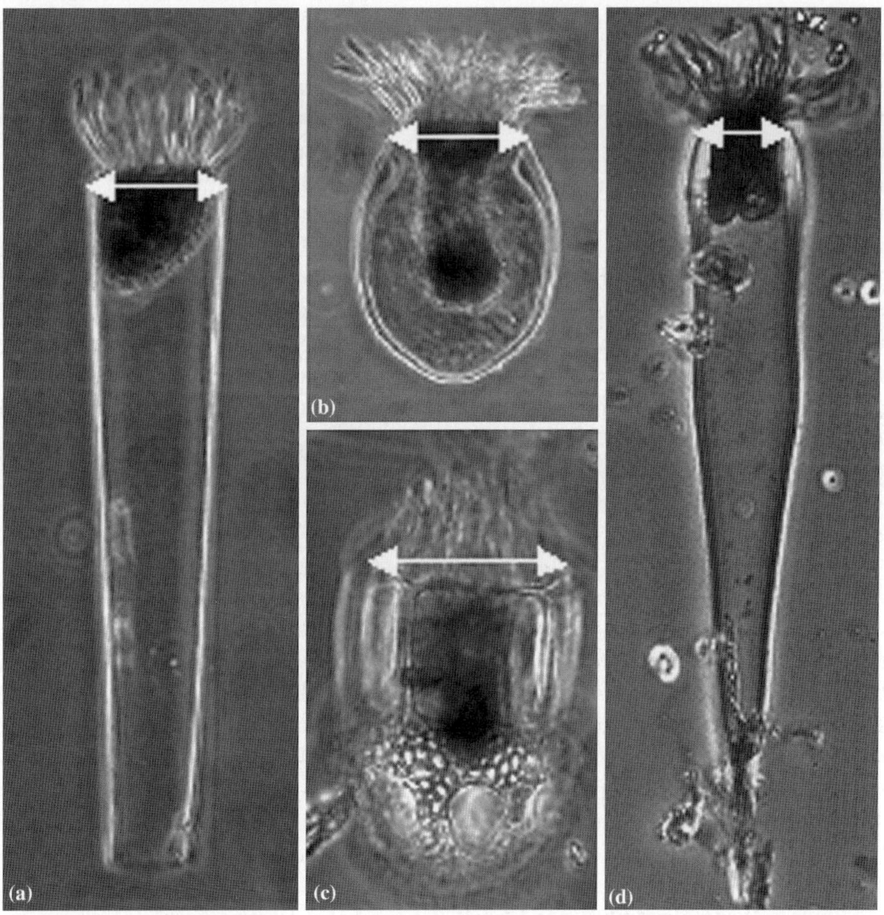

Figure 1. Photomicrographs of Tintinnids. Lugol's -preserved specimens collected from the Bay of Villefranche, N. W. Mediterranean Sea. Arrows indicate the oral opening, lorica oral diameter (LOD) of the lorica or shell into which the ciliate cell may withdraw. (a) *Eutintinnus fraknoii*, LOD = 40 μm. (b) *Proplectella fastigata*, LOD = 40 μm. (c). *Dictyocysta lepida*, LOD = 45 μm., (d) *Xystonellopsis brandti* LOD = 25 μm.

Gallegos, 2001; Modigh & Castaldo, 2002). Tintinnids have the further advantage of representing a monophyletic group, even among competing ciliate classification schemes (e.g., Petz & Foissner, 1992; Lynn & Small, 1997). Recent molecular work (Snoeyenbos-West et al., 2002; Agatha et al., 2005) has largely confirmed the classical (based on morphology) taxonomic classification of tintinnids. Tintinnids are united ecologically as microzooplankters, morphologically as loricate ciliates, and phylogenetically as members of the order Tintinnida.

Before examining patterns of diversity in tintinnids, the first question one might ask is whether tintinnids are a typical or atypical group of plankters. That is, do tintinnids show a global pattern of diversity similar to other planktonic groups such as foraminifera and copepods and phytoplankton taxa or a 'microbial' pattern of "all species are everywhere" *sensu* Fenchel & Findlay (1997). Thus, this will be the first issue considered. Secondly, the relationship between water column structure and diversity among planktonic organisms will be considered. Temporal trends of diversity relative to water column structure in a system characterized by seasonal changes in water column stratification – the Mediterranean Sea, will be explored with regard to foraminifera, tintinnids and dinoflagellates of the genus *Ceratium*. Thirdly, changes in diversity will be examined at an oceanographic sampling station in the N. W. Mediterranean where resources, as chlorophyll, appear distinctly unrelated to changes in water column structure.

Methods

Global distribution of tintinnid species abundance

Literature reports furnishing species abundance for single points in time and space were used. When species abundance were given for more than one date, the date with maximum number of species was taken. Use of a single date from reports giving more than one measure was used to make such reports comparable to those of oceanographic campaigns, which formed the majority of the data points, giving a single estimate per location. No attempt was made to sort data by the sampling technique employed (i.e., plankton net tows or whole water collection). The oceanographic campaign reports and research reports (Table 1) yielded a large number of data points (n = 451) with a near complete latitudinal coverage of 81°N to 75°S. The data base assembled represents a greatly expanded version of that previously presented (Dolan & Gallegos, 2001) which consisted of 168 data points.

Water column stratification and Mediterranean plankton

The link between water column structure and diversity was considered by examining reports of foraminifera community composition across the Mediterranean in the autumn when the Mediterranean is stratified and during winter mixis. Seasonal trends were considered by examining reports of species abundance at a single location over an annual cycle for tintinnids and dinoflagellates of the genus *Ceratium*. The relationship of tintinnid diversity with the size of the mixed layer, as indicated by the depth of the chlorophyll maximum layer, at the beginning and end of the stratified period was examined.

Data on foraminifera (species richness and ln-based H′ values) were based on a report of foraminifera community composition derived from sampling at 13 stations across the Mediterranean in autumn 1986 and 14 stations in winter 1988 (Pujol & Vergnaud Grazzini, 1995: Table 2). Data on species richness by month for tintinnids and a group of phytoplankters-dinoflagellates of the genus *Ceratium* was obtained from Rampi (1948: p. 53) based on material collected from 1938 to 1939 in the N. W. Mediterranean Sea near San Remo, Italy. The relationship between the depth of the chlorophyll maximum layer and tintinnid species richness or diversity (as H′) at the beginning of the stratified period was examined using data from Dolan (2000: Fig. 1) based on samples obtained in June 1993 (Dolan & Marrasé, 1995) and May–June 1996 (Dolan et al., 1999). The relationship at the end of the stratified period was explored using data from Dolan et al., (2002: Figs. 2 and 3) based on sampling conducted in September 1999.

Temporal trends of tintinnid diversity

We examined tintinnid populations in the Bay of Villefranche in samples from a standard station

Table 1. Sources of data used to construct plots of latitude vs. species abundances

Study site	Latitude range	*N*	Reference
Barents Sea	72–70° N	6	Jensen & Hansen (2000)
Barents Sea	73–81° N	12	Boltovskoy et al. (1991)
Chesterfield Inlet Estuary	64–63° N	12	Rogers et al. (1981)
Skagerak (N Sea)	58° N	1	Hedin (1974)
Bedford Basin, NS	44° N	1	Paranjape (1987)
Bay of Fundy	45–42 N	3	Middelebrook et al. (1987)
Damariscotta estuary	44° N	1	Sanders (1987)
Akkeshi Bay, JP	43° N	1	Taguchi (1976)
Coastal Mediterranean	43° N	1	Cariou et al. (1999)
Narragansett Bay	42° N	1	Hargraves (1981)
Long Island Sound	41° N	1	Gold & Morales (1975)
Long Island Sound	41° N	1	Capriuolo & Carpenter (1983)
Open Mediterreanean	41–34°N	23	Dolan (2000)
Atlantic-Mediterreanean	43–31° N	11	Dolan (2002)
Lagoa de Obidos, PT	40° N	1	Silva (1953)
Chesapeake Bay	39–37° N	9	Dolan & Gallegos (2001)
Atlantic	29–60° N	62	Garder (1946)
South Pacific Coastal	12° N	1	Gold & Morales (1977)
Sub & Tropical Pacific	34° N–25° S	62	Kofoid & Campbell (1939)
Atlantic & Pacific	64° N–39° S	160	Campbell (1942)
New Zealand Coastal	42–50° S	4	James & Hall (1995)
Bahia Blanca Estuary	38° S	2	Barria de Cao (1992)
SW Atlantic Open	34–60°S	18	Thompson et al. (1999)
SW Atantic Shelf	40–56° S	16	Thompson et al. (2001)
S Atlantic	59–60° S	23	Wasik & Mikolajczk (1990)
Ross Sea	75° S	1	Monti & Fonda Umani (1995)

N value indicates the number of data points obtained. Note that out of a total of 451 points, the majority are from reports of oceanographic expeditions in which open water sites dominated (i.e., Kofoid & Campbell, 1939; Campbell, 1942; Garder, 1946).

'Point B' (43°41′10″ N, 7°19′00″ E). Sampling was conducted, in principle, weekly. Discrete depth samples from 0, 10, 20, 30, 40, and 50 m were obtained using 5 l Niskin bottles. Temperature and salinity were measured using a Seabird CTD. Chlorophyll *a* samples were analyzed fluorometrically following acetone extraction. For tintinnids, a composite integrated water column sample of 4 l was concentrated using a 20 μm Nitex screen to 20 ml. This method yields tintinnid numbers as high as settling whole water samples (Pierce & Turner, 1994). The entire concentrate, in 3–10 ml aliquots, was settled in sedimentation chambers and examined with an inverted microscope at 200×. Based on previous studies, examining material from 4 l of water likely yielded species abundances of about 50% of those found through examining large volumes (80 l) of water (Cariou et al., 1999). Tintinnids were identified using lorica morphology and the species descriptions found in Campbell (1942), Jörgensen (1924) and Kofoid and Campbell (1929, 1939). Here data from the 45 samples taken in 2002 (seven samples missing from equipment failure or bad weather) are presented.

Results

Global distribution of tintinnid species abundance

Plotting species richness, averaged over increments of 5° latitude or individual points estimates (Fig. 2), gave the commonly found relationship between latitude and species abundance. In both

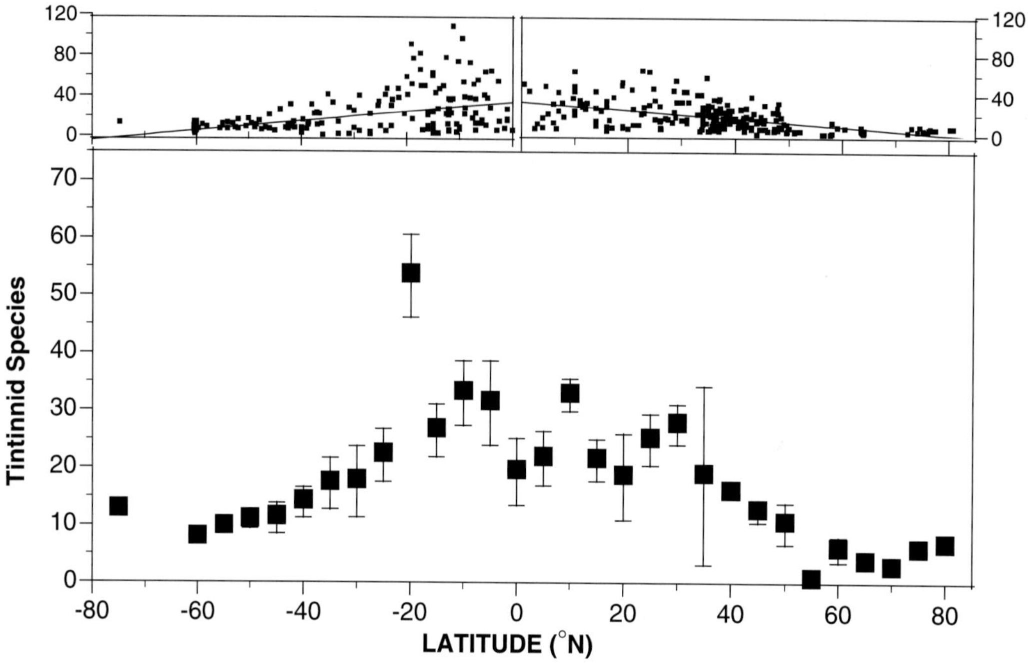

Figure 2. Latitudinal distribution of tintinnid species abundance based on reports given in Table 1. Top panels show individual data points. Bottom panel shows species abundances averaged within bands of 5° of latitude, error bars indicate SE of the averages. Linear regression equation for the southern latitude estimates (individual data points) is $r = 0.45$, $x = 0.48 \times \text{latitude} + 0.37$. Linear regression equation for the nothern latitude estimates (individual data points) is $r = 0.53$, $x = 0.41 \times \text{latitude} + 0.33$.

the southern and northern hemispheres, species richness increases from the poles to lower latitudes with a peak around 20–30° followed by a slight decline toward the equator. Quite similar equations describe the relationships between latitude and species richness for the northern and southern hemispheres (Fig. 2).

Water column stratification and Mediterranean plankton

Water column dynamics are well characterized for the Mediterranean Sea (e.g., Bethoux, 1989). The seasonal stratification of the water column begins in the spring and ends with mixis in autumn; the cool Mediterranean deep water is formed at the surface in winter. Superimposed on the seasonal changes in water column structure throughout the Mediterranean is a general west to east gradient of declining nutrient and chlorophyll concentrations as well as primary production. During the stratified period, there is also a marked gradient of increasing mixed layer depth from west to the east (e.g., Dolan et al., 1999). Thus, oligotrophy increases and when the surface mixed layer is present, from late spring to early autumn, the surface layer is larger and less productive going from the west to the east.

Comparing mixed and stratified conditions, for the foraminifera species richness appears higher in the stratified period comparing estimates across the Mediterranean for February and September (Fig. 3). However, the differences are neither large nor consistent and diversity appears about the same in the autumn compared to the winter. Thus, on a seasonal time scale, changes in water column structure do not appear to be directly relatable to changes in the diversity of planktonic foraminifera.

Now consider temporal changes at a single site. Based on monthly sampling at a coastal station, species richness of two trophically distinct populations, tintinnids and *Ceratium* of the phytoplankton, show similarities (Fig. 4). For both groups, species richness was highest during the unstratified winter period and lowest during the summer corresponding with the stratified period.

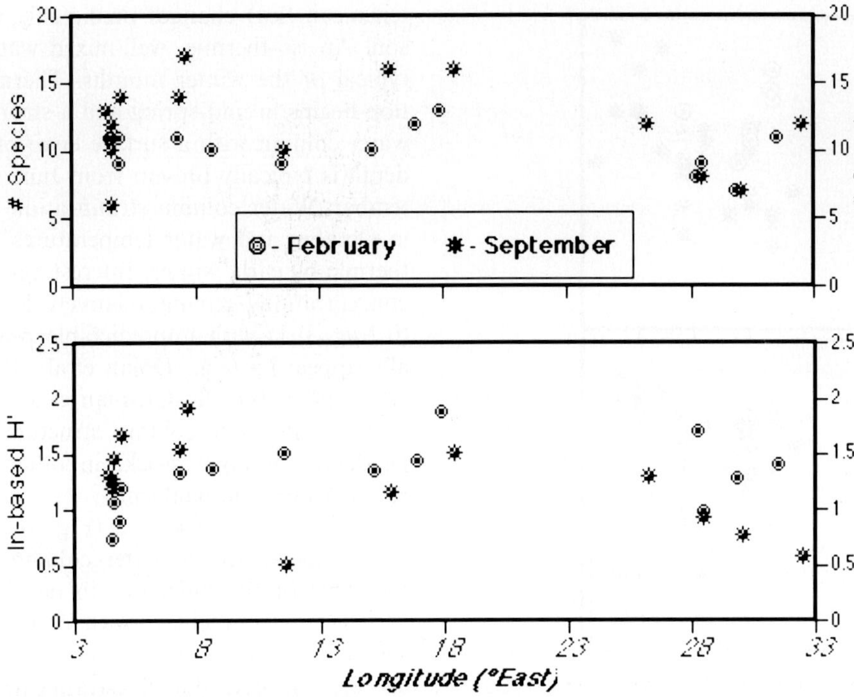

Figure 3. Estimates of the diversity of foraminifera across the Mediterrean Sea in February when the water column is mixed and September when the water colmumn is stratified based on data presented by Pujol & Vergnaud Granzzini (1995). Note that species richness was generally higher in the September compared to February samples. However, this was not true of H′ values.

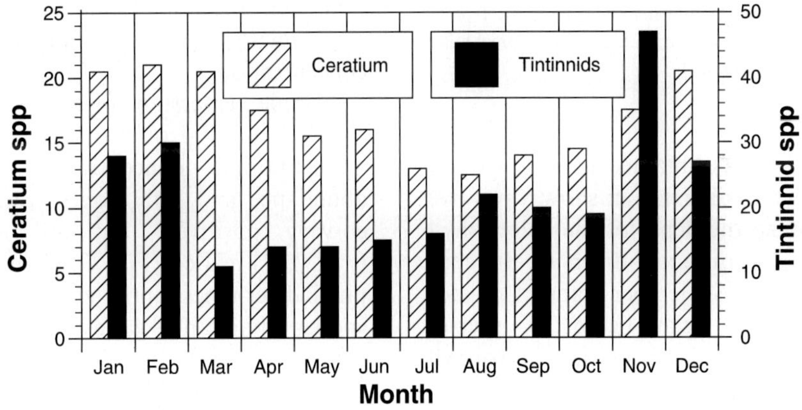

Figure 4. Species richness data by month for the phytoplankton genus Ceratium and tintinnids for waters off San Remo, N. W. Mediterrean Sea, based on a report by Rampi (1948). Note that species richness appears lowest during the summer and highest in the winter for the two trophically distinct groups of organisms.

Based on the admittedly weak evidence of monthly sampling, diversity as species richness (in two distinct groups) appears lowest in the structured, relatively stable water column of the summer.

A separate, and the last question with regard to the effect of structure of the water column, is that of comparing the characteristics of water columns structured to different degrees and at different times. A comparison of species richness and diversity (H′) as a function of the depth of the chlorophyll maximum layer at the beginning and end of the stratified period shows very different

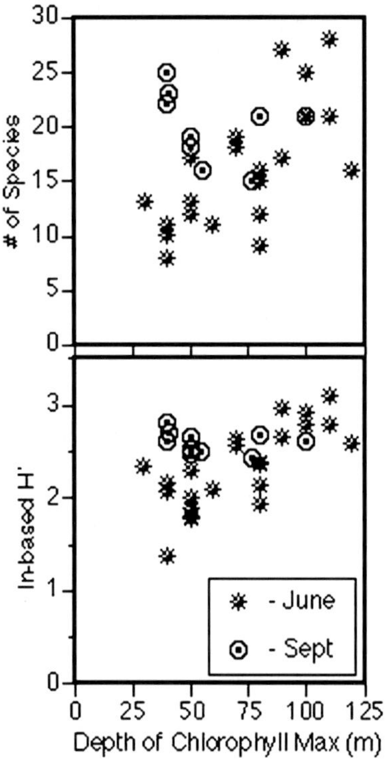

Figure 5. The relationship between species richness (top panel) or diversity as H′ (bottom panel) and the depth of the chlorophyll maximum layer in June, early in the season of the stratified water column, and in September, near the end of water colulmn stratification in the Mediterrean Sea. Data are from a variety of sites across the Mediterrean Sea taken from Dolan (2000) and Dolan et al., (2002).

trends (Fig. 5). Samples obtained in June from a transect across the Mediterranean show a positive relationship between the depth of the chlorophyll maximum layer and tintinnid diversity. In contrast, samples from a September transect do not show any particular relationship of species richness or H′ with the depth of the chlorophyll maximum layer. However, tintinnid diversity (both taxonomic and morphological) was relatable to the size structure of the phytoplankton community. Diversity of resources, roughly considered as size-classes of chlorophyll, was correlated with both morphological and species diversity of tintinnids (Dolan et al., 2002).

Temporal trends of tintinnid diversity

Water column stratification at "Point B" at the entrance to the Bay of Villefranche (N.W. Mediterranean Sea) changes predictably with the season. An iso-thermic, well-mixed water column is typical of the winter months. Thermal stratification begins in mid-spring and a strongly stratified water column with a surface layer of about 10 m depth is typically present from June through September. Water column stratification breaks down in October and water temperatures are near isothermic by early winter. Interestingly, chlorophyll concentrations remain relatively low year-round (0.2 μg 1–1) with unpredictable peaks occasionally appearing (e.g., Dolan et al., 1995; Mostajir et al., 1995; Bustillo-Guzman et al., 1995). Thus, at Point B water column structure varies independently of algal stock, in other terms, food resources for zooplankton.

Data gathered in 2002 (Fig. 6) clearly shows the independence of water column structure, in the form of the difference in density between 0 and 75 m, from chlorophyll concentration. Casual inspection of the graphs showing temporal changes in chlorophyll, stratification, tintinnid diversity and concentration reveals that water column stratification is unrelated to tintinnid concentrations and diversity. Chlorophyll trends however parallel those of tintinnid concentration, and species abundance appears more closely related to tintinnid than chlorophyll concentration.

Simple regression analysis confirms a significant relationship (r = 0.53) between chlorophyll and tintinnid concentration (Fig. 7). Species abundance and total concentration of tintinnids co-vary positively but are not significantly (r = 0.23) related (Fig. 7). Given that (1) chlorophyll determines tintinnid concentration and (2) species abundance varies with tintinnid concentrations, changes in chlorophyll should, through changing tintinnid concentrations, influence diversity. This can be examined by plotting weekly changes in chlorophyll against weekly changes in tintinnid diversity and this indeed shows that the two parameters are tightly related (r = 0.74). Thus, as a mechanism explaining temporal changes in diversity, chlorophyll or resources appears a likely mechanism, especially compared to water column structure. While we can not exclude other mechanisms, such as predation, the close relationship with changes in chlorophyll suggests a direct link.

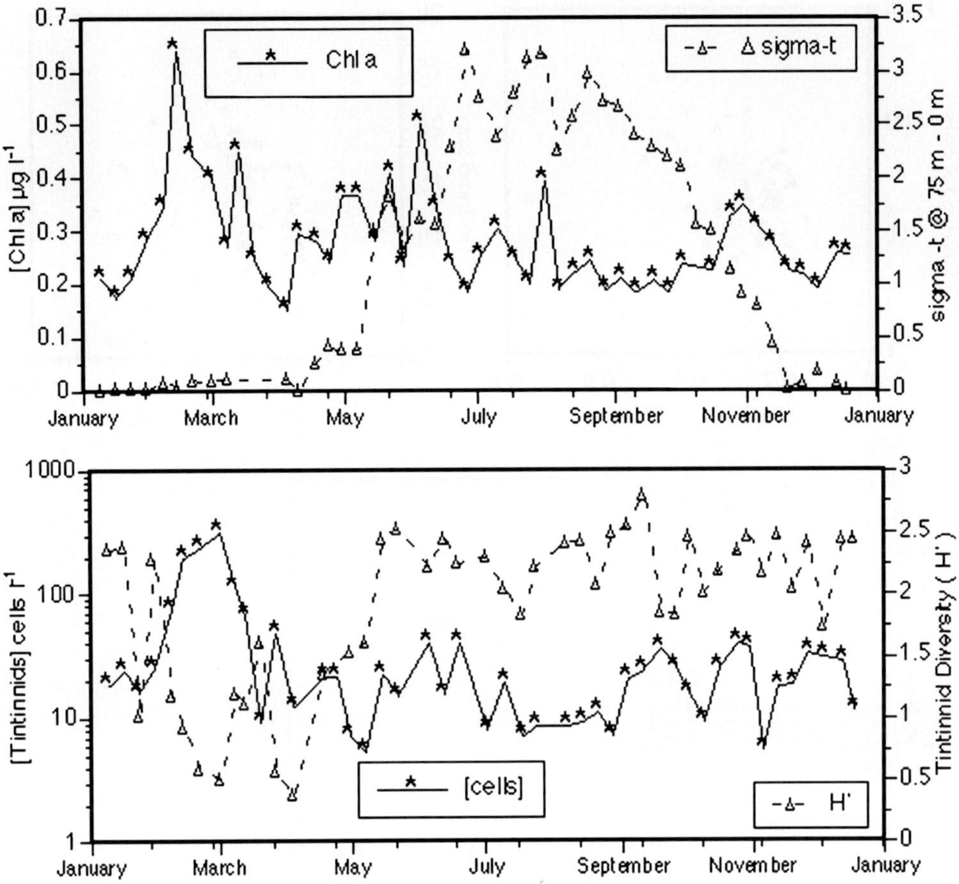

Figure 6. Temporal changes of water column characteristics and the tintinnid community at Point B, Bay of Villefranche in 2002, based on weekly sampling. Top panel shows changes in water column stratification and average chlorophyll concentration. Bottom panel shows species diversity and tintinnid concentrations.

Discussion

Global distribution of tintinnid species abundance

The pattern we found for tintinnid ciliates (species richness increases from the poles to lower latitudes with a peak around 20°–30° followed by a slight decline toward the equator) has been described for a very large variety of marine pelagic taxa ranging from tuna to foraminifera (Angel, 1993; Rutherford et al., 1999; Worm et al., 2003). Latitudinal diversity gradients are, of course, very well known among terrestrial taxa as well. Identification of factors underlying latitudinal diversity gradients has long been, and remains, a very active area of research (e.g., Gaston, 2000). The apparently common peaks of species richness at about 25° N or 25° S, for example recently described for tintinnids through the Indian Ocean (Modigh et al., 2003), are intriguing. However, here we are not concerned with determining the underlying cause(s) of a latitudinal diversity gradient among tintinnid ciliates but rather simply demonstrating its existence.

The existence of a latitudinal gradient of species richness in tintinnids is of some significance because in other groups of ciliates, specifically benthic ciliates, global and local species richness are thought to be equal; that is all species are cosmopolitan and present in all locales, albeit in perhaps very difficult to detect concentrations (Fenchel et al., 1997; Finlay et al., 1998, 1999; Finlay, 2002). This conclusion has been drawn from the results of intensive examination and treatment (incubation of material to provoke excystment of

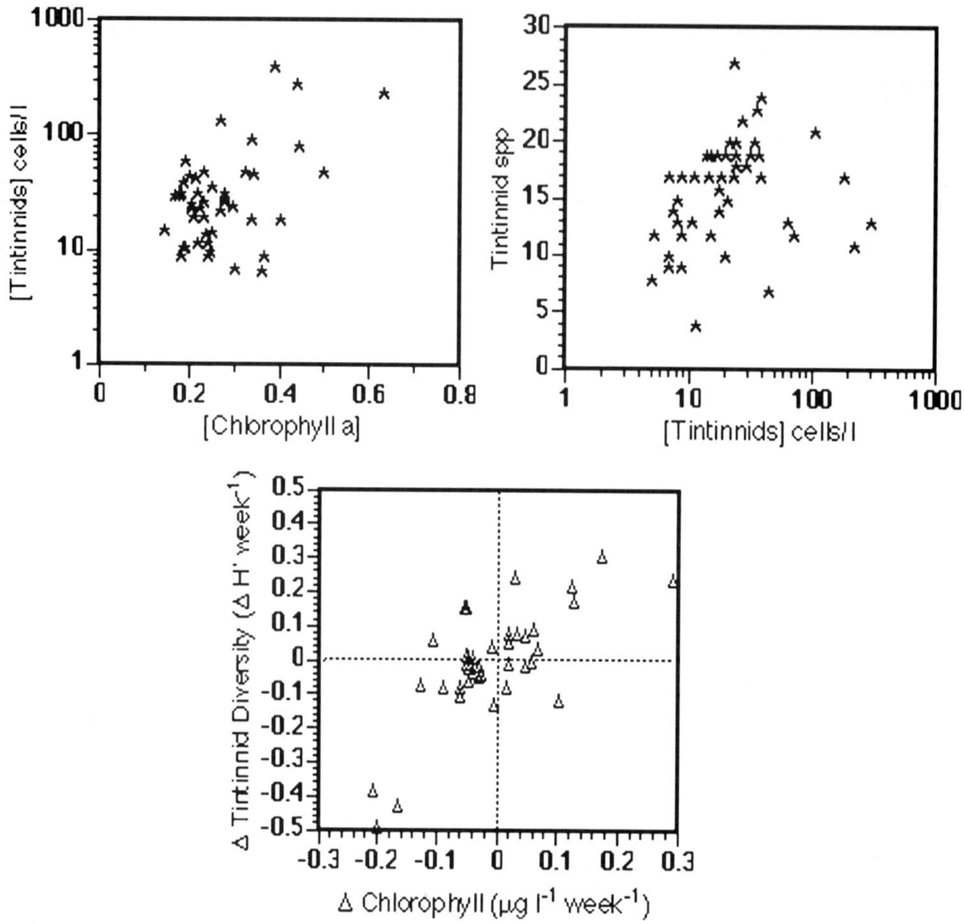

Figure 7. Scatter plots of the relationships between chlorophyll and tintinnid concentrations, tintinnid concentrations and species abundance, weekly changes in chlorophyll and tintinnid diversity as ln-based H′ Shannon values. Chlorophyll and tintinnid concentrations are significantly related ($r = 0.53$, $n = 45$): but tintinnid species richness is not related to tintinnid concentration. Weekly changes in chlorophyll are significantly related to weekly shifts in tintinnid diversity values ($r = 0.74$, $n = 40$).

cysts, etc.) of samples yielding large and similar assemblages, regardless of sample origin. Thus, a latitudinal gradient in the species richness of benthic ciliates should not exist.

The mechanism behind ubiquitous distribution of species is ubiquitous dispersal which effectively prevents endemism (Finlay & Esteban, 2004). Evidence at present can be found to support both global dispersal as well as endemism among marine microbes. For protists, there is evidence of wide dispersal as well as genetic isolation . For example, some taxa of foraminifera appear to experience genetic exchange between arctic and antarctic populations (Darling et al., 2000). In other species (defined morphologically), the quite similar morphologies mask genetic divergence between arctic and antarctic populations (Darling et al., 2004). With regard to prokaryotes, the pattern of 'all species are everywhere' should presumably exist yet there is ample evidence of the existence of endemism among some taxa of free-living prokaryotes (Hedlund & Staley, 2004). One may conclude then, that while many microbes are everywhere not all are.

With regard to tintinnids, while many species are cosmopolitan, Figure 2 shows that global and local diversity do seem very different. Furthermore, tintinnids show a pattern of latitudinal diversity common among pelagic marine taxa. Confirmation of latitudinal changes in species richness of tintinnid communities was recently reported in a study of tintinnids between 29° South and 60° South in the southwestern Atlantic in which latitude was found to be associated with about 50% of the variability in diversity (Thompson, 2004). Nonetheless, it must be admitted that an apparent latitudinal gradient of species abundance does not disprove a cosmopolitan distribution of all species. A gradient in species abundance may simply reflect a gradient in the abundance of more or less distinct niches permitting the development and co-existence of more species.

What defines these niches is unknown. For example, it has been claimed that zooplankton diversity is not related to phytoplankton diversity but rather to phytoplankton biomass (Irigoien et al., 2004). However, these conclusions have been challenged (Dolan, 2005). Thus, even the relationship between zooplankton and phytoplankton diversity is unclear.

Water column stratification and Mediterranean plankton

In the Atlantic, the depth of the mixed layer has been proposed as a correlate of foraminferan species richness (Rutherford et al., 1999). Our examination of water column structure as an influence (direct or indirect) on diversity in Mediterranean plankton failed to provide evidence of a major rôle. Firstly, for Mediterranean populations of foraminifera, there appears to be little difference between species richness in February when the water column is mixed compared to September with a well-defined surface layer (Fig. 3). Secondly, historical reports based on monthly sampling for dinoflagellates (Ceratium) and tintinnids actually suggest the opposite. In coastal waters of the N. W. Mediterranean, species richness appears highest during late autumn and winter when the water column is subject to the most turbulent mixing (Fig. 4). Thirdly, data from transects across the Mediterranean Sea in June compared to September, showed distinct trends in tintinnid diversity despite similarities in water column structure (Fig. 5). Perhaps then, not unexpectedly, planktonic diversity does not show a simple relationship with water column structure.

Temporal trends of tintinnid diversity

The overall lack of a relationship between stratification of the water column and diversity was also seen in the temporal series from the Rade de Villefranche. At 'Point B' stratification follows regular seasonal trends while algal stock varies irregularly (Fig. 6). We found that diversity in tintinnids was loosely related to tintinnid abundance which was, in turn, loosely related to algal stock, measured as chlorophyll concentration. These relationships suggested that changes in chlorophyll (food resources) may be relatable to changes in diversity in tintinnids. Indeed, we found that the weekly changes in chlorophyll concentration were tightly correlated with shifts in tintinnid diversity (Fig. 7). Thus, as a mechanism controlling temporal changes in diversity, chlorophyll or resources appears a likely mechanism, especially compared to water column structure. While we can not exclude other mechanisms, such as predation, the close relationship with changes in chlorophyll suggests a direct link.

We should point out that the Rade de Villefranche may be an unusually dynamic system. For each season, the assemblage is generally dominated by 1–3 species representing about 50% of the population and the identity of the seasonally dominate species has apparently varied comparing reports based on samples taken in the 1950s (Balech, 1959) and the 1970s (Rassoulzadegan, 1979). This is in contrast to the Bay of Naples in which species assemblages appear consistent since the 1930s despite considerable changes in nutrient input into the bay (Modigh & Castalado, 2002). Thus, one might conclude that the Bay of Villefranche is an unusually dynamic system. However, the cycle of seasonal abundance in the bay (distinct peaks in late winter and late autumn separated by a summer minimum) appears common to the Western Mediterranean as it has been found in the Bay of Algers (Vitello, 1964) and the Gulf of Marseille (Travers, 1973). Large interannual (1952–1959) differences in the identity of the dominant tintinnid species have been found for waters off Blanes, Spain (Margalel, 1957, Margalef & Morales, 1960). Thus, the relationships described for the Bay of Villefranche may be common. These relationships suggest that diversity appears more easily relatable to shifts in resources than the physical structure of the environment.

Diversity issues have taken center stage in many areas of biology. Fundamental points of view such as 'diversity is a characteristic of ecosystem function' vs. 'diversity governs ecosystem function' are currently subjects of debate (Naeem, 2002). With regard to marine systems, a few years ago it was remarked that our understanding of marine pelagic biodiversity appears vague (Smetacek, 1996). This is probably due in part to the fact that different mechanisms proposed as important in maintaining diversity in pelagic populations are often closely related. Thus, a structured water column coincides with diverse and stable communities of phytoplankton, herbivorous copepods and predacious copepods (McGowan & Walker, 1980, 1985; Venrick, 1990, 1999).

Here we have attempted to sort out possible mechanisms using field survey data established on different spatial and temporal scales. Tintinnid ciliates appear to be a reasonable model of marine pelagic organisms as their latitudinal diversity gradient pattern is similar to those found in a large variety of taxa. Considering a variety of data from the Mediterranean Sea, the seasonal appearance of a structured water column does not appear to be positively related to planktonic diversity. We showed that temporal changes in tintinnid diversity appeared unrelated to water column structure but could be linked to changes in resources. Thus, overall, resources appear to directly influence diversity.

Acknowledgements

We are grateful to the organizers of the 38th EMBS, H. Queiroga, M. Ribeiro da Cunha, M.-H. Moreira, V. Quintino, A.-M. Rodrigues, J. Serôdio, and M.-A. Cunha. A special thanks to H. Queiroga for an invitation to speak. Financial support for the work described here was provided by INSU of the CNRS through the project CNRS-INSU ATI #01N50/0388: "Effet du Mesozooplancton sur la Diversité des Compartiments Phyto et Microzooplanctonique d'une Zone Cotière Mediterraneene", the shared cost research project NTAP (contract no. EVK3-CT-2000-00022) of the EU RTD Programme "Environment and Sustainable Development" and forms part of the ELOISE projects cluster. This is ELOISE contribution no. 459/40 and a contribution of the PROOF program PROSOPE. Point B hydrological data were provided by the SOMLIT program (CNRS-INSU) and are available through "www.obs-vlfr.fr/rade". Comments of the reviewers led to substantial improvements in this paper.

References

Agatha, S., M. C. Strûder-Kypke, A. Beran & D. H. Lynn, 2005. Pelagostrobilidium neptuni (Montagnes and Taylor, 1994) and Strombidium biarmatum nov. spec. (Ciliophora, Oligotrichea): phylogenetic position inferred from morphology, ontogenesis, and gene sequence data. European Journal of Protistolology 41: 65–83.

Angel, M. V., 1993. Biodiversity of the Pelagic Ocean. Conservation Biology 7: 760–772.

Balech, E., 1959. Tintinnoinea del Mediterraneo. Trabajo Instituto Espanol de Oceanografia 28: 1–88.

Bethoux, J. P., 1989. Oxygen consumption, new production, vertical advection and environmental evolution in the Mediterranean Sea. Deep-Sea Research 36: 769–781.

de Barrio Cao, M. S., 1992. Abundance and species composition of Tintinnina (Ciliophora) in Bahia Blanca Estuary, Argentina. Estuarine Coastal Shelf Science 34: 295–303.

Bolotovskoy, D., S. M. Vivequin & N. R. Swanberg, 1991. Vertical distribution of tintinnids and associated microplankton in the upper layer of the Barents Sea. Sarsia 76: 141–151.

Buckling, A., R. Kassen, G. Bell & P. B. Rainey, 2000. Disturbance and diversity in experimental microcosms. Nature 408: 961–964.

Bustillo-Guzman, J., H. Claustre & J. -C. Marty, 1995. Specific phytoplankton signatures and their relationship with hydrographic conditions in the coastal northwestern Mediterranean Sea. Marine Ecology Progress Series 124: 247–258.

Campbell, A. S., 1942. The Oceanic Tintinnoina of the Plankton Gathered During the Last Cruise of the Carnegie. Carnegie Institution of Washington, Publication 537, Washington D.C.

Capriulo, G. M. & E. J. Carpenter, 1983. Abundance, species composition and feeding impact of tintinnid micro-zooplankton in central Long Island Sound. Marine Ecology Progress Series 10: 277–288.

Cariou, J.-B., J. R. Dolan & S. Dallot, 1999. A preliminary study of tintinnid diversity in the NW Mediterranean Sea. Journal of Plankton Research 21: 1065–1075.

Caron, D. A. & N. R. Swanberg, 1990. The ecology of planktonic sarcodines. Reviews in Aquatic Sciences 3: 147–180.

Caron, D. A., A. F. Michaels, N. R. Swanberg & F. A. Howes, 1995. Primary productivity by symbiont-bearing planktonic sarcodines (Acantharia, Radiolaria, Foraminifera) in surface waters near Bermuda. Journal of Plankton Research 17: 103–129.

Darling, K. F., C. M. Wade, I. A. Stewart, D. Kroon, R. Dingle & A. I. Brown, 2000. Molecular evidence for genetic mixing of artic and antarctic subpolar populations of planktonic foraminifers. Nature 405: 23–24.

Darling K. F., M. Kucera, C. J. Pudsey & C. M. Wade, 2004. Molecular evidence links cryptic diversification in ploar planktonic protists to Quaternary climate dynamics. Proceeding of the National Academy of Sciences USA 101: 7657–7662.

Darwin, C., 1859. On the Origin of Species by Means of Natural Selection. Facsimile edition. 1964. Harvard University Press, Boston, USA.

Dodson, S. I., S. E. Arnott & K. L. Cottingham, 2000. The relationship in lake communities between primary productivity and species richness. Ecology 81: 2662–2679.

De Vargas, C., R. Norris, L. Zaninetti, S. W. Gibb, & J. Pawlowski, 1999. Molecular evidence of cryptic speciation in planktonic foraminifera and their relation to ocean provinces. Proceedings of the National Academy of Sciences USA 96: 2864–2868.

De Vargas, C., M. Bonzon, N. W. Tess, J. Pawlowski & L. Zaninetti, 2002. A molecular approach to biodiversity and biogeography in the planktonic foraminifera *Globigerinella siphonifera* (d'Orbigny). Marine Micropaleontology 45: 101–116.

Dolan, J. R., 2000. Tintinnid ciliate diversity in the Mediterranean Sea: longitudinal patterns related to water column structure in late spring-early summer. Aquatic Microbial Ecology 22: 20–30.

Dolan, J. R., 2005. Marine ecology – different measures of biodiversity. Nature 433: E9 (Jan 27 2005).

Dolan, J. R & C. L. Gallegos, 2001. Estuarine diversity of tintinnids (planktonic ciliates). Journal of Plankton Research 23: 1009–1027.

Dolan, J. R. & C. Marassé, 1995. Planktonic ciliate distribution relative to a deep chlorophyll maximum: Catalan Sea, N. W. Mediterrean, June 1993. Deep-Sea Research 42: 1965–1987.

Dolan, J. R., T. F. Thingstad, & F. Rassoulzadegan, 1995. Phosphate transfer between microbial size-fractions in Villefranche Bay (N. W. Mediterranean Sea) France in autumn 1992.

Dolan, J. R., F. Vidussi & H. Claustre, 1999. Planktonic ciliates in the Mediterranean Sea: longitudinal trends. Deep-Sea Research I 46: 2025–2039.

Dolan, J. R., H. Claustre, F. Carlotti, S. Plounevez & T. Moutin, 2002. Microzooplankton diversity: relationships of tintinnid ciliates with resources, competitors and predators from the Atlantic coast of Morocco to the eastern Mediterranean. Deep-Sea Research 1(49): 1217–1232.

Emmerson, M. C., M. Solan, C. Emes, D. M. Paterson & D. Raffaell, 2001. Consistent patterns and the idiosyncratic effcts of biodiversity in marine ecosystems. Nature 411: 73–77.

Fenchel, T., G. F. Esteban & B. Finlay, 1997. Local vs. global diversity of microorganisms: cryptic diversity of ciliated protozoa. Oikos 80: 220–225.

Finlay, B. J., 2002. Global dispersal of free-living microbial eukaryote species. Science 296: 1061–1063.

Finlay, B. J. & G. F. Esteban, 2004. Ubiquitous dispersal of free-living microorganisms. In Bull, A. T. (ed.) Microbial Diversity and Bioprocessing. ASM Press, Washington D.C., USA: 216–224.

Finlay, B. J., G. F. Esteban & T. Fenchel, 1998. Protozoan diversity: converging estimates of the global number of free-living ciliate species. Protist 149: 29–37.

Finlay, B. J., G. F. Esteban, J. L. Olmo & P. A. Tyler, 1999. Global distribution of free-living microbial species. Ecography 22: 138–144.

Garder, K. R., 1946. Tintinnoinea from the "Michael Sars" North Atlantic Deep-Sea Expedition, 1910. In Murray, J. & J. Hjort, Report on the Scientific Results of the "Michael Sars" North Atlantic Deep-Sea Expedition, 1910. Bergen Musuem, Bergen, Norway.

Gaston, K. J., 2000. Global patterns in biodiversity. Nature 405: 220–227.

Gold, K. & E. A. Morales, 1975. Seasonal changes in lorica sizes and the species of Tinntinida in the New York Bight. Journal of Protozoology 22: 520–528.

Gold, K. & E. A. Morales, 1977. Studies on the Tintinnida of Enewetak Atoll. Journal of Protozoology 24: 580–587.

Hargraves, P. E., 1981. Seasonal variations of tintinnids (Ciliophora: Oligotrichida) in Narragansett Bay, Rhode Island. Journal of Plankton Research 3: 81–91.

Hedlin, B. P. & J. T. Staley, 2004. Microbial endemism and biogeography. In Bull, A. T. (ed.) Microbial Diversity and Bioprocessing. ASM Press, Washington D.C, pp. 225–231.

Hedin, H., 1974. Tintinnids on the swedish west coast. Zoon 2: 123–133.

Huisman, J. & J. F. Weissing, 1999. Biodiversity of plankton by species oscillations and chaos. Nature 402: 407–410.

Hulot, F. D., G. Lacroix, F. Lescher-Moutoué & M. Loreau, 2000. Functional diversity governs ecosystem response to nutrient enrichment. Nature 405: 340–344.

Hutchinson, G. E., 1961. The paradox of the plankton. American Naturalist 95: 137–145.

Irigoien, X., J. Huisman & R. P. Harris 2004. Global biodiversity patterns of marine phytoplankton and zooplankton. Nature 429: 863–867.

James, M. R. & J. A. Hall, 1995. Planktonic ciliated protozoa: their distribution and relationship to environmental variables in a marine coastal ecosystem. Journal of Plankton Research 17: 659–683.

Jensen, F. & B. W. Hansen, 2000. Ciliates and heterotrophic dinoflagellates in the marginal ice zone of the central Barents Sea during spring. Journal of the Marine Biological Association United Kingdom 18: 45–54.

Jörgensen, E., 1924. Mediterranean Tintinnidae. Report of the Danish Oceanographical Expeditions 1908–1910 to the Mediterranean and adjacent seas. Vol. II. Biology, No. 8, J.3 (Thor expedition), Copenhagen.

Kassen, R., A. Buckling, G. Bell & P. B. Rainey, 2000. Diversity peaks at intermediate productivity in a laboratory microcosm. Nature 406: 508–512.

Kofoid, C. A. & A. S. Campbell, 1929. A conspectus of the marine and fresh-water ciliate belonging to the suborder tintinnoinea, with descriptions of new species principally from the Agassiz expedition to the eastern tropical Pacific 1904–1905. University of California Publications in Zoology 34: 1–403.

Kofoid, C. A. & A. S. Campbell, 1939. The tintinnoinea of the eastern tropical Pacific. Bulletin of the Museum of Comparative Zoology of Harvard College 84: 1–473.

Longhurst, A. R., 1985. Relationship between diversity and the vertical structure of the upper ocean. Deep-Sea Research 1(32): 1535–1570.

Loreau, M. & A. Hector, 2001. Partitioning selection and complementarity in biodiversity experiments. Nature 412: 72–76.

Lynn, D. H. & E. B. Small, 1997. A revised classification of the phylum Ciliophora Doflein, 1901. Revista de la Sociadad de la Historia Naturale de Mexico 47: 65–78.

Margalel, R., 1957. Fitoplancton de las costas de Blanes (Gerona) de agosto de 1952 a junio de 1956. Investigaciones Pesqueras 8: 89–95.

Margalef, R. & E. Morales, 1960. Fitoplancton de las costas de Blanes (Gerona) de julio de 1956 a junio de 1959. Investigaciones Pesqueras 16: 3–31.

McGowan, J. A. & P. W. Walker, 1980. Structure in the copepod community of the North Pacific Central grye. Ecological Monographs 49: 195–226.

McGowan, J. A. & P. W. Walker, 1985. Dominance and diversity maintenance in an oceanic ecosystem. Ecological Monographs 55: 103–118.

Middlebrook, K., C. W. Emerson, J. C. Roff & D. H. Lynn, 1987. Distribution and abundance of tintinnids in the Quoddy region of the Bay of Fundy. Canadian Journal of Zoology 65: 594–601.

Modigh, M. & S. Castaldo, 2002. Variability and persistence in tintinnid assemblages in a Mediterranean coastal site. Aquatic Microbial Ecology 28: 299–311.

Modigh, M., S. Castalado, M. Saggiomi & I. Santarpia, 2003. Distribution of tintinnid species from 42° N to 43° S through the Indian Ocean. Hydrobiologia 503: 251–262.

Monti, M. & S. Fonda Umani, 1995. Tintinnids in Terra Nova Bay – Ross Sea during two austral summers (1987/88 and 1989/90). Acta Protozoologica 34: 193–201.

Mostajir, B., J. R. Dolan & F. Rassoulzadegan, 1995. Seasonal variations of pico- and nano-detrital particles (DAPI Yellow Particles DYP) in the Ligurian Sea (NW Mediterranean). Aquatic Microbial Ecology 9: 267–277.

Naeem, S., 2002. Ecosystem consequences of biodiversity loss: the evolution of a paradigm. Ecology 83: 1537–1552.

Naeem, S. & S. Li, 1998. Consumer species richness and autrophic biomass. Ecology 79: 2603–2615.

Naeem, S., D. R. Hahn & G. Schuurman, 2000. Producer-decomposer co-dependency influences biodiversity effects. Nature 403: 762–764.

Ottens, J. J. & A. J. Nederbragt, 1992. Planktic foraminferal diversity as an indicator of ocean environment. Marine Micropaleontology 19: 13–28.

Paranjape, M., 1987. Grazing by microzooplankton in the Eastern Canadian Arctic in summer 1983. Marine Ecology Progress Series 40: 239–246.

Petz, W. & W. Foissner, 1992. Morphology and morphogenesis of *Strobilidium caudatum* (Fromentel), *Meseres corlissi* N. Sp., *Halteria graninella* (Müller), and *Strombidium rehwaldi* N. Sp., and a proposed phylogenetic system for oligotrich ciliates (Protozoa, Ciliophora). Journal of Protozoology 39: 159–176.

Pierce, R. W. & J. T. Turner, 1994. Plankton studies in Buzzards Bay, Massachusetts, USA. IV. Tintinnids, 1987 to 1988. Marine Ecology Progress Series 112: 235–240.

Pujol C. & C. Vergnaud Grazzini, 1995. Distribution patterns of live planktic foraminifers as related to regional hydrography and productive systems of the Mediterranean Sea. Marine Micropaleontology 25: 187–217.

Rassoulzadegan, F., 1979. Evolution annuelle des ciliés pélagiques en Méditerranée nord-occidentale. II. Cilés oligotriches. Tintinnides (Tintinnia). Investigaciones Pesqueras 43: 417–488.

Rampi, L., 1948. Tinttinnoidi delle acque di San Remo. Bollettino di Pesca, di Pisciocoltura e di Idrobiologia 24: 5–11.

Rogers, G. F., J. C. Roff & D. H. Lynn, 1981. Tintinnids of Chesterfield Inlet, Northwest Territories. Canadian Journal of Zoology 59: 2360–2364.

Rutherford, S., S. D'Hondt & W. Prell, 1999. Environmental controls on the geographic distribution of zooplankton diversity. Nature 400: 749–752.

Sanders, R. W., 1987. Tintinnids and other microzooplankton – seasonal distributions and relationships to resources and hydrography in a Maine estuary. Journal of Plankton Research 9: 65–77.

Smetacek, V., 1996. Biodiversity and production in the water mass. In Hempel, G. (ed.) The Ocean and the Poles. Gustav Verlag, , Jena: 207–216.

Snoeyenbos-West, O. L. O., T. Salcedo, G. B. McManus & L. A. Katz, 2002. Insights into the diversity of choreotrich, oligotrich ciliates (Class: Spirotrichea) based on genealogical analyses of multiple loci. International Journal of Systematic and Evolutionary Microbiology 52: 1901–1913.

Silva, E. S., 1953. Estudos de plancton na Lagoa de Obidos II. Tintinnoinea. Revista de Faculdadae de Ciencias de Lisboa 2a Serie C 11: 97–116.

Stoecker, D. K. & J. M. Capuzzo, 1990. Predation on protozoa: its importance to zooplankton. Journal of Plankton Research 12: 891–908.

Swanberg, N. R. & D. A. Caron, 1991. Patterns of sarcodine feeding in epipelagic oceanic plankton. Journal of Plankton Research 13: 287–312.

Taguchi, S., 1976. Microzooplankton and seston in Akkeshi Bay, Japan. Hydrobiologia 50: 195–204.

Thompson, G. A., 2004. Tintinnid diversity trends in the southwestern Atlantic Ocean (29–60°S). Aquatic Microbial Ecology 35: 93–130.

Thompson, G. A., V. A. Alder, D. Boltovskoy & F. Brandini, 1999. Abundance and biogeography of tintinnids (Ciliophora) and associated microzooplankton in the Southwestern Atlantic Ocean. Journal of Plankton Research 21: 1265–1298.

Thompson, G. A., V. A. Alder & D. Boltovskoy, 2001. Tintinnids (ciliophora) and other net microzooplankton (>30 μm) in southw-western Atlantic Shelf break waters. P.S.Z.N. Marine Ecology 22: 343–355.

Travers, M., 1973. Le microplancton du Golfe de Marseille: variations de la composition systématique et de la densité des populations. Tethys 5: 31–53.

Turner, J. T., 1984. The Feeding Ecology of Some Zooplankters that are Important Prey Items of Larval Fish. NOAA Technical Report NMFS 7. NOAA, Washington D.C 28 pp.

Venrick, E. L., 1990. Phytoplankton in an oligotrophic ocean: species structure and interannual variability. Ecology 71: 1547–1563.

Venrick, E. L., 1999. Phytoplankton species structure in the central North Pacific, 1973–1996: variability and persistence. Journal of Plankton Research 21: 1029–1042.

Vitiello, P., 1964. Contribution à l'étude des tintinnides de la baie d'Alger. Pelagos 2: 5–41.

Wasik, A. & E. Mikolajczyk, 1990. Tintinnids near pack-ice between South Shetland and the South Orkney Islands (26 Dec. 1988 – 18 Jan. 1989). Acta Protozoologica 29: 229–244.

Wilf, P., K. R. Johnson & B. T. Huber, 2003. Correlated terrestrial and marine evidence for global climate changes before mass extinction at the Cretaceous-Paleogene boundary. Proceedings of the National Academy of Sciences, USA 100: 599–604.

Williams, D. F. & W. C. Johnson II, 1975. Diversity of recent planktonic foraminifera in the southern Indian Ocean and late Pleistocene paleotemperatures. Quaternary Research 5: 237–250.

Worm, B., H. K. Lotze, H. Hillebrand & U. Sommer, 2002. Consumer versus resource control of species diversity and ecosystem functioning. Nature 417: 848–851.

Worm, B., H. K. Lotze & R. A. Myers, 2003. Predator diversity hotspots in the blue ocean. Proceedings of the National Academy of Sciences, USA 100: 9884–9888.

Hydrobiologia (2006) 555:159–174
H. Queiroga, M.R. Cunha, A. Cunha, M.H. Moreira, V. Quintino, A.M. Rodrigues, J. Serôdio & R.M. Warwick (eds),
Marine Biodiversity: Patterns and Processes, Assessment, Threats, Management and Conservation
DOI 10.1007/s10750-005-1113-5

Fish assemblages in different shallow water habitats of the Venice Lagoon

Anita Franco*, Piero Franzoi, Stefano Malavasi, Federico Riccato & Patrizia Torricelli
Department of Environmental Sciences, University of Venezia, Castello 2737/B, 30122, Venezia, Italy
*(*Author for correspondence: Tel.: +39-0412347739; Fax: +39-0415281494; E-mail: afranco@unive.it)*

Key words: fish assemblages, shallow waters, seagrass beds, marsh creeks, Mediterranean, Venice Lagoon

Abstract

The small-sized fish assemblages of the Venice Lagoon were investigated and compared among five shallow subtidal habitats (seagrass beds, sparsely vegetated habitats, unvegetated sand bottoms, mudflats and saltmarsh creeks) in the Northern lagoon basin. Sampling was carried out seasonally (Spring, Summer and Autumn of 2002) in 4–7 stations for each habitat type, by means of a fine-mesh, small beach seine. Two-way analysis of variance was applied to assess the differences in species richness, fish diversity, density and standing stock amongst habitats, whereas fish assemblage composition was investigated by using multivariate analyses (MDS, ANOSIM, SIMPER). The analyses indicated that seagrass beds and saltmarsh creeks are relevant shallow habitats in structuring the small-sized fish assemblages of the Venice Lagoon, supporting specialized and recognizable fish assemblages. Those in seagrass beds, in particular, were characterized by higher species richness and standing stock with respect to all the others. The structuring role of these habitats was discussed in terms of both habitat complexity and degree of confinement. In contrast, sandy bottoms, mudflats and sparsely vegetated habitats were identified as "transition" habitats, with highly variable fish assemblages, influenced by the contribution of the adjacent habitats, and acting probably as both 'buffer zones' between the other habitats and migration routes for many fish species in the lagoon.

Introduction

Estuaries and coastal lagoons show high levels of habitat heterogeneity and support a large fish production (Elliott & Hemingway, 2002). Estuarine fish assemblages change significantly in relation to habitat characteristics, in particular to the degree of habitat complexity, often due to the level of vegetation coverage (macrophytes or macroalgae), as shown by many studies carried out along the Australian, South African and American coasts (Orth & Heck, 1980; Heck et al., 1989; Sogard & Able, 1991; Connolly, 1994; West & King, 1996; Jenkins & Wheatley, 1998; Paterson & Whitfield, 2000). These studies indicate that vegetated habitats support relatively higher fish diversity, abundance and biomass than the unvegetated ones.

Many studies have been also conducted on fish assemblages of European estuaries, concerning mainly estuarine and lagoonal environments from the Atlantic and Baltic areas (Muus, 1967; Arruda et al., 1988; Nash, 1988; Elliott & Dewailly, 1995; Marshall & Elliott, 1998; Mathieson et al., 2000; Gordo & Cabral, 2001; Elliott & Hemingway, 2002; Lobry et al., 2003) and from the Western Mediterranean Sea (Huvé et al., 1973; Hervé & Buslé, 1980; Amanieu & Lasserre, 1982; Quignard, 1984; Bouchereau et al., 2000; Mariani, 2001; Pérez-Ruzafa et al., 2004; Poizat et al., 2004). However, few papers have been published on estuarine fish assemblages from the Adriatic Sea (Gandolfi et al., 1985; Franzoi et al., 1989, 1992).

The present study investigates the fish assemblages of the Venice Lagoon (North Adriatic Sea), the largest in the Mediterranean basin. This lagoon is characterized by high levels of environmental heterogeneity resulting in several types of habitat suitable for fish species, such as seagrass meadows, bare sand areas, intertidal flats, sandy and muddy subtidal beds, and tidal marshes. Several gradients in the physico-chemical parameters can be detected in the Venice Lagoon, such as the decrease in the sediment grain size from the sea-inlets into the internal salt marsh areas (Sacchi, 1985). The biological communities are therefore expected to be regulated by the confinement gradient, as suggested by Guelorget & Perthuisot (1983). The main aim of the present work is to compare the fish assemblages inhabiting the shallow waters of the Northern basin of the Venice Lagoon, in terms of taxonomic composition, species richness, fish diversity, fish density and standing stock, among some of these habitats, using a combination of multivariate and univariate methods.

Materials and methods

Study site

The Venice Lagoon covers an area of about 540 km^2. It is mostly composed of shallow water areas traversed by shipping canals of different sizes and is connected to the sea by three wide inlets. Due to the shallow water nature of the lagoon and to the tidal range of up to 1 m, which is the largest in the Mediterranean basin, tides flowing through the inlets strongly influence the Venice Lagoon environment. From a hydrological point of view, the lagoon is divided into three main basins: Northern, Central and Southern (Avanzi et al., 1979; Fig. 1). This system is subjected to strong temporal and spatial changes with regard to its morphological and physico-chemical parameters (Sacchi, 1985), which determine high levels of spatial heterogeneity.

Five main types of habitat can be identified in the subtidal shallow water areas (maximum depth of 1.2–1.5 m during high tide) of the Northern basin: (1) seagrass beds (Sb), which are represented by both homospecific and heterospecific meadows composed of *Zostera marina* Linnaeus, *Nanozostera noltii* Honermann (Tomlinson & Posluzny) and *Cymodocea nodosa* (Ucria) Ascherson; (2) sparsely vegetated habitats (sV), with patches of seagrasses covering less than 30% of the sampling area; (3) unvegetated (i.e. without seagrass) sand habitats (Us); (4) unvegetated mudflats (Um); (5) unvegetated saltmarsh creeks (Uc). Seasonally, some presence of macroalgae (mainly *Ulva rigida* C. Agardh) can be detected in these habitats, but their presence is occasional, as they have almost completely disappeared from the lagoon as a result of a declining trend that began in the 1990s (Sfriso et al., 2003). The vegetated and sand habitats are closer to the sea inlets and deeper canals connecting the lagoon to the sea, whereas the other unvegetated habitats (mudflats and saltmarsh creeks) are located farther from the sea inlets, in the internal areas of the lagoon (see Fig. 1), where bottom sediments are finer.

Field sampling

Four to seven replicates were collected in each habitat (stations) on a seasonal basis (Spring, Summer and Autumn), from March to December of 2002, giving a total number of 73 replicates. Fishes were sampled during daylight hours with a seine net of 10-m length, which had a 2 m drop and a knot-to-knot mesh size of 2 mm. Fish were sorted and identified by species *in situ*, and then they were rapidly counted and weighted (± 0.1 g) before they were released in the water. When *in situ* identification was difficult, fish were anaesthetized with MS222 diluted in water, preserved in 10% formalin solution and brought to the laboratory for identification. Three tows (each of them about 15 m in length and 8 m in width) were conducted in each station, so that a total area of about 360 m^2 was sampled.

Data analysis

Fish density and standing stock were calculated by dividing the total number and biomass of each species by the area sampled (360 m^2). Species richness was measured by the total number of species caught in each station, and the

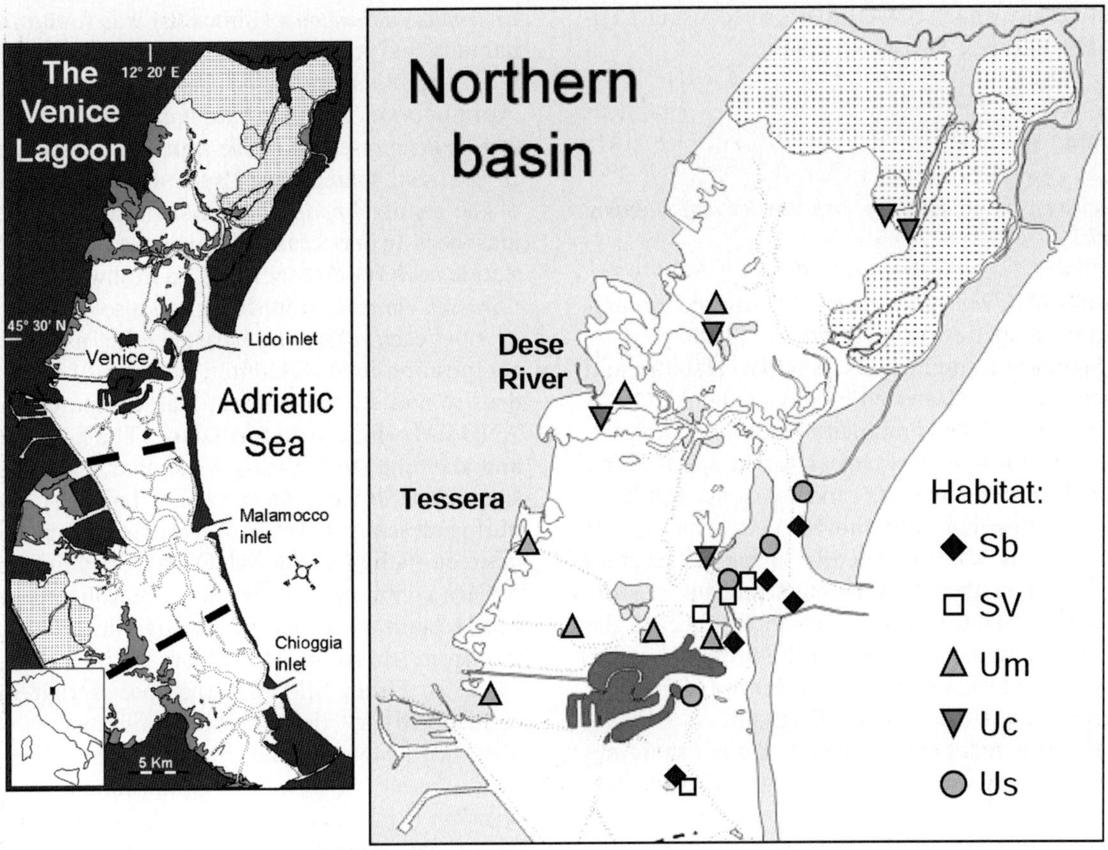

Figure 1. Sampling stations in the Northern basin of the Venice Lagoon, divided by type of habitat (Sb = seagrass bed; sV = sparsely vegetated habitat; Um = unvegetated mudflat; Uc = unvegetated saltmarsh creek; Us = unvegetated sand habitat). Hatched lines in the map of the whole lagoon represent the hydrological boundaries dividing the lagoon in Northern (upper), Central and Southern (lower) basins.

Shannon–Wiener diversity index was calculated (Krebs, 1999).

Two-way analysis of variance (ANOVA) was used to assess habitat differences in total fish density, standing stock, species richness, and fish diversity. Habitat and season were considered to be fixed factors. Fish density and standing stock data were transformed to $\log(x+1)$ in order to meet the assumption of homogeneity of variances (Underwood, 1997), whereas species richness and diversity data were analyzed without any transformation, as the variances were homogeneous (Cochran test: $p > 0.05$). Least significant difference (LSD) test was used for post hoc comparisons. Statistical analyses were carried out by means of STATISTICA 5 (Statsoft, 2004).

Multivariate analyses were carried out on the fish density and standing stock data. These data were transformed by fourth-root and the Bray–Curtis similarity coefficient was calculated to generate similarity matrices (Clarke, 1993). For each season, an analogue of multivariate analysis of variance with a randomization test for significance (ANOSIM) was used to compare habitats in terms of their fish assemblage composition. Where significant differences occurred, multidimensional scaling (MDS) was used to plot the relationships amongst habitats for each season. Similarity matrices used for MDS were also analyzed for the individual species making the greatest contribution to between-group differences (SIMPER; Clarke, 1993). Seasonal differences were investigated by performing ANOSIM for each habitat. All multivariate analyses were performed using PRIMER (Clarke & Warwick, 1994).

Results

Species richness, diversity, fish density and standing stock

A total of 17933 individuals were sampled, representing 49 species from 22 families (Table 1).

Significant interactions between Habitat and Season were not detected by any of the ANOVAs performed on the community variables (Tables 2, 3, 4 and 5). Species richness varied significantly amongst both habitats and seasons (Table 2, Fig. 2a). Post hoc tests showed that, overall, species richness was significantly higher in seagrass beds than in the other habitats and, as regards seasonal variations, was significantly higher in the summer than in the spring, with autumn showing intermediate values. A similar result was found for fish standing stock (Table 5, Fig. 2d).

Diversity index and fish density did not show any significant variation across habitats, whereas they differed significantly amongst seasons, with higher values in summer (Tables 3 and 4, Fig. 2b, c).

Fish assemblages composition

The fish assemblage in seagrass beds was dominated by two pipefishes of the genus *Syngnathus* (*Syngnathus typhle* Linnaeus and *Syngnathus abaster* Risso), which accounted for about 64% of the total number of individuals caught in this habitat. The grass goby, *Zosterisessor ophiocephalus* (Pallas), was also abundant in seagrass beds and was well represented in the sparsely vegetated habitat and unvegetated mudflats too.

The marbled goby, *Pomatoschistus marmoratus* (Risso), was the dominant taxon in all other habitats, with percent abundances ranging from 42 to 63%. The sand smelt *Atherina boyeri* Risso was ubiquitous and abundant across habitats. The grey mullets were abundant mainly in unvegetated saltmarsh creek (*Liza ramada* (Risso)) and sand habitat (*Liza saliens* (Risso)), whereas *Liza aurata* (Risso) and *Mugil cephalus* Linnaeus were restricted to the latter. The European anchovy *Engraulis encrasicolus* (Linnaeus) was found in all habitats, but it was primarily collected in unvegetated mudflats and sand habitats, whereas the cyprinodontid *Aphanius fasciatus* (Valenciennes) was well represented in the saltmarsh creeks only. In contrast, some species, such as the two species of the genus *Hippocampus*, were restricted to seagrass beds. In one occasion, in an unvegetated mudflat station near Tessera (see Fig. 1), a freshwater species, *Carassius carassius* (Linnaeus), was also caught.

For each sampling season, the assemblage composition in each habitat, in terms of both fish density and standing stock, was compared using ANOSIM with Habitat as factor. The fish density and standing stock data gave similar results, with significant differences amongst habitats found during all seasons (Table 6). Pairwise comparisons between habitats (by ANOSIM) highlighted a pattern common to all the three seasons: seagrass beds sustain a fish assemblage significantly different from those of the other habitats (Fig. 3, Table 7). Using SIMPER the species that most contributed to these habitat differences were identified (Table 7). The three pipefishes, *Syngnathus abaster*, *Syngnathus typhle* and *Nerophis ophidion* (Linnaeus), and the grass goby, *Zosterisessor ophiocephalus*, discriminated the seagrass beds from all other habitats, being preferably found in the former during all three seasons. Also the black goby, *Gobius niger* Linnaeus, contributed to separate the seagrass fish community from the others, but its contribution was evident in the spring, due to the low density of this species during the other seasons. Also the sand smelt *Atherina boyeri* and the peacock blenny *Salaria pavo* (Risso) contributed to characterize the fish assemblages in seagrass beds with respect to the other habitats, especially in summer. Further species, which were preferably caught in the other habitats, were found to contribute to the differences with the seagrass beds, but they varied over seasons and habitats (Table 7). Significant differences in the fish community structure were detected also between the saltmarsh creeks and some of the other habitats, although these differences were particularly evident only during summer and autumn (Fig. 3, Table 7). However, a common pattern could be recognized: saltmarsh creek assemblages were distinguished from those of the other habitats due to the presence of two species, the lagoon goby

Table 1. Mean fish density (number of individuals 100 m^{-2}), percentage of the total individuals (%) and standing stock (g 100 m^{-2}) for fish taxa collected from March to December 2002 in the five habitat types (Sb = seagrass bed; sV = sparsely vegetated habitat; Um = unvegetated mudflat; Uc = unvegetated saltmarsh creek; Us = unvegetated sand habitat)

Species	Sb			sV			Um			Uc			Us		
	Fish density	%	Standing stock	Fish density	%	Standing stock	Fish density	%	Standing stock	Fish density	%	Standing stock	Fish density	%	Standing stock
Atherinidae															
Atherina boyeri	10.07	11.03	16.33	5.44	14.11	10.13	3.79	5.77	9.84	10.50	15.26	15.34	8.36	11.50	9.33
Belonidae															
Belone belone	0.31	0.34	0.21	0.02	0.06	0.02	0	0	0	0	0	0	0.28	0.38	0.46
Blennidae															
Salaria pavo	1.07	1.18	8.27	0.09	0.24	0.02	0.04	0.06	0.07	0.42	0.61	1.25	0.19	0.25	0.91
Parablennius sanguinolentus	0.28	0.30	0.67	0	0	0	0	0	0	0	0	0	0.39	0.54	0.55
Parablennius tentacularis	0.07	0.08	0.74	0	0	0	0	0	0	0	0	0	0	0	0
Callionymidae															
Callionymus risso	0.22	0.24	0.16	0.02	0.06	0.01	0.03	0.04	0.01	0	0	0	0.07	0.10	0.04
Clupeidae															
Sardina pilchardus	0	0	0	0	0	0	0	0	0	0	0	0	0.02	0.03	0.04
Sprattus sprattus	0.02	0.02	0.03	0	0	0	0	0	0	0	0	0	0.05	0.06	0.06
Cyprinidae															
Carassius carassius	0	0	0	0	0	0	0.10	0.15	29.86	0	0	0	0	0	0
Cyprinodontidae															
Aphanius fasciatus	0	0	0	0	0	0	0	0	0	7.22	10.50	3.98	0.02	0.03	0.002
Engraulidae															
Engraulis encrasicolus	0.48	0.53	0.27	0.07	0.18	0.04	6.78	10.32	4.82	0.75	1.10	0.24	4.93	6.79	1.64
Gadidae															
Merlangius merlangus	0	0	0	0.02	0.06	0.01	0	0	0	0	0	0	0	0	0
Gobiidae															
Gobius cobitis	0.22	0.24	0.55	0.02	0.06	0.14	0	0	0	0	0	0	0.88	1.21	0.94
Zosterisessor ophiocephalus	11.89	13.01	65.11	2.45	6.36	20.41	3.94	6.01	10.67	0.46	0.66	2.50	1.25	1.72	1.06
Gobius niger	0.87	0.95	4.78	0.88	2.28	5.07	0.13	0.19	1.01	0.02	0.03	0.06	0.07	0.10	0.45

Continued on p. 164

Table 1. (Continued)

Species	Sb			sV			Um			Uc			Us		
	Fish density	%	Standing stock	Fish density	%	Standing stock	Fish density	%	Standing stock	Fish density	%	Standing stock	Fish density	%	Standing stock
Knipowitschia panizzae	0	0	0	0.23	0.60	0.09	2.42	3.68	0.56	5.44	7.90	0.78	0	0	0
Pomatoschistus canestrinii	0	0	0	0	0	0	3.13	4.76	1.50	2.42	3.52	1.06	0	0	0
Pomatoschistus minutus	0	0	0	0.28	0.72	0.45	1.76	2.69	3.49	0.06	0.09	0.11	0.02	0.03	0.002
Pomatoschistus marmoratus	0.89	0.97	0.36	24.44	63.38	8.58	40.14	61.12	9.84	29.25	42.51	8.16	32.01	44.06	8.33
Labridae															
Symphodus sp.1	0.04	0.04	1.33	0	0	0	0	0	0	0	0	0	0	0	0
Symphodus sp.2	0.19	0.20	0.25	0	0	0	0	0	0	0	0	0	0.02	0.03	0.04
Symphodus roissali	0.07	0.08	0.93	0.02	0.06	0.35	0	0	0	0	0	0	0	0	0
Moronidae															
Dicentrarchus labrax	0	0	0	0	0	0	0.03	0.04	0.01	0	0	0	0	0	0
Mugilidae															
Liza aurata	0	0	0	0	0	0	0	0	0	0	0	0	1.16	1.59	0.38
Liza ramada	0.06	0.06	1.65	0	0	0	0.44	0.68	1.21	9.72	14.13	4.01	0	0	0
Liza saliens	0.09	0.10	0.40	0.09	0.24	0.38	0.10	0.15	5.32	1.61	2.33	22.37	17.36	23.89	7.32
Mugil cephalus	0	0	0	0	0	0	0	0	0	0	0	0	0.02	0.03	0.002
Mullidae															
Mullus surmuletus	0.02	0.02	0.01	0.05	0.12	0.06	0.01	0.02	0.03	0	0	0	0.30	0.41	0.28
Pleuronectidae															
Platichthys flesus	0.13	0.14	1.23	0.02	0.06	0.33	1.36	2.07	13.91	0.28	0.40	0.90	0.07	0.10	0.32
Poeciliidae															
Gambusia holbrooki	0	0	0	0	0	0	0	0	0	0.04	0.06	0.004	0.12	0.16	0.03
Scianeidae															
Sciaena umbra	0.06	0.06	0.01	0	0	0	0.01	0.02	0.01	0	0	0	0.07	0.10	0.01
Scophthalmidae															
Scophthalmus rhombus	0	0	0	0	0	0	0	0	0	0	0	0	0.02	0.03	0.03

Soleidae															
Solea vulgaris	0.02	0.02	0.34	0.63	1.62	0.23	0.40	0.61	0.72	0.40	0.58	0.97	0.67	0.92	0.28
Solea impar	0	0	0	0	0	0	0.01	0.02	0.14	0	0	0	0.02	0.03	0.002
Sparidae															
Boops boops	0.07	0.08	0.04	0	0	0	0	0	0	0.02	0.03	0.05	0	0	0
Diplodus puntazzo	0.02	0.02	0.002	0	0	0	0	0	0	0	0	0	0	0	0
Diplodus annularis	0.04	0.04	0.06	0	0	0	0	0	0	0	0	0	0.12	0.16	0.01
Lithognathus mormyrus	0.04	0.04	0.04	0	0	0	0	0	0	0	0	0	0	0	0
Sparidae *n.i.*	0.02	0.02	0.002	0	0	0	0	0	0	0	0	0	0	0	0
Syngnathidae															
Hippocampus guttulatus	0.09	0.10	0.62	0	0	0	0	0	0	0	0	0	0	0	0
Hippocampus hippocampus	0.02	0.02	0.002	0	0	0	0	0	0	0	0	0	0	0	0
Nerophis ophidion	4.98	5.45	2.38	0.05	0.12	0.01	0.04	0.06	0.04	0	0	0	0.19	0.25	0.06
Syngnathus abaster	17.81	19.50	9.12	2.69	6.96	1.30	0.49	0.74	0.34	0.14	0.20	0.07	1.23	1.69	0.36
Syngnathus acus	0.06	0.06	0.53	0	0	0	0	0	0	0	0	0	0	0	0
Syngnathus taenionotus	0.13	0.14	0.15	0.07	0.18	0.07	0.38	0.57	0.18	0.02	0.03	0.01	0.07	0.10	0.03
Syngnathus tenuirostris	0.04	0.04	0.74	0	0	0	0	0	0	0	0	0	0.05	0.06	0.01
Syngnathus typhle	40.91	44.78	44.07	0.97	2.52	1.95	0.10	0.15	0.23	0.04	0.06	0.03	2.57	3.54	2.22
Triglidae															
Trigla lucerna	0	0	0.15	0	0	0	0.03	0.04	0.10	0	0	0	0.07	0.10	0.08
Mean density	1.86		3.29	0.79		1.01	1.34		1.92	1.40		1.26	1.48		0.72
Number of species	35			21			25			19			32		

Table 2. Analysis of variance of fish species richness (mean number of species) in the five habitat types over three seasons

Source	SS	DF	MS	*F*	*p*	LSD test
Season	128.755	2	64.378	7.442	<0.01	Su > Au = Sp
Habitat	120.516	4	30.129	3.483	<0.05	Sb > Uc = sV = Um = Us
Season × Habitat	69.902	8	8.738	1.010	ns	
Error	501.712	58	8.650			

LSD test: $p<0.05$. Sb = seagrass beds; sV = sparsely vegetated habitats; Um = unvegetated mudflats; Uc = unvegetated saltmarsh creeks; Us = unvegetated sand habitats; Sp = Spring; Su = Summer; Au = Autumn; SS = sums of squares; DF = degrees of freedom; MS = mean square; ns = not significant.

Table 3. Analysis of variance of fish diversity (Shannon–Wiener index) in the five habitat types over three seasons

Source	SS	DF	MS	*F*	*p*	LSD test
Season	2.132	2	1.006	4.860	<0.05	Su > Au = Sp
Habitat	0.877	4	0.219	1.059	ns	
Season × Habitat	1.935	8	0.242	1.170	ns	
Error	12.004	58	0.207			

Sp = Spring; Su = Summer; Au = Autumn; SS = sums of squares; DF = degrees of freedom; MS = mean square; ns = not significant.

Table 4. Analysis of variance of fish density (number of individuals/m^2) in the five habitat types over three seasons

Source	SS	DF	MS	*F*	*p*	LSD test
Season	0.570	2	0.285	13.112	<0.001	Su > Au = Sp
Habitat	0.110	4	0.028	1.275	ns	
Season × Habitat	0.180	8	0.023	1.038	ns	
Error	1.261	58	0.022			

LSD test: $p<0.05$. Sp = Spring; Su = Summer; Au = Autumn; SS = sums of squares; DF = degrees of freedom; MS = mean square; ns = not significant.

Table 5. Analysis of variance of fish standing stock (g/m^2) in the five habitat types over three seasons

Source	SS	DF	MS	*F*	*p*	LSD test
Season	0.208	2	0.104	3.729	<0.05	Su > Au = Sp
Habitat	0.656	4	0.164	5.881	<0.001	Sb > Um = Uc = sV = Us
Season × Habitat	0.267	8	0.033	1.199	ns	
Error	1.618	58	0.028			

LSD test: $p<0.05$. Sb = seagrass beds; sV = sparsely vegetated habitats; Um = unvegetated mudflats; Uc = unvegetated saltmarsh creeks; Us = unvegetated sand habitats; Sp = Spring; Su = Summer; Au = Autumn; SS = sums of squares; DF = degrees of freedom; MS = mean square; ns = not significant.

Knipowitschia panizzae (Verga) and the leaping mullet *Liza saliens*, which were preferably found in this habitat (Table 7).

Significant seasonal fluctuations in the fish assemblage composition were also detected for each habitat by means of ANOSIM, as shown in Table 6.

Discussion

The fish assemblages changed significantly over time within each habitat type. The degree of the seasonal variation was similar, or higher in some cases, with respect to the spatial variation, as revealed by the results of both univariate and

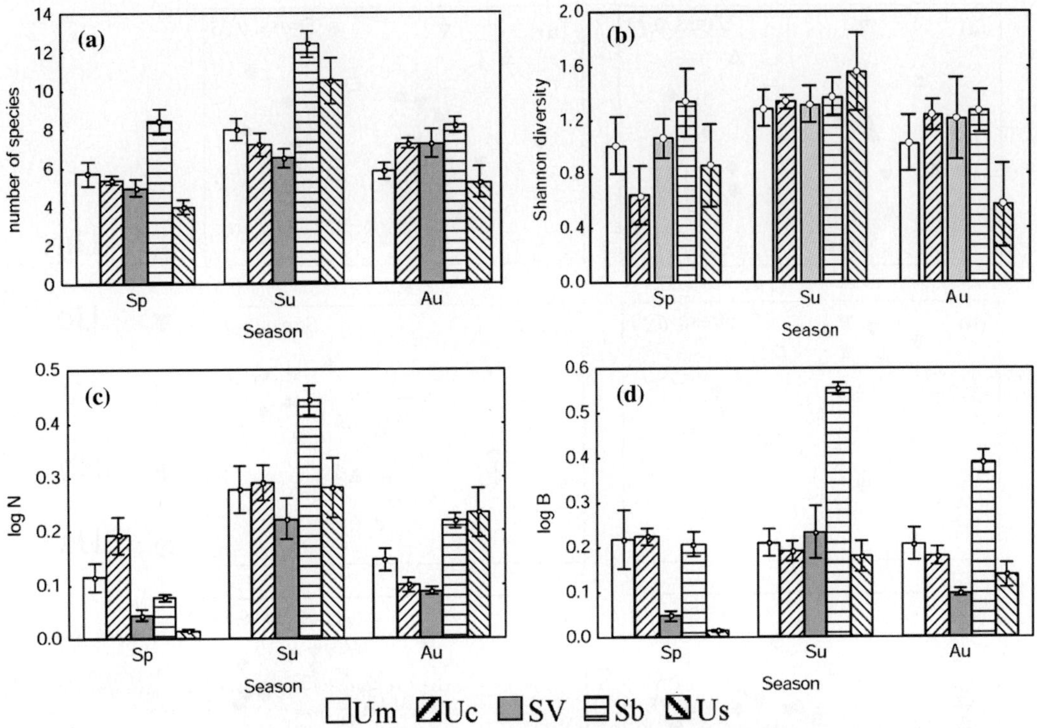

Figure 2. (a) Mean species richness (number of species per station), (b) mean fish diversity (Shannon–Wiener index), (c) mean fish density (log transformed data of the number of individuals/360 m^2, log N), and (d) mean fish standing stock (log transformed data of the fish biomass, g/360 m^2, log B) at the five types of habitats in the three sampling seasons. Bars indicate standard errors.

multivariate analyses. The seasonal variation showed the same direction in each habitat type, with higher values of total fish density, standing stock, species diversity and richness found during summer than in spring and autumn. A high level of seasonal variation is a recurrent pattern of the estuarine fish communities of temperate systems, mainly due to recruitment of the resident species

Table 6. ANOSIM results (global R-values and p-level) on fish density and standing stock among habitats and seasons (within each season and habitat, respectively)

	Fish density		Fish standing stock	
	Global-R	p-level	Global-R	p-level
Among habitats				
Spring	0.342	0.001	0.403	0.001
Summer	0.412	0.001	0.383	0.002
Autumn	0.52	0.001	0.423	0.001
Among seasons				
Sb	0.579	0.001	0.512	0.001
sV	0.314	0.009	0.311	0.015
Us	0.503	0.011	0.49	0.006
Um	0.251	0.001	0.206	0.006
Uc	0.196	0.031	0.156	0.054

Sb = seagrass beds; sV = sparsely vegetated habitats; Um = unvegetated mudflats; Uc = unvegetated saltmarsh creeks; Us = unvegetated sand habitats.

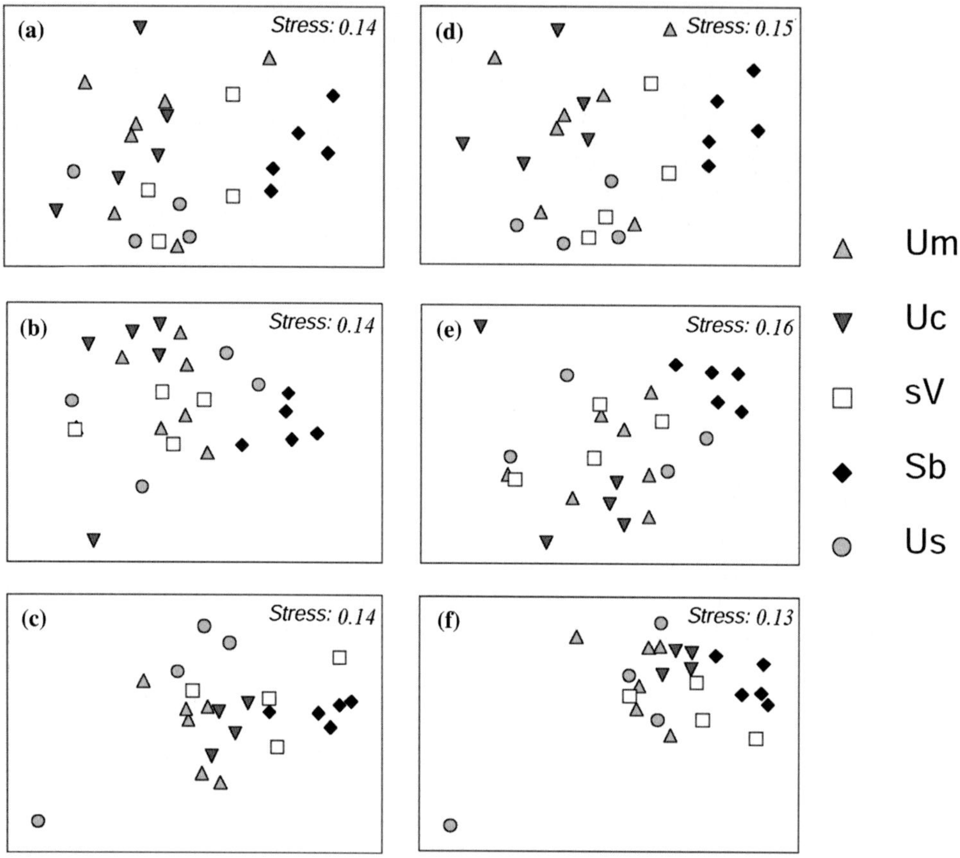

Figure 3. MDS plots of relationships amongst fish assemblages associated with the habitats in terms of fish density (a, b, c) and standing stock (d, e, f), for each season (a and d = Spring; b and e = Summer; c and f = Autumn).

and to the seasonal migrations of the marine transient fishes (McErlean et al., 1973; Potter et al., 1986).

The fish assemblages of seagrass beds were distinctly different from those of the other habitats, and these differences persisted over time, at least within the investigated period (spring– autumn). The fish community in the seagrass beds was dominated by estuarine resident fishes (Paterson & Whitfield, 2000). This assemblage was typified by high densities of syngnathids (*Syngnathus abaster*, *Syngnathus typhle*, and *Nerophis ophidion*), and the grass goby *Zosterisessor ophiocephalus*, which significantly contributed to discriminate the seagrass habitat from the others. These are specialized species, which are closely associated with the seagrass due either to their specific morphological and behavioral adaptations (as for Syngnathids; Howard & Koehn, 1985; Lourie et al., 1999) or to their reproductive demands (as for the grass goby; Malavasi et al., 2002). Although the composition of the fish community in the seagrass beds changed significantly over time, the stations belonging to this habitat type were highly similar to each other, in terms of fish assemblage structure, within each season. This indicates that the simply presence of the vegetation coverage contributes to make the fish assemblage more homogeneous within a given seasonal period.

According to many authors (Kikuchi & Peres, 1977; McRoy & Helfferich, 1980; Rozas & Odum, 1987; Bell & Pollard, 1989; Heck et al., 1989; Lubbers et al., 1990; Jenkins & Wheatley, 1998; Guidetti, 2000; Lazzari & Tupper, 2002), seagrass beds support higher number of fish species, higher abundance and higher standing stock than other habitats. However, a few other works did not detect differences between the seagrass beds and the

Table 7. Pairwise comparisons (by ANOSIM) amongst habitats in terms of fish assemblage composition, for each season

	sV	Us	Um	Uc
Spring				
Sb	***Nerophis ophidion*** ***Syngnathus typhle*** *Solea vulgaris*	***Zosterisessor ophiocephalus*** ***Nerophis ophidion*** ***Syngnathus typhle*** ***Syngnathus abaster*** ***Gobius niger*** *Solea vulgaris*	***Nerophis ophidion*** ***Zosterisessor ophiocephalus*** ***Syngnathus typhle*** ***Gobius niger*** ***Syngnathus abaster*** *Pomatoschistus marmoratus* *Platichthys flesus*	***Nerophis ophidion*** ***Syngnathus abaster*** ***Syngnathus typhle*** ***Zosterisessor ophiocephalus*** ***Gobius niger*** *Platichthys flesus*
SV		ns	ns	ns
Us			ns	*Platichthys flesus* *Pomatoschistus marmoratus*
Um				ns
Summer				
	sV	Us	Um	Uc
Sb	***Syngnathus typhle*** ***Nerophis ophidion*** ***Syngnathus abaster*** ***Salaria pavo*** ***Z. ophiocephalus*** *P. marmoratus* *Gobius niger*	***Atherina boyeri*** ***Syngnathus typhle*** ***Z. ophiocephalus*** ***Salaria pavo*** *P. marmoratus*	***Syngnathus typhle*** ***Atherina boyeri*** ***Nerophis ophidion*** ***Syngnathus abaster*** ***Z. ophiocephalus*** ***Salaria pavo*** *P. marmoratus* *Engraulis encrasicolus*	***Syngnathus typhle*** ***Z. ophiocephalus*** ***Syngnathus abaster*** ***Nerophis ophidion*** ***Gobius niger*** *P. marmoratus* *Knipowitschia panizzae* *Liza saliens*
sV		ns	ns	***P. marmoratus*** ***Z. ophiocephalus*** *Liza saliens* *Knipowitschia panizzae*
Us			ns	ns
Um				***Engraulis encrasicolus*** *Liza saliens* *Knipowitschia panizzae*

Continued on p. 170

Table 7. (Continued)

	sV	Us	Um	Uc
Autumn				
Sb	***Z. ophiocephalus*** ***Syngnathus typhle*** ***Nerophis ophidion*** ***Atherina boyeri*** ***Syngnathus abaster*** *Salaria pavo*	***Z. ophiocephalus*** ***Syngnathus abaster*** ***Nerophis ophidion*** ***Syngnathus typhle*** *P. marmoratus*	***Syngnathus abaster*** ***Z. ophiocephalus*** ***Syngnathus typhle*** ***Nerophis ophidion*** *P. marmoratus* *Engraulis encrasicolus*	***Syngnathus abaster*** ***Syngnathus typhle*** ***Z. ophiocephalus*** ***Nerophis ophidion*** ***Atherina boyeri*** *Knipowitschia panizzae* *Engraulis encrasicolus* *Liza saliens*
sV		ns	***Syngnathus abaster*** ***Salaria pavo*** *Engraulis encrasicolus*	***Syngnathus abaster*** ***Salaria pavo*** *Engraulis encrasicolus* *Liza saliens* *Knipowitschia panizzae*
Us			ns	***P. marmoratus*** *Knipowitschia panizzae*
Um				ns

ns = not significant difference. Where the differences resulted significant ($p < 0.05$) the discriminant species (as resulted from SIMPER analysis) are indicated: in bold italic for the species which were more abundant in the habitat in row, and in italic for the species which were more abundant in habitat in column. (Sb = seagrass bed; sV = sparsely vegetated habitat; Um = unvegetated mudflat; Uc = unvegetated saltmarsh creek; Us = unvegetated sand habitat).

unvegetated habitats in fish abundance and species richness (Hanekom & Baird, 1984; Heck & Thoman, 1984). Our results showed that in the seagrass beds of the Venice Lagoon, fish species richness and standing stock were higher than in the other habitats, whereas the fish abundance did not change significantly across habitats. Several studies (Kikuchi & Peres, 1977; McRoy & Helfferich, 1980; Jenkins & Wheatley, 1998; Guidetti, 2000) suggest that high numbers of species are attracted to seagrass beds due to the higher structural complexity of such vegetated systems. However, the higher species richness we found in seagrass beds of the Northern basin of the Venice Lagoon could be due to other local factors, such as the proximity of these habitats to the sea inlets and colonization processes (Garcia-Charton & Pérez-Ruzafa, 2001; Mariani, 2001; Pérez-Ruzafa et al., 2004). Results concerning fish abundance and standing stock may be explained considering the different dominant species characterizing the different habitats. In fact, the higher fish standing stock found in seagrass beds, if compared with other habitats, results mainly from the high abundance of a larger species, the grass goby. On the other hand, the across habitat homogeneity in total fish abundance could be due to the high contribution from small-sized epibenthic species, such as those belonging to the genus *Pomatoschistus*, which use the unvegetated habitats as feeding, reproductive and nursery grounds (Elliott & Hemingway, 2002). These species, given the high reproductive and recruitment rates (Miller, 1984), contribute strongly to maintain fish densities similar to that found in the seagrass beds.

Seagrass beds, sustaining high abundances of grass goby, resulted also to be important habitats in supporting the local artisanal fishery in the Venice Lagoon, being the grass goby one of the target species of this kind of fishery (Granzotto et al., 2001). Moreover, seagrass beds hosted some threatened species, like *Hippocampus hippocampus* (Linnaeus) and *Hippocampus guttulatus* Cuvier (IUCN, 2002), which, although with low abundances, were restricted to this habitat alone in the lagoon shallow waters (see Table 1).

In contrast, the fish assemblages in sparsely vegetated (with less than 30% of the seagrass cover) and unvegetated habitats were constituted by relatively smaller-sized species, such as small-sized residents and marine transient fishes, caught mainly as juveniles, in accordance with the results of Paterson & Whitfield (2000). The marbled goby *Pomatoschistus marmoratus*, dominated these assemblages, accounting for 42–63% of the total catch. Other small-sized resident species, such as *Knipowitschia panizzae* and *Aphanius fasciatus*, were also abundant in these assemblages (in particular in the saltmarsh creeks and mudflats). Migrant fish, such as the common sole, *Solea vulgaris* (Linnaeus), and the flounder, *Platichthys flesus* (Linnaeus), during spring, and the leaping mullet and the European anchovy during summer and autumn, are also elements of these assemblages. Despite the common features of the fish assemblages in the sparsely vegetated and unvegetated habitats, a difference could be detected amongst them in terms of species composition, which was more pronounced between the unvegetated saltmarsh creeks and the seagrass sparsely vegetated and unvegetated sand habitats. In fact, two species, *Knipowitschia panizzae* and *Liza saliens* discriminated the fish assemblage of saltmarsh creeks from the others. The difference relative to the unvegetated mudflat assemblages was less apparent, especially during autumn, due to the shared abundance of *Engraulis encrasicolus* amongst the two unvegetated mud habitats. The differences between the saltmarsh creeks and other habitats in terms of fish assemblage composition were evident particularly during summer and autumn, as in this period the recruitment of *K. panizzae* (summer) and the migration of *E. encrasicolus* (summer and autumn) and *L. saliens* (autumn) from the sea towards the inner areas of the lagoon determine increased abundances of these species in that habitat. Saltmarsh creeks were also important habitats for species like *Aphanius fasciatus* and *Pomatoschistus canestrinii* (Ninni) (see Table 2), which, together with the abundant *Knipowitschia panizzae*, are "species of community interest whose conservation requires designation of special areas of conservation", as reported in the "Habitat and Species Directive" 92/43/EEC (EEC, 1992).

A characteristic fish assemblage could not be recognized in the sparsely vegetated and unvegetated sand habitats. The fish assemblages there were made up of species which were found both in seagrass beds and in unvegetated mud habitats (see Tables 1 and 2), suggesting that sparsely vegetated and unvegetated sand habitats act probably as 'buffer zones' between the other ones and as migration routes for many fish species.

On the whole, these results suggest that two habitats – seagrass beds and saltmarsh creeks – are relevant in structuring the small-sized fish assemblages of the Venice Lagoon, due to different ecological reasons. The seagrass beds sustain high diversity not only because of the three-dimensional structure of the plant canopy and its associated specialized fish community, but also because these habitats are under a strong influence of the communication channels with open sea. The marine influence highly contributes to increase species diversity in the lagoonal systems, according to the confinement theory (Guelorget & Perthuisot, 1983). Also, the saltmarsh creeks offer a complex topography in the saltmarsh environment creating protection and a number of ecological niches available for fish species (Desmond et al., 2000). This habitat is usually located at the other extreme of the confinement gradient with respect to the seagrass beds, but it does not seem to support higher fish abundance as predicted by the theory of the paralic domain (Guelorget & Perthuisot, 1983). Our analysis seems therefore to suggest that both habitat complexity and degree of confinement are important determinants of the structure of the fish assemblages.

These results suggest therefore that basically two categories of habitats contribute to sustain the highly diversified fish assemblages in shallow waters of the Venice Lagoon: "structuring" habitats, like seagrasses and saltmarsh creeks, where a specialized and recognizable fish assemblage can be found, and "transition" habitats (sandy bottoms, mudflats and sparsely vegetated habitats), where fish assemblages are highly variable and influenced by the contribution of the adjacent habitats. The management and conservation of the fish communities of the lagoon probably requires a constant monitoring of both these categories of habitats.

Acknowledgments

The present work was financed by the Co.Ri.La. Project: "Lo studio della biodiversità per un equilibrio fra conservazione e sfruttamento in laguna di Venezia".

References

Amanieu, M. & G. Lasserre, 1982. Organisation et évolution des pueplements lagunaires. Oceanologica Acta. Proceedings International Symposium on coastal lagoons, Bordeaux France: pp. 201–213.

Arruda, L. M., J. P. Andrade & M. M. Cunha, 1988. Abundance, diversity, and community structure of the fish population in the Ria de Aveiro (Portugal). Oceanologica Acta 11: 235–240.

Avanzi, C., V. Fossato, P. Gatto, R. Rabagliati, P. Rosa Salva & A. Zitelli, 1979. Ripristino, conservazione ed uso dell'ecosistema lagunare veneziano. Comune di Venezia, Venezia: 197 pp.

Bell, J. D. & D. A. Pollard, 1989. Ecology of fish assemblages and fisheries associated with seagrasses. In Larkum, A. W. D. A. J. McComb, & S. A. Shepard (eds.) Biology of Seagrasses. Elsevier, Amsterdam: 565–609.

Bouchereau, J.-L., O. Guelorget, Y. Vergne & J.-P. Perthuisot, 2000. The fish fauna in the biological organization of a lagoonar paralic system: the Prévost–Arnel lagoonal complex (Languedoc, France). Vie Milieu 50: 19–27.

Clarke, K. R., 1993. Non-parametric multivariate analyses of changes in community structure. Australian Journal of Ecology 18: 117–143.

Clarke, K. R. & R. M. Warwick, 1994. Change in Marine Communities: An Approach to Statistical Analysis and Interpretation. Natural Environment Research Council, UK: 144 pp.

Connolly, R. M., 1994. A comparison of fish assemblages from seagrass and unvegetated areas of a southern Australian estuary. Australian Journal of Marine and Freshwater Research 45: 1033–1044.

Desmond, J. S., J. B. Zedler & G. D. Williams, 2000. Fish use of tidal creek habitats in two southern California salt masrhes. Ecological Engineering 14: 233–252.

EEC, 1992. Council Directive 92/43/EEC of 21 May 1992 on the conservation of natural habitats and of wild fauna and flora. Official Journal L206, 22/07/1992, pp. 7–50.

Elliott, M. & F. Dewailly, 1995. The structure and components of European estuarine fish assemblages. The Netherlands Journal of Aquatic Ecology 29: 397–417.

Elliott, M. & K. L. Hemingway, 2002. Fishes in Estuaries. Blackwell Science, UK: 636 pp.

Franzoi, P., V. U. Ceccherelli & R. Rossi, 1992. Differenze di abitudini alimentari tra novellame e specie ittiche residenti in una laguna del Delta del Po. Oebalia XVII181–186.

Franzoi, P., R. Trisolini, A. Carrieri & R. Rossi, 1989. Caratteristiche ecologiche del popolamento ittico ripario della sacca di Scardovari (Delta de Po). Nova Thalassia 10: 399–405.

Gandolfi, G., E. Ioannilli & R. Vitali, 1985. Caratteristiche biologiche delle comunità ittiche, studi sulle migrazioni ed aspetti quantitativi delle attività alieutiche nel Delta del Po. Nova Thalassia 7: 281–309.

Garcia-Charton, J. A. & A. Pérez-Ruzafa, 2001. Spatial pattern and the habitat structure of a Mediterranean rocky fish local reef assemblage. Marine Biology 138: 917–934.

Gordo, L. S. & H. N. Cabral, 2001. The fish assemblage structure of a hydrologically altered coastal lagoon: the Obidos lagoon (Portugal). Hydrobiologia 459: 125–133.

Granzotto, A., P. Franzoi, A. Longo, F. Pranovi & P. Torricelli, 2001. La pesca nella laguna di Venezia un percorso di sostenibilità nel recupero delle tradizioni. Lo stato dell'arte. Rapporto sullo sviluppo sostenibile 2.2001 – Foundation ENI Enrico Mattei.

Guelorget, O. & J.-P. Perthuisot, 1983. Le domaine paralique.Expressions geologiques, biologiques et economiques du confinement. Travaux du Laboratoire de Geologie 16: 1–136.

Guidetti, P., 2000. Differences among fish assemblages associated with nearshore *Posidonia oceanica* seagrass beds, rocky-algal reefs and unvegetated sand habitats in the Adriatic Sea. Estuarine, Coastal and Shelf Science 50: 515–529.

Hanekom, N. & D. Baird, 1984. Fish community structures in *Zostera* and non-*Zostera* regions of the Kromme estuary, St Francis Bay. South African Journal of Zoology 19: 295–301.

Heck, K. L. & T. A. Thoman, 1984. The nursery role of seagrass meadows in the upper and lower reaches of the Chesapeake Bay. Estuaries 7: 70–92.

Heck, K. L., K. W. Jr. Able, M. P. Fahay & C. T. Roman, 1989. Fishes and decapod crustaceans of Cape Cod eelgrass meadows: species composition, seasonal abundance patterns and comparison with unvegetated substrated. Estuaries 12: 59–65.

Hervé, P. & J. Bruslé, 1980. L'étang de salses-leucate écologie générale et ichtyofaune. Vie Milieu 30: 275–283.

Howard, R. K. & J. D. Koehn, 1985. Population dynamics and feeding ecology of pipefish (syngnathidae) associated with eelgrass beds of Western Port, Victoria. Australian Journal of Marine and Freshwater Research 36: 361–370.

Huvé, H., A. Kiener & R. Riouall, 1973. Modifications de la flore et des populations ichthyologiques des etangs de berre et de vaine en fonction des conditions hydrologiques crees par le deversement de la durance. Bulletin du Musée d'Histoire Naturelle de Marseille 33: 123–134.

IUCN, 2002. 2002 IUCN Red List of threatened species. http://www.redlist.org. Downloaded on 14 August 2002.

Jenkins, G. P. & M. J. Wheatley, 1998. The influence of habitat structure on nearshore fish assemblages in a southern Australian embayment: comparison of shallow seagrass, reef-algal and unvegetated sand habitats, with emphasis on their importance to recruitment. Journal of Experimental Marine Biology and Ecology 221: 147–172.

Kikuchi, T. & J. M. Peres, 1977. Consumer ecology of seagrass beds. In McRoy, C. P. & C. Helfferich (eds.) Seagrass Ecosystems: A Scientific Perspective. Dekker, New York: 147–193.

Krebs, C. J., 1999. Ecological Methodology – 2nd ed. Addison-Wesley, California, USA:: 620 pp.

Lazzari, M. A. & B. Tupper, 2002. Importance of shallow water habitats for demersal fishes and decapod crustaceans in Penobscot Bay, Maine. Environmental Biology of Fishes 63: 57–66.

Lobry, J., L. Mourand, E. Rochard & P. Elie, 2003. Structure of the Gironde estuarine fish assemblages: a comparison of European estuaries perspective. Aquatic Living Resources 16: 47–58.

Lourie, S. A., A. C. J. Vincent & H. J. Hall, 1999. Seahorses: An Identification Guide to the World's Species and their Conservation. Project Seahorse, London U.K: 214 pp.

Lubbers, L., W. R. Boynton & W. M. Kemp, 1990. Variations in structure of estuarine fish communities in relation to abundance of submersed vascular plants. Marine Ecology Progress Series 65: 1–14.

Malavasi, S., J. Coppola, F. Pranovi, A. Granzotto, A. Franco & P. Torricelli, 2002. Habitat riproduttivo di Zosterisessor ophiocephalus Pall. (Pisces, Gobiidae) in laguna di Venezia e osservazioni sulle caratteristiche dei riproduttori. Lavori della Società Veneziana di Scienze Naturali 27: 47–56.

Mariani, S., 2001. Can spatial distribution of ichthyofauna describe marine influence on coastal lagoons? A central Mediterranean case study. Estuarine and Coastal Shelf Science 52: 261–267.

Marshall, S. & M. Elliott, 1998. Environmental influences on the fish assemblage of the Humber estuary. U.K. Estuarine Coastal and Shelf Sciences 46: 175–184.

Mathieson, S., A. Cattrijsse, M. J. Costa, P. Drake, M. Elliott, J. Gardner & J. Marchand, 2000. Fish assemblages of European tidal marshes: a comparison based on species, families and functional guilds. Marine Ecology Progress Series 204: 225–242.

McErlean, A. J., S. G. O'Connor, J. A. Mihursky & C. I. Gibson, 1973. Abundance, diversity and seasonal patterns of estuarine fish populations. Estuarine and Coastal Marine Science 1: 19–36.

McRoy, C. P. & C. Helfferich, 1980. Applied aspects of seagrasses. In Phillips, R. C. & C. P. McRoy (eds.) Handbook of Seagrass Biology: an Ecosystem Perspective. Garland STM Press, New York: 297–343.

Miller, P. J., 1984. The tokology of gobioid fishes. In Potts, G. W. & R. J. Wootton (eds.) Fish Reproduction: Strategies and Tactics. Academic Press, London: 119–153.

Muus, B. J., 1967. The Fauna of Danish Estuaries and Lagoons. Distribution and ecology of dominating species in the shallow reaches of the mesohaline zone, Kobenhavn: 269 pp.

Nash, R. D., 1988. The effects of disturbance and severe seasonal fluctuations in environmental conditions of north temperate shallow-water fish assemblages. Estuarine Coastal and Shelf Sciences 26: 123–135.

Orth, R. J. & K. L. Jr. Heck, 1980. Structural components of the eelgrass (*Zostera marina*) meadows in the lower Chesapeake Bay – fishes. Estuaries 3: 278–288.

Paterson, A. W. & A. K. Whitfield, 2000. The ichthyofauna associated with an intertidal creek and adjacent eelgrass beds in the Kariega Estuary, South Africa. Environmental Biology of Fishes 58: 145–156.

Pérez-Ruzafa, A., J. I. Quispe-Becerra, J. A. Garcia-Charton & C. Marcos, 2004. Composition, structure and distribution of the ichthyoplancton in a Mediterranean coastal lagoon. Journal of Fish Biology 64: 1–17.

Poizat, G., E. Rosecchi, P. Chauvelon, P. Contournet & A. J. Crivelli, 2004. Long-term fish and macro-crustacean community variation in a Mediterranean lagoon. Estuarine Coastal and Shelf Science 59: 615–624.

Potter, I. C., P. N. Claridge & R. M. Warwick, 1986. Consistency of seasonal changes in an estuarine fish assemblage. Marine Ecology Progress series 32: 217–228.

Quignard, J. P., 1984. The biological and environmental characteristics of lagoons as the biological basis of fisheries management (in French). GFCM – Studies and Reviews, FAO 61: 3–38.

Rozas, L. P. & W. E. Odum, 1987. fish and macrocrustacean use of submerged plant beds in tidal freshwater marsh creecks. Marine Ecology Progress Series 38: 101–108.

Sacchi, C. R., 1985. Le sel de La Palice: réflexions sur le paralin méditerranéen. Memoires de Biologie Marine Oceanografique 15: 71–89.

Sfriso, A., C. Facca & P. F. Ghetti, 2003. Temporal and spatial changes of macroalgae and phytoplankton in a Mediterranean coastal area: the Venice lagoon as a case study. Marine Environmental Research 56: 617–636.

Sogard, S. M. & K. W. Able, 1991. A comparison of eelgrass, sea lettuce macroalgae, and marsh creeks as habitats for epibenthic fishes and decapods. Estuarine Coastal and Shelf Sciences 33: 501–519.

StatSoft. 2004. Electronic Statistics Textbook StatSoft Inc., Tulsa OK USA Web: http://www.statsoft.com/textbook/stathome.html.

Underwood, A. J., 1997. Experiments in Ecology The Logical Design and Interpretation using Analysis of Variance University Press, Cambridge: 504 pp.

West, R. J. & R. J. King, 1996. Marine, brackish, and freshwater fish communities in the vegetated and bare shallows of an Australian coastal river. Estuaries 19: 31–41.

Hydrobiologia (2006) 555:175–183
H. Queiroga, M.R. Cunha, A. Cunha, M.H. Moreira, V. Quintino, A.M. Rodrigues, J. Serôdio & R.M. Warwick (eds),
Marine Biodiversity: Patterns and Processes, Assessment, Threats, Management and Conservation
DOI 10.1007/s10750-005-1114-4

The effect of *Zostera noltii*, *Spartina maritima* and *Scirpus maritimus* on sediment pore-water profiles in a temperate intertidal estuary

A.I. Lillebø[1,]*, M.R. Flindt[2], M.A. Pardal[1] & J.C. Marques[1]
[1]*IMAR – Institute of Marine Research, Department of Zoology, University of Coimbra, 3004-517, Coimbra, Portugal*
[2]*Biological Institute, SDU – University of Odense, Campusvej 55, 5230, Odense M, Denmark*
*(*Author for correspondence: Tel.: +351-239-837797; Fax: +351-239-823603; E-mail: lillebo@ci.uc.pt)*

Key words: salt marsh, nutrient profiles, estuary, nitrogen, phosphorus

Abstract

The objective of the present work was to study the effect of plants common in temperate latitudes (*Zostera noltii, Spartina maritima* and *Scirpus maritimus*) on sediment nutrient profiles, and to compare it to sand- and mud-flats without vegetation. The study focused on the organic matter contents, the concentration of dissolved inorganic nutrients (PO_4–P, NH_3–N, NO_3–N), an on the estimation of the total amount of these nutrients during day and night conditions and their potential net-fluxes. It was also hypothesised that in an estuarine system, different plants may have specific effects, and consequently different contributions to the system nutrient dynamics as a whole. Sediment profiles of loss on ignition (LOI) showed an increase of the organic matter contents from sand-flat, to *Zostera, Spartina*, mud-flat and *Scirpus*. Statistically, there were significant differences between sediment profiles of phosphate, ammonia and nitrate (Mann-Whitney test, $p < 0.05$), during day and night periods. These results suggest that there is an intense mobility of nutrients in the sediment, showing a day-night variation of nutrient concentrations in the pore-water. In the plants' rhizosphere, the day-night variation of nutrients seemed dependent on plant biomass and penetration of the roots. Additionally, coupling between plant and sediment seems to be a species-specific process. In spring, *Scirpus* salt marsh reaches the maximum density and biomass, and despite the higher organic matter contents in the plant covered sediment, *Scirpus* acts as a sink of nutrients. In contrast, the top 10 cm of the sediment in the *Spartina* salt marsh and in the *Zostera* beds may contribute to the efflux of nutrients during the night period, especially phosphate.

Introduction

Salt marshes and sea-grass beads have been characterised as efficient sinks for nutrients, and buffering the effects of nutrient inputs, (e.g., Anderson et al., 1997; Andersen & Ring, 1999; Flindt et al., 1999). The root-sediment interaction is complex and covers a wide range of biogeochemical processes, (e.g., Caçador et al., 1996; Wigand et al., 1997). Vegetation may also act as sediment traps playing an important role in the settling of suspended matter (Kamp-Nielsen & Flindt, 1993; Caçador et al., 1996, 2000; Andersen & Ring, 1999; Flindt et al., 1999). Several papers have reported chemical changes in the rhizosphere of several plants, including the redox potential (Eh), organic matter contents, metal availability and oxygen and nutrient profiles, (e.g. Kamp-Nielsen & Flindt, 1993; Caçador et al., 1996, 2000; Anderson et al., 1997; Cartaxana & Lloyd, 1999; Flindt et al., 1999; Azzoni et al., 2001). Nonetheless, few papers have addressed comparative studies of different plants in natural estuarine environments (Wigand et al., 1997; Sánchez et al., 1998). It was hypothesised that different plants may have different specific effects, and consequently different contributions to the system nutrient dynamics as a whole.

The Mondego estuary is located on the Atlantic coast of Portugal (40°08′ N, 8°50′ W). It is about 7 km long and 2–3 km across at its widest part. Wetland habitats occupy about 18% of the south arm of the Mondego estuary, where the *Scirpus maritimus* L. population occupies the inner mud-flat areas, *Spartina maritima* (Curtis) Fernald occurs in the higher downstream mud-flat and sand-flat areas, and the *Zostera noltii* Hornem beds occupy the downstream mud-flats (Fig. 1). In the 1980s, *Z. noltii* beds occupied a broad expanse along the southern arm reaching the inner most parts of the estuary. Nowadays, this grass-like flowering plant has become restricted to a small patch located downstream, having been replaced elsewhere by green macroalgae (Marques et al., 1997, 2003; Cardoso et al., 2002). These three species also differ in their annual dynamics. New leaves of *Z. noltii* appear in late winter (late February/March) and the eelgrass meadows develop over the intertidal mud-flats in spring/summer. Leaf cover begins to decline during autumn/winter. *S. maritima* is a rhizomatous grass forming extensive monotypic stands (Sánchez et al., 1997), with a continuous but very slow growth (Adams & Bate, 1995). *S. maritimus* is a stoutly rhizomatous perennial sedge (Karagatzides & Hutchinson, 1991), and usually forms similar dense monospecific stands in shallow brackish marshes (Lieffers & Shay, 1982). In the Mondego estuary this species has a particularly aboveground life cycle with a growing season from January to April/May (Lillebø et al., 2003).

The objective was to study the sediment profiles at *Z. noltii* meadows and on the *S. maritima* and *S. maritimus* salt marshes in comparison with sand and mud-flats without vegetation, concerning: the organic matter contents; the dissolved inorganic nutrients (PO_4–P, NH_3–N, NO_3–N) followed by an estimation of the total amount of these nutrients during day and night conditions; and finally compare the potential net-fluxes. This study took place in May, because spring corresponds to the season where *S. maritimus* and *Z. noltii* reaches maximum

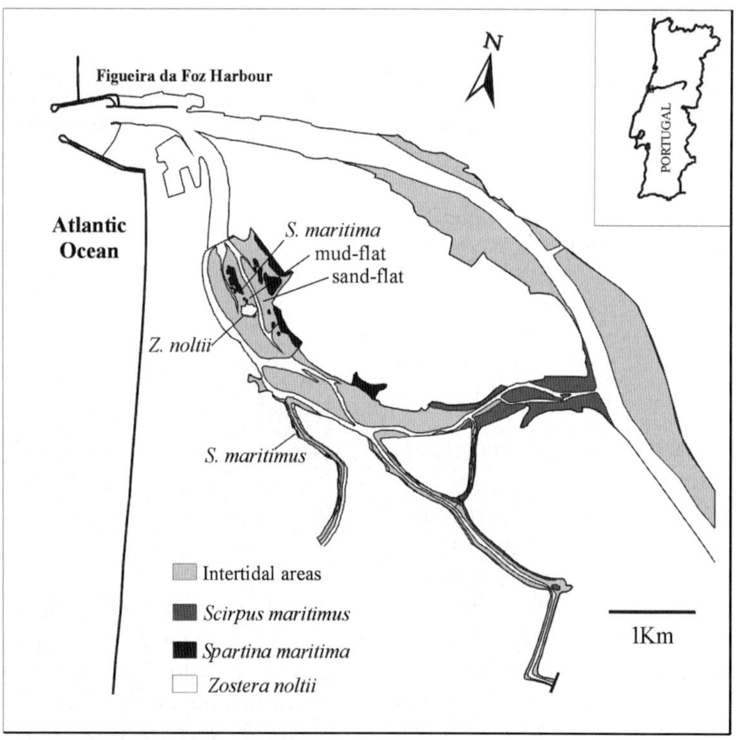

Figure 1. The location of the Mondego estuary and location of the study sites: *Spartina* and *Scirpus* salt marshes, *Zostera* meadows and mud- and sand-flats without vegetation.

aboveground biomass. The mean above and belowground biomass (Mean ± STD) in spring is, respectively, 51 ± 29 g Dwt m^2 and 25 ± 8 g Dwt m^2 for *Zostera*, 1866 ± 679 g Dwt m^2 and 2183 ± 1119 g Dwt m^2 for *Spartina*, and 435 ± 273 g Dwt m^2 and 3374 ± 890 g Dwt m^2 for *Scirpus*, in IMAR-Data base and Lillebø et al. (2003).

Material and methods

Five different study areas were established in the southern arm of the Mondego estuary on the 18th of May 2000: a bare sediment mud-flat; a sand-flat area; the *Zostera* beds; the *Spartina* salt marsh; and the *Scirpus* salt marsh. In order to characterise each area, triplicate sediment cores (10 cm depth) were analysed (1 cm slices) for percentage of water contents (Wwt–Dwt, 105 °C for 24 h), and loss on ignition (LOI) (Dwt-AFDW, 550 °C for 6 h). Sediment nutrient profiles were studied by placing vertically dialysis chambers in the muddy and sandy bare sediments and in the plants' rhizospheres; for a more detailed description of the technique see Kamp-Nielsen & Flindt (1993). In our case, the water samples were collected with a syringe (2.5 ml, corresponding to each chamber volume), and simultaneously the chamber was filled up with distilled water. To evaluate the effect of plants, samples were taken after 12 h exposure under dark conditions (night period) and 12 h under light conditions (day).

All water samples were analysed for dissolved inorganic phosphate, ammonia and nitrate in a rapid flow autoanalyser (RFA 300 Alpkem) according to Alpkem methodologies (1990). A non-parametric Mann-Whitney test was applied to determine the significance of differences (95% confidence level) between the sediment nutrient profiles (Zar, 1996).

Estimates of the amount of nutrients in the sediment at each depth were performed considering the LOI, the nutrient concentration in the interstitial water, the sediment sample specific mass and the water volume fraction. The total amount was obtained by integration of the amounts through depth.

Possible night efflux rates were calculated taking into account the difference between the night and day total amount plus the estuarine area occupied by each species population (*Zostera* = 8850 m^2, *Spartina* = 104720 m^2 and *Scirpus* = 105863 m^2).

Results

The profiles of percentage of LOI in the sediment indicate an increase of organic matter from sand-flat, to *Zostera, Spartina* and to mud-flat (Fig. 2). *Scirpus* marshes, located in the most inner estuarine areas, with lower hydrodynamics and higher sedimentation of fine sediments, showed a higher percentage of LOI.

Phosphate and ammonia concentrations in sediment pore-water increased with depth, especially in *Zostera* (Fig. 3a and b) and in the mud-flat (Fig. 4a and b) profiles, and during the night. On the other hand, nitrate concentrations were

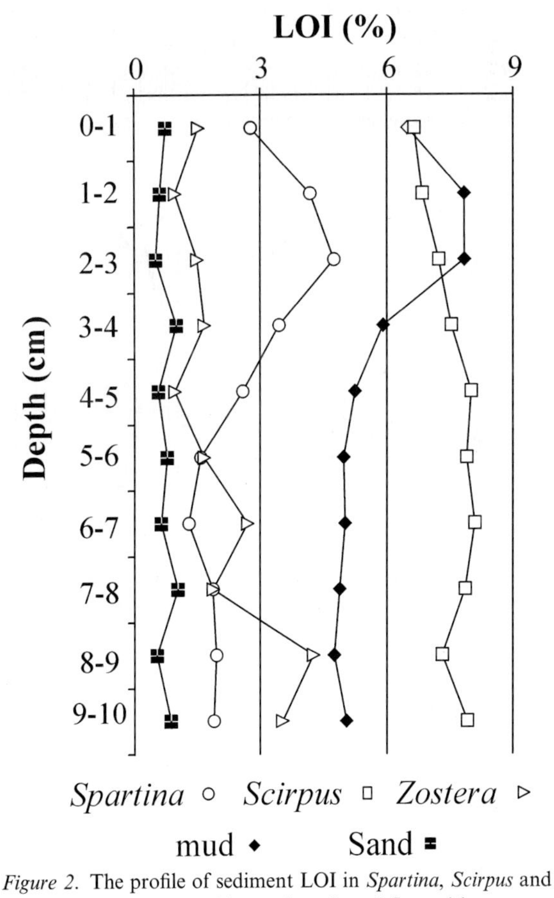

Figure 2. The profile of sediment LOI in *Spartina, Scirpus* and *Zostera* rhizosphere, and in mud- and sand-flats without vegetation.

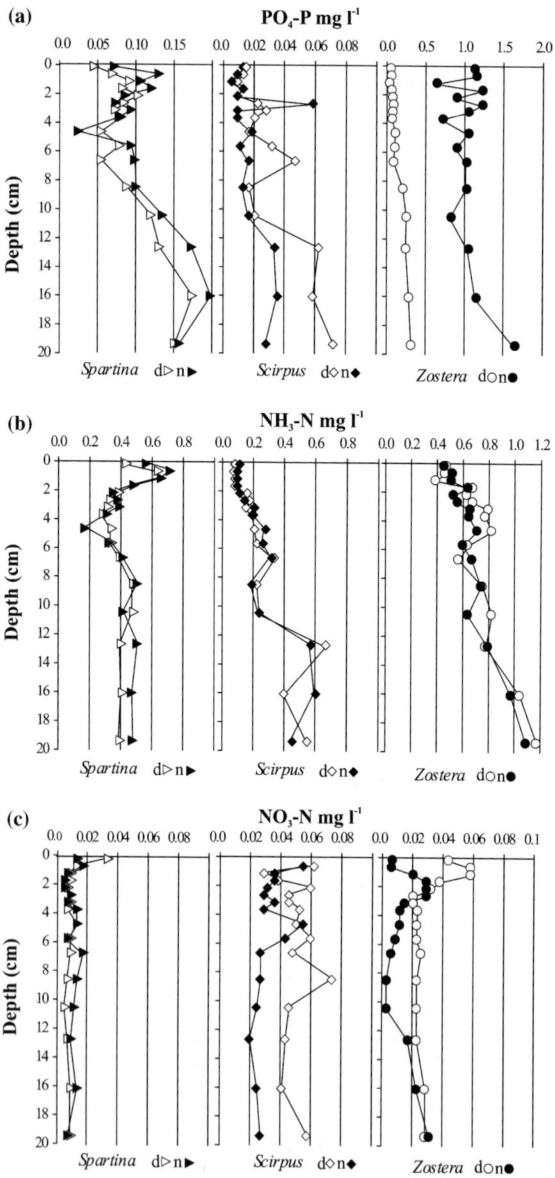

Figure 3. Nutrient concentrations in sediment pore-water profiles in *Spartina*, *Scirpus* and *Zostera* rhizosphere, during day and night: (a) Phosphate; (b) Ammonia; (c) Nitrate.

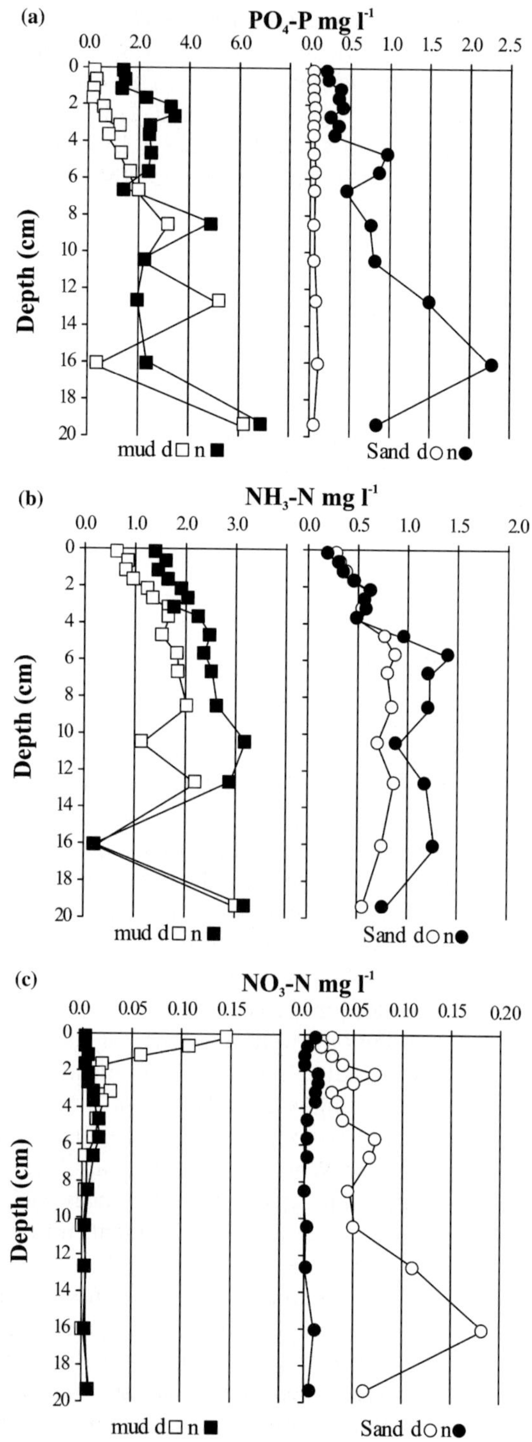

Figure 4. Nutrient concentrations in sediment pore-water profiles mud- and sand-flats without vegetation, during day and night: (a) Phosphate; (b) Ammonia; (c) Nitrate.

higher during the day and tended to decrease with depth, except for *Scirpus* rhizosphere were it was quite constant (Fig. 3c), and in the sand-flat were it only slightly increased (Fig. 4c).

The non-parametric Mann-Whitney tests showed that day-night profiles were significantly different in the mud and sand-flats without vegetation (Table 1A). In the vegetated flats, day-night profiles were statistically different for PO_4–P and NO_3–N in *Z. noltii* rhizosphere and for NO_3–N in *S. maritimus* rhizosphere. In the *Spartina* rhizosphere there were no statistically significant differences between day and night

Table 1. Results from the non-parametric Mann-Whitney expressing the significance of differences (95% confidence level) in sediment PO_4–P, NH_3–N and NO_3–N profiles

		Mud-flat	Sand-flat	*Spartina maritima*	*Scirpus maritimus*	*Zostera noltii*
A: Between day and night concentrations						
PO_4–P	Mud-flat	0.0079	–	–	–	–
	Sand-flat	–	0.0000	–	–	–
	Spartina maritima	–	–	n.s.	–	–
	Scirpus maritimus	–	–	–	n.s.	–
	Zostera noltii	–	–	–	–	0.0000
NH_3–N	Mud-flat	0.0136	–	–	–	–
	Sand-flat	–	n.s.	–	–	–
	Spartina maritima	–	–	n.s.	–	–
	Scirpus maritimus	–	–	–	n.s.	–
	Zostera noltii	–	–	–	–	n.s.
NO_3–N	Mud-flat	n.s	–	–	–	–
	Sand-flat	–	0.0000	–	–	–
	Spartina maritima	–	–	n.s.	–	–
	Scirpus maritimus	–	–	–	0.0017	–
	Zostera noltii	–	–	–	–	0.0033
B: Between concentrations during day						
PO_4–P	Sand-flat	0.0000	–	–	–	
	Spartina maritima	0.0000	0.0003	–	–	
	Scirpus maritimus	0.0000	0.0034	0.0000	–	
	Zostera noltii	0.0000	0.0001	n.s.	0.0000	
NH_3–N	Sand-flat	0.0002	–	–	–	
	Spartina maritima	0.0000	0.0109	–	–	
	Scirpus maritimus	0.0000	0.0001	0.0007	–	
	Zostera noltii	0.0006	n.s.	0.0002	0.0000	
NO_3–N	Sand-flat	0.0022	–	–	–	
	Spartina maritima	n.s.	0.0000	–	–	
	Scirpus maritimus	0.0014	n.s.	0.0000	–	
	Zostera noltii	0.0129	0.0033	0.0000	0.0002	
C: Between concentrations during night						
PO_4–P	Sand-flat	0.0000	–	–	–	
	Spartina maritima	0.0000	0.0000	–	–	
	Scirpus maritimus	0.0000	0.0000	0.0000	–	
	Zostera noltii	0.0000	0.0021	0.0000	0.0000	
NH_3–N	Sand-flat	0.0000	–	–	–	
	Spartina maritima	0.0000	0.0167	–	–	
	Scirpus maritimus	0.0000	0.0001	0.0016	–	
	Zostera noltii	0.0000	n.s.	0.0004	0.0000	
NO_3–N	Sand-flat	n.s.	–	–	–	
	Spartina maritima	0.0205	0.0065	–	–	
	Scirpus maritimus	0.0000	0.0000	0.0000	–	
	Zostera noltii	0.0134	0.0038	n.s.	0.0003	

profiles considering the three nutrients analysed (Table 1A).

The nutrient profiles in the mud- and sand-flat without vegetation were statistically different from each other and from all other studied areas during day (Table 1B) and night situations (Table 1C). There were statistically significant differences concerning the concentrations of the three nutrients analysed between *Spartina* and *Scirpus* rhizosphere profiles and between *Spartina* and *Zostera* rhizosphere profiles, under day and night situations. *Scirpus* and *Zostera* rhizosphere were also statistically different for PO_4–P, NH_3–N and NO_3–N during the day (Table 1B) and night (Table 1C).

The total amount of phosphate in the first 10 cm of sediment was lower in the sand-flat, increasing in *Scirpus*, *Spartina* and in *Zostera* rhizosphere, and was much higher in the mud-flat without vegetation (Fig. 5a). The total amount of ammonia was lower in *Scirpus* rhizosphere, increasing in the sand-flat, *Zostera* rhizosphere, mud-flat and was highest in *Spartina* rhizosphere (Fig. 5b). Nitrate total amount was higher in *Scirpus* rhizosphere, decreasing in *Spartina* and *Zostera* rhizosphere, in sand-flat, and was much lower in the mud-flat without vegetation (Fig. 5c).

Zostera roots penetrate approximately 5 cm while *Scirpus* and *Spartina* may have active roots down to 20 cm in depth. The potential nutrient efflux from plants' rhizospheres during the night shows that at this time of the year *Scirpus* salt marsh acts as a sink of nutrients, *Spartina* salt marsh may contribute to the efflux of nutrients during the night period, especially phosphate, and *Zostera* beds may have a comparatively strong contribution to phosphate efflux during the night (Fig. 6).

Discussion

The day-night amount of nutrients exchanged in the sediment pore-water without vegetation and the day-night sediment-water exchanges in plants' rhizospheres seem quite different. However, LOI can express the potential availability of nutrients in the sediment, explaining the low phosphorus and nitrogen concentration in the sand-flat profiles compared with the high concentration in the

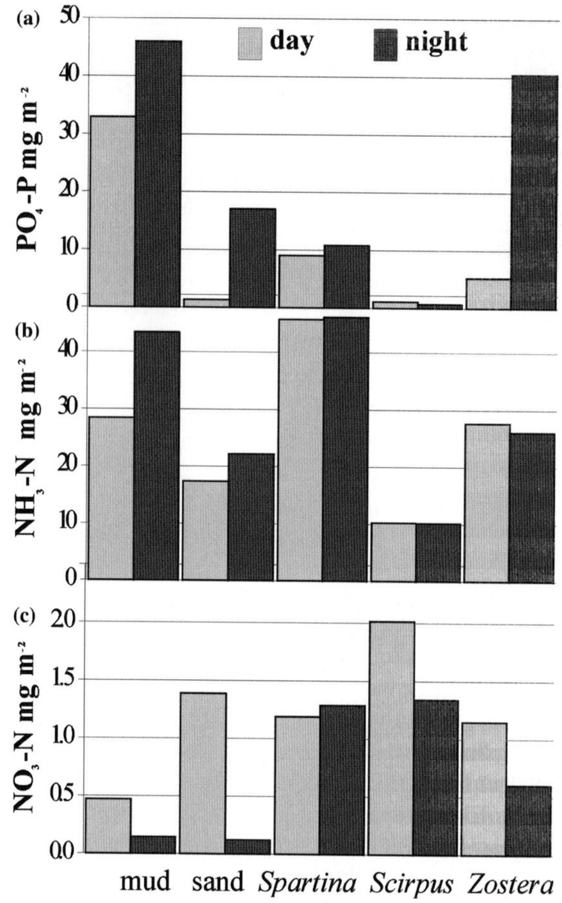

Figure 5. Total amount of nutrients in *Spartina*, *Scirpus* and *Zostera* rhizosphere, and in mud- and sand-flats without vegetation, during day and night: (a) Phosphate; (b) Ammonia; (c) Nitrate.

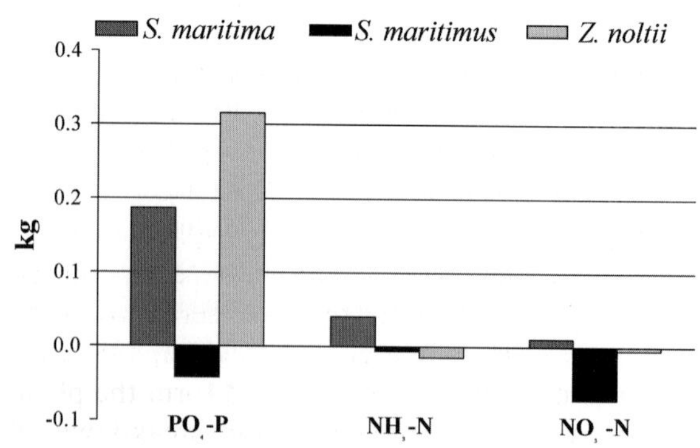

Figure 6. The potential night efflux of phosphate, ammonia and nitrate from *Spartina*, *Scirpus* and *Zostera* rhizospheres.

mud-flat (Falcão & Vale, 1998), as particle-water equilibrium and reactions between these two phases can influence the nutrient availability for plant uptake. On the other hand, in the plant rhizosphere, nutrients may diffuse towards the roots and the uptake of water by plants creates advective transport of water and salt in the direction of the roots. Additionally, the sediment phosphate adsorption capacity and the form of dissolved inorganic nitrogen will be dependent upon the oxic state of the sediment, (e.g., Flindt, 1994; Cowan et al., 1996). In the areas without vegetation, the significant differences between day and night profiles, with higher phosphate and ammonia concentrations during the night coupled with the extremely low nitrate concentration, may reflect oxygen depletion due to respiration, which during the day may be compensated by autotrophic oxygen production from associated microalgae. In addition, the integrated low amount of nitrate and the high ammonia pool most likely also reflect the oxygen limitation of the nitrification process (Kamp-Nielsen & Flindt, 1993; Flindt, 1994; Flindt et al., 1999). In the colonised sediments day-night variation of nutrient concentrations in sediment pore-water seemed to be very much dependent on plant root biomass and penetration in depth. In spring, the average biomass of *Zostera* roots in the top 5 cm of sediment is 25 ± 8 g Dwt m^2, while salt marsh plants have active roots down to 20 cm in depth and the average roots biomass is respectively 2183 ± 1119 g Dwt m^2 for *Spartina* and 3374 ± 890 g Dwt m^2 for *Scirpus*. These results suggest that in spring these species may enhance the nutrient uptake for growth proposes and promote a more oxidized rhizosphere, which explains the lower phosphate and ammonia concentrations in the interstitial water during day and night situations, when compared with bare muddy sediments. Nutrient profiles in *Zostera* beds should be compared to sandy-muddy bare sediment.

Macrophytes can influence the nitrogen cycling directly by taking up dissolved inorganic nitrogen for growth purposes. Additionally, ammonification may be enhanced by re-mineralisation of particulate organic matter (Kemp et al., 1984), by dissolved organic nitrogen released form the plant roots (Smith et al., 1988) and from breakdown of senescent plant material (Kenworthy & Thayer, 1984). The nitrification process is generally limited by low oxygen and ammonia concentrations (Henriksen & Kemp, 1988).

The sediment pore-water profiles also showed that plant coverage influences differently the nutrient profiles. Results show that *Z. noltii, S. maritima* and *S. maritimus* have a specific interaction with the sediment, as has been suggested for other species (Wigand et al., 1997). Significant differences between various plant rhizosphere nutrient profiles may result from the depth of root penetration, plant biomass and plant physiology.

Zostera noltii is able to take-up nutrients from the water column through the leaves and from the interstitial water *via* the rhizomes (Short, 1987). However, it has been shown that rooted submersed macrophytes are able to take up almost all phosphorus needed from the interstitial water (Flindt et al., 1999). The penetration of *Zostera* roots into the sediment aerates the upper layers, and allows oxygen to penetrate into the top few centimetres (Short, 1987). The low phosphate concentrations during the day, especially in the top 5 cm of the sediment, may express the increased P-adsorption due to the release of oxygen, and the consumption for growth proposes. However, during the night, and considering the top 10 cm of the sediment, *Z. noltii* beds may make a comparatively strong contribution to the phosphate efflux, due to the prevalence of heterotrophic processes and the fact that the rooting zone only covers 50% of the sediment column. In *Zostera* beds, nitrogen is usually the limiting element and is most easily absorbed as ammonium (Short, 1987). Relatively high nitrification rates in *Zostera* beds have been attributed to the release of oxygen from roots (Barko et al., 1991; Flindt, 1994). Additionally, stimulated denitrification has been explained by enhanced nitrification and leakage of easily degradable dissolved organic carbon (Flindt, 1994).

Both salt marsh species have the ability to transport oxygen to the belowground parts, down to 20 cm depth into otherwise impermeable sediments, where it is used for root respiration and oxidation of the rhizosphere (Adams & Bate, 1995; Cleavering et al., 1995). This process may lead to the precipitation of iron oxides around the roots (Vale et al., 1990), and ultimately increase the phosphate adsorption capacity of the sediment

(Berner & Berner, 1996). Moreover, rhizosphere oxidation can provide aerobic micro-sites with available nitrate, and mitigate the likely predominance of ammonium over much of the marshes (Adam, 1990; Kamp-Nielsen & Flindt, 1993). The oxygen diffusion to the sediment is driven by photosynthetic oxygen production (Azzoni et al., 2001), and the fact that the *Scirpus* population is reaching maximum density and biomass, may enhance both biotic and abiotic reoxidation of reduced compounds. Additionally, the likely increase in oxidised surface area of *Scirpus* compared to *Spartina* suggests a higher nitrification rate in the root rhizosphere and subsequently denitrification. Further on, denitrification may be stimulated by high quality dissolved organic carbon released by the roots (Kamp-Nielsen & Flindt, 1993; Flindt et al., 1999). On the other hand, *Scirpus* is a C3 photosynthetic mechanism type (Boschker et al., 1999), while *Spartina* is an alternative (C4) type (Carter, 1988), or just a C4 type (Benito & Onaindia, 1991). It has been suggested that plants with a C4 photosynthetic pathway have, at least theoretically, a number of competitive advantages over C3 species (Adam, 1990), namely a higher potential productivity, higher water-use efficiency and also a more efficient use of available nitrogen (Adam, 1990). Furthermore, *Spartina* has continuous growth, and considering the top 10 cm were the rooting zone covers 100% of the sediment, this salt marsh area may contribute to the efflux of phosphate during the night period.

Our results show that nutrients are intensely exchanged inside the colonised sediments, and suggest that coupling between plant and sediment is a species-specific process, allowing a more comprehensive and environmentally correct understanding of estuarine systems functioning.

Acknowledgements

This study was carried out in the scope of the POCTI – Formar e Qualificar – Medida 1.1 programme (Portugal) through a Post-Doctoral grant (SFRH/BPD/5650/2001). Furthermore, it was supported by the research project DYNAMOD (POCTI/M6S137431/2001) funded by the Portuguese FCT. The authors are indebted to all colleagues from IMAR-Coimbra who assisted in the field and laboratory work. The authors would like to thank the Freshwater Biological Laboratory, University of Copenhagen, Denmark, for the support of this study.

References

Adam, P., 1990. Plants and Salinity in Salt Marsh Ecology. Cambridge University Press, Cambridge, pp. 208–277.

Adams, J. B. & G. C. Bate, 1995. Ecological implications of tolerance of salinity and inundation by *Spartina maritima*. Aquatic Botany. 52: 183–191.

Alpkem, 1990. Rapid Flow Analyser – Methodology A303–S202, A303–S170, A303–S020. ALPKEM Corporation, Clackamas, Oregon, USA.

Andersen, F. Ø. & P. Ring, 1999. Comparison of phosphorus release from littoral and profundal sediments in a shallow eutrophic lake. Hydrobiologia 408/409: 175–183.

Anderson, I. C., C. R. Tobias, B. B. Neikirk & R. L. Wetzel, 1997. Development of a process-based nitrogen mass balance model for Virginia (USA) *Spartina alterniflora* salt marsh: implications for net DIN flux. Marine Ecology Progress Series 29: 13–27.

Azzoni, R. G., M. Giordani, D. T. Bartoli & P. Welsh Viaroli, 2001. Iron, sulphur and phosphorus cycling in the rhizosphere sediments of an euthrophic *Ruppia cirrhosa* meadow (Valle Smarlacca, Italy). Journal of Sea Research 45: 15–26.

Barko, J. W., D. Gunnison & S. R. Carpenter, 1991. Sediment interactions with submersed macrophyte growth and community dynamics. Aquatic Botany 41: 41–65.

Benito, I. & M. Onaindia, 1991. Biomass and aboveground production of four angiosperms in Cantabrian (N. Spain) salt marshes. Vegetario 96: 165–175.

Berner, E. K. & R. Berner, 1996. Marginal marine environments, estuaries. Global Environment, Water, Air and Geochemical Cycles. Chapter 7. Estuaries, Prentica-Hall Inc, New Jersey: 284–311.

Boschker, H. T. S., J. F. C. de Brouwer & T. E. Cappenberg, 1999. The contribution of macrophyte-derived organic matter to microbial biomass in salt-marsh sediment: Stable isotope analysis of microbial biomarkers. Limnology and Oceanography 44: 309–319.

Caçador, I., C. Vale & F. Catarino, 1996. Accumulation of Zn, Pb, Cr and Ni in sediments between roots of the Tagus estuary salt marshes, Portugal. Estuarine, Coastal and Shelf Science 42: 393–403.

Caçador, I., C. Vale & F. Catarino, 2000. Seasonal variation of Zn, Pb, Cu and Cd concentrations in the root-sediment system of *Spatina maritima* and *Halimione portugaloides* from Tagus estuary salt marshes. Marine Environment Research 49: 279–290.

Cardoso, P. G., A. I. Lillebø, M. A. Pardal, S. M. Ferreira & J. C. Marques, 2002. The effect of different primary producers on *Hydrobia ulvae* population dynamics: a case study in a temperate intertidal estuary. Journal Experimental Marine Biology Ecology 277: 173–195.

Cartaxana, P. & D. Lloyd, 1999. N_2, N_2O and O_2 profiles in a Tagus estuary salt marsh. Estuarine, Coastal Shelf Science 48: 751–756.

Carter, R. W. G., 1988. Coastal wetlands, salt marshes. Coastal Environments. An Introduction to the Physical, Ecological and Cultural Systems of Coastlines. Ac. Press, London: 335–346.

Cowan, J. L. W., J. R. Pennock & W. R. Boynton, 1996. Seasonal and interannual patterns of sediment nutrient and oxygen fluxes in Mobile Bay, Alabama (USA): regulating factors and ecological significance. Marine Ecology Progress Series 141: 229–245.

Falcão, M & C. Vale, 1998. Sediment–water exchanges of ammonium and phosphate in intertidal and subtidal areas of a mesotidal coastal lagoon (Ria Formosa). Hydrobiologia 373/374: 193–201.

Flindt, M. R., 1994. Measurements of nutrient fluxes and mass balances by on-line *in situ* dialysis in a Zostera marina bed culture. Verhandlungen Internationale Vereinigung für theoretische und angewandte Limnologie 25: 2259–2264.

Flindt, M. R., M. A. Pardal, A. I. Lillebø, I. Martins & J. C. Marques, 1999. Nutrient cycling and plant dynamic in estuaries: a brief review. Acta Oecologica 20: 237–248.

Henriksen, K. & W. M. Kemp, 1988. Nitrification in estuarine and coastal marine sediments: methods, patterns and regulating factors. In Blackburn, T. H. & J. Sorensen (eds.) Nitrogen Cycling in Coastal Marine Environments. John Wiley and sons, New York: 207–250.

Kamp-Nielsen, L. & M. R. Flindt, 1993. On-line recording of porewater profiles from *in situ* dialysis. Verhandlungen Internationale Vereinigung für theoretische und angewandte Limnologie 25: 151–156.

Karagatzides, J. D. & I. Hutchinson, 1991. Intraspecific comparison of biomass dynamics in *Scirpus americanus* and *Scirpus maritimus* on the Fraser river delta. Journal of Ecology 79: 459–476.

Kemp, W. M., W. R. Boynton, R. R. Twilley, J. C. Stevenson & L. G. Ward, 1984. Influeces of submersed vascular plants on ecological processes in upper Chesapeake Bay. In Kennedy, V. S. (ed.) The Estuary as a Filter. Academic Press, New York: 367–394.

Kenworthy, W. J. & G. W. Thayer, 1984. Production and decomposition of the roots and rhizomes of seagrasses, *Zostera marina* and *Thalassia testudinum*, in temperate and subtropical marine ecosystems. Bulletin of Marine Science 35: 364–379.

Lieffers, V. J. & J. M. Shay, 1982. Seasonal growth and standing crop of Scirpus maritimus var. paludosus in Saskatchewan. Canadian Journal of Botany 60: 117–125.

Lillebø, A. I., M. A. Pardal, J. M. Neto & J. C. Marques, 2003. Salinity as the major factor affecting *Scirpus maritimus* annual dynamics. Evidence from field data and greenhouse experiment. Aquatic Botany 77: 111–120.

Marques, J. C., M. A. Pardal, S. N. Nielsen & S. E. Jørgensen, 1997. Analysis of the properties of exergy and biodiversity along an estuarine gradient of eutrophication. Ecological Modelling 62: 155–167.

Marques, J. C., S. N. Nielsen, M. A. Pardal & S. E. Jørgensen, 2003. Impact of eutrophication and river management within a framework of ecosystem theories. Ecological Modelling 166: 147–168.

Sánchez, J. M., X. L. Otero, J. Izco & F. Marcías, 1997. Growth form and population density of *Spartina maritima* (Curtis) Fernald in Northwest Spain. Wetlands 17: 368–374.

Sánchez, J. M., X. L. Otero & J. Izco, 1998. Relation between vegetation and environmental characteristics in a salt-marsh system on the coast of Northwest Spain. Plant Ecology 136: 1–8.

Short, F. T., 1987. The effects of sediment nutrients on eelgrasses; literature review and mesocosm experiment. Aquatic Botany 27: 41–57.

Smith, R. D., A. M. Pregnall & R. S. Alberte, 1988. Effects of anaerobiosis on root metabolism of *Zostera marina* (eelgrass): Implications for survival in reducing sediments. Marine Biology 98: 131–141.

Vale, C., F. M. Catarino, C. Cortesão & M. I. Caçador, 1990. Presence of metal-rich rhizoconcretions on the roots of *Spartina maritima* from the salt marshes of the Tagus estuary, Portugal. Science Total Environment 97/98: 617–626.

Wigand, C., J. C. Stevenson & J. C. Cornwell, 1997. Effects of different submersed macrophytes on sediment biogeochemistry. Aquatic Botany 56: 233–244.

Zar, J., 1996. Biostatistical Analysis, 3rd edn. Prentice-hall International Inc., Simon and Schuster/A viacom Company, Upper Saddle River, NJ, USA, 662 pp.

Hydrobiologia (2006) 555:185–192
H. Queiroga, M.R. Cunha, A. Cunha, M.H. Moreira, V. Quintino, A.M. Rodrigues, J. Serôdio & R.M. Warwick (eds),
Marine Biodiversity: Patterns and Processes, Assessment, Threats, Management and Conservation
DOI 10.1007/s10750-005-1115-3

Biogeographic patterns of intertidal macroinvertebrates and their association with macroalgae distribution along the Portuguese coast

Sílvia G. Pereira[1], Fernando P. Lima[1,2], Nuno C. Queiroz[1,2], Pedro A. Ribeiro[2] & António M. Santos[1,2,*]

[1]*Departmento de Zoologia-Antropologia, Faculdade de Ciências da Universidade do Porto, 4099-002 Porto, Portugal*
[2]*CIBIO, Centro de Investigação em Biodiversidade e Recursos Genéticos, Campus Agrário de Vairão, 4485-661 Vairão, Portugal*
*(*Author for correspondence: E-mail: amsantos@fc.up.pt)*

Key words: Amphipoda, biogeography, Isopoda, macroalgae, Portugal, Tanaidacea

Abstract

Geographical patterns in the distribution of epifaunal crustaceans (Amphipoda, Isopoda and Tanaidacea) occurring with dominant macroalgal species were investigated along the Portuguese rocky coast. Three regions, each encompassing six shores, were studied. Algal species were selected according to their geographical distribution: *Mastocarpusstellatus* and *Chondrus crispus* (north); *Bifurcariabifurcata* (north-centre); *Plocamiumcartilagineum* and *Cystoseiratamariscifolia* (centre-south); *Corallina* spp. and *Codiumtomentosum* (entire coast). Multivariate techniques were used to test for differences in crustacean assemblage composition between sub-regions and host algal species. A clear gradient of species substitution was observed from north to south. Differences in abundance and diversity of epifaunal crustaceans were observed between southern locations and the remaining sites. Four species were recorded for the first time in the Portuguese coast. Among the 57 *taxa* identified, southern distribution limits were observed for three species and northern distribution limits were observed for four species. Interestingly, the observed geographical patterns in epifaunal abundance and diversity were not related with geographical changes in the indentity of the dominant algal species.

Introduction

Understanding the causes driving species distribution is a major challenge of modern biogeography. The analysis of the relation between patterns of distribution of organisms and physical or biological factors is usually the first step towards this goal (Hoffman & Blows, 1994). In the marine environment, most large to medium scale studies were primarily focused on the importance of physical factors in the distribution of species (van den Hoek, 1984; Cambridge et al., 1987; Zacharias & Roff, 2001).

Marine algae are known to provide habitats for a wide range of animal species (Williams & Seed, 1992). Several studies addressed the role of seaweeds as determinants of epifauna diversity (Kitching, 1981; Kelaher et al., 2001; Parker et al., 2001; Chemello & Milazzo, 2002; Christie et al., 2003). However, few attempts have been made to investigate the relation between geographical changes in algae composition and patterns of macroinvertebrate diversity (Arrontes & Anadón, 1990a; Russo, 1997).

A considerable number of both cold- and warm-water algal species reach their distributional limits within the Portuguese coast (Ardré, 1971). In the lower eulittoral, northern shores are dominated by large macrophytes, such as *Himanthalia elongata* (Linnaeus) S.F. Gray, and the red algae *Chondrus crispus* Stackhouse and *Mastocarpus stellatus* (Stackhouse). Towards the south, these species are replaced by *Cystoseira tamariscifolia* (Hudson) and other red algae become dominant, especially *Plocamium cartilagineum* (Linnaeus) (Ardré, 1970, 1971). Despite these differences, no attempts have been made to couple such information with the patterns of distribution of the associated fauna. Recent biogeographical studies in the Portuguese coast addressed solely the distribution and abundance of conspicuous animals (see Boaventura et al., 2002). In contrast, smaller organisms, including those inhabiting seaweeds, have been poorly studied. The information available on these *taxa* often comes from unpublished academic theses, and in most cases is of limited geographic scope.

The purpose of this study was to: (1) provide a description of the patterns of distribution of algal-dwelling crustaceans (Amphipoda, Isopoda and Tanaidacea) along the Portuguese rocky coast and (2) make a preliminary attempt to test the relationship between patterns of distribution of crustacean epifauna and conspicuous macroalgal species.

Material and methods

Study area

The study area encompassed the whole continental Portuguese coast and was divided in three regions (North, Centre and South), corresponding to the main stretches of rocky coastline (Fig. 1). Within each region, six shores with similar geomorphology and wave exposure were selected. Samples of the dominant macroalgae were collected from each shore.

Sampling procedures

Algal species were selected according to their geographical distribution. *M. stellatus* and *C. crispus* are abundant in the north of Portugal, becoming rare or absent towards the south. *Bifurcaria bifurcata* R. Ross is characteristic of northern and central shores. *P. cartilagineum* and *C. tamariscifolia* are abundant in the centre and south. *Corallina* spp. (*C. elongata* J. Ellis & Solander and *C. officinalis* Linnaeus) and *Codium tomentosum* Stackhouse are common along the entire study area. For further details see Figure 2.

Algae were collected either individually (*Cystoseira*, *Codium* and *Bifurcaria*) or by scraping 20 × 20 cm quadrats of monospecific stands (*Corallina*, *Chondrus*, *Mastocarpus*, *Plocamium*). Samples were preserved in formalin (10%). Extraction of animal species was made by washing the algae in flow water through a set of sieves (5 mm to 250 μm mesh). The fraction retained in the 250 μm sieve was sorted under the binocular miscroscope (10× magnification). In addition, the algae were also inspected under a binocular microscope to pick any remaining organisms. All animals were counted and identified to the lowest possible *taxon*. Due to the different structure of the selected algal species, the density of animals was expressed as number of individuals per volume of algae. After removal of macroinvertebrates the algae were dried for 2 h. They were then inserted into a graduated cylinder with a fixed amount of water, and the volume of algae was estimated as the difference between the initial and final volume.

Data analysis

Multivariate analyses were carried out with the PRIMER package (Clarke & Warwick, 1994). Non-parametric multidimensional scaling (nMDS) ordination of samples (alga/site) was performed using Bray–Curtis similarity coefficient. Transformation of animal abundances into presence–absence was chosen in order to minimise data variability. To determine a possible correlation between epifaunal assemblages and algal species a Mantel test (Sokal & Rohlf, 1995) was carried out using distance matrices for sites based on presence-absence of animal and algal species. One-way ANOSIM analyses to test for differences in epifaunal assemblages of algal species were made separately for each region to overcome the problem related to the lack of orthogonality of the two factors involved (regions and algal species).

One-way ANOSIM analyses were also employed to test for differences in species composition between the three regions. In order to avoid confounding effects between regions and algae, only species which span along the entire coast were used (*Corallina* spp. and *C. tomentosum*). To achieve an overall Type I error rate of $\alpha = 0.05$ in multiple tests, a Bonferroni correction was used (Quinn & Keough, 2002).

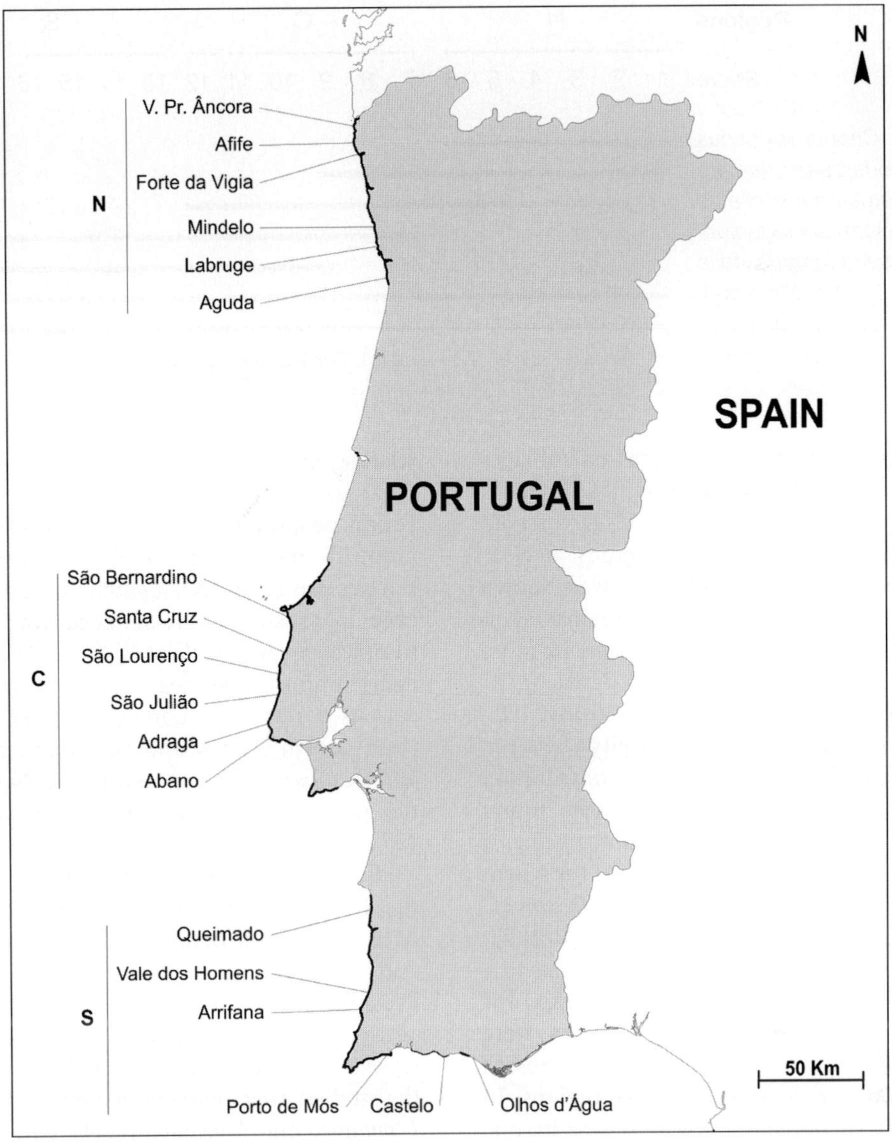

Figure 1. Location of the studied shores in the Portuguese coast. The main stretches of rocky coastline are depicted in black. sampled location. N: 1 – Vila Praia de Âncora, 2 – Afife, 3-Forte da Vigia, 4 – Mindelo, 5 – Labruge and 6 – Aguda; C: 7 – São Bernardino, 8 – Santa Cruz, 9 – São Lourenço, 10 – São Julião, 11 – Adraga and 12 – Abano; S: 13 – Queimado, 14 – Vale dos Homens, 15 – Arrifana, 16 – Porto de Mós, 17 – Castelo and 18 – Olhos d'Água.

Results

A total of 57 *taxa* were identified (34 amphipods, 19 isopods and four tanaids). Most *taxa* occurred with more than one algal species (32% of the *taxa* were observed at least in five algae). Seven species were seaweed-specific (five were found in *Corallina* and two in *Cystoseira*). From these, only *Biancolina algicola* Della Valle (found in *Cystoseira*) occurred consistently in more than one site.

The nMDS ordination (Fig. 3) revealed a significant interspersion of samples. No clear pattern between regions or algal species was observed. Furthermore, the stress value (0.2) suggests that too much reliance should not be placed on the detail of the plot (Clarke & Warwick, 1994).

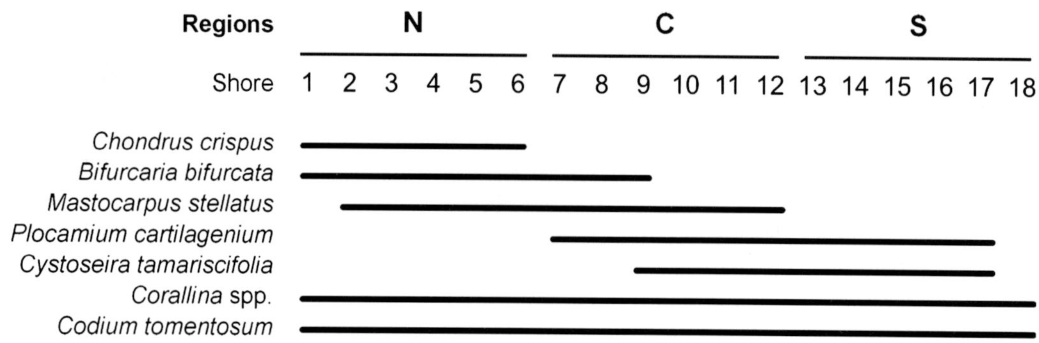

Figure 2. Shores where the algae species were sampled. For locations names see Figure 1.

Differences in macroinvertebrate assemblages were found between regions (ANOSIM $R = 0.209$; $p < 0.05$). Multiple tests revealed significant differences between all regions. A clear latitudinal gradient of substitution was shown plotting macroinvertebrate species against sampled sites. Figure 4 depicts this gradient in more detail.

Low similarity within regions was found (32–45%), denoting a high heterogeneity between replicate shores. The *taxa* with higher contribution to the dissimilarity between regions were found after SIMPER analysis: *Jassa* spp., *Idotea pelagica* Leach and *Dynamene bidentata* (Adams) for North vs. Centre; *Idotea granulosa* Rathke, *Dexamine spiniventris* (Costa) and *Synisoma capito* (Rathke) for North vs. South; *Dexamine spiniventris* (Costa), *I. pelagica* Leach and *Jassa* spp. for Centre vs. South. Although most species were found along the entire coast, several were present exclusively in the northern or southern regions. In most cases, these species were represented by one or few individuals found in one type of seaweed at a single shore and did not contribute in a consistent way to the differences between sub-regions.

ANOSIM analyses did not reveal any significant differences in epifaunal assemblages between algal species for the southern and central regions (ANOSIM $R = 0.048$, $p > 0.05$ and $R = 0.306$, $p > 0.05$ respectively). For the northern region differences were found between *B. bifurcata*, *Corallina* spp. and the remaining algae (ANOSIM $R = 0.445$, $p < 0.05$). The Mantel test revealed a low, but statistically significant, correlation between algae and epifauna ($r = 0.2429$; $p < 0.05$).

Discussion

Broad geographical differences in crustacean epifauna diversity were found along the continental Portuguese coast. A statistically significant difference in crustacean assemblage composition was found between the three regions. The species gradient supports the idea that the Portuguese coast acts as a region of contact between warm-water (from north Africa and the Mediterranean Sea) and cold-water species (from the North Sea and the Arctic) as described in earlier works (Ardré, 1970, 1971).

Four species were recorded for the first time in the Portuguese coast: two amphipods, *Caprella mitis* Mayer and *B. algicola*, one isopod, *S. capito*, and one tanaid, *Leptochelia savignyi* (Kroyer). Previously, the first three species were only recorded in the Mediterranean Sea. Not surprisingly, southern distribution limits were also detected in the studied area: *Amphitolina cuniculus* (Stebbing), *I. pelagica* and *I. granulosa*. The two idoteids were thought to be limited to northern Spain (Arrontes & Anadón, 1990b), but their range extends as far as southern Portugal.

Apparently, the observed geographical differences in epifaunal crustaceans were not related with differences in dominant algal species. These observations differ from those described by Arrontes & Anadón (1990a) for northern Spain, who found that several isopod species responded to geographical changes in algal composition. Interestingly, the three idoteid species were closely associated with algae that displayed geographical changes in abundance. This is certainly not the case in the Portuguese coast. *Idotea granulosa* and

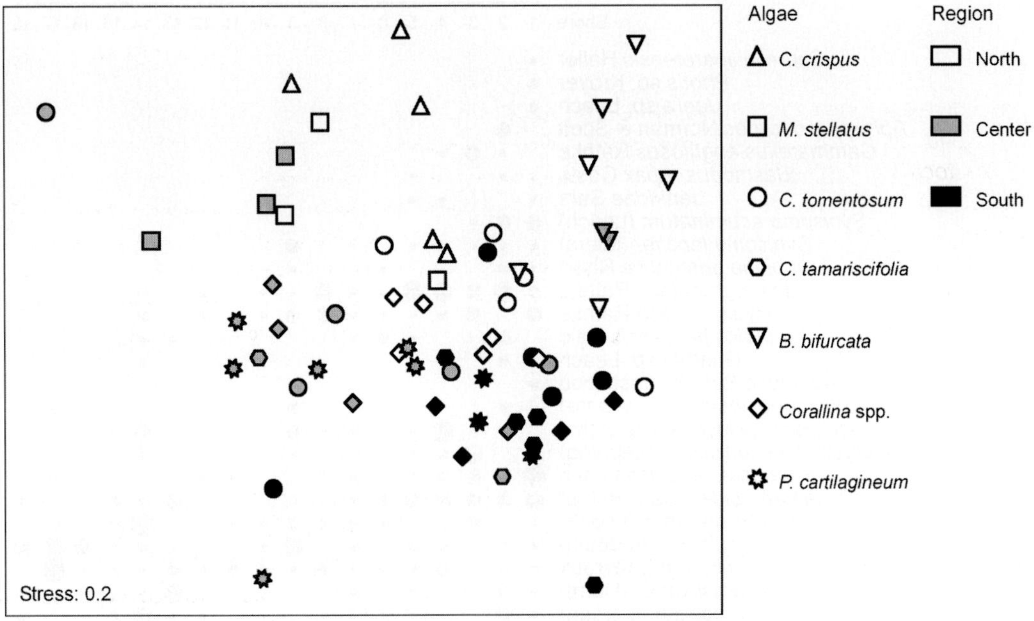

Figure 3. nMDS ordination of the samples (algae/site). Regions are depicted by different grades of shade, and algal species by different shapes.

I. pelagica were commonly found in *Corallina* (which ranges from north to south) and in both northern (*Chondrus* and *Mastocarpus*) and southern (*Cystoseira* and *Plocamium*) algal species.

Most *taxa* displayed no preference for a particular alga, occurring in association with different macroalgae species at different geographical locations. Only *B. algicola* was found consistently in *Cystoseira* at two different sites. The lack of a strict host specialisation by epifauna has been reported by several authors (Russo, 1990; Parker et al., 2001).

Algal architecture might play a more important role in determining epifaunal diversity and abundance (e.g. Crisp & Mwaiseje, 1989; Chemello & Milazzo, 2002; Kelaher, 2003). In the present study, finely branched macroalgae (*Cystoseira*, *Plocamium* and *Corallina*) exhibited a more diversified and abundant epifauna, when compared with the other five species, which have a less complex architecture. *Mastocarpus* and *Chondrus*, the only algae with blade-like structure, presented the lowest diversity values. These findings prompt for a more detailed analysis with a sampling strategy aimed towards algal complexity rather than algal species. In the present study, many non-dominant, but common, algal species were left out. Yet, when contiguous to the dominant plants, they may also account for a large proportion of the epifauna, especially when highly mobile organisms are considered (Gunnill, 1982).

Some of the species recorded in this study are probably transient in the lower shore. This is the case of *Phtisica marina* Slabber, *Synisoma acuminatum* (Leach), *Idotea baltica* (Pallas) and *Campecopea hirsuta* (Montagu), which are known to occur along the entire Portuguese coast, but appeared sporadically on some samples. The first three are mainly sublittoral species, whilst the latter is typically found in the upper midshore among barnacles and in *Lichina pygmaea* (Lightfoot) Agardh (Arrontes & Anadón, 1990b, for the isopod species). Other species are also abundant in non-algal substrates. In the present study *I. pelagica* was more common in the centre and southern Portuguese coast, but in fact it is one of the most abundant species found among mussels in the northern shores (A. M. Santos, unpublished data). Removal of such species from the data did not alter the final outcome of the analysis.

Analysis of temporal and small-scale spatial variability was not addressed in the present study due to logistic constraints given the large sampling area. However, both sources of variation are known

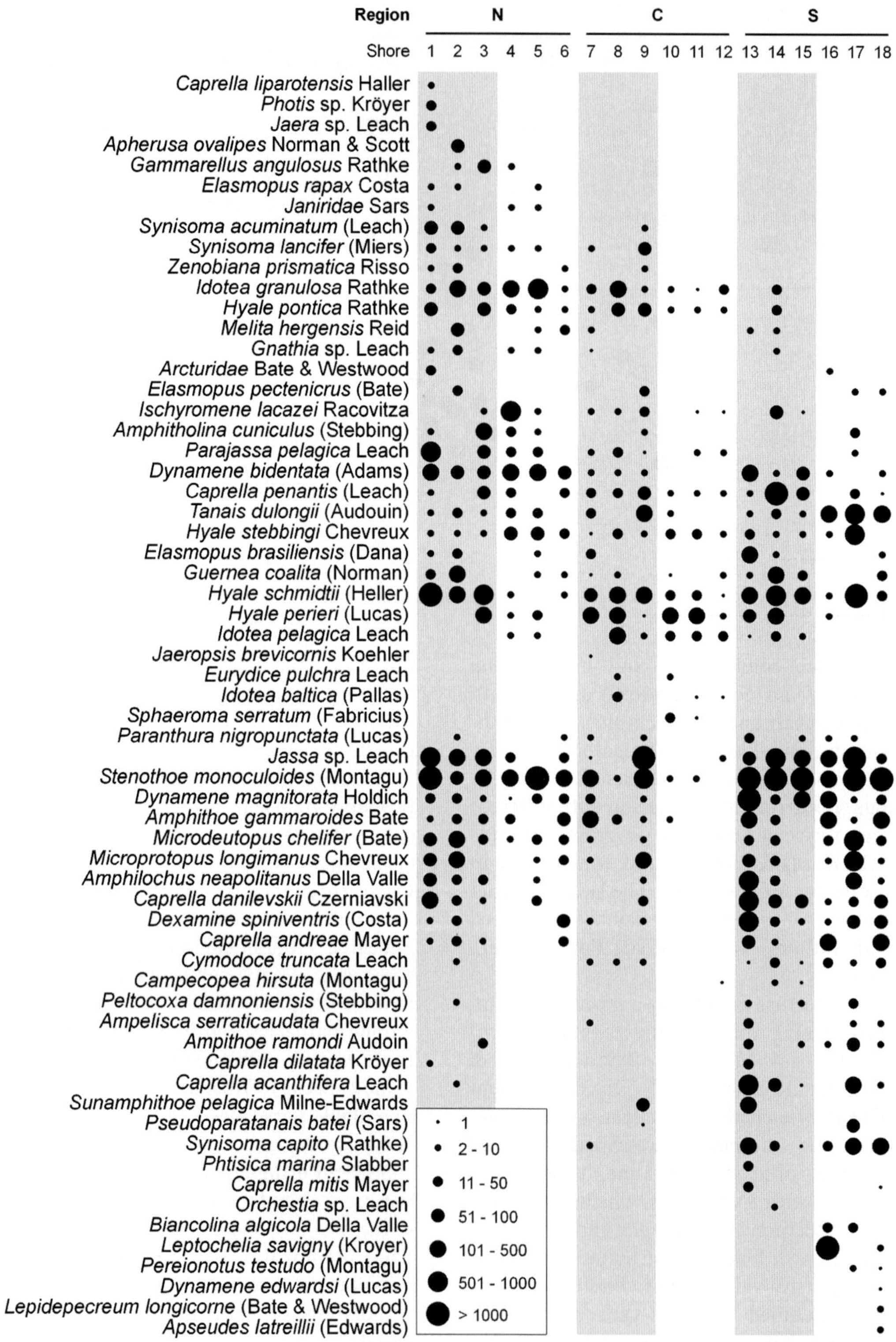

Figure 4. Graphical representation of macroinvertebrate abundance vs. sampled location. Abundances are expressed as number of individuals per 1000 cm^3 of algae. For locations names see Figure 1.

to be important (Kelaher et al., 2001). Futhermore, some species may exploit distinct habitats in different phases of their life-cycle. For example, Arrontes & Anadón (1990a) found that many isopod species displayed marked seasonal variation in algal occurrence, with macroalgae providing a habitat for juvenile stages, while adults occurred elsewhere. Other traits, such as feeding behaviour, vagility or response to predators, may also contribute to the high variation in epifaunal abundance and composition observed among different algal species (Williams & Seed, 1992) or even different thalli of the same or similar species (Gunnill, 1982; Kelaher et al., 2001). Therefore, replication at these small-scale levels is mandatory to distinguish wandering organisms from true host-plant specialists, which are more likely to be affected by changes in the distribution of their host.

The present study found that epifaunal crustacean assemblages display marked geographical differences along the Portuguese coast. Northern and southern limits of distribution were detected for several of the 57 *taxa* identified. In addition, four new species were recorded outside their previously known distribution range. Biogeographical patterns of epifauna seem not to be related with geographical changes in dominant algal species. Instead, physical factors, such as temperature, water currents and wave exposure, or biological factors, such as algal architecture, might play a more important role as determinants of epifaunal distribution and should be investigated in detail.

Acknowledgements

F.P. Lima and P.A. Ribeiro were funded by Fundação para a Ciência e Tecnologia (FCT) grants ref. SFRH/BD/8730/2002 and ref. SFRH/BD/8232/2002, respectively. We also would like to thank the contribution of two anonymous referees and Prof. Vitor Quintino for their helpful comments on the manuscript.

References

Ardré, F., 1970. Contribution à l'étude des algues marines du Portugal I. La Flore. Portugalia Acta Biologica 10: 1–423.

Ardré, F., 1971. Contribution à l'étude des algues marines du Portugal II. Ecologie et Chorologie. Bulletin du Centre d'Etudes et de Recherches Scientifiques. Biarritz 8: 359–574.

Arrontes, J. & R. Anadón, 1990a. Seasonal variation and population dynamics of isopods inhabiting intertidal macroalgae. Scientia Marina 54: 231–240.

Arrontes, J. & R. Anadón, 1990b. Distribution of intertidal isopods in relation to geographical changes in macroalgal cover in the bay of Biscay. Journal of the Marine Biological Association of the United Kingdom 70: 283–293.

Boaventura, D., P. Ré, L. C. Fonseca & S. J. Hawkins, 2002. Intertidal rocky shore communities of the continental portuguese coast: analysis of distribution patterns. P. S. Z.: N. Marine Ecology 23(1): 69–90.

Cambridge, M. L., A. M. Breeman, S. Skraak & C. van den Hoek, 1987. Temperature responses of tropical to warm temperate *Cladophora* species in relation to their diistribution in the North Atlantic Ocean. Helgolaender Wissenschaftliche Meeresuntersuchungen 41: 329–354.

Chemello, R. & M. Milazzo, 2002. Effect of algal architecture on associated fauna: some evidence from phytal molluscs. Marine Biology 140: 981–990.

Christie, H., N. M. Jörgesen, K. M. Norderhaug & E. Waage-Nielsen, 2003. Species distribution and habitat exploitation of fauna associated witk kelp (*Laminaria hyperborea*) along the Norwegian coast. Journal of the Marine Biological Association of the United Kingdom 83: 687–699.

Clarke, K. R. & R. M. Warwick, 1994. Similarity-based testing for community pattern: the two-way layout with no replication. Marine Biology 118: 167–176.

Crisp, D. J. & B. Mwaiseje, 1989. Diversity in intertidal communities with special reference to the *Corallina officinalis* community. Scientia Marina 53(2–3): 365–372.

Gunnill, F. C., 1982. Effects of plant size and distribution on the numbers of invertebrate species and individuals inhabiting the brown alga *Pelvetia fastigiata*. Marine Biology 69: 263–280.

Hoffman, A. A. & M. W. Blows, 1994. Species borders: ecological and evolutionary perspectives. Trends in Ecology and Evolution 9: 223–227.

Kelaher, B. P., 2003. Changes in habitat complexity negatively affect diverse gastropod assemblages in coralline algal turfs. Oecologia 135: 431–441.

Kelaher, B. P., M. G. Chapman & A. J. Underwood, 2001. Spatial patterns of diverse macrofaunal assemblages in coralline turf and their associations with environmental variables. Journal of the Marine Biological Association of the United Kingdom 81: 917–930.

Kitching, J. A., 1981. The flora and fauna associated with *Himanthalia elongata* (L.) S. F. Gray in relation to water current and wave action in the Lough Hyne Marine Nature Reserve. Estuarine, Coastal and Shelf Science 25: 663–676.

Parker, J. D., J. E. Duffy & R. J. Orth, 2001. Plant species diversity and composition: experimental effects on marine epifaunal assemblages. Marine Ecology Progress Series 224: 55–67.

Quinn, G. P. & M. J. Keough, 2002. Experimental Design and Data Analysis for Biologists Cambridge University Press, 537 pp.

Russo, A. R., 1990. The role of seaweed complexity in structuring Hawaiian epiphytal amphipod communities. Hydrobiologia 194: 1–12.

Russo, A. R., 1997. Epifauna living on sublittoral seaweeds around Cyprus. Hydrobiologia 344: 169–179.

Sokal, R. R. & F. J. Rohlf, 1995. Biometry (3rd ed.). W. H. Freeman & Co, New York 887 pp.

van den Hoek, C., 1984. World wide latitudinal and longitudinal seaweed distribution patterns and their possible causes, as illustrated by the distribution of Rodophytan genera. Helgolaender Wissenschaftliche Meeresuntersuchungen 38: 227–257.

Williams, G. A. & R. Seed, 1992. Interactions between macrofaunal epiphytes and their host algae. In John, S. H. D. M. & J. H. Price (eds.) Plant–Animal Interactions in the Marine Benthos. Clarendon Press, Oxford: 189–211.

Zacharias, M. A. & J. C. Roff, 2001. Explanations of patterns of intertidal diversity at regional scales. Journal of Biogeography 28: 471–483.

Hydrobiologia (2006) 555:193–206

H. Queiroga, M.R. Cunha, A. Cunha, M.H. Moreira, V. Quintino, A.M. Rodrigues, J. Serôdio & R.M. Warwick (eds),
Marine Biodiversity: Patterns and Processes, Assessment, Threats, Management and Conservation
DOI 10.1007/s10750-005-1116-2

Seasonal variations of *Cystoseira barbata* (Stackhouse) C. Agardh frond architecture

Annalisa Falace* & Guido Bressan
Department of Biology, University of Trieste, Via L. Giorgieri 10, 34127, Trieste, Italy
*(*Author for correspondence: E-mail: falace@univ.trieste.it)*

Key words: canopy algae, *Cystoseira barbata*, three-dimensional complexity, habitat architecture, phenology, seasonality, Mediterranean

Abstract

The practical assessment of the biogenic structural complexity poses some problems in non-symmetrical or three-dimensional macroalgae. This study represents the first attempt to measure quantitatively the phenology and the plasticity of *Cystoseira barbata* (Stackhouse) C. Agardh in the Mediterranean Sea. This canopy-forming brown alga is characterised by a ramified monopodial cauloid and a dendroid frond. To analyse the morphological-spatial changes in time and the plasticity of *C. barbata* an experiment was conducted in a protected sea area of the Northern Adriatic Sea on thalli taken from a natural population and a transplanted one, re-located in deeper water. The three-dimensional complexity of thalli and the frond architecture were analysed by means of quantitative structural (total area, perimeter) and spatial (intercepting area, volume and interstitial area) attributes. A preliminary assessment of the primary production of the basiphyite-epiphyte system was investigated by a volumetric analysis. The thalli showed a seasonal pattern characterised by a marked phenological variation of the fronds, both in size and in shape. The seasonal trend was comparable in the both populations studied, with fronds largest in spring–summer, whereas the fall of phylloids and branchelets occurs in autumn–winter. Also the abundance of algal epibionts on the host varied seasonably and depended mainly on the life-form of *Cystoseira* fronds, where complex fronds encouraged more abundant associated epibionts. The morphological variability of different individuals was also investigated using some bio-structural indices. These indices were proposed as useful for a more detailed description of the phenology of *C. barbata* and for a better evaluation of the potential micro-spatial and structural habitat available inside the fronds. The proposed protocol and the quantitative descriptors analysed may be employed in physiological or ecological studies, being useful for a standardisable classification of a habitat's complexity. The analysed spatial and structural attributes, as well as the definition of the complexity of *C. barbata*, enable the evaluation of the area available for the attachment and shelter of epibionts together with the number and type of potential habitats.

Introduction

The observation that the structural complexity of habitats plays an important role in determining the abundance and diversity of species has been recorded by many researchers in a number of different ecological systems, both terrestrial and aquatic (Hicks, 1980; O'Connor, 1991; Gee & Warwick, 1994; Jacobi & Langervin, 1996). Moreover biodiversity does not only vary depending on the physical nature of the environment, but the biotic community itself may influence the physical features of the habitat and, therefore, the distribution and the abundance of species (Hicks, 1980; Edgar, 1983; Dean & Connell, 1987b; Menge & Sutherland, 1987; Gibbons, 1988; Jones et al., 1994; Russo, 1997; Bulleri et al., 2002; Chemello & Milazzo, 2002; Piraino et al., 2002). Different authors have proposed several non-independent mechanisms by means of which the geometric complexity of substrata can enhance colonization rates (i.e., protecting the organisms from the environment, providing an increased number of resources and niches and increasing the area available for settlement) (Hicks, 1980; Dean & Connell, 1987a, b; Gibbons, 1988; O'Connor, 1991; Jacobi & Langervin, 1996; Chemello & Milazzo, 2002).

In particular, at a small spatial scale of observation (10^{-2}, 10^{-3} m), the development of canopy-forming algae can promote spatial heterogeneity of coastal rocky bottoms, leading to an increase of additional habitats and providing three-dimensional complexity (Hicks, 1980; Dean & Connell, 1987b; Gibbons, 1988; Jenkins et al., 1999; Benedetti-Cecchi et al., 2001; Chemello & Milazzo, 2002). Canopy algae (referred as keystone species *sensu* Paine, 1969 or ecological engineering species *sensu* Jones et al., 1994) play a fundamental role in affecting the understory assemblages, by modifying physical factors such as light (Reed & Foster, 1984; Kennelly, 1989; Figueiredo et al., 2000; Melville & Connell, 2001) and desiccation (Menge, 1978; McCook & Chapman, 1991), and by the sweeping effects of the fronds (Hawkins, 1983; Jenkins et al., 1999). Furthermore, a good relationship exists between algal structural complexity and epibiotic diversity/abundance (Hicks, 1980; Edgar, 1983; Dean & Connell, 1987b; Gibbons, 1988; Russo, 1997; Aarnio & Mattila, 2000; Hernández-Carmona et al., 2000; Chemello & Milazzo, 2002).

Seaweed size and phenology are important attributes not only in macroalgal ecology (Dudgeon et al., 1995; Hernàndez et al., 1999), but also in eco-physiological studies (Kübler & Dudgeon, 1996; Hernàndez et al., 1999). For instance, macroalgal photosynthesis, nutrient uptake and growth have been demonstrated to be size-dependent and, in particular, to be correlated to some allometric ratios (e.g., surface area:volume) (Hein et al., 1995; Kübler & Dudgeon, 1996; Hernàndez et al., 1999).

Until recent times morphological studies on macrophytes have been mainly descriptive and concern problems of systematic, evolutionary biology and cytology, but quantitative descriptors are needed to solve problems of physiology or ecology (Kailov & Firsov, 1976).

In terrestrial habitats Lawton (1994, in Gee & Warwick, 1994) has defined plant architecture with the aid of two characteristics: the size and the variety of plant structures, both in form and persistence. In marine habitats Hacker & Steneck (1990) referred to 'habitat architecture' as the 'structural and spatial attributes that define a habitat'. In the majority of benthic systems, algal structural complexity has been defined by the simple morphological description of the thallus (i.e., sheet-like, branched blades, distichous branchelets), or by using traditional morphometric features such as biomass, length, branch number, etc. (Khailov & Firsov, 1976; Edgar, 1983; Dean & Connell, 1987b; Russo, 1997; Hull, 1997; Chemello & Milazzo, 2002). Recently, algal architecture has been described by using quantitative descriptors such as the area and the volume (Hicks, 1980; Stoner & Lewis, 1985; Dean & Connell, 1987b; Hacker & Steneck, 1990; O'Connor, 1991; Jacobi & Langevin, 1996) or the fractal dimensions (Gee & Warwick, 1994). Some of these studies have been conducted in detail by means of mathematical algorithms applied to artificial substrata, which therefore present a definite structural smoothness and regular shapes; other researches have focalised on algal species with simplified morphological shapes (i.e., laminar or dichotomously fronds) (Dean & Connell, 1987b; Edgar, 1991; Jacobi & Langervin, 1996).

At present in the Mediterranean Sea very few studies have aimed to highlight the importance of marine algae as biological formers of habitat complexity and to measure appropriately the spatial and structural characteristics of the biological substratum (Chemello & Milazzo, 2002). Canopy-forming algae of the genus *Cystoseira* (Fucales, Phaeophyta) are widely distributed, colonizing from shallower water to the sublittoral in the Mediterranean Sea (Boudouresque, 1971; Ribera et al., 1992; Russo, 1997; Bulleri et al., 2002; Otero-Schmitt & Pérez-Cirera, 1996). Different species of *Cystoseira* may present different phenologies that can also vary within the same species following the geographical distribution (Gomez et al., 1982; Cormaci et al., 1992; Otero-Schmitt & Pérez-Cirera, 1996).

C. barbata (Stackhouse) C. Agardh, a brown macroalga endemic to the Mediterranean, has a thallus which consists of a perennial cauloid and a dendroid frond. Within its broad geographical range *C. barbata* may show considerable morphological variation, related to environmental conditions. Kailov & Firsov (1976) conducted some studies on this species using traditional morphometric measures (length, biomass, age) in the Black Sea area. Russo (1997), in research conducted around Cyprus on the epifauna living on *C. barbata* fronds, has defined the complexity of thalli as biomass of fronds, but these measures have no relation to either plant fine structure or surface area (Gee & Warwick, 1994; Chemello & Milazzo, 2002). Recent advances in image analysis have used the fractal dimension of a substratum as an index of habitat complexity (Gee & Warwick, 1994; Kübler & Dudgeon, 1996). However the applicability of this index depends on the morphological structure of the studied macroalga (Gee & Warwick, 1994). The analysis may be relatively easy to apply to simple broad and two-dimensional flat or regularly dichotomously branched thalli, while it is more difficult to apply and less precise for non-symmetrical or three-dimensional plants.

The biogenic structural complexity of habitats poses some problems for their practical assessment, particularly in comparative studies (Jacobi & Langevin, 1996). The quantitative analysis of macroalgal phenological variation is a necessary precursor to linking morphological diversity and physiological or ecological processes (Ralph et al., 1988; Hernàndez et al., 1999). Given the importance of these parameters it is necessary to propose an operative standardisable protocol useful for the evaluation of the complexity of habitat-forming species and for possible application to the morphologies of simplified algae (i.e., laminar fronds) and to more complex structures like those in the present study. The aim of this paper is to measure the architecture of *C. barbata* fronds, analysing different morphological spatial and structural attributes.

Materials and methods

The study site was a protected sea area at Izola (Slovenia–Northern Adriatic Sea) (latitude 45°32′41.3″N; longitude 13°40′39.2″E), where the rocky wave exposed shores are characterised by a continuous natural population of *C. barbata* that extends down to 2 m deep. The experiment was conducted on thalli of this natural population and on a transplanted one. To analyse the morphological plasticity of *C. barbata*, that may be related to the environmental stress, transplantation units were used. Fifty modules (transplantation units) were deployed, comprising four bricks fixed inside a plastic box. On every transplantation unit, 6 thalli of *C. barbata* were fixed to the bricks using polyurethane foam. These modules were then interspersed with natural substrata in the study site area at a depth of 3–4 m. The modules were placed at random, usually 20–25 cm apart. Unfortunately many of these long-term samples were destroyed before their scheduled removal time.

Samples were taken monthly for one year (April 02–March 03), apart from June, August and December, due to international bureaucratic impediments. During each sampling 3 thalli were taken randomly from the natural and transplanted population. The samples collected were fixed in a 4% formalin seawater solution.

To quantify the substratum geometry some architectural characteristics were measured as structural (total area, perimeter) and spatial (intercepting area, volume and interstitial area) attributes. According to Jacobi and Langevin (1996) the *intercepting area* of a three-dimensional substratum can be calculated by averaging the area of several projections on a plane. To calculate the intercepting area of a three-dimensional substratum of irregular shape we used an original method deploying a square-based glass prism, on the base of which a framework was placed; this permitted rotation of the thallus by 45° each time with respect to the longitudinal axis. The shape of the thalli was then analysed in three-dimensions, from four non-specular perspective angles. A reticulum, with a mesh dimension of 1 cm, was placed on two sides of the prism and the percentage cover in each grid was estimated. The data from the four readings of each sample were averaged. The matrices obtained were elaborated using the software Surfer and the Kringing algorithm interpolation to generate images that visualised the areas with the same probability of presence of the object in the space. Thus, the three-dimensional complexity of the sample and the intercepting area was summarised in a plane-image. These images were further elaborated, grouping the areas of equal probability of presence in five classes of cover: 1 (20%), 2 (21–40%), 3 (41–60%), 4 (61–80%) and 5 (≥81%). The percentage of each class in each image was calculated using Adobe Photoshop 6.0 software. It must be emphasised, however, that the total sum of contributions is never 100% for this study (this theoretical value may be reached only for a thallus that occupies the whole area of the prism).

The matrices, obtained by the reading of covers in the prism, were also elaborated using the Stax 2000 program for multivariate analysis. The Soerensen index and group average linkage were used to produce the dendrogram.

The *areas* and the *perimeters* of the collected thalli were measured by scanning images of each specimen. Two images were scanned for each sample by rotating the thallus by 90° in the prism in order to collect more detailed information. The images were acquired at 600 dpi, on a grey scale to avoid possible interference due to the water column located between the scanner sensor and the object. The digital images were elaborated by means of Adobe Photoshop 6.0. In order to calculate the *smallest available area* for the colonisation of epiphytes during the phase of vegetative rest only the area of the cauloid was estimated. The calculation was achieved by summing the length of every branching of the perennial cauloid. Since the diameter of the stipe was almost constant (2.7 mm on average), the whole cauloid was approximated to a cylinder, as high as the sum of lengths of the branching. Therefore, the lateral surface area of the cylinder was calculated.

The *volume* of each thallus was measured by placing the alga in a graduated cylinder containing water, and determining the volume of water displaced. This measure was conducted firstly by analysing together the basiphyte–epiphyte system and then the single components, after careful removal of any visible epiphytes with forceps. The volume provided information on the size of a body and an indirect appraisal of the primary production of the investigated populations using a non-destructive methodology. The thalli sampled are preserved in the collection of the Department of Biology for further researches concerning light penetration in the frond, epiphyte distribution and capture of sediment (works in progress).

Using the collected data, the following biostructural indices of the thalli were analysed, each being considered as useful for the evaluation of potential microhabitats available for the colonisation of epibionts and substrate microtopography:

Fragmentation Index (CP/A) was proposed by Austin (1984) to compare figures that do not present a regular geometrical shape. The correction of the ratio Perimeter/Area (CP/A = 0.282 perimeter/√area) enables the comparison of the irregular surfaces.

Interstitial Area Index (cm^2) (IA) is a second index applicable to the scanned images of the thalli, that is related to the amount of free space within the fronds. It was calculated as: IA = total area/sum of the interstitial areas.

Surface Complexity Index (SC) refers to the degree of roughness or irregularity of the surface of a structure. It was calculated, from the digital images, as: SC = Perimeter/Tangent Rectangle Perimeter.

Results

The morphological-spatial changes in time of *C. barbata* frond architecture were analysed in three-dimensions. Only the most representative seasonal phenological typologies are discussed here.

The intercepting areas of September thalli are shown in Figure 1a and b. During the vegetative resting period the thalli showed reduced dimensions in both height and width. The density, however, reached values of cover ranging between 20 and 60% (class of cover 2 and 3) in the central basal area of the thallus, due to the perennial cauloid. In

Figure 1. Images elaborated using the software Surfer that visualise areas with the same probability of the object's presence in space. In each image the classes of cover calculated using Adobe Photoshop 6.0 software are reported ((a) thallus no. 1 of September transplanted population; (b) no. 2 of September natural population; (c) no. 2 of March transplanted population; (d) no. 3 of March natural population; (e) no. 1 of May transplanted population; (f) no. 3 of April natural population).

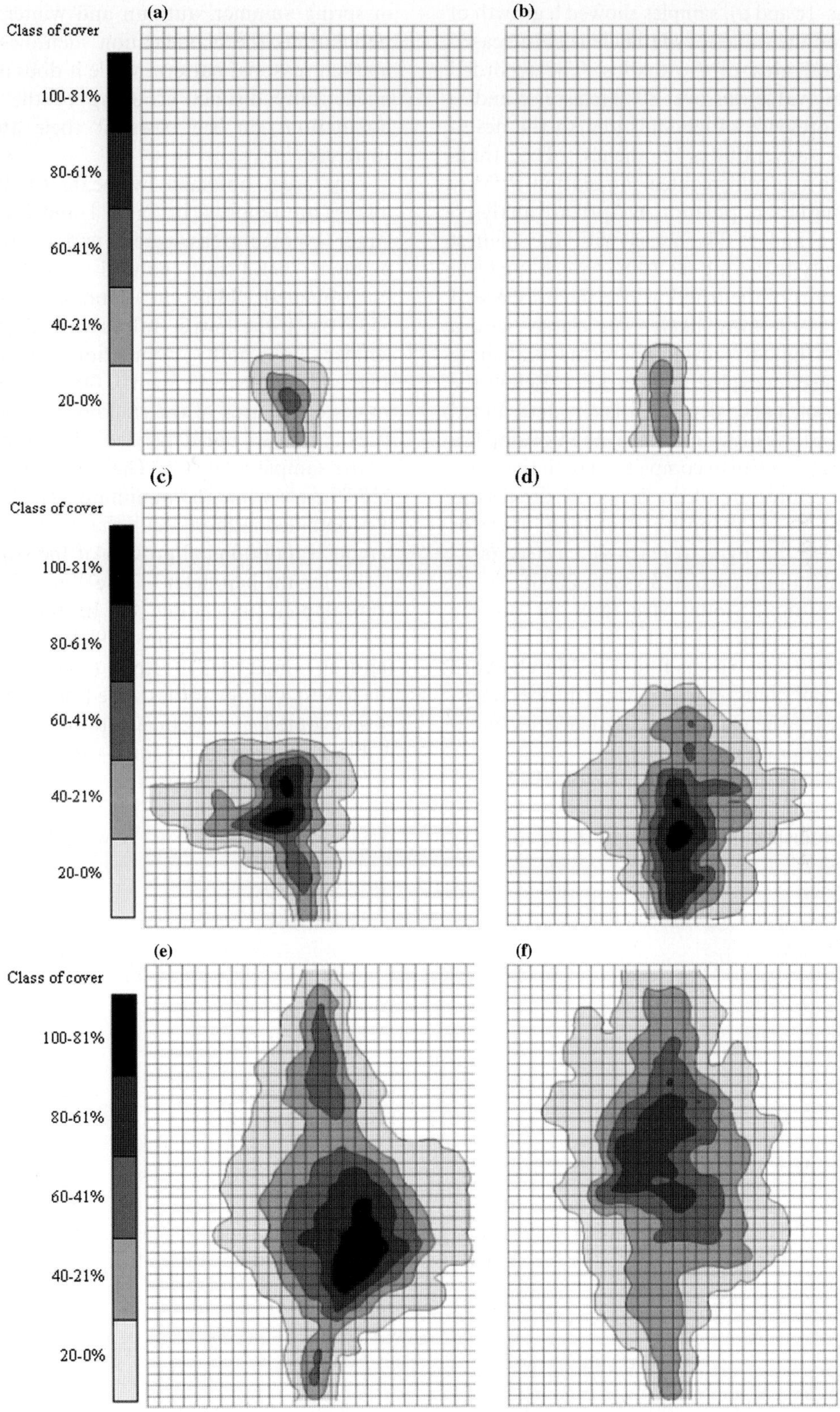
(a)
(b)
Class of cover
100-81%
80-61%
60-41%
40-21%
20-0%
(c)
(d)
Class of cover
100-81%
80-61%
60-41%
40-21%
20-0%
(e)
(f)
Class of cover
100-81%
80-61%
60-41%
40-21%
20-0%

March (Fig. 1c and d), samples showed a growth of branches and the values of spatial density increased in the central area of the thallus. The recorded cover density value was over 80% (classes 4 and 5) due to the presence of cauloids and to the development of adventitious branches. The thalli belonging to the transplanted population in May and to the natural population in April are analysed in Figure 1e and f. The areas with the highest spatial density (covering 60–100%) were found in the central part of the frond, where the presence of branches was the most likely. The classes with a lower cover (lighter greys), where the branching was less dense, occupied a wider area. Below the central area of higher density, a layer, determined by the cauloid, was found towards the bottom at a relatively high density in comparison with the outer layer. The architecture of the fronds of these samples seemed to be complex, stretching upwards which enabled the algae to reach 40 cm (see morphometric data in Falace & Bressan, 2003).

The three-dimensional structures of all the samples are summarised in Figure 2. The larger class of cover (class 5) was found in March–April in both populations, and in May in the transplanted thalli, when the species reached its maximum architectural complexity.

The seasonal trend in the phenological variability is highlighted by the dendogram (Fig. 3) which distinguishes three groups of thalli collected in spring–summer, autumn and winter. For both populations the classification identifies the same marked seasonal pattern, while it does not identify phenetically distinct grouping of the two plant populations on the basis of their architectural structure.

The areas obtained by means of the scanned images are reported in Table 1 and Figure 4. The larger average areas were found in April (transplanted: 391.3 cm^2; natural: 443.4 cm^2), May (transplanted: 436.8 cm^2) and March (natural: 553.3 cm^2); the lowest values were found in September in both populations (transplanted: 45.6 cm^2; natural: 43.5 cm^2). On the contrary, the values of the area of the cauloid did not vary on a seasonal basis, but depended on the morphology of the sample analysed. The maximum value was 126.79 cm^2; while the minimum value (10.50 cm^2) was obtained in those thalli that were less ramified, having a monopodial growth of the stipe.

As for the perimeter (Table 1 and Fig. 4) in the transplanted population, the highest average values were measured in May (192.9 cm), and for the natural population in April (253.2 cm), whereas the lowest average values were found in September (transplanted: 47.8 cm; natural: 47.1 cm). In Table 1 the monthly average values of the volume of the 'basiphyte–epiphyte' system are reported. The monthly averages of the volumes of the basiphyte (Table 1 and Fig. 5) were higher in April

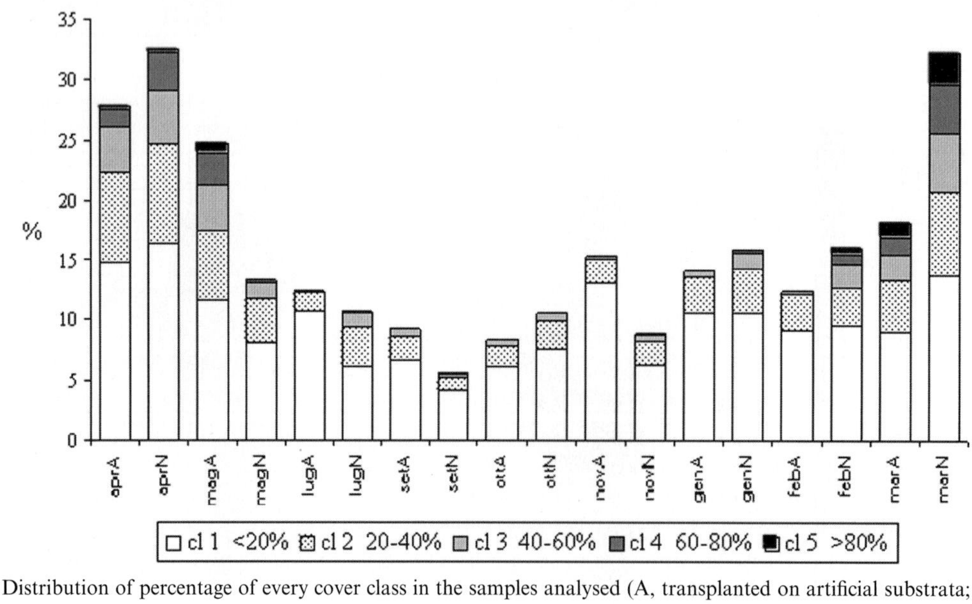

Figure 2. Distribution of percentage of every cover class in the samples analysed (A, transplanted on artificial substrata; N, natural).

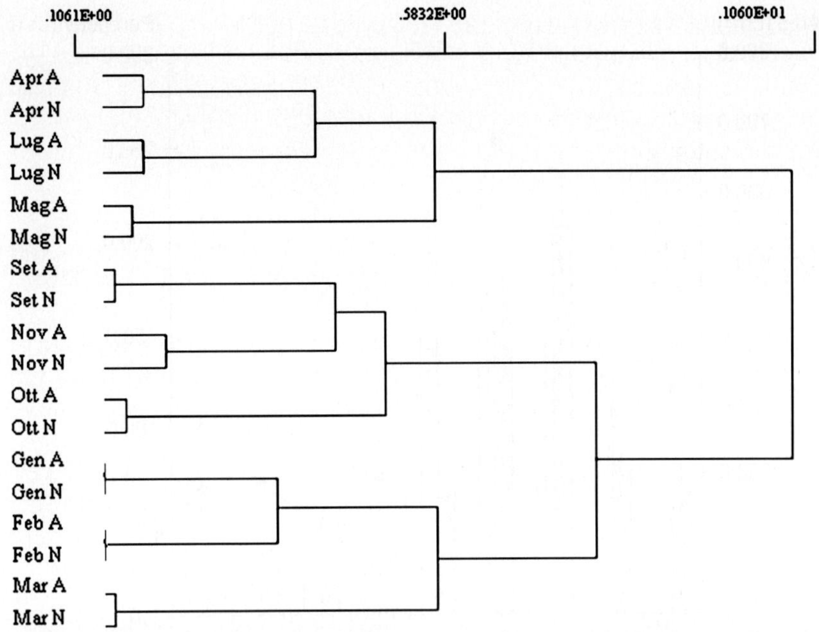

Figure 3. Seasonal trend of the phenological variability (A, transplanted; N, natural).

(34.1 cm^3) and May (39.9 cm^3) in the transplanted population, and in April (38.3 cm^3) and March (62.7 cm^3) in the natural population. As far as epiphytes were concerned, the highest average values were recorded (Table 1) for both populations in the spring period with a peak in March in

Table 1. Collection date, area (cm^2), perimeter (cm), volume basiphyte (cm^3), volume epiphyte (cm^3), Fragmentation index (CP/A) values of samples (A, transplanted on artificial substrata; N, natural)

	Area (cm^2)	SEM	Perimeter (cm)	SEM	Volume basiphyte (cm^3)	SEM	Volume epiphyte (cm^3)	SEM	CP/A	SEM
April A	391.3	79.6	148.7	21.3	34.1	6.4	31.0	11.9	2.15	0.12
April N	443.4	36.7	253.2	7.4	38.3	5.1	45.0	15.2	3.41	0.21
May A	436.8	152.7	192.9	48.8	39.9	15.1	19.4	6.6	2.64	0.19
May N	173.5	14.8	139.5	12.4	15.8	4.0	31.6	9.2	2.98	0.14
July A	75.3	15.8	67.8	9.8	8.2	1.8	6.2	3.0	2.21	0.09
July N	108.2	58.1	72.6	22.1	8.2	1.4	9.9	2.0	2.11	0.11
September A	45.6	12.1	47.8	10.3	5.8	1.9	3.4	1.8	1.99	0.17
September N	43.5	10.5	47.1	3.0	6.1	2.5	10.0	4.3	2.08	0.20
October A	74.5	9.9	70.0	5.9	9.2	1.0	12.8	9.9	2.30	0.13
October N	94.6	11.8	70.7	5.3	12.6	2.0	16.4	2.3	2.06	0.09
November A	94.6	20.3	100.9	4.9	8.8	2.0	14.0	6.7	3.00	0.26
November N	81.8	39.6	66.9	24.6	12.3	4.4	22.1	3.6	2.08	0.25
January A	131.6	27.2	109.5	14.5	9.8	1.4	10.9	1.5	2.71	0.16
January N	177.1	58.7	119.8	27.9	16.8	6.6	31.9	2.6	2.57	0.19
February A	107.9	33.4	88.3	24.5	8.2	1.8	9.3	3.3	2.37	0.35
February N	209.7	26.2	108.1	14.4	14.3	4.2	32.4	5.8	2.16	0.39
March A	249.4	136.9	113.3	33.9	26.2	15.3	32.7	3.4	2.20	0.03
March N	553.3	114.8	198.2	14.8	62.7	19.4	79.6	13.6	2.47	0.31

Values are mean ratios and the Standard Error of the Mean (SEM).

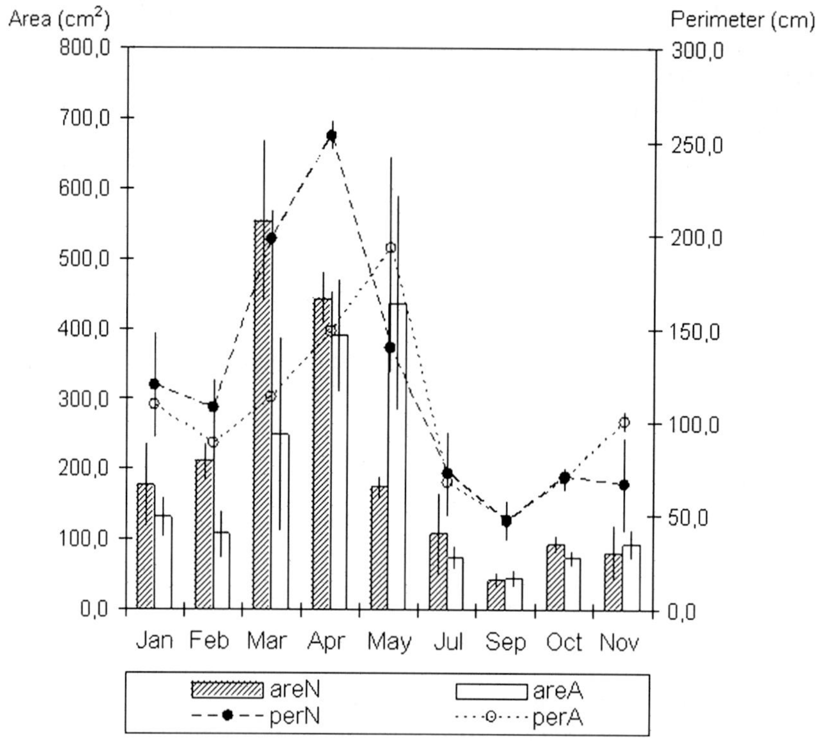

Figure 4. Average monthly area and perimeter values in the natural and transplanted populations.

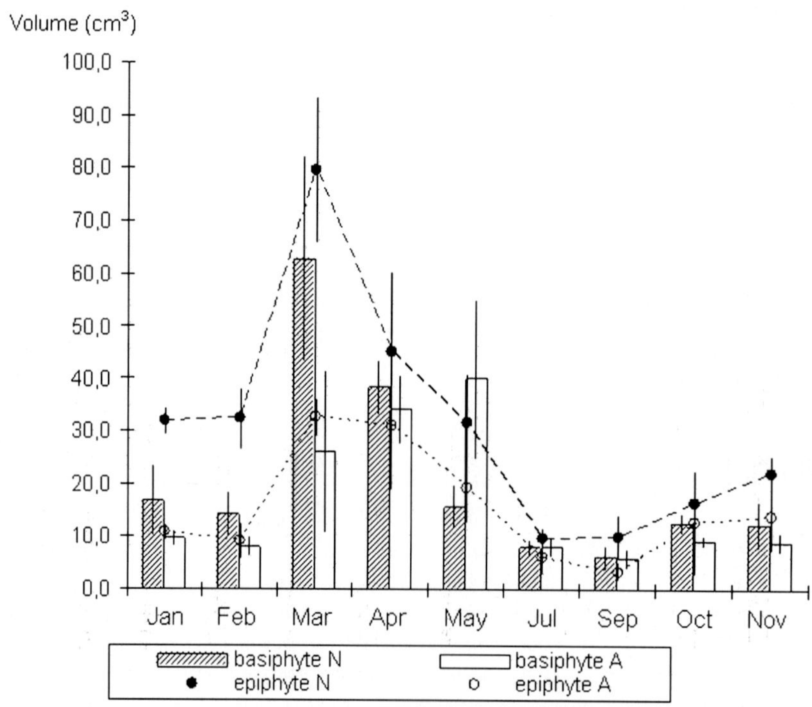

Figure 5. Average monthly volume values of the basiphyte and epiphyte system in the natural and transplanted populations.

the natural population (79.6 cm^3). The contributions to the basiphyte–epiphyte system of the epiphytes varied between 40 and 70%. In February, for example, the average volume of epiphytes (32.4 cm^3) removed from the thalli of the natural population was almost double that of the basiphyte (14.3 cm^3).

In view of the fact that the values of the biostructural indices did not vary seasonally, but depended exclusively on the structure of the frond and on the morphological variability of different individuals, only a few cases are given below as examples:

- Fragmentation Index (CP/A): the highest monthly average values (Table 1 and Fig. 6) of this index were associated with the samples with an irregular thallus; the lowest values were reached when the shape was more compact. The lowest average values were recorded in samples taken in October for the natural population (CP/A = 2.06) and in September for the transplanted population (CP/A = 1.99), respectively.
- Interstitial Area Index (IA): to explain the range of this index two characteristic examples of the process of evolution of the frond were chosen and compared (Fig. 7). In the case of September thalli (Fig. 7a), characterised by the absence of the frond, therefore not revealing any holes IA = 113. On the contrary, February thalli of the natural population (Fig. 7b) showed a developed frond with several spaces between fronds (IA = 11).
- The Surface Complexity Index provides the degree of roughness or irregularity of thallus outline, and also in this case two representative thalli, describing the individual variability of the frond attained by *C. barbata*, are shown (Fig. 8). In September (Fig. 8b), in thalli characterised by the absence of the frond and by the presence of the ramified perennial cauloid, the surface complexity value was 0.13, while in April (Fig. 8a), when the thallus had the typical dendroid branched frond arising from the cauloid, it reached 0.50.

Discussion

Although the effects of algal complexity have been the focus of several studies (Edgar, 1991; Russo, 1997), previous research has employed traditional analysis, using morphometric attributes such as biomass, height, width, degree of branching, stem width and number of branches emerging from the stem (Edgar, 1983; Russo, 1997; Chemello & Milazzo, 2002).

This study represents a first attempt to quantitatively assess the phenology and the morphological plasticity of *C. barbata*. The protocol proposed and the analysis conducted permitted us to quantify and to test the main development phases of the thalli and to assess in detail the

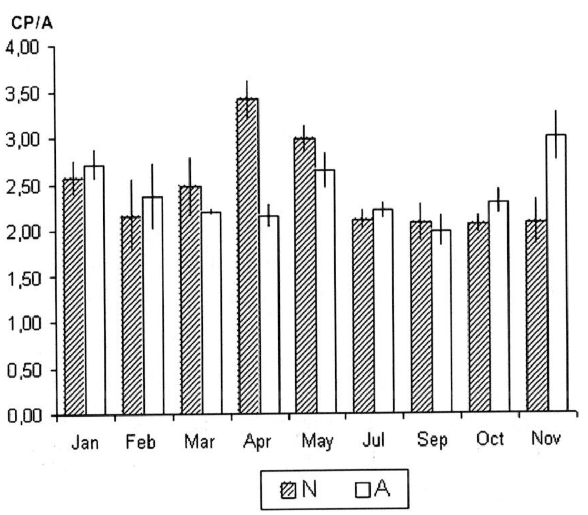

Figure 6. Variability of Fragmentation Index (CP/A) in natural and transplanted thalli.

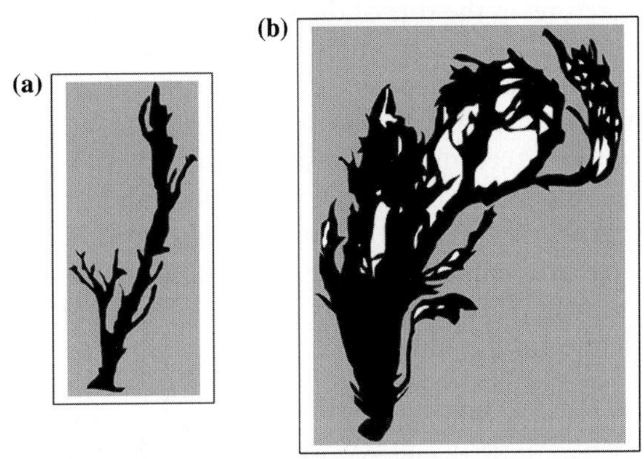

Figure 7. Interstitial Area Index (IA). Two representative digital images employed to evaluate the index are shown ((a) thallus no. 2 of September natural population; (b) thallus no. 3 of February natural population). The thallus area is shown in black and the areas between fronds in white.

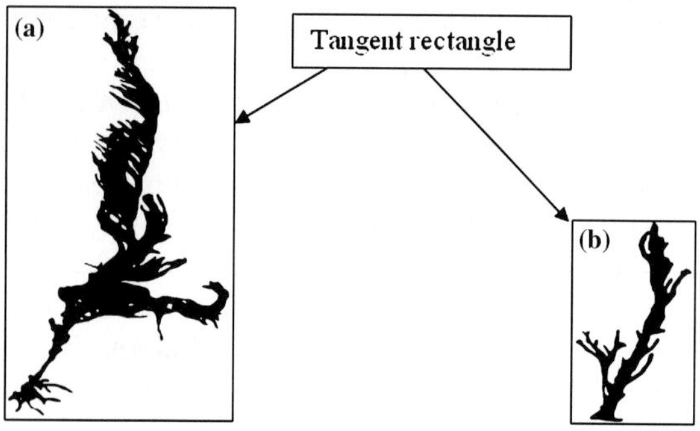

Figure 8. Surface Complexity Index (SC). Two digital images: (a) thallus no. 2 of April natural population; (b) thallus no. 2 of September natural population.

morphological and seasonal variation of fronds by means of different structural and spatial attributes.

The intercepting area, calculated by the use of the prism, allowed us to analyse this spatial characteristics in three-dimensions as 'probability of presence of the alga' in space. The seasonal variation of *C. barbata* is characterised by marked changes in the size of the thallus and shows a considerable degree of morphological variability throughout its range. The seasonal phenological changes of this Hemiphanerophyceae, the spatial organisation and the luxuriance of the thallus are determined by the model of expansion and reduction of the frond. The latter is largest in spring–summer, whereas the fall of phylloids and branchelets occurring in autumn–winter results in important variations in morphometric measures (Falace & Bressan, 2003), as well as in the surface area and, therefore, in the role of the species within the community. The intercepting area may be a fundamental parameter in studies concerning the efficiency of this species as biological former of canopy. According to Jacobi & Langevin (1996), the settlement of an organism depends on the probabilities that it locates a substratum, and therefore on its intercepting area. The assessment of this spatial attribute allows to examine in detail the architectural complexity, that in turn may be related to spatial heterogeneity, chemical-physical gradients and availability of secondary substratum. Furthermore, the increasing frond complexity may also significantly influence the passage of currents and waves and, therefore, the rates of transport and deposition of suspended particles (Dean & Connell, 1987b; Gibbon, 1988; Duggins et al., 1990).

From a functional point of view, the light regime and other physical and chemical parameters in the understory assemblages are not modified significantly during the vegetative resting period of *C. barbata*, when the abscission of phylloids and branchlets takes place and the perennial stipe is the only surface that can be colonised. With the progressive growth of branches and adventitious ramifications, the highest spatial complexity of the frond implies an enrichment of available habitats and a great variability of chemical-physical gradients.

The intercepting area has been estimated by Jacobi & Langevin (1996) using artificial substrata that mimic macroalgae, to evaluate the role of geometric characteristics on settlement of benthic organisms. However, this methodology could hardly be applicable to biological substrata with an irregular shape and densely ramified frond such as *C. barbata*. Scanning the thalli in the prism to assess *C. barbata* surface area values proves to be very useful when applied to dendritic structures of this kind. On fronds with a laminar structure, other researchers have directly scanned or photocopied the algae to calculate this structural attribute (Kübler & Dudgeon, 1996; Hernàndez et al., 1999). Dean & Connell (1987a) and Hicks (1980) estimated the surface area on the basis of biomass. For thalli that can be approximated to geometric figures (cylinders, cones or plane figures) other authors have used a calibration curve of portions of thalli versus surface or the formulae of geometric objects closest in shape (Jacobi & Langervin, 1996; Hernàndez et al., 1999).

The surface area of *C. barbata* here is considered, according to Hicks (1980), as a biological index of the potential habitable space of the algal substratum. The average surface area values (Fig. 4) recorded during the experiment, show a seasonal harmonic pattern. It must be pointed out that the data analysed show high variability, coinciding with evident morphological diversity of the fronds and, in particular, of the relative maximum area values. This evidence may be indicative of a heterogeneous population characterized by the coexistence of individuals in different morphogenetic phases. In addition, the recorded values of surface area for the transplanted thalli do not achieve the relative maximum values of the natural ones and present a temporal lag in the pattern values. The delay in the development of the transplanted thalli may be related to the higher depth of submersion of the artificial modules.

The same observations can be made for the perimeter maximum average values (Fig. 4). On the contrary, the perimeter minimum average values were similar in both populations because of the persistence of only the perennial cauloid during the resting period. By comparing the area and perimeter temporal patterns, it may be observed that in the natural population the perimeter values decrease while the area values increase in April, associated with the development of the adventitious branches on the cauloid surface. The perimeter may thus give important information on the morphogenesis and can contribute to a better definition of the substratum complexity with respect to the structural luxuriance of the frond. Furthermore by analysing the time-course of the values of the perimeter, in order to assess the functional role of *C. barbata* as a habitat-forming species, it is possible to interpret the spring values as having the highest available space and variability of microhabitats. This structural parameter has been also used by Gee & Warwick (1994) who measured the length of the outline of the frond drawn under a binocular dissection microscope fitted with a camera lucida.

The area of the cauloid, as well as representing the potential available area during the vegetative resting period when there is no frond, can also furnish useful information in taxonomic studies. In fact, *C. barbata* is characterised by a monopodial cauloid that may be more or less ramified. Unlike other biogeographical areas, in the Northern Adriatic Sea it is often found in a highly ramified form. The area of the cauloid could thus be used in comparative studies for the characterisation of ecoforms.

The proposed bio-structural indices contribute to a more detailed description of the species, but also enable a better assessment of the micro-spatial and structural habitat complexity. According to Jacobi & Langevin (1996) the processes of colonisation and the survival of epibionts might depend on different structural aspects of the substratum and can be related to the number and type of microhabitats that must be carefully described and quantitatively evaluated.

The CP/A index provides information on the variability of potential microhabitats. In fact higher values of this index identify thalli with an irregular surface and outline, and therefore having higher structural and spatial morphological frond complexity. The highest average values are found in April in the natural population (Fig. 6). Kübler & Dudgeon (1996) have used a similar index (the ratios of the perimeter to the square root of the surface area) to evaluate the plasticity and the complexity of *Chondrus crispus* Stackhouse fronds under different temperature conditions.

The Interstitial Area index refers to the amount of free space, permitting a detailed evaluation of the potential ecological niches presented inside the fronds. The lowest values are associated with fronds rich in holes, in relation to the area of the thallus. Hacker & Steneck (1990) and Jacobi & Langevin (1996) have employed the interstitial volumes to obtain similar information. However to study macroalgal epibiont cover, sediment trap capability and light penetration on canopy macroalgae the area measure between fronds may be more useful than volumes.

Regarding the Surface Complexity index, previous studies carried out on the degree of roughness or irregularity have concerned artificial substrata (Bortone et al., 2000). Despite the few studies that have been conducted on this parameter, the general finding is that epibiont species richness is higher on complex surfaces than on simple ones (O'Connor, 1991).

In this study, a preliminary assessment of the autogenic engineering role of *C. barbata* and therefore of its efficiency as a substratum for algal epibionts has been estimated by volumetric analysis (Fig. 5). The analysis of the average monthly volume values of the 'basiphyte–epiphyte' system indicates that the abundance of algal epibionts depends mainly on the life-form of the *Cystoseira* fronds. *C. barbata*, by vastly increasing the area of potential suitable substrata, may hold a volume of algal epiphytes equal to or even higher than its own volume. This observation is in accordance with published data that have demonstrated that the microenvironments, created by different algal shapes and blade packing, are important in determining patterns of abundance of associated species, and that complex macroalgae encourage more abundant and well diversified epibionts than simple and flat thalloid ones (Edgar, 1983, 1991; Gibbons, 1988; Russo, 1997; Gee & Warwick,

1994; Chemello & Milazzo, 2002). Nevertheless, in our research the pattern of epiphyte abundance also depends on the natural seasonal variability of epiphyte spores. Furthermore, the volume of epiphytes on transplanted thalli, even though it follows the seasonal pattern of the natural ones, is lower, basiphyte volume being equal (Fig. 5). The reduced light penetration in relation to the deeper location of the modules could have reduced the colonisation rate and the development of epiphytes.

Concerning the morphological plasticity of *C. barbata* in response to environmental stress, the identification of structural differences between populations on natural and artificial substrata is mainly limited to single monthly cases, without any remarkable or substantial morphological changes due to the transplantation. The algae of both natural and artificial substrata were characterised by a similar seasonal cycle and by the same phenological variation of the thalli. However, in the transplanted thalli there is a temporal lag in the frond's development probably due to the higher depths of immersion. This could be linked to different light penetration and temperature, particularly when abiotic signals determine the more significant phenetic transformations.

The spatial and structural attributes analysed, as well as the definition of the architectural complexity of *C. barbata*, permit the evaluation of the area available for the attachment and shelter of epibionts together with the number and type of potential habitats. The data suggest that the potential habitat of *C. barbata* varies seasonally during the lifetime of individual fronds, and the quantification of the structural complexity may also allow a better prediction of the physiological consequences of this morphological plasticity. The representation of three-dimensional plant architecture provides replicated and standardisable measures of algal structure, in order to develop a broader understanding of the role of canopy species in the creation and maintenance of habitat diversity. Furthermore the indices proposed, besides allowing us to model the development of plant architecture over time, may enable the evaluation of degrees of complexity and a classification of habitat-former species complexity in comparative studies. The information and data obtained in this study will be correlated in the course of the research with the other levels of structural complexity added to the system, like epiphyte colonisation and sediment capture. Finally the analysis conducted on the habitat architecture may also find a useful application in the design of artificial plants that mimic these morphologies (artificial algae) for projects aimed at saving and restoring the environment.

Acknowledgements

We thank Dr Elisa Zanelli and Dr Enrico Bressan for their helpful cooperation and assistance in field and laboratory work. Special thanks to Dr Antony Jensen for valuable comments and revision on the manuscript. Furthermore the authors wish to thank the two anonymous proof-readers for their attentive reading of the work and for the valuable suggestions that have permitted us to edit and improve the manuscript.

References

Aarnio, K. & J. Mattila, 2000. Predation by juvenile *Platichthys flesus* (L.) on shelled prey species in a bare sand and a drift algae habitat. Hydrobiologia 440: 347–355.

Austin, R. F., 1984. Measuring and comparing two-dimensional shapes. In Gaile, G. L. & C. J. Willmott (eds.) Spatial Statics and Models. Reidel Publ. Co, Boston: 293–312.

Benedetti-Cecchi, L., F. Pannacciulli, F. Bulleri, P. S. Morchella, L. Airoldi, G. Relini & F. Cinelli, 2001. Predicting the consequences of anthropogenic disturbance: large-scale effects of loss of canopy algae on rocky shores. Marine Ecology Progress Series 214: 137–150.

Boudouresque, C. F., 1971. Contribution à l'étude phytosociologique des peuplements algaux des côtes varoises. Vegetation 22: 83–184.

Bortone, S. A., M. A. Samoilys & P. Francour, 2000. Fish and Macroinvertebrates Evaluation Methods. In Seaman W. J. (ed.), Artificial Reef Evalution with Application to Natural Marine Habitats. CRC Marine Science Press: 128–159.

Bulleri, F., L. Benedetti-Cecchi, S. Acunto, F. Cinelli & S. J. Hawkins, 2002. The influence of canopy algae on vertical patterns of distribution of low-shore assemblages on rocky coasts in the northwest Mediterranean. Journal of Experimental Marine Biology and Ecology 267: 89–106.

Chemello, R. & M. Milazzo, 2002. Effect of algal architecture on associated fauna: some evidence from phytal molluscs. Marine Biology 140: 981–990.

Cormaci, M., Furnari, G., Giaccone, G., Scammacca, B. & Serio, D., 1992. Observation taxonomiques et biogéographiques sur quelques espèces du genre *Cystoseira* C. Agardh. Bulletin de l'Institut océanographique, Monaco 9: 21–35.

Dean, R. L. & J. H. Connell, 1987a. Marine invertebrates in an algal succession. II. Tests of hypothesis to explain changes in diversity with succession. Journal of Experimental Marine Biology Ecology 109: 217–247.

Dean, R. L. & J. H. Connell, 1987b. Marine invertebrates in an algal succession. III. Mechanisms linking habitat complexity with diversity. Journal of Experimental Marine Biology Ecology 109: 249–273.

Dudgeon, S. R., J. E. Kübler, R. L. Vadas & I. R. Davidson, 1995. Physiological responses to environmental variation in intertidal red algae: does thallus morphology matter?. Marine Ecology Progress Series 117: 193–206.

Duggins, D. O., J. E. Eckman & A. T. Sewell, 1990. Ecology of understory kelp environment. 2. Effects of kelps on recruitment of benthic invertebrates. Journal of Experimental Marine Biology and Ecology 143: 27–45.

Edgar, G. J., 1983. The ecology of S. E. Tasmanian phytal animal communities. IV. Factors affecting the distribution of amphithoids among algae. Journal of Experimental Marine Biology and Ecology 70: 205–225.

Edgar, G. J., 1991. Artificial algae as habitat for mobile epifauna: factors affecting colonization in a Japanese Sargassum bed. Hydrobiologia 226: 111–118.

Falace A. & G. Bressan, 2003. Intervento pilota di restauro biologico mediante tecniche di traspianto algale. Biologia Marina Mediterranea 11(2): 499–503.

Figueiredo, O. M. A., J. M. Kain & T. A. Norton, 2000. Responses of crustose corallines to epiphyte and canopy cover. Journal of Phycology 1: 17–24.

Gee, J. M. & R. M. Warwick, 1994. Body-size distribution in a marine metozoan community and the fractal dimensions of macroalgae. Journal of Experimental Marine Biology Ecology 178: 247–259.

Gibbons, M. J., 1988. The impact of sediment accumulations, relative habitat complexity and elevation on rocky shore meiofauna. Journal of Experimental Marine Biology Ecology 122: 225–241.

Gomez, A., M. A. Ribera & J. A. Seoane Camba, 1982. Estudio fenológico de varias especies del género *Cystoseira* en Mallorca. Collectanea Botanica 13: 841–855.

Hacker, S. D. & R. S. Steneck, 1990. Habitat architecture and the abundance and body-size-dependent habitat selection of a phytal amphipod. Ecology 71(6): 2269–2285.

Hawkins, S. J., 1983. Interaction of *Patella* and macroalgae with settling *Balanus balanoides* (L.). Journal of Experimental Marine Biology and Ecology 71: 55–72.

Hein, M., M. F. Pedersen & K. Sand-Jensen, 1995. Size-dependent nitrogen uptake in micro- and macroalgae. Marine Ecology Progress Series 164: 21–34.

Hernández, I., J. R. Andria, M. Christmas & B. A. Whitton, 1999. Testing the allometric scaling of alkaline phosphatase activity to surface/volume ratio in benthic marine macrophytes. Journal of Experimental Marine Biology Ecology 241: 1–14.

Hernández-Carmona, G., O. García, D. Robledo & M. Foster, 2000. Restoration techniques for *Macrocystis pyrifera* (Phaeophyceae) populations at the Southern limit of their distribution in México. Botanica Marina 43: 273–284.

Hicks, G. R. F., 1980. Structure of phytal Harpacticoid Copepod assemblages and the influence of habitat complexity and turbidity. Journal of Experimental Marine Biology Ecology 44: 157–192.

Hull, S. L., 1997. Seasonal changes in diversity and abundance of ostracods on four species on intertidal algae with different structural complexity. Marine Ecology Progress Series 161: 71–82.

Jacobi, C. L. & R. Langevin, 1996. Habitat geometry of benthic substrata: effects on arrival and settlement of mobile epifauna. Journal of Experimental Marine Biology Ecology 206: 39–54.

Jenkins, S. R., S. J. Hawkins & T. A. Norton, 1999. Interaction between a fucoid canopy and limpet grazing in structuring a low shore intertidal community. Journal of Experimental Marine Biology Ecology 233: 41–63.

Jones, C. G., J. H. Lawton & M. Shachak, 1994. Organisms as ecosystem engineers. Oikos 69: 373–386.

Kennelly, S. J., 1989. Effects of kelp canopies on understory species due to shade and scour. Marine Ecology Progress Series 50: 215–224.

Khailov, K. M. & Yu. K. Firsov, 1976. The relationships between weight, length, age and intensity of photosynthesis and organotrophy in the thallus of *C. barbata* from the Black Sea. Botanica Marina XIX: 329–334.

Kübler, J. E. & S. R. Dudgeon, 1996. Temperature dependent change in the complexity of form of *Chondrus crispus* fonds. Journal of Experimental Marine Biology Ecology 207: 15–24.

McCook, L. J. & A. R. O. Chapman, 1991. Community structure following massive ice scour on an exposed rocky-shore: effects of *Fucus* canopy algae and of mussels during late succession. Journal of Experimental Marine Biology and Ecology 154: 137–169.

Melville, A. J. & S. D. Connell, 2001. Experimental effects of kelp canopies on subtidal coralline algae. Australian Ecology 26: 102–108.

Menge, B. A., 1978. Predation intensity in a rocky intertidal community Relation between predator foraging activity and environment harshness. Oecologia 34: 1–16.

Menge, B. A. & J. P. Sutherland, 1987. Community regulation: variation in disturbance, competition, and predation to gradients of environmental stress and recruitment. American Naturalist 130: 730–757.

O'Connor, N. A., 1991. The effects of habitat complexity on the macroinvertebrates colonising wood substrates in a lowland stream. Oecologia 85: 504–512.

Otero-Schmitt, J. & J. L. Pérez-Cirera, 1996. Epiphytism on *Cystoseira* (Fucales, Phaeophyta) from the Atlantic Coast of Northwest Spain. Botanica Marina 39: 445–465.

Paine, R. T., 1969. A note on trophic complexity and community stability. American Naturalist 103: 91–93.

Piraino, S., G. Fanelli & F. Boero, 2002. Variability of species' roles in marine communities: change of paradigm for conservation priorities. Marine Biology 140: 1067–1074.

Stoner, A. W. & F. G. Lewis, 1985. The influence of quantitative and qualitative aspects of habitat complexity in tropical seagrass meadows. Journal of Experimental Marine Biology Ecology 94: 19–40.

Ralph, P. J., D. A. Morrison & A. Addison, 1998. A quantitative study of the patterns of morphological variation within *Hormosira banksii* (Turner) Decaisne (Fucales: Phaeophyta) in South-eastern Australia. Journal of Experimental Marine Biology Ecology 225: 285–300.

Reed, D. C. & M. S. Foster, 1984. The effect of canopy shading on algal recruitment and growth in a giant kelp forest. Ecology 65: 937–948.

Ribera, M. A., A. Gómez-Garreta, T. Gallardo, M. Cormaci, G. Furnari & G. Giaccone, 1992. Check-list of Mediterranean Seaweeds Fucophyceae I., (Warming, 1884). Botanica Marina 35: 109–130.

Russo, A. R., 1997. Epifauna living on sublittoral seaweeds around Cyprus. Hydrobiologia 344: 169–179.

Hydrobiologia (2006) 555:207–219
H. Queiroga, M.R. Cunha, A. Cunha, M.H. Moreira, V. Quintino, A.M. Rodrigues, J. Serôdio & R.M. Warwick (eds),
Marine Biodiversity: Patterns and Processes, Assessment, Threats, Management and Conservation
DOI 10.1007/s10750-005-1117-1

Epifaunal inventory of two shipwrecks from the Belgian Continental Shelf

Vincent Zintzen[1,2,*], Claude Massin[2], Alain Norro[3] & Jérôme Mallefet[1]
[1]*Catholic University of Louvain, Laboratory of Marine Biology, Place Croix du Sud 3, Kellner's building, B-1348, Louvain-la-Neuve, Belgium*
[2]*Royal Belgian Institute of Natural Sciences, Malacology Section, 29 Vautier street, B-1000, Brussels, Belgium*
[3]*Royal Belgian Institute of Natural Sciences, Management Unit of Mathematical Model of the North Sea, 100 Gulledelle, B-1200, Brussels, Belgium*
*(*Author for correspondence: E-mail: vincent.zintzen@naturalsciences.be)*

Key words: shipwrecks, Belgian Continental Shelf, hard substrata, epifauna

Abstract

Shipwrecks are almost the only subtidal substrata available for epifaunal colonization along the Belgian coastal waters and have never been scientifically prospected up to now. Two shipwrecks have been investigated during the summers of 2001 and 2002. *De-visu* observations and examination of 14 scraped surfaces of 0.0625 m^2 allowed the identification of a total of 121 macrofauna species. Species richness cumulative curves provide an estimated number of 150–280 species. Both shipwrecks are dominated by Cnidarians. The tube-dwelling amphipod *Jassa herdmani* was also particularly abundant. Striking differences were observed between shipwrecks. Different faunal assemblages were observed on vertical and horizontal surfaces. Species richness could not been correlated with diversity indices. Samples with high species diversity were observed when the Hydrozoan *Tubularia indivisa* was dominant. On the contrary, when the Anthozoan *Metridium senile* was dominant, samples showed a very low species richness. Finally, the Poriferan *Dysidea fragilis* has to be considered as a new species for the Belgian fauna.

Introduction

Since the second part of the 19th century, the benthos of the Southern Bight of the North Sea has been intensively studied. The first intensive campaigns performed by Gilson (1900) gave an insight on the faunal diversity of the Belgian coasts but it is only since the early seventies that various researchers described the benthic communities of that region (Heip & Decraemer, 1974; Jensen, 1976; Govaere et al., 1980; Vincx, 1981). Focussing respectively on the macro and meiobenthos, Vanosmael et al. (1982) and Willems et al. (1982) stressed the ecological value of the Belgian Continental Shelf (BCS) sandbanks, Southwest – Northeast oriented. They suggested that those banks act as possible islands and generate a range of habitats for marine fauna. More recently, Beyst et al. (2001) listed the species of the surf zone occurring on Belgian sandy beaches and raised the question of their nursery function. Dewicke et al. (2003) emphasized an onshore–offshore gradient in the density and biomass of the hyperbenthos, here defined as the small animals living close to the sea bed. In 2004, Van Hoey et al. summarized the large-scale spatial distribution of the macrobenthos of the BCS by combining a large amount of data collected from 1994 to 2000. All these researches directly concern the benthic fauna of soft bottoms. Despite the studies of Daro (1969, 1970), De Pauw & Van Damme (1992) and Volckaert et al. (2002) on the faunal and floral assemblage of intertidal structures along the coast, there is a paucity of available data devoted to communities of hard substrata on the BCS. This lack of information is a consequence of hydrodynamism schemes. Indeed fine sediments dominate the Southern part of the North Sea due to a hydrodynamism decrease after the Dover Strait (Prygiel et al., 1988) while epifauna associated with pebbles is still well represented in the Strait (Davoult, 1990).

Assuming that *Asterias rubens* mainly feeds on molluscs (*Mytilus edulis* L.) that need hard substrata to develop (Castilla & Crisp, 1973), the findings of this echinoderm could indirectly reveal the presence of hard substrata. Only a few limited places where concentrations of the starfish *A. rubens* occur on the BCS were documented by De Clerck et al. (1973, 1974a, b, 1975) and therefore could be assigned to hard substrate regions. Moreover, Maertens (1989) and Deleu (2002) confirmed the presence of such regions on the BCS but so far studies on the fauna of these areas have never been performed (Kerckhof & Houziaux, 2003). Finally, Degraer (1999) and Van Hoey et al. (2004) mention a very special macrobenthic community with high density of the Bivalve *Barnea candida* (L.) in outcropping tertiary clay layers near Oostende. As a consequence, the question of epifaunal assemblages on hard substrata for the Belgian waters has been hardly addressed. Although unnatural, shipwrecks represent another type of hard substrate and more than 200 recent shipwrecks which are a potential threat to navigation or fisheries are recorded along the Belgian coast (Norro, pers. communication). Moreover, it has been recently estimated that 10 000 wrecks lie in the Dutch sector of the North Sea (Leewis et al., 2000). Theses structures represent substrata available for the colonization of subtidal sessile epibenthic communities in Belgian coastal waters. Even if the fauna living on these structures could be regarded as exotic, shipwrecks represent nevertheless a part of the habitat diversity that cannot be neglected. Technical problems are certainly the main reason why the fauna of shipwrecks have scarcely been studied (Massin et al., 2002). This fauna has been prospected along the Dutch Continental Shelf (Leewis et al., 2000) and on a single shipwreck near the Isle of Lundy (Bristol Channel, England; Hiscock, 1980) but this has never been done for the BCS. This paper presents the first results of a detailed study aiming to understand the possible role of such structures in the biological diversity of the Southern Bight of the North Sea. These preliminary results present a first estimation of the fauna of two shipwrecks from the BCS with a distinction between horizontal and vertical surfaces in order to determine whether different associations could be due to sedimentation. We intend to describe the basic features of the dominant communities found on shipwrecks with emphasis on some species of special interest.

Material and method

The macrofauna (fauna retained on a 1 mm sieve) of two BCS shipwrecks was investigated from July to the beginning of September during the years 2001 and 2002 on board of the A.962 'RV Belgica.' These two shipwrecks were chosen according to the following criteria: (i) large size thus ease of location, (ii) good state allowing safe working conditions, (iii) location out of navigation roads, and (iv) sunk for at least 10 years to reach a mature community (Leewis et al., 2000). The first shipwreck investigated was the Birkenfels, 156 m long, 42 m depth, sunk in 1966 and lying 30 nautical miles from the coast (WGS-84 coordinates: N 51°38′,989 – E 02°32′,268). The second one was the Bourrasque, 106 m long, 16 m depth, sunk in 1940 and lying 8 nautical miles from the coast (WGS-84 coordinates: N 51°14′,964 – E 02°33′,026). The Birkenfels is located in open sea conditions with a maximum current speed during neap and spring tides of 0.9 and 1.7 knots, respectively. The Bourrasque lies between the Buiten Ratel sandbank and the Kwintebank close to the shore. The maximum current speeds during neap and spring tides are 0.8 and 1.3 knots, respectively. These two wrecks lie in different water conditions according to the distance from the coast, since the BCS displays a gradient from turbid, nutrient rich and well-mixed inshore waters towards more oceanic, less turbid and less productive offshore waters (Cattrijsse & Vincx, 2001). Figure 1 shows the location of the two chosen shipwrecks among the other wrecks of the BCS.

Teams of three divers sampled vertical and horizontal oriented surfaces in order to document the possible faunal differences due to sedimentation process. During a tidal cycle (≅12 h), currents turn around wrecks so that all external surfaces are exposed. Due to the large size of the wrecks and the

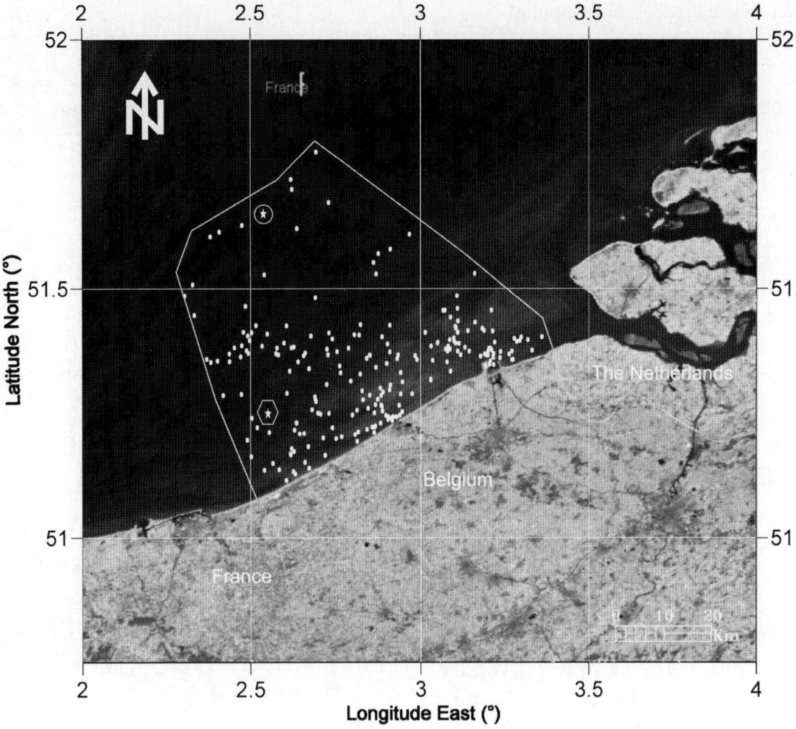

Figure 1. Location of the 231 inventoried wrecks of the Belgian Continental Shelf. The shipwreck of the Birkenfels is outlined by a circle; the shipwreck of the Bourrasque by a hexagon. Coordinates are WGS-84. Source: Afdeling Waterwegen Kust (AWK), Belgium. Map background by Image courtesy of MODIS Rapid Response Project at NASA/GSFC.

poor underwater visibility (1–4 m), it was impossible to sample the same locations twice; therefore, all the sampling spots must be considered as differently exposed areas. As a consequence, on these exposed surfaces the most important factor for the settlement of sessile epifauna should be governed by the orientation (horizontal/vertical) of surfaces to the current. A total of 7 samples has been taken on each shipwreck (Table 1) representing a total surface of 0.875 m^2.

Sampling of protected surfaces such as overhangs and more sheltered areas inside the shipwrecks was not undertaken. Moreover, abiotic parameters being quite similar for all collection times (July and September 2001, 2002), it is assumed that no seasonal or annual effects influence the observed variations in fauna.

A three step procedure was adopted in order to assess the macrofauna of the shipwrecks. First, each team randomly placed a 50 × 50 cm frame on a vertical or horizontal surface and digital pictures of the frame were taken. Then, a *de-visu* inventory of the dominant visible species was performed within the 50 × 50 cm frame. Finally, a subsurface of 25 × 25 cm was scraped and transferred into plastic bags for further analysis. Divers also identified fish and jellyfish around the shipwrecks but these were not included in the analysis although they are listed for the reader's information.

On board, animals were relaxed in 3.5% $MgCl_2$ solution for two hours and then transferred to buffered formalin solution (final concentration 4%, pH 8.2–8.4). Later on, specimens were transferred to 70% buffered alcohol for permanent storage. The samples were then sorted and species counted and identified to the lowest possible taxonomic level. All the collected material is deposited in the collections of the Royal Belgian Institute of Natural Sciences under the IG number 29462. Additional species were identified by examination of digital pictures. Colonies of taxa such as Hydrozoans or Bryozoans were each counted as one individual.

Species richness was estimated through species accumulation curves based on two non-parametric estimators: the Jack2 estimator (Burnham & Overton, 1978, 1979) and ACE (Abundance based

Table 1. Date, depth and orientation of the samplings

Bourrasque			Birkenfels		
Date	Depth (m)	Orientation	Date	Depth (m)	Orientation
04/07/02	17.8	Horizontal	11/07/01	29	Horizontal
04/07/02	17.8	Vertical	11/07/01	27	Horizontal
05/07/02	20.0	Vertical	11/07/01	22	Horizontal
05/07/02	18.4	Horizontal	04/09/02	23	Horizontal
05/07/02	20.0	Vertical	04/09/02	23	Vertical
05/07/02	18.0	Vertical	04/09/02	23	Vertical
05/07/02	20.0	Horizontal	04/09/02	23	Vertical

Coverage Estimator) (Chazdon et al., 1998). ACE is a modified version of the Chao2 estimator developed by Chao & Lee (1992) that takes into account the overestimation in species richness of the Chao2, especially when a small number of samples is used. The estimates were computed with EstimateS using 100 randomizations without replacement (Colwell, 1997). The sample's diversity was described using Hill numbers N_0, N_1, N_2 and N_∞. N_0 is the species richness, N_1 the exponential Shannon–Wiener index, N_2 the reciprocal of Simpson's index and N_∞ the reciprocal of the proportional abundance of the most common species (reciprocal of Berger–Parker index) (Hill, 1973).

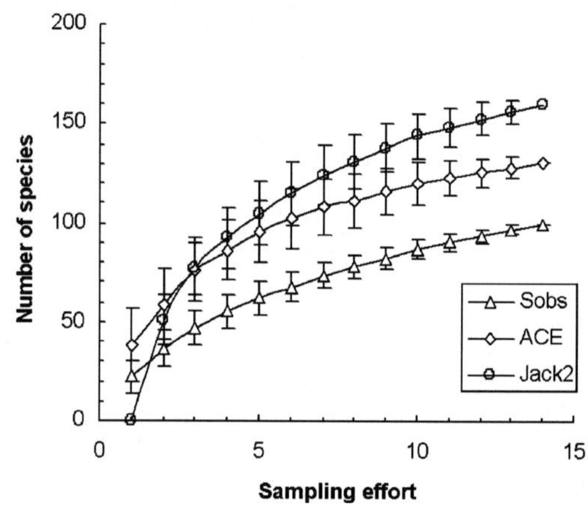

Figure 2. Species accumulation curves. Estimators of the species richness are the total number of observed species (Sobs), the Abundance based Coverage Estimator (ACE) and the Jacknife 2 Estimator of true richness (Jack2). Plotted values are mean ± SD of 100 estimates based on 100 randomization.

Results

Species richness, abundance and diversity

The total number of taxa identified is 121 (see Table A in Appendix). Among them 22 were identified *de-visu* by divers: three Scyphozoans, two Actinarians, one Hydrozoan, three Opisthobranchia, two Decapoda and 11 species of fish. The other 99 taxa were identified from the scraped surface, with 65 and 74 taxa for the Birkenfels and Bourrasque, respectively. Estimates of cumulative species richness against sampling effort showed no sign of approaching asymptotic values (Fig. 2). Nearly 100 species were observed while the estimated number of species for ACE and Jack2 are respectively 131 and 160. Fitting a log regression on the number of species ($R^2 \geq 0.99$), predicted values for 100 samples are respectively 156 for Sobs, 205 for Jack2 and 280 for ACE.

The mean, minimum and maximum number of species for the shipwrecks is presented in Table 2. The number and density of species for both shipwrecks are similar with more variation among samples for the Bourrasque. The huge mean and standard error for the density is due to a single species, the tube-dwelling amphipod *Jassa herdmani* (Walker). Its density ranges from 100 to 101,120 ind/m^2.

Figure 3 shows the rank/abundance graph by shipwreck. Both are dominated by a small number of species with many 'rare' species. The dominance of *J. herdmani* reaches 76% when all samples are pooled together. Table 3 shows the cumulative dominance for pooled samples when *J. herdmani* is excluded: 81% of the dominance is due to the 10 most abundant species. The most represented

Table 2. Number of species (N_0) and density on the two shipwrecks samples (25 × 25 cm)

Shipwreck	N_0	Density (ind/m^2)
Birkenfels		
Mean	25 ± 5	18 460 ± 28 930
Min	18	3950
Max	33	77 056
Bourrasque		
Mean	19 ± 11	20 300 ± 43 810
Min	11	1870
Max	41	119 540

Mean is expressed with it standard error.

phyla in terms of species diversity for both shipwrecks are Polychaets, Crustaceans, Molluscs and Cnidarians (Table 4). Less than 50% of the species are common to both wrecks (Table 5). The Spearman rank correlation between species richness (N_0) of samples and the other Hill numbers (N_1, N_2 and N_∞) is not significant ($p > 0.05$) (Table 6). The low value for the dominance index indicates a high degree of unevenness but this does not hamper high species richness. On the contrary, the samples containing most species (sample 4 and 14, with 33 and 41 species, respectively) have low values for the N_1, N_2 and N_∞. A dense mat of a

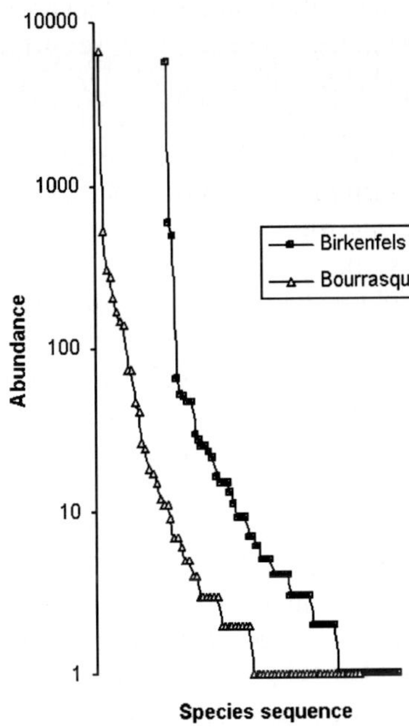

Figure 3. Rank abundance plot for the onshore (Bourrasque) and offshore (Birkenfels) shipwrecks.

Table 3. Main dominant species of the complete dataset

Species	Dominance (%)	Cumulative dominance (%)
Pisidia longicornis	19.1	19.1
Phtisica marina	13.8	32.9
Ophiotrix fragilis	12.6	45.6
Phyllodoce mucosa	8.2	53.8
Lanice conchilega	8.2	61.9
Diadumene cincta	6.3	68.2
Metridium senile	4.8	73.1
Balanus crenatus	3.5	76.6
Stenothoe valida	2.6	79.3
Eumida sanguinea	1.9	81.2

Colonial species (Hydrozoa and Bryozoa) and the amphipod *Jassa herdmani* are excluded.

Table 4. Distribution of the organisms among taxonomic groups for both shipwrecks (nr: not represented)

Phylum	Birkenfels (%)	Bourrasque (%)
Polychaeta	33.3	37.8
Crustacea	21.2	21.6
Mollusca	19.7	13.5
Cnidaria	9.1	17.6
Echinodermata	6.1	4.1
Bryozoa	4.5	2.7
Nemertina	1.5	nr
Porifera	1.5	1.4
Sipunculia	1.5	1.4
Tunicata	1.5	nr

Table 5. Shared species based on taxonomic groups by the two shipwrecks

Phylum	Number of species		% shared
	Birkenfels	Bourrasque	
Sipunculia	1	1	100
Mollusca	13	10	53
Crustacea	14	16	43
Echinodermata	4	3	40
Polychaeta	22	28	39
Cnidaria	6	13	36
Bryozoa	3	2	25
Nemertina	1	0	0
Porifera	2	1	0
Tunicata	1	0	0

Table 6. Hill numbers values for the 14 samples with indication on the orientation of the sampling surface. N_0 is the species richness, N_1 the exponential Shannon-Wiener index, N_2 the reciprocal of Simpson's index and N_∞ the reciprocal of the proportional abundance of the commonest species (reciprocal of Berger-Parker index)

	N_0	N_1	N_2	N_∞
Birkenfels				
Horizontal	11	4.23	3.10	2.00
	15	6.22	4.18	2.30
	21	7.59	5.27	3.16
	23	4.63	2.14	1.47
Vertical	11	5.43	4.04	2.66
	13	5.59	4.00	2.58
	41	2.14	1.39	1.18
Bourrasque				
Horizontal	18	2.01	1.43	1.21
	23	5.00	2.85	1.77
	25	10.07	6.51	3.15
	33	1.37	1.10	1.05
vertical	22	5.27	3.80	2.79
	28	6.08	3.97	2.50
	28	8.58	5.91	3.69

Hydrozoan species, *Tubularia indivisa* L. is the common feature of these two samples. The erect perisarc of this species, which can reach 15–20 cm long, creates a third dimension onto the two-dimensional shipwreck surface.

Faunal assemblage description

We observed two main communities on both shipwrecks. The first one is dominated by the hydrozoan *Tubularia indivisa* associated with *Jassa herdmani* (Amphipoda). The second one is dominated by *Metridium senile* (L.) (Anthozoa). Cnidarians are the dominant phylum represented particularly by three very abundant species: *T. indivisa* (Hydrozoa), *M. senile* (Anthozoa) and *Diadumene cincta* Stephenson (Anthozoa). *T. indivisa* is always associated with *Jassa herdmani*. This amphipod builds a tube made of sediment that aggregates around the perisarc and sometimes even completely covers it. Samples with a high species diversity are, as previously mentioned, linked to an important development of *T. indivisa* colonies. One can assume that the perisarc acts as a new colonization surface for a wide range of animals. On the contrary, concentration of *M. senile* does not permit the development of other species and therefore surfaces where *M. senile* is dominant are characterized by very few species.

At the prospected depth there was no algal development. The animal community was dominated by carnivores/scavengers and passive suspension-feeders. For the active suspension-feeders, only a few bivalve species were found as juveniles and in low densities; moreover, only three species of sponges and one tunicate were identified, all in low densities. The ophiuroid *Ophiotrix fragilis* (Abildgaard) is mainly found offshore on the Birkenfels where it can form dense aggregations up to 2000 ind/m^2. The Serpulid *Pomatoceros triqueter* (L.) covers the metal sheet of that shipwreck while *Balanus crenatus* Bruguiére is found on all onshore samples (Bourrasque). *Phyllodoce mucosa* (Oersted) occurred at a density ranging between 16 to 900 ind/m^2 on the Birkenfels and between 160 to 1680 ind/m^2 on the Bourrasque. The Anomuran crab *Pisidia longicornis* (L.) is common at both sites, being represented by juveniles with densities ranging between 32 to 4860 ind/m^2. Only juveniles of *Psammechinus miliaris* (Gmelin) were found (maximum test length of 10 mm). Differences between the faunal associations on vertical and horizontal surfaces were observed. Sedimentation on horizontal surface favors some soft sediment species. On these surfaces tubes of the Polychaeta *Lanice conchilega* (Pallas) were observed at a maximum density of 3680 ind/m^2. *Nassarius incrassatus* (Ström) was only found on horizontal surface of the Birkenfels. The Polychaet *Phyllodoce mucosa* was preferentially found on horizontal surfaces.

The mollusc *Epitonium clathratulum* (Kanmacher) has been found in densities ranging from 32 to 144 ind/m^2 on horizontal surfaces on both shipwrecks but more frequently on the Birkenfels. The Poriferan *Dysidea fragilis* (Montagu) is recorded for the first time in Belgian waters.

Discussion

In this study the total sampled area on both shipwrecks only represents 0.875 m^2, nevertheless 65 and 74 macrofauna species have already been identified on the offshore and onshore sites respectively. The species accumulation curves indicate that more intensive sampling will certainly add more species to the list. With only two shipwrecks and 14 samples we already reached the species richness mentioned by Leewis et al. (2000) (127 species) for a 5 year study covering 21 Dutch shipwrecks. The low species richness observed by Leewis et al. (2000) might be related to the fact that the sampling was mainly made by *de-visu* methods with a potential information loss on small macrofauna species. On the other hand, Hiscock (1980) identified a total of 187 taxa on a 1.4 m^2 horizontal surface of a shipwreck from the isle of Lundy (SW of England) sunk 5 years previously. The expected species richness of the two shipwrecks prospected in the present study is closer to the value given by Hiscock (1980). Cattrijse & Vincx (2001) reviewed the Belgian data on macrobenthos for soft bottoms and found that species richness for a site varied from 4 to 33 and that species richness for a zone (sites pooled for 9 defined zones on the BCS) ranged from 19 to 87. Comparison with our data indicates that average macrobenthos biodiversity on the BCS is at least ten times higher on hard substrata than on soft substrata.

The pebble communities in the Dover Straits have a pooled species diversity of 211 species for the coastal zone and 128 species for the offshore zone (Davoult, 1990). The observed faunal assemblage on shipwrecks shows some resemblance to sites subjected to high hydrodynamics in the Dover Strait. In this community, epifauna is the dominant group of organisms. Four dominant species are common to pebble communities and studied shipwrecks: *Ophiotrix fragilis*, *Pisidia longicornis*, *Lepidonotus squamatus* (L.) and *Psammechinus miliaris*. On the contrary, pebble communities are not dominated by Hydrozoan and Anthozoan species.

The absence of macro algal development at the studied depth is probably due to the low light intensity reaching the wrecks. The animal community is dominated by carnivores/scavengers and passive suspension-feeders and only very few active suspension-feeding species were observed. This situation can be compared with what has been described on hard substrata in the Oosterschelde (SW Netherlands). Leewis & Waardenburg (1990) studied the faunal assemblages after the construction of a storm surge barrier in the mouth of the Oosterschelde estuary. They found that *Metridium senile* and *Tubularia indivisa* communities dominate the area close to the dam, where the current velocity is the strongest. Away from the dam, under low current conditions, communities with Porifera and Tunicata develop. Indeed, these active filter-feeders are more competitive when the current velocity is low (de Kluijver & Leewis, 1994). In another study in the Adriatic Sea (Italy, Mediterranean Sea) a community on outcrops and shipwrecks in low current conditions was also dominated by active suspension filter-feeders: Tunicata, Bivalvia and Porifera accounted for 55% of the species diversity while passive suspension filter-feeders represented only 10% of the species (Gabriele et al., 1999). It is likely that the more sheltered parts of the shipwrecks (mainly inside the shipwrecks themselves) might harbor active suspension filter-feeders but this part of the shipwrecks has not been investigated so far. In both shipwrecks, the amphipod *Jassa herdmani* was found in huge densities up to 100 000 ind/m^2. They inhabit self-constructed tubes from which they extend the anterior half of their bodies to gather detritus and filter suspended particles and plankton (Conlan, 1989). The construction of these tubes prevents them from being washed out by strong currents (Ulrich et al., 1995). They are known as important fouling organisms. This species seems to out-compete most of the other amphipods that are represented by three species of the genus *Stenothoe* and five species of Caprellidea. The members of the genus *Stenothoe* are filter-feeders (Barnard, 1969) and some are known to live on Hydrozoa ramifications (Lewis, 1992). Caprellidea are mainly filter-feeders but some species adopt a scraping feeding strategy (Guerra-Garcia, 2002). Our results indicated that poor species assemblages are associated with the huge development of the Anthozoan *M. senile* in some places; a phenomenon already described by de Kluijver & Leewis (1994). One possible reason is the asexual mode of reproduction of this anthozoan by basal laceration (Manuel, 1981) that tends to overgrow other organisms (Bucklin, 1987). Another reason could be toxin production (Anderluh & Macek, 2002) and exudation preventing larval settlement (Koh, 1997).

The hard substrata provide attachment sites and microhabitats for a range of species typically not found in the surrounding soft sediments. As with other artificial structures, shipwrecks raise the question of the 'attraction versus production' debate (see among others Bohnsack et al., 1997; Pickering & Whitmarsh, 1997; Bortone, 1998; Page et al., 1999; Hall et al., 2000; Jensen et al., 2000). Although not quantified, the increased density of fish around the shipwreck was probably important since 11 species were encountered; among them, the commercially important species *Gadus morhua* L.

Finally, some species have to be considered as having conservation value because of their scarcity or rareness. Several living specimens of the amphipod *Caprella tuberculata* Guérin were found on the Birkenfels. This species does not seem to be uncommon but is scarce along the British coasts (Isaac et al., 1990) and has been mentioned only very recently for the Belgian marine fauna (Massin et al., 2002). *Epitonium clathratulum* is a rare species for the Belgian marine fauna; it was only known from a few stranded specimens (Balckeljau, 1986; Vanhaelen, 1989) and one single living specimen collected on a breakwater (Jonckheere, 2001). In the Netherlands, the first record of a living specimen was made by Eisma (1966) and more recently it was found in the Osterschelde (The Netherlands) (Wetsteyn & Nieuwenhuize, 1996). During the present study, *E. clathratulum* was observed on both shipwrecks, the species being more frequent on the Birkenfels. *E. clathratulum* was found only on horizontal surfaces were a thin layer of sediment occurs. This observation highlights the importance of sedimentation on horizontal surfaces that seems to favour some soft sediment molluscs and polychaeta species such as *Lanice conchilega*. The Epitonidae are carnivorous and known to feed on Anthozoa (Graham, 1988). But compared to other horizontal samples, relatively few Anthozoa (*M. senile* and *D. cincta*) were present in the samples where *E. clathratulum* was found. The Poriferan *Dysidea fragilis* is recorded for the first time in the Belgian waters. This species is widespread but seldom dominant in the British Isles and can be found along the Atlantic coast of Europe and in the Mediterranean Sea (Ackers et al., 1992).

As a conclusion, these preliminary results emphasize spots of high species richness for the sessile and slow moving epifauna. The two investigated sites show striking differences in terms of species assemblage but more intensive sampling is needed in order to discern ecological patterns. The possible role of such 'hard substrata islands' in an environment dominated by soft bottom sediments needs further research in order to evaluate their impact on the reproduction, dissemination and protection of North Sea species. In this respect, two separate questions need to be addressed. First, what is the origin of the species present on shipwrecks in the Belgian waters? Second, do shipwrecks allow these species to develop, reproduce and disseminate? We do not know anything about natural hard substrata available on the BCS for the colonization of subtidal epifauna. It seems that if present, they occur in very localized and limited areas. A comparison between the fauna encountered on these natural and unnatural outcrops could be helpful in order to know if shipwrecks are discrete spots where a natural fauna occurs or if shipwrecks play an important role for the dissemination of some species that would otherwise not be present. In this later case, shipwrecks, even if hotspots for diversity, could favor the dissemination of possibly harmful exotic species and act as stepping-stones to put in touch different populations of a single species. This could lead to a loss of genetic diversity and fitness for all or part of these populations.

Acknowledgements

It is our pleasure to thank Drs Y. Laitat, A. Simon, D. Delbare and Mrs J.P. De Blauwe, J. Haelters, A. Vanhaelen & P. Van de Steen for their help on the field. Drs J. Bouillon (Hydrozoa), Cl. De Broyer (Crustacea), Th. Backeljau (Mollusca), Ph. Willenz (Porifera) & K. Wouters (Crustacea) and Mrs G. Van Hoey (Polychaeta) and Fr Kerckhof (Mollusca, Crustacea) for their help in species identification; our sincere thanks to officers and crew of the RV Belgica for their help during the 2001 and 2002 cruise, to RBINS (malacology section and MUMM department) for logistical support. Special thanks to Ir Vincent Tigny for his help in the mapping process, to two anonymous reviewers who made suggestions in order to improve our paper and to C. Vanderlinden for English revision. This study has been supported by a 'plan Action II' grant from the

Federal Office for Scientific, Technical and Cultural Affairs (OSTC).

References

Ackers, R. G., D. Moss & B. E. Picton, 1992. Sponges of the British Isles: A Colour Guide and Working Document. Marine Conservation Society, Herefordshire 175 pp.

Anderluh, G. & P. Macek, 2002. Cytolytic peptide and protein toxin from sea anemones (Anthozoa: Actiniaria). Toxicon 40: 111–124.

Backeljau, T., 1986. Lijst van de recente mariene mollusken van België. Studiedocumenten Koninklijk Belgisch Instituut voor Natuurwetenschappen 29: 1–106.

Barnard, J. L., 1969. The family and genera of marine gammaridean amphipod. U.S. National Museum Bulletin 271: 1–535.

Beyst, B., K. Hostens & J. Mees, 2001. Factors influencing fish and macrocrustacean communities in the surf zone of sandy beaches in Belgium: temporal variation. Journal of Sea Research 46: 281–294.

Bohnsack, J. A., A. M. Ecklund & A. M. Szmant, 1997. Artificial reef research: is there more than the attraction-production issue?. Fisheries 22: 14–16.

Bortone, S. A., 1998. Resolving the attraction-production dilemma in artificial reef research: some yeas and nays. Fisheries 23: 6–10.

Bucklin, A., 1987. Adaptative advantages of patterns of growth and asexual reproduction of the sea anemone *Metridium senile* (L.) in intertidal and submerged populations. Journal of Experimental Marine Biology and Ecology 110: 225–243.

Burnham, K. P. & W. S. Overton, 1978. Estimation of the size of a closed population when capture probabilities vary among animals. Biometrika 65: 623–633.

Burnham, K. P. & W. S. Overton, 1979. Robust estimation of population size when capture probabilities vary among animals. Ecology 60: 927–936.

Castilla, J. C. & D. J. Crisp, 1973. Responses of *Asterias rubens* to water currents and their modification by certain environmental factors. Netherlands Journal of Sea Research 7: 171–190.

Cattrijsse, A. & M. Vincx, 2001. Sustainable Management of the North Sea: Biodiversity of the Benthos and the Avifauna of the Belgian Coastal Waters. Federal Office for Scientific, Technical and Cultural Affairs, Brussels 48 pp.

Chao, A. & S.-M. Lee, 1992. Estimating the number of classes via sample coverage. Journal of the American Statistical Association 87: 210–217.

Chazdon, R. L., R. K. Colwell, J. S. Denslow & M. R. Guariguata, 1998. Statistical methods for estimating species richness of woody regeneration in primary and secondary rain forests of NE Costa Rica. In Dallmeier, F. & J. A. Comiskey (eds.) Forest Biodiversity Research, Monitoring and Modelling: Conceptual Background and Old World Case Studies. Parthenon Publisher, Paris: 285–309.

Colwell, R. K., 1997. EstimatesS: Statistical estimation of species richness and shared species from samples. Version 5. User's Guide and application published at: http://viceroy.eeb.uconn.edu/estimates.

Conlan, K. E., 1989. Delayed reproduction and adult dimorphism in males of the amphipod genus *Jassa* (Corophioidea: Ischyroceridae): an explanation for systematic confusion. Journal of Crustacean Biology 9: 601–625.

Costello, M. J., C. S. Emblow & R. White (eds), 2001. European Register of Marine Species. A check-list of marine species in Europe and a bibliography of guides to their identification, Vol. 50. Patrimoines naturels: 463 pp.

Daro, M., 1969. Etude écologique d'un brise-lame de la côte belge. 1. Description et zonation des organismes. Annales de la Société Royale Zoologique de Belgique 99: 112–152.

Daro, M., 1970. Etude écologique d'un brise-lame de la côte belge. 2. Biologie et développement saisonnier des espèces. Annales de la Société Royale Zoologique de Belgique 100: 159–189.

Davoult, D., 1990. Biofaciès et structure trophique du peuplement des cailloutis du Pas-de-Calais (France). Oceanologica Acta 13: 335–348.

De Clerck, R., N. Cloet & N. Redant, 1973. Bestandopname langsheen de Belgische kust. 7. Mededelingen van het Rijkstation voor Zeevisserij, Ghent.

De Clerck, R., N. Cloet & N. Redant, 1974a. Bestandopname langsheen de Belgische kust. 9. Mededelingen van het Rijkstation voor Zeevisserij, Ghent.

De Clerck, R., N. Cloet & N. Redant, 1974b. Bestandopname langsheen de Belgische kust. 8. Mededelingen van het Rijkstation voor Zeevisserij, Ghent.

De Clerck, R., N. Cloet & N. Redant, 1975. Bestandopname langsheen de Belgische kust. 10. Mededelingen van het Rijkstation voor Zeevisserij, Ghent.

Degraer, S., M. Vincx, P. Meire & Offringa, 1999. The macrozoobenthos of an important wintering area of the common scoter (*Melanitta nigra*). Journal of the Marine Biological Association of the UK 79: 243–251.

de Kluijver, M. J. & R. J. Leewis, 1994. Changes in the sublittoral hard substrate communities in the Oosterschelde estuary (SW Netherlands), caused by changes in the environmental parameters. Hydrobiologia 283: 265–280.

Deleu, S., 2002. Tide-topography interactions near the kink in the Westhinder sandbank (Southern North Sea). M. Sc. Thesis, Gent University: 43 pp.

De Pauw, N. & D. Van Damme 1992. Mariene Fauna en Flora. MER Study. Elektriciteitcentrale Zeebrugge (N.V. ELECTRABEL - S.P.E.), Zeebrugge.

Dewicke, A., A. Cattrijsse, J. Mees & M. Vincx, 2003. Spatial patterns of the hyperbenthos of subtidal sandbanks in the southern North Sea. Journal of Sea Research 49: 27–46.

Eisma, D., 1966. The distribution of benthic marine molluscs off the main Dutch coast. Netherlands Journal of Sea Research 3: 107–163.

Gabriele, M., A. Bellot, D. Gallotti & R. Brunetti, 1999. Sublittoral hard substrate communities of the northern Adriatic Sea. Cahiers de Biologie Marine 40: 65–76.

Gilson, G., 1900. Exploration de la mer sur les côtes de la Belgique. Mémoire du Muséum d'Histoires naturelles de Belgique 1: 1–81.

Govaere, J. C. R., D. Van Damme, C. Heip & L. A. P. De Coninck, 1980. Benthic communities in the Southern Bight of the North Sea and their use in ecological monitoring. Helgolander Wissenschaftliche Meeresuntersuchungen 33: 507–521.

Graham, A., 1988. Mollusca, Prosobranch and Pyramidellid Gastropods. Synopses of the British fauna (New series), N°2. Academic Press, London 662 pp.

Guerra-Garcia, M., 2002. Re-descriptions of *Caprella linearis* (Linnaeus, 1767) and *C. septentrionalis* Kröyer, 1838 (Crustacea: Amphipoda: Caprellidea) from Scotland, with an ontogenetic comparison between the species and a study of the clinging behaviour. Sarsia 87: 216–235.

Hall, S. J., S. A. Gray & Z. L. Hammett, 2000. Biodiversity-productivity relations: an experimental evaluation of mechanisms. Oecologia 122: 545–555.

Heip, C. & W. Decraemer, 1974. The diversity of nematode communities in the Southern North Sea. Journal of the Marine Biological Association of the UK 54: 251–255.

Hill, M. O., 1973. Diversity and evenness: a unifying notation and its consequence. Ecology 54: 427–431.

Hiscock, K., 1980. Marine life on the shipwreck of the M.V. 'Robert'. Report of the Lundy Field Society 32: 40–44.

Isaac, M. J., P. Making, E. Naylor, G. Smaldon & R. G. Withers, 1990. Crustacea II: Malacostraca Peracarida. In Hayward, P. J. & J. S. Ryland (eds.) The marine fauna of the British Isles and North-West Europe. 1 Oxford University Press, Oxford: 362–488.

Jensen, A. C., K. J. Collins & A. P. M. Lockwood, 2000. Artificial Reefs in European Seas. Kluwer Academic Publishers, Dordrecht, The Netherlands 508 pp.

Jensen, P., 1976. Free-living marine nematodes from a sublittoral station in the North Sea off the Belgian coast. Biologisch Jaarboek 44: 231–255.

Jonckheere, I., 2001. Witte wenteltrap *Epitonium clathratulum* levend op een golfbreker te Koksijde. Strandvlo 21: 53–57.

Kerckhof, F. & J.-S. Houziaux, 2003. Biodiversity of Belgian marine areas. In Peeters, M. A. Franklin, & J.L. Goethem Van(eds.) Biodiversity in Belgium. Royal Belgian Insitute of Natural Sciences, Brussels: 350–385.

Koh, E. G. L., 1997. Secretion of bioactive compounds by a scleractinian coral. Proceedings of the 8th International Coral Reef Symposium, 24–29 June 1996. Lessios, H. A. & MacIntyre, I. G. (eds), Smithsonian Tropical Research Institute, Balboa, Republic of Panamá: 1263–1266.

Leewis, R. J. & H. W. Waardenburg, 1990. Flora and fauna of the sublittoral hard substrata in the Oosterchelde (The Netherlands) – interactions with the North Sea and the influence of a storm surge barrier. Hydrobiologia 195: 189–200.

Leewis, R. J., G. W. N. M. van Moorsel & H. W. Waardenburg, 2000. Shipwrecks on the Dutch continental shelf as artificial reefs. Artificial reefs in European seas. Jensen, A. C, Collins, K. J. & Lockwood, A. P. M. (eds), Kluwer Academic Publishers: 419–434.

Lewis, B. L., 1992. Abundance, distribution and behaviour of a commensal amphipod *Stenothoe valida* Dana on the hydrocoral *Millepora complanata* Lamarck. Bulletin of Marine Science 51: 245–249.

Manuel, R. L., 1981. British Anthozoa. Synopses of the British Fauna (New series), N°18. Academic Press, London 241 pp.

Maertens, D., 1989. Ecological monitoring of the new dumping area on the Belgian Continental Shelf. In Pichot, G. (ed.), Progress in Belgian Oceanographic Research 1989. Proceedings of the North Sea symposium held in Ghent, 14 February 1989. Prime Minister's Services. Science Policy Office, Brussels: 297–324.

Massin, C., A. Norro & J. Mallefet, 2002. Biodiversiy of a shipwreck from the Belgian Continental Shelf: monitoring using scientific diving Preliminary results. Bulletin de l'Institut Royal des Sciences Naturelles de Belgique, Biologie 72: 67–72.

Page, H. M., J. E. Dugan, D. S. Dugan, J. B. Richards & D. M. Hubbard, 1999. Effects of an offshore oil platform on the distribution and abundance of commercially important crab species. Marine Ecology – Progress Series 185: 47–57.

Pickering, H. & D. Whitmarsh, 1997. Artificial reefs and fisheries exploitation: a review of the 'attraction versus production' debate, the influence of design and its significance for policy. Fisheries Research 31: 39–59.

Prygiel, J., D. Davoult, J. M. Dewarumez, R. Glaçon & A. Richard, 1988. Description et richesse des peuplements benthiques de la partie française de la Mer du Nord. Comptes Rendus de l'Académie des Sciences, Série III 306: 5–10.

Ulrich, I., K. Anger & U. Schöttler, 1995. Tube-building in two epifaunal amphipod species, *Corophium insidiosum* and *Jassa falcata*. Helgolander Wissenschaftliche Meersuntersuchungen 49: 393–398.

Vanhaelen, M.-T., 1989. Enkele waarnemingen van het strand tussen Koksijde en Oostduinkerke tijdens en na de warme zomer van 1989. Strandvlo 9: 117–120.

Van Hoey, G., S. Degraer & Vincx, 2004. Macrobenthic community structure of soft-bottom sediments at the Belgian Continental Shelf. Estuarine, Coastal and Shelf Science 59: 599–613.

Vanosmael, C., K. A. Willems, D. Claeys, M. Vincx & C. Heip, 1982. Macrobenthos of a sublittoral sandbank in the Southern Bight of the North Sea. Journal of the Marine Biological Association of the UK 62: 521–534.

Vincx, M., 1981. New and little known nematodes from the North Sea. Cahiers de Biologie Marine 22: 431–451.

Volckaert, A., S. Engledow, S. Degraer, M. Vincx, E. Coppejans & M. Hoffmann, 2002. The epilithic macrofauna and macroalgae of the hard substrates along the Belgian coast. Bulletin de l'Institut Royal des Sciences Naturelles de Belgique, Biologie 72 (Suppliment): 13–15.

Wetsteyn, L. P. M. J. & J. Nieuwenhuize, 1996. *Epitonium clathratulum* (Kanmacher, 1798) (Gasteropoda, Prosobranchia: Epitoniidae), new for the Oosterschelde (The Netherlands), and some remarks on its presence in the Dutch Delta area. Basteria 59: 239–240.

Willems, K. A., C. Vanosmael, D. Claeys, M. Vincx & C. Heip, 1982. Benthos of a sublittoral sandbank in the southern bight of the North Sea: general considerations. Journal of the Marine Biological Association of the UK 62: 549–557.

Appendix

Table A. List of species found only in the Birkenfels, Bourrasque and common to both shipwrecks. Species designated by 'photo' were not collected but identified by mean of digital pictures. Species designated by * were identified in situ by divers. The taxonomy is based on the 'European Register of Marine Species' (Costello et al., 2001)

		Birkenfels specific taxa	Bourrasque specific taxa	Common taxa to both shipwrecks
PORIFERA		*Dysidea fragilis* (Montagu, 1818)	*Haliclona* sp.	
		Halichondria cfr panicea (Pallas, 1766)		
CNIDARIA	Hydrozoa		*Bougainvillia muscus* (Allman, 1863)	*Hydractinia echinata* (Flemming, 1828)
			Campanularia volubilis (Linnaeus, 1758)	*Obelia bidentata* Clarke, 1875
			Clytia gracilis (Sars, 1850)	*Sarsia eximia* (Allman, 1859)
			Clytia hemisphaerica (Linnaeus, 1767)	*Tubularia indivisa* Linnaeus, 1758
			Hydrallmania falcata (Linnaeus, 1758)	
			Laomedea flexuosa Alder 1857	
			Nemertesia antennina (Linnaeus, 1758)	
			Obelia dichotoma (Linnaeus, 1758)	
			Sertularia cupressina Linnaeus, 1758	
			Tubularia larynx Ellis & Solander, 1786	
	Schyphozoa	*Aequorea vitrina* Gosse, 1853*		*Cyanea lamarcki* Péron & Lesueur,1810*
		Chrysaora hysoscella (Linnaeus, 1767)*		
	Anthozoa	*Alcyonium digitatum* Linnaeus, 1758	*Sagartia* sp.*	*Diadumene cincta* Stephenson, 1925
		Anemonia sp. (photo)		*Metridium senile* (Linnaeus, 1767)
		Sagartia sp. (photo)		*Sagartia troglodytes* (Price *in* Johnston, 1847)
		Urticina felina (Linnaeus, 1767) (photo)		
NEMERTA		Nemertinea spp.		
ANNELIDA	Polychaeta	*Alentia gelatinosa* (M. Sars, 1835)	Pectinariidae	*Autolytus* sp.
		Eulalia cfr fucescens de Saint Joseph, 1888	*Cirratulus filiformis* Keferstein, 1862	*Eumida sanguinea* (Oersted, 1843)
		Nereis zonata Malmgren, 1867	*Eteone picta Quatrefages, 1865*	*Eupolymnia nesidensis* (Delle Chiaje, 1828)
		Poecilochaetus serpens Allen, 1904	*Eulalia* spp.	*Eusyllis blomdstrandi* Malmgren, 1867
		Procerastea halleziana Mallaquin, 1893	*Eulalia viridis* (Linnaeus, 1768)	*Harmothoe extenuata* (Grube, 1840)
		Syllis armillaris (O.F. Müller, 1776)	*Gattyana cirrhosa* (Pallas, 1766)	*Harmothoe* spp.
		Thelepus cincinnatus (Fabricius, 1780)	*Nereimyra punctata* (O.F. Müller, 1788	*Kefersteinia cirrata* (Keferstein, 1862)
			Nicomache sp.	*Lanice conchilega* (Pallas, 1766)
			Ophelia sp.	*Lepidonotus squamatus* (Linnaeus, 1758)
			Phyllodoce mucosa (Linnaeus, 1767)	*Nereis pelagica* Linnaeus, 1758
			Phyllodoce spp.	*Nereis* sp.

Continued on p. 218

Table A. (Continued)

		Birkenfels specific taxa	Bourrasque specific taxa	Common taxa to both shipwrecks
			Polydora sp.	*Pomatoceros triqueter* (Linnaeus, 1758)
			Sabellaria spinulosa Leuckart, 1849	*Phyllodoce mucosa* (Öersted, 1843)
			Scoloplos armiger (O.F. Müller, 1776)	*Procerastea perieri* Gravier, 1900
			Thelepus setosus (Quatrefages, 1865)	*Syllis gracilis* Grube, 1840
MOLLUSCA	Bivalvia	*Heteranomia squamula* (Linnaeus, 1758)		*Aequipecten opercularis* (Linnaeus, 1758)
		Venerupis geographica (Gmelin, 1791)		*Mytilus edulis* Linnaeus, 1758
				Mysella bidentata (Montagu, 1803)
	Gasteropoda	*Archidoris pseudoargus* (Rapp, 1827)	*Cuthona* sp.	Aeolidiidae
		Coryphella verrucosa (M. Sars, 1829)	*Nassarius reticulatus* (Linneaus, 1758)	*Crepidula fornicata* (Linnaeus, 1758)
		Dendronotus frondosus (Ascanius, 1774)		*Epitonium clathratulum* (Kanmacher, 1798)
		Euspira pulchella (Risso, 1826)		*Eubranchus* sp.
		Flabellinidae sp. (photo)		Rissoidae
		Nassarius incrassatus (Ström, 1768)		
SIPUNCULIA				*Sipunculus* sp.
CRUSTACEA	Decapoda	*Hias araneus* (Linneaus, 1758)		*Cancer pagurus* Linneaus, 1758
		Macropodia rostrata (Linnaeus, 1761)		*Liocarcinus holstatus* (Fabricius, 1798)
		Paguridae		*Liocarcinus* sp.
				Necora puber (Linnaeus, 1767)*
				Pagurus berhnardus (Linnaeus, 1758)
				Pilumnus hirtellus (Linnaeus, 1761)
				Pisidia longicornis (Linneaus, 1767)
	Amphipoda	*Caprella tuberculata* Guérin, 1836	*Pseudoprotella phasma* (Montagu, 1804)	*Caprella linearis* (Linnaeus, 1767)
			Pariambus typicus (Kroyer, 1844)	*Jassa herdmani* (Walker, 1893)
			Stenothoe marina (Bate, 1856)	*Phtisica marina* Slabber, 1769
			Stenothoe sp.	*Stenothoe monoculoides* (Montagu, 1815)
				Stenothoe valida Dana, 1855
	Cirripedia		*Balanus crenatus* Bruguiére, 1789	
	Cumacea		Cumacea	
	Copepoda		Copepoda	
BRYOZOA		*Disporella hispida* (Fleming, 1828)	*Conopeum seurati* (Canu, 1928)	*Electra pilosa* (Linnaeus, 1767)
ECHINODERMATA	Asteroidea	*Asterias rubens* Linnaeus, 1758		
	Echinoidea		Clypeasteroidea	*Psammechinus miliaris* (Gmelin, 1778)
	Ophiuroidea	*Ophiura albida* Forbes, 1839		*Ophiotrix fragilis* (Abildgaard, 1789)

TUNICATA	*Molgula cfr occulta* Kupffer, 1875		
PISCES	*Gadus morhua* Linnaeus, 1758* *Pollachius pollachius* (Linnaeus, 1758)* *Scomber scombrus* Linnaeus, 1758* *Trisopterus luscus* (Linnaeus, 1758)* *Trisopterus minutus* (Linnaeus, 1758)* *Trachurus trachurus* (Linnaeus, 1758)*	*Dicentrarchus labrax* (Linnaeus, 1758)* *Myoxocephalus scorpius* (Linnaeus, 1758)* *Parablennius gattorugine* (Linnaeus, 1758)*	*Pollachius virens* (Linnaeus, 1758)* *Pomatoschistus* sp. (photo)*

Assessment

Hydrobiologia (2006) 555:223–227
H. Queiroga, M.R. Cunha, A. Cunha, M.H. Moreira, V. Quintino, A.M. Rodrigues, J. Serôdio & R.M. Warwick (eds),
Marine Biodiversity: Patterns and Processes, Assessment, Threats, Management and Conservation
DOI 10.1007/s10750-005-1118-0

Using asymmetrical designs for environmental impact assessment of unplanned disturbances

Nuno C. Queiroz[1,2], Fernando P. Lima[1,2], Pedro A. Ribeiro[1], Sílvia G. Pereira[2] & António M. Santos[1,2,*]

[1]*CIBIO, Centro de Investigação em Biodiversidade e Recursos Genéticos, Campus Agrário de Vairão, 4485-661 Vairão, Portugal*

[2]*Departmento de Zoologia-Antropologia, Faculdade de Ciências da Universidade do Porto, 4099-002 Porto, Portugal*

*(*Author for correspondence: E-mail: amsantos@fc.up.pt)*

Key words: beyond BACI, impact assessment, oil spill, rocky shores, macrobenthic fauna

Abstract

Environmental impact assessment of unplanned disturbances is often difficult to accomplish due to the absence of 'before' data for the impacted sites. In an attempt to overcome this problem, a beyond BACI model is used in order to detect possible changes in the temporal patterns of variation when no previous data are available. The model attempted to detect changes in the abundance of macroinvertebrate species inhabiting the intertidal mussel matrix after an oil spill which occurred in northern Portugal. The detection of a significant impact failed, most probably due to low temporal replication. An extension of the analysis, including the hierarchical arrangement of temporal variability in periods, suggests that increasing the number of sampling times may result in a higher efficiency of the model.

Introduction

The unambiguous detection of environmental impacts in natural ecosystems requires the statistical comparison of differences in temporal changes of a certain variable (for example, the mean abundance of a species) from before to after the disturbance, between the putatively impacted site and several control locations. Beyond BACI analysis (Underwood, 1991, 1992, 1993, 1994) was specifically developed to address this problem, and is still the most logically robust method available to date.

Monitoring studies aiming to assess the effect of unexpected anthropogenic activities on natural populations require careful planning, especially in the selection of the number and location of sampling sites. In the case of oil spills, a reasonable procedure consists in the identification of 'risk areas', like the vicinity of harbours and oil refineries, where sampling should be focused. Unfortunately, the logistic constraints common to the vast majority of medium and long-term monitoring programmes make it almost impossible to survey more than a few control locations.

Such difficulty was felt after a small scale oil spill which occurred in northern Portugal in December 2000. A monitoring programme had been underway on northern Portuguese rocky shores since 1997. Its main goal was to acquire data on macrobenthic assemblages from a number of sites within a stretch of 90 km of coastline for before/after impact assessment. Although intensively studied, the only rocky shore affected by the oil spill had not been regularly monitored. Hence, no comparable data were available before the spill for that particular site. In this work, a beyond BACI model was used in order to detect environmental impacts when no 'before' data are available for the impacted site. Given the fact that there were many sampled sites (eight) which could serve as controls, these were divided into two sets. One of them provided 'after' data, as in usual ACI analysis, and the other was used to estimate 'before' variability as suggested by Underwood (1994).

Material and methods

Data acquisition

The study was carried out at eight locations in northern Portugal between August 1999 and June 2002. Sampling was conducted on randomly assigned dates and during low-tide by scraping over 20 × 20 cm replicate areas in the mussel matrix. The collected material was preserved in formalin (4%). Macroinvertebrate species found in the samples were identified and counted. Only species consistently present across sampling times were used in the analyses: *Brachystomia rissoides* (Hanley), *Littorina neritoides* (L.), *Gibbula umbilicallis* (da Costa), *Nucella lapillus* (Röding), *Rissoa parva* (da Costa), *Lasaea rubra* (Montagu), *Hyale* Rathke spp., *Idotea pelagica* Leach and *Jaera albifrons* Leach.

Statistical analysis

The model adopted in this work (Table 1) is a modified beyond BACI, since the factor Locations (L) is nested in Before-After (B). From the available set of sampled locations, four were randomly assigned to provide an estimate of natural variability before the incident. These were contrasted with 'after' data from a different set of three controls and the putatively impacted location. By explicitly including 'before' data for control locations, changes unrelated to the putative impact can be specifically tested.

The analysis is based on two assumptions. First, the patterns of temporal change within controls must not be different from before to after the disturbance, which means that the interactions T (Aft) × C (Aft) and T (Bef) × C (Bef) should not be significant. Second, if they are, their magnitude should not be different (two tailed test T (Aft) × C (Aft) vs. T (Bef) × C (Bef) not significant). Denominators for the *F*-ratios were identified following the logic of beyond-BACI designs (Underwood, 1992, 1993).

Homogeneity of variances was tested using Cochran's test (Winer et al., 1991) and when necessary, data were transformed to log $(x+1)$. Missing values were compensated by averaging the remaining replicates and reducing the degrees of freedom of the Residual term (Underwood, 1997; Quinn & Keough, 2002).

Results

The beyond BACI analyses failed to detect a significant impact of the oil spill on the studied species. As an example, Table 2 depicts the results for *Lasaea rubra, Hyale* spp. and *Idotea pelagica*. For *Brachystomia rissoides*, *Nucella lapillus*, *Rissoa parva* and *Lasaea rubra* the

Table 1. General beyond BACI model used in the present study

Source of variation	DF	Estimates
B	1	Before vs. After impact
L (B)	$2(l-1)$	Differences between locations
L (Bef)	$l-1$	Differences between locations sampled before the impact
L(Aft)	$l-1$	Differences between locations sampled after the impact
C (Aft)	$l-2$	Differences between controls sampled after the impact
I	1	Differences between impacted and control locations after the impact
T (B)	$2(t-1)$	Differences between sampling times
T (Bef) × L (Bef)	$(t-1)(l-1)$	Variability patterns of locations sampled before the impact
T (Aft) × L (Aft)	$(t-1)(l-1)$	Variability patterns of locations sampled after the impact
T (Aft) × C (Aft)	$(t-1)(l-2)$	Variability patterns of control locations sampled after the impact
T (Aft) × I	$t-1$	Variability patterns of the impacted site vs. controls sampled after the impact
Residual	$2tl(n-1)$	–
Total	$2tln-1$	–

Table 2. Selected examples of the beyond BACI results

Source of variation	df	*Lasaea rubra*			*Hyale* spp.			*Idotea pelagica*		
		SS	MS	*F*	SS	MS	*F*	SS	MS	*F*
B	1	1.645	1.645		1.127	1.127		19.130	19.130	
L (B)	6	43.677	7.279		2.864	0.477		15.000	2.500	
L (Bef)	3	26.082	8.694		1.221	0.407		11.417	3.806	
L (Aft)	3	17.595	5.865		1.643	0.548		3.584	1.195	
C (Aft)	2	4.655	2.327		0.424	0.212		2.426	1.213	
I	1	12.940	12.940		1.219	1.219		1.157	1.157	
T (B)	4	2.933	0.733		2.826	0.707		4.057	1.014	
T (Bef) × L (Bef)	6	10.752	1.792		2.622	0.437		1.383	0.231	
T (Aft) × L (Aft)	6	7.896	1.316		1.382	0.230		3.936	0.656	
T (Aft) × C (Aft)	4	2.616	0.654	1.453	1.096	0.274	2.192	3.066	0.766	3.563*
T (Aft) × I	2	5.279	2.640	5.867*	0.287	0.143	1.144	0.870	0.435	
Residual	93	41.888	0.450		11.63	0.125		20.009	0.215	
Total	116	108.791			22.453			63.515		

Data sets were homoscedastic after log $(x+1)$ transformation. Missing values were compensated by averaging the remaining replicates and reducing the degrees of freedom. B = Before vs. After; L = locations; I = impacted; C = control; T = sampling times. *Significant ($p < 0.05$).

analyses ended up in non-significant T (Aft) × I vs. T (Bef) × C (Bef), meaning that the patterns of variability in the impacted site were not different from those measured at the controls before the impact. In the case of *Hyale* spp. and *Rissoa parva* the interaction T (Aft) × I was not significant. For *Idotea pelagica*, the patterns of variability in the impacted site were not different from those measured at the control sites after the impact (T (Aft) × I vs. T (Aft) × C (Aft) not significant). It is noteworthy that in all eight analyses the test T (Aft) × C (Aft) vs. T (Bef) × C (Bef) was non-significant.

Discussion

The failure to detect differences in the patterns of variability of the selected species may indicate that these were not significantly affected by the oil spill. On the other hand, it could be a consequence of the low overall efficiency of the analyses which, in turn, may result from high spatial and temporal variability of the data (Underwood, 1994). The main challenge in environmental impact assessment studies is to isolate the effect of interest from natural variability. In highly variable systems, this is seldom achievable with low spatial and/or temporal replication.

Small-scale oil spills like the present one are classified as pulse events (Bender et al., 1984), which normally cause short-term disturbances. According to Underwood & Chapman (2003), beyond BACI designs which use nested sampling times within periods are more powerful in the detection of such disturbances. It was not possible to include periods in the model employed in this study due to insufficient temporal replication prior to the disturbance. However, 'after' data were not limited by such constraint. A new approach was attempted, using an after control/impact (ACI) design, which allowed the inclusion of one more control location and another sampling time. Increasing replication in time enabled the hierarchical arrangement of temporal variability, by nesting sampling times within an extra factor (periods).

Results revealed significant differences in the patterns of temporal variation in abundance between impacted and control shores for three of the nine studied taxa (Table 3). Those were detected at different spatial scales – T (P) × L or P × L depending on the species. It is very likely that the increase of small-scale temporal variability

Table 3. Results of the asymmetrical ANOVAs to test for differences in temporal patterns of variation after the spill, between the impacted location and several controls

Source of variation	df	*Lasaea rubra*			*Hyale* spp.			*Jaera albifrons*		
		SS	MS	*F*	SS	MS	*F*	SS	MS	*F*
L	4	30.029	7.507		798.548	199.637		5.205	8.801	
P	1	7.205	7.205		937.996	937.996		0.786	0.786	
T(P)	2	7.249	3.624		154.548	77.274		1.693	0.846	
P × L	4	6.558	1.640		824.820	206.205		2.596	0.649	
P × I	1	2.243	2.243		520.596	520.596	10.635*	0.697	0.697	
P × C	3	4.315	1.438		304.224	101.408	2.074	1.899	0.633	
T (P) × L	8	7.876	0.985		495.420	61.927		9.125	1.141	
T (P) × I	2	3.066	1.533	3.992*	162.299	81.149	1.660	0.173	0.087	†
T (P) × C	6	4.810	0.802	2.089	333.121	55.520	1.135	8.952	1.492	2.908*
Residual	80	30.697	0.384		3911.567	48.895		41.073	0.513	
Total	99	89.614			117.258			90.478		

†2-tailed test: T (P) × I vs. T(P) × C ($F=17.149$*). Data sets were homoscedastic after square root transformation for *Hyale* spp. and *log* $(x+1)$ transformation for the remaining species. L = locations; I = impacted; C = control; P = periods; T = sampling times. *Significant ($p<0.05$).

detected at the impacted site was related to the effects of the spill and also to distinct rates of recovery of the different species, after the oil was removed or washed away.

ACI models have been used by several authors (e.g. Chapman et al., 1995; Roberts, 1996; Glasby, 1997; Lardicci et al., 1999; Bishop et al., 2002; Underwood et al., 2003). By excluding 'before' data from the model a relationship between the human activity in question and observed effects can only be presumed. However, the beyond BACI model used in this study showed no significant differences between patterns of variation among controls from before to after the putative impact. Therefore, it is safe to assume that eventual differences in patterns of variation detected by the ACI analyses between the impacted and control shores are likely to have been caused by the oil spill.

The present work demonstrated the applicability of asymmetrical designs in the detection of unplanned anthropogenic disturbances. As demonstrated by this study, sampling as many sites as possible at the expense of temporal replication should be avoided, especially if it can be anticipated that the disturbance is of small scale type and thus unlikely to affect any of the selected sites. The inclusion of factor 'periods' seems to be mandatory to detect small scale temporal effects, particularly when organisms that exhibit seasonal variation in abundance are monitored. The apparent complexity of the analysis does not imply a sophisticated sampling strategy. Thus, monitoring programs targeted towards the detection of environmental disturbances should be easily set up according to beyond BACI models.

Acknowledgements

This work was funded by Direcção Regional do Ambiente e Ordenamento do Território – Norte. F.P. Lima and P.A. Ribeiro were funded by Fundação para a Ciência e Tecnologia (FCT) grants (refs. SFRH/BD/8730/2002 and SFRH/BD/8232/2002, respectively). We thank two anonymous referees for their comments, which contributed to the improvement of the manuscript.

References

Bender, E. A., T. J. Case & M. E. Gilpin, 1984. Perturbation experiments in community ecology: theory and practice. Ecology 65: 1–13.

Bishop, M. J., A. J. Underwood & P. Archambault, 2002. Sewage and environmental impacts on rocky shores: necessity of identifying relevant spatial scales. Marine Ecology Progress Series 236: 121–128.

Chapman, F. S., A. J. Underwood & G. A. Skilleter, 1995. Variability at different spatial scales between a subtidal assemblage exposed to the discharge of sewage and two control assemblages. Journal of Experimental Marine Biology and Ecology 189: 103–122.

Glasby, T. M., 1997. Analysing data from post-impact studies using asymmetrical analyses of variance: a case study of epibiota on marinas. Australian Journal of Ecology 22: 448–459.

Lardicci, C., F. Rossi & F. Maltagliati, 1999. Detection of thermal pollution: variability of benthic communities at two different spatial scales in an area influenced by a coastal power station. Marine Pollution Bulletin 38: 296–303.

Quinn, G. P. & M. J. Keough, 2002. Experimental Design and Data Analysis for Biologists. Cambridge University Press, Cambridge 537 pp.

Roberts, D. E., 1996. Effects of the North Head deep-water sewage outfall on nearshore coastal reef macrobenthic assemblages. Marine Pollution Bulletin 33: 303–308.

Underwood, A. J., 1991. Beyond BACI: experimental designs for detecting human environmental impacts on temporal variations in natural populations. Australian Journal of Marine and Freshwater Research 42: 569–587.

Underwood, A. J., 1992. Beyond BACI: the detection of environmental impacts on populations in the real, but variable, world. Journal of Experimental Marine Biology and Ecology 161: 145–178.

Underwood, A. J., 1993. The mechanics of spatially replicated sampling programmes to detect environmental impacts in a variable world. Australian Journal of Ecology 18: 99–116.

Underwood, A. J., 1994. On Beyond BACI: sampling designs that might reliably detect environmental disturbances. Ecological Applications 4: 3–15.

Underwood, A. J., 1997. Experiments in ecology: their logical design and interpretation using analysis of variance. Cambridge University Press, Cambridge 504 pp.

Underwood, A. J. & M. G. Chapman, 2003. Power, precaution, Type II error and sampling design in assessment of environmental impacts. Journal of Experimental Marine Biology and Ecology 296: 49–70.

Underwood, A. J., M. G. Chapman & D. E. Roberts, 2003. A practical protocol to assess impacts of unplanned disturbance: a case study in Tuggerah Lakes Estuary, NSW. Ecological Management and Restoration 4S4–S11.

Winer, B. J., D. R. Brown & K. M. Michels, 1991. Statistical Principles in Experimental Design. McGrawHill, New York 1057 pp.

Threats

Hydrobiologia (2006) 555:231–240
H. Queiroga, M.R. Cunha, A. Cunha, M.H. Moreira, V. Quintino, A.M. Rodrigues, J. Serôdio & R.M. Warwick (eds),
Marine Biodiversity: Patterns and Processes, Assessment, Threats, Management and Conservation
DOI 10.1007/s10750-005-1119-z

Mass mortality of coral reef ascidians following the 1997/1998 El Niño event

Francisco Kelmo[1,2], Martin J. Attrill[1,*] & Malcolm B. Jones[1]
[1]*School of Biological Sciences, University of Plymouth, Drake Circus, Plymouth, PL48AA, United Kingdom*
[2]*Brazilian Research Council–CNPq Brazil & Departamento de Zoologia, Instituto de Biologia, , Universidade Federal da Bahia, Campus Universitário de Ondina, 40210-340, Salvador, Brazil*
(**Author for correspondence: Tel.: +44-1752-232916; Fax: +44-1752-232970; E-mail: mattrill@plymouth.ac.uk*)
Received 21 April 2005; accepted in revised form 17 October 2005

Key words: ENSO, climate, Ascidiacea, community, bioerosion, Brazil

Abstract

In April/May each year from 1995 to 2000, ascidians were sampled randomly with 35 1m^2 quadrats from three different reef habitats (intertidal reef tops, coastal reef walls and shallow-bank reefs) at four replicate localities (Praia do Forte, Itacimirim, Guarajuba and Abai) in northern Bahia (Brazil). As the sampling period included the 1997/1998 El Niño event, the most severe on record, for the first time these results allow a quantitative assessment of the impact of this major environmental stressor on the biodiversity of associated coral reef ascidians. Across all reef habitats, 22 ascidian species were recorded from three different orders (Aplousobranchia, Phlebobranchia and Stolidobranchia). After El Niño, all species showed significantly altered densities (ANOVA, $F=602.90$, $p<0.0001$); many species were absent from the reefs within 2 years of the El Niño period, but densities of *Lissoclinum perforatum* (all reefs) and *Echinoclinum verrilli* (subtidal reefs) increased significantly from 1998 onwards. Univariate and multivariate analyses confirmed that significant changes in assemblage composition had occurred. BIOENV analysis identified turbidity, mean temperature and cloud cover as the main factors best explaining these assemblage changes. Our results suggest that although the 1997/1998 El Niño had a differential effect on the species contributing to the ascidian assemblage of Brazilian coral reefs, most species disappeared and those remaining are likely to enhance reef degradation through their bioeroding activities.

Introduction

Ascidians are conspicuous, sessile benthic marine invertebrates, relatively abundant in shallow-water ecosystems wherever a suitable substratum for attachment is present (Millar, 1971). Whilst some aspects of their biology, including embryology (Albrieux & Villaz, 2000; Bishop et al., 2000; Nixon et al., 2000), reproductive physiology (Levasseur & McDougal, 2000), evolution (Cameron et al., 2000; Swalla et al., 2000), immunology (Findlay & Smith, 1995), genetics (Fujimura & Takamura, 2000), toxicology (Searle & Molinski, 1994) and chemistry (Wipf & Uto, 2000) have received recent attention, ascidians remain an enigmatic group in respect to their role in marine ecosystem functioning.

For some geographical regions, the ascidian fauna is well documented (Millar, 1971), but the same is not true for all parts of the globe (particularly South America) nor for all ecosystems (particularly coral reefs) (Fairfull & Harriott, 1999; Mundy & Babcock, 2000). Of the few Brazilian reports on ascidians (e.g. da Rocha, 1991; da Rocha & Monniot, 1993; da Rocha & Nasser, 1998; da Rocha et al., 1999), none includes the coral reef fauna. To address these deficiencies, a long-term study (initiated in 1995) was established to record the temporal and spatial variation in diversity and assemblage composition of ascidians from coral reefs in northern Bahia (Brazil). During the sampling programme, a marked increase of seawater temperature occurred globally (Pittock, 1999; Kumar et al., 2001; Pezzi & Cavalcanti, 2001) associated with the 1997/1998 El Niño event, the warming component of the El Niño Southern Oscillation (ENSO). This had severe consequences for coral reefs worldwide, including bleaching and mass mortalities of coral (Glynn et al., 2001; Vargas-Angel et al., 2001). Previously, we reported the effects of the 1997/1998 El Niño on the cnidarians (Kelmo & Attrill, 2001; Kelmo et al., 2003), bryozoans (Kelmo et al., 2004) and echinoderms (Attrill et al., 2004) from these same reefs in Brazil. In this paper, we assess the impact of the 1997/1998 El Niño on the ascidian biodiversity of Bahian coral reefs.

Material and methods

Study site

In Brazil, coral reefs occur throughout the northeastern coast and constitute the only coral constructions in the South Atlantic. The reefs studied here, in northern Bahia, are the most extensive and richest coral systems in Brazil (Leão, 1986). They exist on the narrowest part of the Eastern Brazilian Continental Shelf (average width 15 km between the São Francisco and Doce Rivers) and extend 20 km between the beaches of Abaí and Praia do Forte (see Fig. 1 in Kelmo & Attrill, 2001). A full description of the geological history and morphology of the reefs, plus the climate of the region, has been published previously in Leão et al. (1997) and Kelmo et al. (2003).

Sampling

In April/May each year from 1995 to 2000, ascidians were collected from three different reef habitats (coastal emergent reef tops [intertidal], coastal reef walls and shallow-bank reefs [both subtidal]) at four replicate localities in northern Bahia (Praia do Forte, Itacimirim, Guarajuba and Abai). Samples were collected by hand at low tide either by walking on the reefs (reef tops) or by SCUBA divers (reef walls and shallow-bank reefs). All samples are based upon 35 1m^2 quadrats placed randomly on each reef location; quadrats were randomly positioned each year, not fixed in position and repeat-sampled. Ascidians were counted *in situ* (number of colonies/individuals per quadrat) and small specimens or fragments (compound species) were collected for identification in the laboratory using mainly the illustrated guide of Rodrigues et al. (1998) and other relevant texts. Voucher specimens, collected each year, are stored at the scientific collection of the Department of Zoology, University Federal of Bahia, Brazil (UFBA).

Environmental measurements

Information on surface seawater temperature, sunlight irradiation, air temperature, rainfall and cloud cover were obtained from the Brazilian Meteorological Institute (INMET). In addition, on each sampling visit to each of the three reef habitats at the four sites, seawater temperature, salinity and pH were measured with a pH/conductivity/TDS/temperature meter (Hanna® model HI 991300); turbidity was estimated with a Secchi disc.

Data analysis

ANOVA and Kruskall–Wallis tests were used to identify significant variations in environmental and biotic univariate variables between pre- and post-El Niño periods. In addition to density, three univariate indices of the ascidian assemblage were calculated [Margalef's species richness (d), Shannon–Wiener diversity ($H'e$) and Pielou's evenness (J)], with data normalised and $\log(x+1)$ transformed where necessary prior to analysis. *Post-hoc* pairwise comparisons were undertaken using either Tukey–Kramer or Dunn's multiple comparison tests.

A triangular matrix of similarities between the ascidian samples was constructed using the Bray–Curtis similarity coefficient. The similarity matrix was subjected to clustering and ordination analyses using the program PRIMER [Plymouth Routines in Multivariate Ecological Research (Carr, 1996)]. Clustering was by an hierarchical agglomerative method using group-average linking, resulting in a

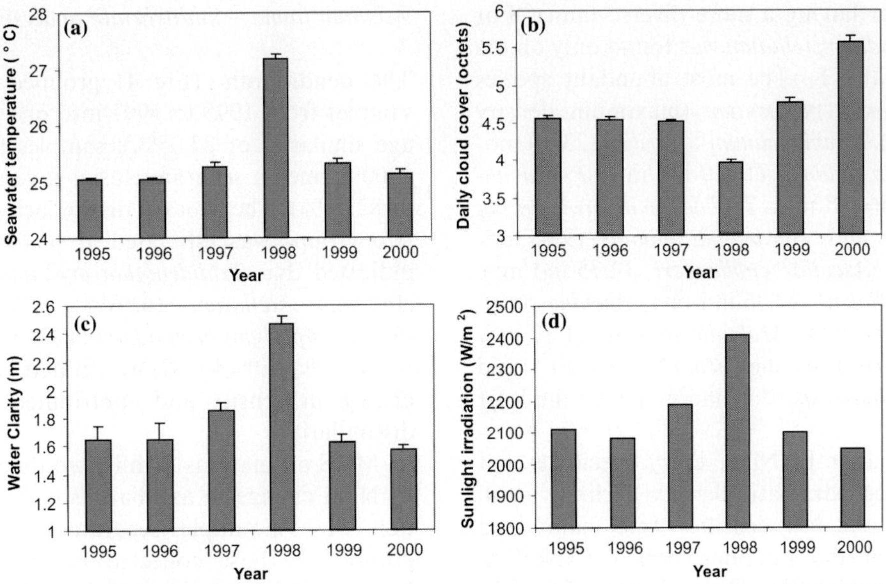

Figure 1. Examples of environmental measurements (means ± SE for (a–c), annual total for (d)) taken from the sampled Bahian reefs illustrating the impact of 1997–1998 El Niño on local conditions. (a) seawater temperature, (b) water clarity (using a Secchi disk), (c) cloud cover, (d) sunlight irradiation. Differences in 1998 are significant for all variables (see Kelmo et al., 2003, 2004 for full details).

dendogram. Ordination was by non-metric multidimensional scaling (MDS). Formal significance testing of differences in assemblage composition between years was tested using ANOSIM (Clarke & Green, 1988). Contribution of species to the groupings observed in the cluster and ordination analyses was examined using the SIMPER procedure ['similarity percentages' (Clarke, 1993)]. The BIOENV method (Clarke & Ainsworth, 1993), used to investigate the relationship between environmental variables and ascidian assemblage data, correlates the similarity matrix derived for the ascidian communities with an equivalent for the suite of environmental measurements taken at each site, defining which variables 'best explain' (Clarke & Ainsworth, 1993) the observed changes in assemblage composition.

Results

Environmental measurements

The physico-chemical parameters varied little between reefs and habitats, and were combined for analysis. Full data analysis has been previously published (e.g. Kelmo et al., 2003); results showed clearly that all means of physical factors were significantly (all $p < 0.05$) different in 1998 compared with other years (Fig. 1) with the exception of pH and the salinity of the subtidal sites (see Kelmo et al., 2003). In summary, water temperatures were the highest on record for these reef systems (with mean temperatures in the intertidal pools 0.5–1.0 °C higher than the main seawater), influenced by changing weather patterns. El Niño conditions result in reduced rainfall in the Bahia region, with the consequence of decreased land run-off and input from rivers and reduced cloud cover (Fig. 1). These changes resulted, respectively, in increased clarity of the naturally turbid waters around the reef (Fig. 1b, Kelmo et al., 2003) and elevated sunlight reaching the reef systems.

Ascidian fauna – univariate analysis

A total of 22 ascidian species was recorded. All species found on the intertidal reef tops were common to the other reef sites; however, there were differences in the fauna between the subtidal reef walls and the offshore shallow-bank reefs, with the latter having a more diverse fauna. For example, *Aplidium stellatum* was found only on the reef walls (Table 1). The most abundant species were *Lissoclinum perforatum* (maximum density 26.92 ind/m^2), *Echinoclinum verrilli* (23.30 ind/m^2), *Clavelina oblonga* (16.19 ind/m^2), *Pyura antillarum* (13.46 ind/m^2), *Trididemnum orbiculatum* (9.67 ind/m^2) and *Herdmania momus* (9.65 ind/m^2), whilst *Ascidia challengeri* (2.75 ind/m^2), *Ascidia meridionalis* (2.36 ind/m^2), *Ascidia coralloides* (2.35 ind/m^2), *Molgula robusta* (1.82 ind/m^2), *Polyandrocarpa anguinea* (1.15 ind/m^2) and *Molgula manhattensis* (0.38 ind/m^2) were the least abundant.

One year after El Niño, every species except two experienced dramatic density declines in all reef habitats and low densities were maintained to the end of the sampling period (ANOVA, $F=602.90$, $p<0.0001$) (Table 1, Fig. 2). Ten species were unrecorded during 1998 and 16 species were locally extinct within 2 years of the El Niño event. Conversely, the densities of two species, *Lissoclinum perforatum* and *Echinoclinum verrilli*, increased significantly post-El Niño in at least two habitats (Table 1). For example, in 2000 *L. perforatum* was the only ascidian recorded from the intertidal reef top habitat and was at its highest density for 6 years. *E. verrilli* density increased at the subtidal sites but, in common with all species apart from *L. perforatum*, a decline in density was apparent on the intertidal reef tops.

Comparisons of univariate measures of the ascidian assemblages pre- and post-El Niño showed significant differences in species richness (ANOVA, $F=60.506$, $p<0.0001$), diversity (ANOVA, $F=416.95$, $p<0.0001$; Fig. 3) and evenness (ANOVA, $F=18.543$, $p<0.0001$), with consistent trends at each of the four locations.

Patterns of diversity changes were similar for each reef habitat (Fig. 3). For the reef tops, diversity values declined significantly post-El Niño (ANOVA, $F=124.40$, $p<0.0001$) with zero diversity recorded in 2000 when the reef tops were inhabited by only one species (Fig. 3a). Similar significant declines post-El Niño were measured for the reef walls (ANOVA, $F=588.04$, $p<0.0001$), with the lowest mean values recorded in 2000 (Fig. 3b), and for the shallow-bank reefs (ANOVA, $F=424.18$, $p<0.0001$) (Fig. 3c).

Ascidian fauna – multivariate analysis

The dendrogram (Fig. 4) grouped all the reef samples from 1995 to 1997 into one cluster (average similarity of 87.05%); samples from 1998 to 2000 formed a separate cluster (average similarity of 82.45%). The average dissimilarity between the two groups was estimated at 63.64%. SIMPER indicated that *Trididemnum orbiculatum* (7.18%), *Aplidium stellatum* (6.53%), *Phallusia nigra* (6.5%), *Aplidium bermudae* (6.09%) and *Botryllus schlosseri* (6.08%) demonstrated the greatest change in density and contributed most to the dissimilarities.

MDS ordinations highlighted the marked post-El Niño change in ascidian assemblage composition (Fig. 5). Samples separated into two distinct groups: (i) those collected before 1998 (pre-El Niño period) and (ii) samples taken from 1998 (post-El Niño period). The clustering of samples from 1999 and 2000 with those from 1998 demonstrated that the assemblage composition remained similar for the three years post-El Niño, with no indication of recovery to the pre-El Niño community. The significance of the observed changes was confirmed by ANOSIM (all global $R>0.96$, $p<0.001$).

BIOENV analysis indicated that a combination of turbidity, mean temperature and cloud cover ($r=0.653$) 'best explained' the changes in assemblage patterns post-El Niño (1998) in comparison with pre-El Niño years. For the reef tops, water temperature together with cloud cover ($r=0.642$) gave the best fit with assemblage patterns.

Discussion

Several reports have highlighted that the 1997/1998 El Niño had significant large-scale effects on coral reefs worldwide (Arthur, 2000; Vargas-Angel et al., 2001; Guzman & Cortes, 2001; Kelmo et al., 2003). Such effects include abnormally elevated and prolonged surface seawater temperature (Pezzi & Cavalcanti, 2001), reduced turbidity (Kelmo & Attrill, 2001) and increased solar radiation (Conde et al., 2000). Combination of these variables are known to cause disturbances to coral reef associated organisms (Glynn, 1990), including mortality (Kelmo et al., 2004).

Table 1. Mean densities of ascidian species in each reef habitat from Bahia in the 3 years pre- El Niño (1995–1997) and the 3 years post-El Niño (1998–2000)

Species	Average density m^{-2}		Average term %
	Post-El Niño Years (1998–2000)	Pre-El Niño Years (1995–1997)	
Reef top			
Lissoclinum perforatum	0.56	0.13	21.67
Herdmania momus	0.00	0.30	16.25
Clavelina oblonga	0.01	0.28	14.62
Phallusia nigra	0.02	0.25	12.98
Echinoclinum verrilli	0.02	0.20	10.55
Styella plicata	0.00	0.17	9.11
Botrylloides nigrum	0.02	0.16	8.16
Reef wall			
Styela plicata	0.00	0.73	12.26
Piura antillarum	0.00	0.70	11.69
Aplidium stellatum	0.03	0.74	10.18
Trididemnium orbiculatum	0.04	0.62	8.83
Botryllus schlosseri	0.00	0.38	8.08
Echinoclinum verrilli	1.66	0.58	7.85
Phallusia nigra	0.04	0.39	7.27
Botrylloides nigrum	0.02	0.35	7.27
Herdmania momus	0.00	0.25	7.04
Lissoclinum perforatum	1.40	0.80	5.60
Clavelina oblonga	0.03	0.23	5.07
Ascidia challengeri	0.00	0.15	4.85
Molgula manhanttensis	0.00	0.11	4.03
Shallow banks			
Trididemnium orbiculatum	0.00	0.18	7.18
Amouricium stellatum	0.01	0.75	6.53
Phallusia nigra	0.01	0.61	6.15
Amouricium bermudae	0.00	0.53	6.09
Botryllus schlosseri	0.00	0.55	6.08
Echinoclinum verrilli	1.26	0.68	6.08
Piura antillarum	0.08	1.04	6.07
Botrylloides nigrum	0.00	0.53	5.71
Styella plicata	0.00	0.57	5.62
Eictenascidia conklini	0.00	0.39	5.43
Bostrichobranchus pilularis	0.00	0.40	5.31
Lissoclinum perforatum	1.31	0.94	5.21
Clavelina oblonga	0.02	0.37	4.78
Clavelina fasciculate	0.00	0.26	4.36
Ascidia challengeri	0.00	0.23	3.49
Ascidia meridionale	0.00	0.20	3.27
Ascidia coralloides	0.00	0.19	3.21

The El Niño component of the El Niño Southern Oscillation is the appearance and persistence of anomalously warm water in the low latitude eastern Pacific (Barber & Chavez, 1986). The ENSO cycle is a natural aperiodic, coupled ocean/atmosphere cycle that determines both the

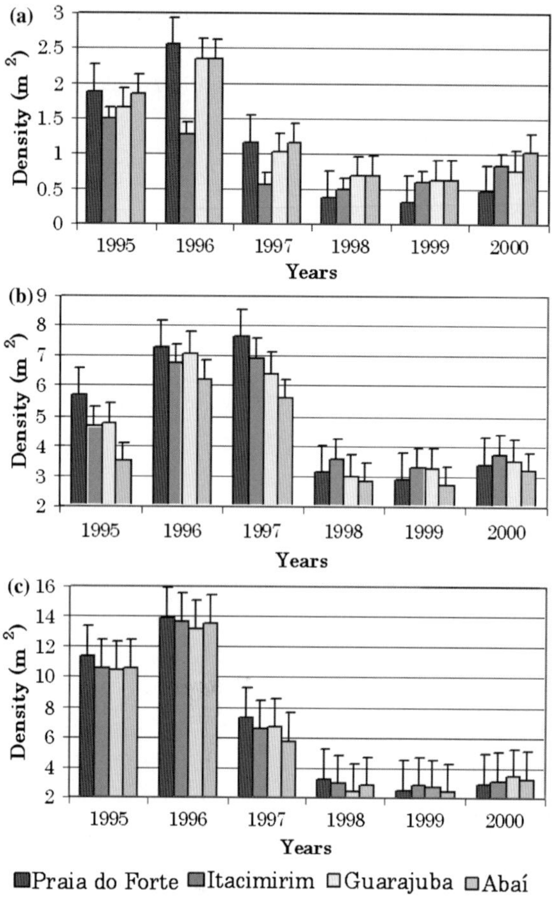

Figure 2. Variation in density (mean ± SE) of reef-associated ascidians recorded from northern Bahian reefs over the 6-year period of investigation. El Niño occurred between 1997 and 1998 samples. (a) intertidal reef tops, (b) subtidal reef walls, (c) shallow-bank reefs.

climatological mean conditions and the major interannual variability of the large-scale heat flux (Barber & Kogelschatz, 1990). In the tropical Atlantic, where the Bahian reefs are situated, a number of perturbations can be observed during El Niño periods (Kumar et al., 2001; Pezzi & Cavalcanti, 2001), including intensification of surface trades over the equatorial Atlantic (Philander, 1986), increased insolation and abnormally increased sea surface temperatures (SSTs) (Ropelewski & Halpert, 1987), depression of the thermocline (McPhaden, 1993), and severe droughts (Horel et al., 1986). As this natural phenomenon is accompanied also by reduced nutrient replenishment to sun-lit surface waters, with further depression of plankton production and disruption of many trophic links (Glynn, 1990), it exerts an important control on the development and sustainability of coral reefs throughout the affected regions. The large-scale El Niño phenomenon results in significant changes in the environmental conditions surrounding the coral reefs of Bahia. In particular, three consequences are most likely to be affecting the reef ascidians. Following the global trend, there was a marked increase in water temperatures across the study region, with record high sea surface temperatures being recorded in 1998. Intertidal pools naturally experience greater variation in environmental conditions than the adjacent sea (Newell, 1979); the highest water temperatures were recorded in the reef top pools, which also had a significantly higher salinity regime than other years, reflecting evaporation. Secondly, reefs were exposed to higher levels of sunlight (and thus potentially damaging UV radiation) due to decreased cloud cover and increased clarity of water, a consequence of reduced land run-off as a result of low precipitation. All habitats therefore received higher than normal levels of UV radiation, the intertidal reef due to the reduced cloud cover and, especially, the subtidal reefs due to additional reduced turbidity of the water. Finally, the disruption of trophic links due to indirect responses related to production, etc. would potentially affect the feeding and survival of adult and larval ascidians.

We suggest, therefore, that the marked reduction in density of the ascidians from the shallow-bank reefs of northern Bahia, and the dramatic disappearance of several species ubiquitous before the 1997/1998 El Niño event (which have not yet reappeared in the community), was associated directly with the 1997/1998 El Niño event. In particular, we propose that these changes were in response to abnormally high temperatures, the impacts of UV radiation (Kelmo et al., 2003) and disruption of the trophic links to adults and larvae, restricting reproduction and settlement of new individuals. da Rocha et al. (1999) proposed that the abundance of ascidians is regulated mostly by seawater temperature, with some influence of reduction of turbidity (Hurlbut, 1993) and increased light intensity (Monniot, 1965), supporting the suggested mechanism of impact in this study. The reproductive biology of ascidians is a relatively well investigated subject (da Rocha et al.,

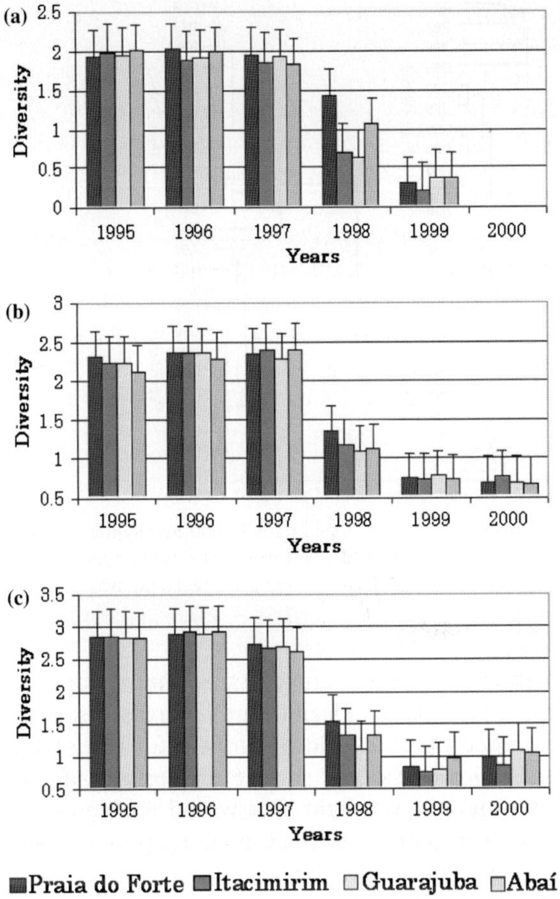

Figure 3. Variation in diversity (*H'e*, mean ± SE) of reef-associated ascidians recorded from northern Bahian reefs over the 6-year period of the investigation. (a–c) as Figure 2.

1999), and temperature fluctuations may influence directly the reproductive cycles of temperate (Dybern, 1965) and tropical species (Goodbody, 1974). Increased temperatures can induce recruitment, but larval settlement seems to be affected by neither increased temperature nor solar radiation, partly explained by the very short pelagic phase (Nielsen, 2001). However, larval attachment is compromised seriously because the increased temperatures (in association with the other El Niño anomalies recorded in the study area) induce mortality of the marine invertebrates that are the preferred substratum for attachment by ascidians such as hydroids, octocorals and bryozoans (Kelmo et al., 2004).

Ascidians are filter feeders, with a diet mostly of phytoplankton and other organic particles in suspension (Millar, 1971). The decline in productivity and subsequent reduction of food supply during the El Niño period disrupts many trophic links and it has been suggested that inappropriate diet in marine invertebrates may lead to increased metabolism with subsequent decline in reserves of protein, lipids and carbohydrates (Glynn, 1993). For ascidians, it is likely that undernourished/starving individuals will have reduced growth rate and impaired reproduction; together with potential UV damage (Kelmo et al., 2003), this may explain their disappearance from the reef environment in years subsequent to 1998. In this context, it is interesting to note that the two species with increased abundances after the 1997/1998 El Niño (*Lissoclinum perforatum* and *Echinoclinum verrilli*) are compound species able to reproduce both sexually and asexually by budding (Nielsen, 2001) and are not obligate associates with any 'live' substrata. To survive, both species must have a comparatively high level of tolerance to increased water temperature, witnessed by their abundance in the naturally stressful intertidal reef top pools. Following El Niño, and in the absence of other ascidians, these two species were able to proliferate through asexual reproduction, spreading onto surrounding areas regardless of substratum type. The negative aspect of such dominance, however, is that these species are bioeroders and significant increase in their numbers could result in intensive reef biodegredation similar to that witnessed for population increases in other bioeroding organisms (e.g. Glynn, 1988).

Lissoclinum perforatum and *Echinoclinum verrilli* responded in a similar manner within subtidal sites, but this was not evident on the intertidal reef tops. Here, *E. verrilli* suffered similar reductions in densities to other ascidian species, whilst *L. perforatum* proliferated to higher densities than recorded previously. By 2000, *L. perforatum* was the only ascidian species present in the intertidal habitat. Out of all three habitats, the intertidal pools present the most stressful conditions due to naturally fluctuating environmental conditions that make them markedly different from subtidal reef walls and shallow-bank reefs. These harsh conditions are reflected in the overall lower diversity in this habitat (Table 1). Whilst *E. verrilli* thrives within the intertidal pools under normal conditions, it would appear that the additional stresses associated with El Niño, including the

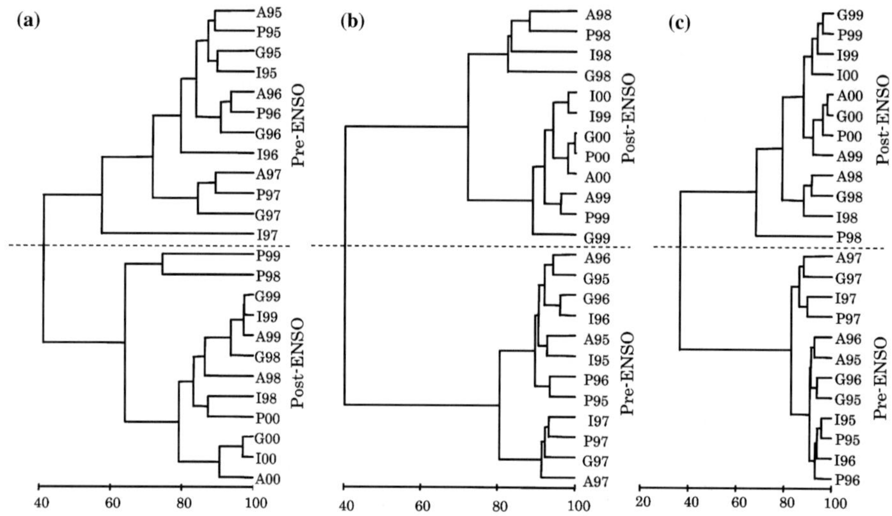

Figure 4. Dendrogram constructed from density data of ascidian assemblages for each reef habitat 1995–2000 highlighting the separation of pre- and post-El Niño samples in terms of the assemblage composition. (a) intertidal reef tops, (b) subtidal reef walls, (c) shallow-bank reefs. Horizontal axis is % similarity. Each sample is represented by a letter for replicate reef and two numerals corresponding to year.

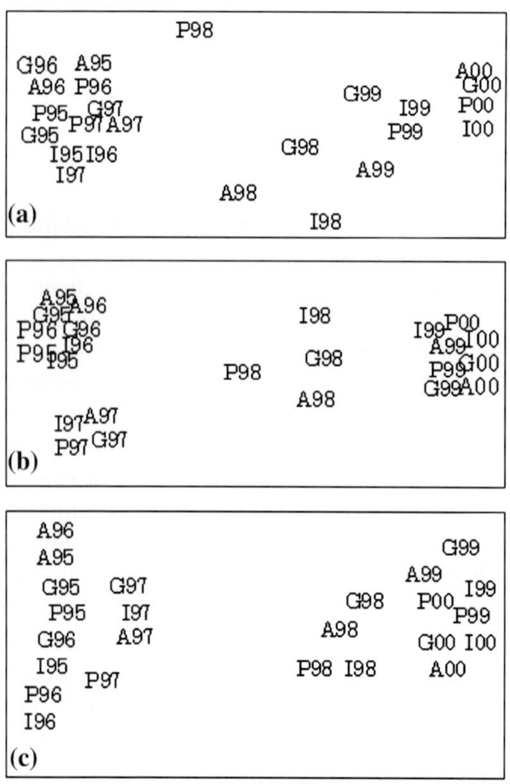

Figure 5. MDS ordinations of ascidian density data from each Bahian reef habitat 1995–2000. Codes as for Figure 4.

highest recorded temperatures (maximum recorded: 31 °C) and significantly increased salinity, exceeded the threshold of survival for this species and colonies did not survive the stresses of temperature and UV light that would enable expansion of the colony through budding post-El Niño. Subsequently, *L. perforatum* increased its density markedly within the pools.

The results of this study demonstrate that the environmental changes associated with ENSO events can have severe consequences for non-coral invertebrates associated with reefs, even in areas such as Brazil where the impact of El Niño is indirect. Additionally, the effect of El Niño continued to be evident for at least 2 years after the end of the episode resulting in the ascidian assemblage on one reef habitat being reduced to a single species. Coral reefs are recognised as being major repositories of biodiversity, with most of this diversity being in groups additional to the corals themselves. It would appear that some sessile invertebrate groups such as ascidians and bryozoans (Kelmo et al., 2004) are exceedingly sensitive to the environmental stresses associated with El Niño, resulting in mass mortalities and local extinctions from large areas of reef. Few such studies on associated coral reef biodiversity have been undertaken and, if the results from Bahia are representative of other reef systems, the global consequences of more severe and frequent ENSO events in a climate change scenario (Timmermann et al., 1999) could be even more dramatic than previously thought and provide a major cause of concern for the conservation of the world's marine life.

Acknowledgements

We would like to thank the Brazilian Research Council (CNPq) for supporting this research and all colleagues of FK in Bahia who helped gather the data for this study.

References

Albrieux, M. & M. Villaz, 2000. Bilateral asymmetry of the inositol triphosphate-mediated calcium signaling in two-cell ascidian embryos. Biology of the Cell 92: 277–284.

Arthur, R., 2000. Coral bleaching and mortality in three Indian reef regions during an El Niño southern oscillation event. Current Science 79: 1723–1729.

Attrill, M .J., F. Kelmo & M. B. Jones, 2004. Impact of the 1997–8 El Niño event on the coral reef-associated echinoderm assemblage from northern Bahia, NE Brazil. Climate Research 26: 151–158.

Barber, R. T. & F. P. Chavez, 1986. Ocean variability in relation to living resources during the 1982/83 El Niño. Nature 319: 279–285.

Barber, R. T. & J. E. Kogelschatz, 1990. Nutrients and productivity during the 1982/83 El Niño. In Glynn, P. W. (ed.) Global Ecological Consequences of the 1982–83 El Niño Southern Oscillation. Elsevier, Amsterdam: 21–53.

Bishop, J. D. D., P. H. Manríquez & R. N. Hughes, 2000. Water-borne sperm triggers vitellogenic egg growth in two sessile marine invertebrates. Proceedings of the Royal Society of London B 267: 1165–1169.

Cameron, C. B., J. R. Garey & B. J. Swalla, 2000. Evolution of the chordate body plan: new insights from phylogenetic analyses of deuterostome phyla. Proceedings of the National Academy of Science USA 97(9): 4469–74.

Carr, M. R., 1996. Plymouth Routines in Multivariate Analysis Primer User Manual. Natural Environment Research Council, UK 45 pp.

Clarke, K. R., 1993. Non-parametric multivariate analysis of changes in community structure. Australian Journal of Ecology 18: 117–143.

Clarke, K. R. & M. Ainsworth, 1993. A method of linking multivariate community structure to environmental variables. Marine Ecology Progress Series 92: 205–219.

Clarke, K. R. & R. H. Green, 1988. Statistical design and analysis for a 'biological effects' study. Marine Ecology Progress Series 46: 213–226.

Conde, D., L. Aubriot & R. Sommaruga, 2000. Changes in UV penetration associated with marine intrusions and freshwater discharge in a shallow coastal lagoon of the Southern Atlantic Ocean. Marine Ecology Progress Series 207: 19–31.

da Rocha, R. M., 1991. Replacement of the compound ascidian species in a southeastern Brazilian fouling community. Annals of Institute of Oceanography, Sao Paulo 39: 141–153.

da Rocha, R. M. & F. Monniot, 1993. *Didemnum rodriguesi* sp. Nov., a new didemnid tunicate common to southern Brazil and New Caledonia. Annals of the Institute of Oceanography, Sao Paulo 69: 261–265.

da Rocha, R. M. & C. M. Nasser, 1998. Some ascidians (Tunicata, Ascidiacea) from Parana State southern Brazil. Revista Brasileira de Zoologia 15: 633–642.

da Rocha, R. M., T. M. C. Lotufo & S. A. Rodrigues, 1999. The biology of *Phallusia nigra* Savigny, 1816 (Tunicata: Ascidiacea) in southern Brazil: spatial distribution and reproductive cycle. Bulletin of Marine Science 64: 77–87.

Dybern, B. I., 1965. The life cycle of *Ciona intestinalis* (L) f. *tipica* in relation to the environmental temperature. Oikos 16: 109–131.

Fairfull, S. J. L. & V. J. Harriott, 1999. Succession, space and coral recuitment in subtropical fouling community. Marine and Freshwater Research 50: 235–242.

Findlay, C. & V. J. Smith, 1995. Antibacterial activity in the blood cells of the solitary ascidian, *Ciona intestinalis*, in vitro. Journal of Experimental Zoology 273: 434–444.

Fujimura, M. & K. Takamura, 2000. Characterization of an ascidian DEAD-box gene, Ci-DEAD1: specific expression in the germ cells and its mRNA localization in the posteriormost blastomeres in early embryos. Development Genes and Evolution 210: 64–72.

Glynn, P. W., 1988. El Niño warming, coral mortality and reef framework destruction by echnoid bioerosion in the eastern Pacific. Galaxea 7: 129–160.

Glynn, P. W., 1990. Coral mortality and disturbances to coral reefs in the tropical eastern Pacific. In Glynn, P. W. (ed.) Global Ecological Consequences of the 1982–83 El Niño Southern Oscillation. Elsevier, Amsterdam: 55–126.

Glynn, P. W., 1993. Coral reef bleaching: ecological perspectives. Coral Reefs 12: 1–17.

Glynn, P. W., J. L. Matt, A. C. Baker & M. O. Calderon, 2001. Coral bleaching and mortality in Panama and Ecuador during the 1997–1998 El Niño Southern Oscillation event: spatial/temporal patterns and comparison with the 1982–1983 event. Bulletin of Marine Science 69: 79–109.

Goodbody, I., 1974. The physiology of ascidians. Advances in Marine Biology 12: 1–149.

Guzman, H. M. & J. Cortes, 2001. Changes in reef community structure after fifteen years of natural disturbances in the Eastern Pacific (Costa Rica). Bulletin of Marine Science 69: 133–149.

Horel, J. D., V. E. Kousky & M. T. Kagano, 1986. Atmospheric conditions in the Atlantic Sector during 1983 and 1984. Nature 322: 248–251.

Hurlbut, C. J., 1993. The adaptive value of larval behaviour of a colonial ascidian. Marine Biology 115: 253–262.

Kelmo, F. & M. J. Attrill, 2001. Cnidarian community structure of coastal reefs from northern Bahia. Bulletin of Marine Science, Brazil 69: 547–557.

Kelmo, F., M. J. Attrill & M. B Jones, 2003. Effects of the 1997–98 El Niño on the cnidarian community of a high turbidity coral reef system (northern Bahia, Brazil). Coral Reefs 22(4): 541–550.

Kelmo, F., M. J. Attrill, M. B. Jones & R. C. T. Gomes, 2004. El Niño induced local extinction of coral reef bryozoan species from northern Bahia, Brazil. Biological Conservation 118: 609–617.

Kumar, A., W. Q. Wang, M. P. Hoerling & A. Leetmaa, 2001. The sustained North American warming of 1997 and 1998. Journal of Climate 14: 345–353.

LeVesseur, M. & A. McDougall, 2000. Sperm-induced calcium oscillations at fertilisation in ascidians are controlled by cyclin B1–dependent kinase activity. Development 127: 631–641.

Leão, Z. M. A. N., 1986. Guia para identificacao dos corais do Brasil, Universidade Federal da Bahia. Salvadoro, Brasil 57 pp.

Leão, Z. M. A. N., R. K. P. Kikuchi, M. P. Maia & R. A. L. Lago, 1997. A catastrophic coral cover decline since 3000 years B.P., Northern Bahia, Brazil. Proc. 8th Int. Coral Reef Symposium. Panama City, Republic of Panama 1: 583–588.

McPhaden, M. J., 1993. TOGA-TAO and the 1991–93 El Niño-Southern oscillation event. Oceanography 6: 36–44.

Millar, R. H., 1971. The biology of ascidians. Advances in Marine Biology 9: 1–100.

Monniot, C., 1965. Les 'Blocks à Microcosmus' des fonds chalutables de al région de Banyuls-sur-Mer. Vie et Milieu 16: 819–849.

Mundy, C. & R. Babcock, 2000. Are vertical distribution patterns of scleractinian corals maintained by pre- or post-settlement processes? A case study of three contrasting species. Marine Ecology Progress Series 198: 109–119.

Newell, R. C., 1979. Biology of Intertidal Animals. 3rd edn. Marine Ecological Surveys Ltd, Faversham, 781 pp.

Nielsen, C., 2001. Animal Evolution, 2nd edn. Oxford University Press, Oxford, 563 pp.

Nixon, V. L., A. McDougall & K. T. Jones, 2000. Ca^{2+} oscillations and the cell cycle at fertilisation of mammalian and ascidian eggs. Biology of the Cell 92: 187–196.

Pezzi, L. P. & I. F. A. Cavalcanti, 2001. The relative importance of ENSO and tropical Atlantic sea surface temperature anomalies for seasonal precipitation over South America: a numerical study. Climatic Dynamics 17: 205–212.

Philander, S. G. H., 1986. Unusual conditions in the tropical Atlantic Ocean in 1984. Nature 322: 236–238.

Pittock, B. A., 1999. Coral reefs and environmental change: adaptation to what?. American Zoologist 39: 10–29.

Rodrigues de , S. A., R. M. da Rocha & T. M. C. Lotufo, 1998. Guia Illustrado para Identificacao das Ascidias do Estado de Sao Paulo. FAPESP, Sao Paulo 190 pp.

Ropelewski, C. F. & M. S. Halpert, 1987. Global and regional scale precipitation patterns associated with the El Niño/Southern Oscillation. Monthly Weather Review 115: 1606–26.

Searle, P. A. & T. F. Molinski, 1994. Five new alkaloids from the tropical ascidian, *Lissoclinum* sp Lissoclinotoxin A is chiral. Journal of Organic Chemistry 59: 6600–6605.

Swalla, B. J., C. B. Cameron, L. S. Corley & J. R. Garey, 2000. Urochordates are monophyletic within deuterostomes. Systematic Biology 49: 52–64.

Timmermann, A., J. Oberhuber, A. Bacher, M. Esch, M. Latif & E. Roeckner, 1999. Increased El Niño frequency in a climate model forced by future greenhouse warming. Nature 398: 694–697.

Vargas-Angel, B., F. A. Zapata, H. Hernandez & J. M. Jimenez, 2001. Coral and coral reef responses to the 1997–98 El Niño event on the Pacific coast of Colombia. Bulletin of Marine Science 69: 111–132.

Wipf, P. & Y. Uto, 2000. Total synthesis and revision of stereochemistry of the marine metabolite Trunkamide A. Journal of Organic Chemistry 65: 1037–1049.

Hydrobiologia (2006) 555:241–251
H. Queiroga, M.R. Cunha, A. Cunha, M.H. Moreira, V. Quintino, A.M. Rodrigues, J. Serôdio & R.M. Warwick (eds),
Marine Biodiversity: Patterns and Processes, Assessment, Threats, Management and Conservation
DOI 10.1007/s10750-005-1120-6

Changes in the range of some common rocky shore species in Britain – a response to climate change?

N. Mieszkowska[1,*], M.A. Kendall[2], S.J. Hawkins[1], R. Leaper[1], P. Williamson[3], N.J. Hardman-Mountford[1,2] & A.J. Southward[1]
[1]*Marine Biological Association of the U.K., Citadel Hill, PL1 2PB Plymouth, England*
[2]*Plymouth Marine Laboratory, Prospect Place, West Hoe, PL1 3DH Plymouth, England*
[3]*School of Environmental Sciences, University of East Anglia, NR4 7TJ Norwich, Norfolk, England*
*(*Author for correspondence: E-mail: nova@mba.ac.uk)*

Key words: climate, change, intertidal, rocky, indicator, MarClim

Abstract

Since the 1990s there has been a period of rapid climate warming in Europe. Long-term broad scale datasets coupled with time series at specific locations for rocky intertidal species dating back to the 1950s have been collected in Britain and Ireland. Resurveys of the original locations in 2001–2003 have been undertaken to identify changes in the biogeographical range and abundance of these species. The results show that some 'southern' species including *Osilinus lineatus* da Costa and *Gibbula umbilicalis* da Costa have undergone north and north-eastern range extensions. Populations have increased in abundance and adult size has decreased since the previous surveys were conducted. These changes have been synchronous throughout Britain, strongly suggesting that climate is responsible. The use of intertidal species as indicators of climate change is proposed.

Introduction

The global average surface temperature has been increasing since records began in 1861. Proxy data for the Northern Hemisphere indicate that the increase in temperature observed during the 20th century is the largest in the last 1000 years (Hulme et al., 2002), with the 1990s the warmest decade and 1998 the warmest year on record (Houghton et al., 2001). The ocean heat content has also increased since modern measurements began in the 1950s. In the western English Channel off Plymouth a 1 °C increased in sea surface temperature has occurred since 1990. This increase exceeds any other change recorded over the past 100 years (Hawkins et al., 2003) and is most apparent in winter months (Fig. 1). Global climate change models predict an acceleration of the current warming trend during the first half of this century as a response to anthropogenic emissions of greenhouse gases. These models also predict that the earth will become warmer than at any period during the past 40 million years (Houghton et al., 2001). The rate of warming cannot be predicted with certainty but models based on medium-high emissions scenarios indicate that sea surface temperatures around Britain will increase between 0.5 and 5 °C by 2080 (Hulme et al., 2002).

Responses of a wide range of marine biota to oceanic warming have already been documented. These include changes in phenology, as observed in the migration patterns of squid in the English Channel (Sims et al., 2001), shifts in abundance and distribution which have been shown in changes to plankton and fish assemblages in the North Sea and English Channel (Southward et al., 1995; Beaugrand et al., 2002; Stebbing et al., 2002) and changes in the community dynamics of intertidal

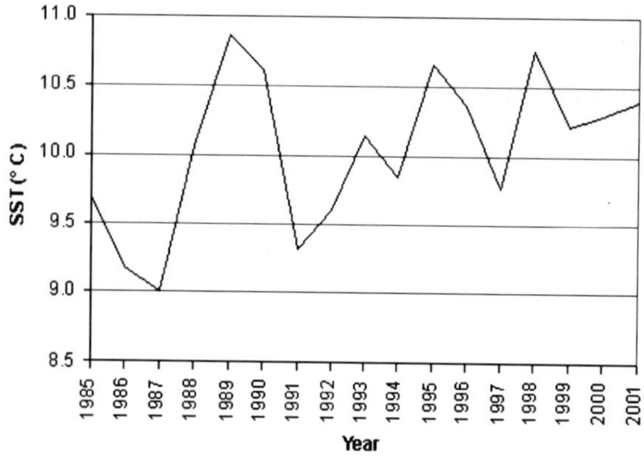

Figure 1. Mean winter (Jan–Mar) SST for the Western English Channel from NOAA Pathfinder AVHRR data for 1° grid square centred on 50° N 4° W. The data were provided by the Earth Observing System Data and Information System (EOSDIS), Distributed Active Archive Center at Goddard Space Flight Center which archives, manages, and distributes this data set.

invertebrates in the North West Atlantic (Barry et al., 1995; Bertness et al., 1999; Sagarin et al., 1999; Murray et al., 2001). As the climate continues to warm it is hypothesised that a poleward shift in the ranges of species will occur (Graham & Grimm, 1990; Fields et al., 1993; Southward et al., 1995; Parmesan, 1996; Sagarin et al., 1999). Climate-induced changes in the range of terrestrial species have already been documented in a wide variety of systems (Jeffree & Jeffree, 1996; Parmesan et al., 1999; Parmesan & Yohe, 2003; Root, 2003) and similar patterns of change are likely in the marine environment.

Britain and Ireland straddle a biogeographic boundary between cold boreal 'northern' waters and warmer lusitanean 'southern' waters (Forbes, 1853; Lewis, 1964) and many intertidal species reach their northern or southern limits of distribution in or close to the British Isles. Thus it is an ideal location for studying the effects of climate change on these species. The Marine Biodiversity and Climate Change Project, MarClim (established in 2001) was set up to assess and forecast the influence of climate change on marine rocky shore biodiversity in Britain and Ireland, and make use of existing long-term and current data.

It is fortunate that extensive historical broadscale baseline of data on intertidal rocky shore fauna and flora has been collected in Europe (Southward & Crisp, 1954a; Fischer-Piette & Prenant, 1956; Crisp & Southward, 1958; Crisp & Fisher-Piette, 1959). Observations at some sites date back to the 1930s, with continuous data since the 1950s for certain species including the barnacles *Chthamalus stellatus* Poli, *Chthamalus montagui* Southward [which were synonymous until 1976 (Southward, 1976)] and *Semibalanus balanoides* Linneaus (Southward & Crisp, 1954b; Crisp & Southward, 1958; Southward, 1967; Southward, 1991). In addition, the population structures of the 'southern' trochid gastropods *Gibbula umbilicalis* da Costa and *Osilinus* (*Monodonta*) *lineatus* da Costa were studied close to the northern limits of distribution in Britain as part of the NERC Rocky Shore Surveillance Group study on processes regulating coastal ecosystems (Kendall & Lewis, 1986; Kendall, 1987).

Methods

The MarClim project collects both broadscale and quantitative data following the methods employed during historical surveys.

Broadscale

Semi-quantitative broadscale surveys of rocky intertidal flora and fauna were undertaken during the 1950s at approximately 400 sites around the coastline of Britain and Ireland (Southward & Crisp, 1954a; Crisp & Southward, 1958). The abundance of each species was assessed using the semi-logarithmic ACFOR abundance scale developed by Crisp & Southward (1958: 160, Table 2) to determine the biogeographic ranges of littoral species. MarClim repeated broad-scale surveys at approximately 300 of the historical sites in 2001–2003 (Fig. 2) using the original ACFOR classification method, including semi-exposed and exposed shores where anthropogenic impacts are often limited in Britain.

Thirty-three of the 39 species that were recorded in the original broadscale surveys of the 1950s were selected for inclusion in the resurveys. An additional 23 species were included in the resurvey list. All species selected reach their northern or southern limits of distribution close to Britain and Ireland and populations are therefore more likely to be sensitive to changes

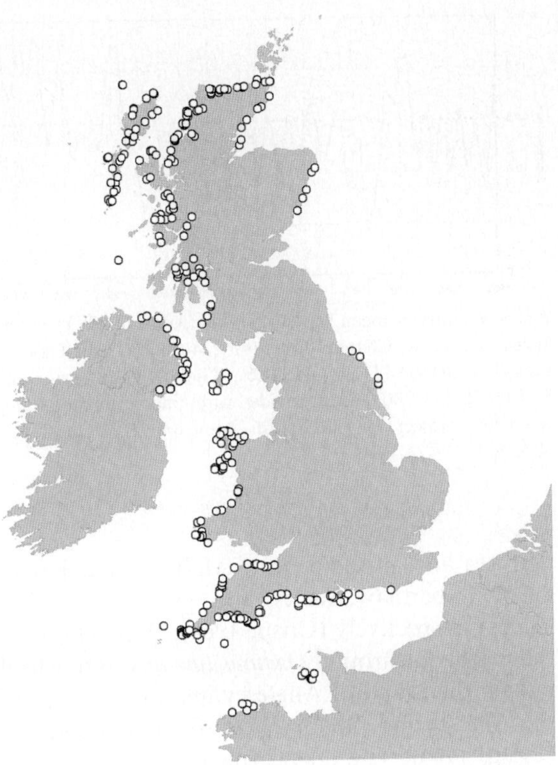

Figure 2. Locations of broadscale resurveys completed in 2002 and 2003.

in climate than those species that are at the centre of their range (Lewis, 1976). Pairs of northern and southern species whose distributions overlap and are competitors on the rocky shores of Britain were included to determine how climate mediates species interactions. Each shore was searched by 2 operators for a duration of 30 min, and the abundance of each species recorded. Target species were also searched for at sites beyond the last documented range edges to determine whether range expansions had occurred.

Quantitative

Southward and Crisp also undertook quantitative counts of adults of the southern barnacle species *Chthamalus stellatus* and *Chthamalus montagui* and the northern barnacle species *Semibalanus balanoides* in 3–4 quadrats at the MHWN, MTL and MLWN shore levels at each sample location (Southward & Crisp, 1954b; Southward, 1991). During the MarClim resurvey, 10 replicated barnacle counts in 25 or 10 cm^2 quadrats, were carried out at the same tidal heights at each site. Digital photographic recording of barnacle quadrats was also trialled as a potential replacement for counts *in situ*.

Quantitative counts of the 'northern' limpet *Patella vulgata* Linneaus and the 'southern' limpet *Patella depressa* Pennant have been collected on shores in the south and south-west of England, Wales and France since 1980 by S.J. Hawkins. These counts have been continued as part of the MarClim project. Ten replicated counts of individuals in a randomly placed 0.25 m^2 quadrat were made within the area of the shore where *P. depressa* was most abundant. These surveys complement the research on barnacles (outlined above) and increase the understanding of the role of climate in 'northern' and 'southern' species competitive interactions.

The population structures of the 'southern' trochid *Gibbula umbilicalis* were recorded at 16 sites close to the northern range limits between 1977 and 1985 (Fig. 3) (Kendall & Lewis, 1986). Approximately 200 animals (where available) were collected during each survey and their basal

Figure 3. Survey locations of *Osilinus lineatus* (closed circles) and *Gibbula umbilicalis* (open circles).

diameters measured to the nearest millimetre. Population structures of another 'southern' trochid *Osilinus lineatus* were also studied at 16 sites close to northern range limits in Britain and France in 1986 (Fig. 3), (Kendall, 1987). Replicated timed collections of individuals involving intensive searching of boulders, underlying gravel and bedrock were carried out in areas of the shores where juvenile abundance was estimated to be greatest. The basal diameter of all individuals was recorded and the animals returned to the shore. At each site resurveyed between 2001 and 2003, five replicated timed counts of 3 min duration were carried out in the area of greatest juvenile abundance for *O. lineatus*. The basal diameter to the nearest millimetre was recorded for each individual and the animals replaced on the shore. The same method was adopted for *G. umbilicalis* at historical sites to make them more amenable to statistical analyses than the original dataset. Quantitative counts were also made at an additional 25 sites for *O. lineatus* and 42 sites for *G. umbilicalis* throughout Britain and northern France to allow analyses of the resurvey data in a wider geographical context.

All of the original datasets including metadata have been extracted from field notebooks and electronically archived in a central database. Resurvey data from 2002 to 2003 have also been archived in database and hard copy format. All data entries have been subject to quality assurance checks involving the original collectors.

Results

Changes in biogeographic range

The MarClim project has completed the second year of its 4-year operational schedule and the results of the biogeographic range studies undertaken to date are reported here. The original broadscale surveys were made in the 1950s during a warm climatic period (Fig. 4). A cool period followed, whose onset was dramatically marked by the severe winter of 1962/1963, the coldest since 1740 (Crisp, 1964). Species close to their northern and north-eastern range edges in Britain were severely affected by the extremely cold weather,

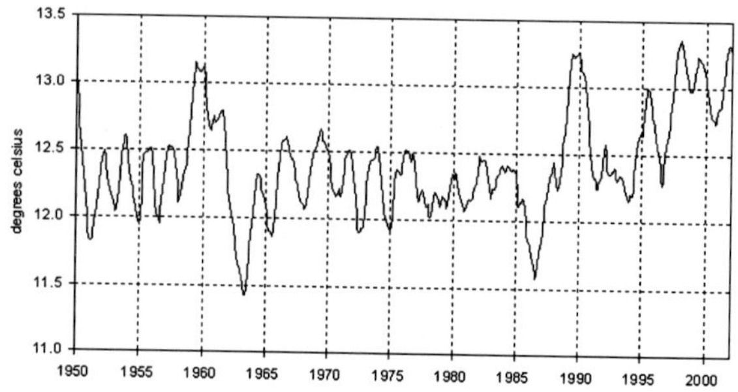

Figure 4. Annual mean sea surface temperature (SST) in the Western English Channel (50–51° N, 4–5° W). Met Office – GISST/MOHMATN4/MOHSST6 – Global Ice coverage and SST (1856–Feb 2003) data provided with kind permission to the MarClim project by the British Atmospheric Data Centre (BADC).

with sea and air temperatures falling to as low as 0.6 °C (Aberthaw, Wales) and −3.4 °C (Swansea, Wales), respectively (Crisp, 1964). Populations of the trochid gastropod *Osilinus lineatus* were wiped out on the island of Anglesey and along much of the Welsh and South West England coastline (Crisp, 1964). The tubeworm *Sabellaria alveolata* Linneaus and the barnacle *Balanus perforatus* Brugière also suffered severe mortalities along the Welsh and southern English coastlines, where they had previously reached their northern and north-eastern range limits (Crisp, 1964). A second abnormally cold winter occurred in 1978/1979. Populations of these southern species again suffered mortalities during this winter, but these were on a smaller scale, as the weather was not as severe as in 1962/1963 (Kendall et al., 1987). The climate remained in this cool phase until the late 1980's, when the current warming trend began.

Data collected by MarClim has shown that the southern species *Osilinus lineatus, Gibbula umbilicalis*, *Sabellaria alveolata* and *Balanus perforatus* have all re-colonised locations close to their northern range limits from where they were lost after the cold winter of 1962/1963. There is also evidence that southern species declined in abundance and northern species increased commensurately in the 1960s, 1970s and early 1980s (e.g., barnacles (Southward, 1967, 1980; Southward et al., 1995; Herbert et al., 2003); limpets (Southward et al., 1995)). This indicates that these re-extensions of range have occurred during the current period of warming since the late 1980s.

Another southern species, the brown algae *Bifurcaria bifurcata* Ross has undergone oscillations in its distribution on the south coast of England. A population was recorded in Dorset during the 1960s (E. Burrows, unpublished), but in recent decades it has been restricted to South Devon, with a sharp range edge between the Salcombe and Dart estuaries in the vicinity of Start Point ((WGS84) 50.22° N, −3.64° W) (JNCC, 1998). Several plants were recorded on Portland Bill (50.51° N, −2.45° W) in 2002 which represent a north-eastern extension in range of approximately 140 km.

The southern trochids *Osilinus lineatus* and *Gibbula umbilicalis* have also extended their northern range edges since the last surveys were conducted in 1985 (Kendall & Lewis, 1986; Kendall, 1987). The previous range edge of *O. lineatus* (defined as the last breeding population) was recorded at Lyme Regis (50.71° N, −2.950° W) on the south coast of England (Kendall, 1987). In 2002 and 2003 a mixed age population including recruits from 2001 and 2002 was found at Osmington Mills (50.63° N, −2.37° W), over 50 km east of the previous range edge population. Individuals had occasionally been recorded at this location in the 1960s (Hawthorne, 1965) and the 1980s (Kendall, pers. obs.), and further east (e.g., Kimmeridge (50.61° N, −2.13° W), Herbert, pers comm.; Preece, 1993) but a multi-cohort population has not been previously documented. The last documented position of the northern range edge of breeding populations of *G. umbilicalis* was Skerray (58.44° N, −4.30° W), on the north coast of Scotland in 1985, although isolated individuals were occasionally found 45 km further north east (Kendall & Lewis, 1986). A breeding population was recorded 55 km north-east of Skerray in 2002, with isolated individuals found up to 20 km past this location (Mieszkowska, unpublished data).

Discussion

Mechanisms setting northern range edges

The resurveys have revealed that in recent years, a number of species of invertebrates and macroalgae with southern distributions have extended their range in the British Isles (Southward & Crisp, 1954a; Hawthorne, 1965; Kendall & Lewis, 1986; Kendall, 1987; JNCC, 1998). The survey sites selected are all semi-exposed or exposed shores, minimising the effects of local environmental and anthropogenic impacts on the system. However, it is very difficult to separate the effects of temperature from other concomitant factors including the effects of coastal and industrial development, and therefore field experiments and laboratory manipulations of temperature are required in order to increase our understanding of the mechanisms driving the observed changes.

The two main biological processes that have been implicated in setting the geographical distribution limits of animals are survival and reproduction (Hutchins, 1947; Lewis et al., 1982). If the limits are determined by survival alone, extreme environmental conditions will cut back the range and less extreme conditions will allow its expansion (Lewis et al., 1982). Adult mortality has been shown to set the range edges of species during two extreme cold periods in Britain (Crisp, 1964; Kendall et al., 1987). However, population and recruitment data from previous studies indicate that in all but the most severe winters low temperatures do not result in a loss of significant adult individuals from populations.

Populations that are limited by poor recruitment success would be expected to have irregular age structures, missing year classes or a bias towards old animals (Lewis et al., 1982). The most northerly populations of *Gibbula umbilicalis* studied between 1976 and 1984 exhibited these characteristics. Partial or full recruitment failure occurred across all of these populations during the 1976–1984 time-series (Kendall & Lewis, 1986). A synchronous, but less well marked pattern of recruitment success/failure was observed at populations further south in Britain, suggesting that a large scale factor such as climate or hydrography may have been responsible (Lewis et al., 1982; Bowman, 1985; Kendall & Lewis, 1986). The size frequencies of all the populations that were revisited in 2002 and 2003 show that recruitment has been more successful in recent years (Mieszkowska & Kendall, unpublished data). As a consequence, population densities and the potential reproductive output of these northern populations has increased as increasing numbers of surviving juveniles become reproductively mature. On the north coast of Scotland, mixed age cohort populations now exist in locations where only occasional individuals were found in the 1980s (Kendall & Lewis, 1986; Bowman, pers comm.). These range extensions are thought to be linked to the increased reproductive/recruitment success of populations close to the range edge.

In both *Gibbula umbilicalis* and *Osilinus lineatus* the length of the breeding period is shorter in populations closer to northern geographic limits than in populations at the centre of the range, suggesting that changes in reproductive activity may be temperature driven (Bode et al., 1986; Kendall & Lewis, 1986; Lewis, 1986; Kendall, 1987). Such temperature sensitivity indicates that the reproductive success of these species may be altered by rapid climate warming. In the cold winter of 1978/1979 Williamson, Kendall and Lewis (Williamson et al., unpublished data) transplanted 1000 adult *O. lineatus* from a breeding population at Aberaeron (52.24° N, −4.26° W), mid Wales to Brighouse Bay (54.78° N, −4.12° W) in South West Scotland and a further 1000 animals to Burniston (54.33° N, −0.42° W) in the North Sea[1].

At Brighouse Bay and Burniston, which are both located north of existing range limits, transplanted animals were still able to successfully develop gonads. Interestingly, the mean peak gonad weight for transplanted animals was greater at both sites than in the source population at Aberaeron. The Aberaeron population spawned in June 1979, but spawning was delayed until July in the Burniston population and August in the Brighouse Bay population (Fig. 5). The mean monthly sea surface temperatures were

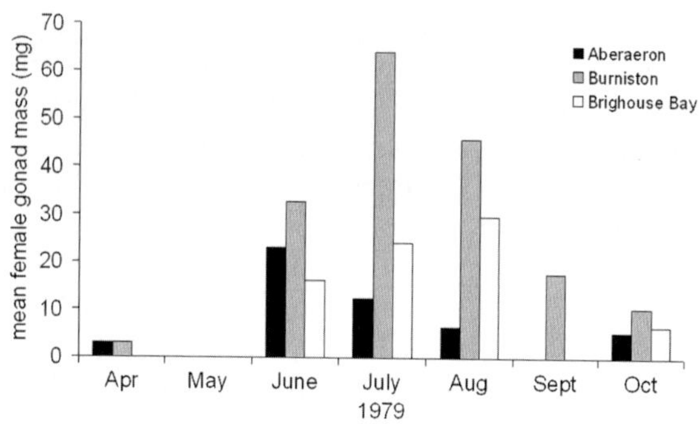

Figure 5. Female gonad cycles of *Osilinus lineatus* in the source population (Aberaeron) and transplant populations (Burniston and Brighouse Bay).

lower at both Brighouse Bay and Burniston than at Aberaeron throughout the period of gonad maturation. These results show that changes in reproductive activity occurred in animals transplanted to a colder temperature regime.

Sea and air temperatures are directly linked to latitude, with a temperature gradient extending from the tropics to the poles, although local modifications can occur due to factors such as upwelling and continental influences. Any general latitudinal cline in the length of the reproductive cycle and spawning periodicity would therefore suggest a correlation with temperature (Hutchins, 1947). To investigate the variation in the onset of the reproductive cycle in *Osilinus lineatus* and *Gibbula umbilicalis* along a latitudinal gradient a sample network extending from sites close the northern range edges in Britain and Ireland to mid range sites in Portugal has been established. Gonad samples from adults of both species are taken at monthly intervals from each site and are analysed to determine the timing and periodicity of gametogenesis and spawning at each location. Laboratory experiments are also underway to determine whether changes in temperature can alter the gametogenic period and extend the duration and frequency of spawning events in *O. lineatus* and *G. umbilicalis*. These experiments will increase our understanding of changes to the potential reproductive output as northern range limits are approached, and of the role that temperature plays in setting northern geographical limits of distribution in these southern species.

[1] The transplant experiments were undertaken in 1978. Since then the scientific community has become fully aware of the implications of moving animals to areas that they had not previously colonised. The data is referred to in this paper as it is of value in determining the mechanisms underpinning species distributions, but the practice of transplants is not supported by the MarClim Project. The animals that were transplanted on the original survey did not colonise on shore and no populations of *Osilinus lineatus* have been recorded at these sites in subsequent years. Such transplants should not be attempted or condoned.

Table 1. Proposed climate change indicator species

Species	Biogeographic range	Data	References
Osilinus lineatus	Morocco to Eastern English Channel, North Wales & North East Ireland	Baseline data 1952, 1954, 1958, 1964, 1986. Resurvey 2002, 2003. Population studies 1987, 2002, 2003. Reproductive studies.	Southward & Crisp (1954a), Southward & Crisp (1954b), Crisp & Southward (1958), Kendall (1987), Kendall et al. in prep.
Gibbula umbilicalis	Morocco to North Scotland	Baseline data 1952, 1954, 1958, 1964, 1974–1986. Resurvey 2002, 2003. Population studies 1974–1986, 2002, 2003. Reproductive studies 2002, 2003.	Lewis (1952), Southward & Crisp (1954a and b), Crisp & Southward (1958), Garwood & Kendall (1986), Kendall & Lewis (1986), Mieszkowska et al. in prep.
Chthamalus montagui	North Africa & Mediterranean to Isle of Wight & Shetland Islands	Baseline data 1954, 1958. Resurvey 2002, 2003. Population studies.	Fischer-Piette (1936 as *C. stellatus*), Southward & Crisp (1954a), Southward & Crisp (1954b), Crisp & Southward (1958)
Semibalanus balanoides	Northern Spain to Spitzbergen	Baseline data 1954. Resurvey 2002, 2003. Recruitment studies 1997.	Fischer-Piette (1936), Southward & Crisp (1954a), Southward & Crisp (1954b), Crisp & Southward (1958), Jenkins et al. (2000)
Balanus perforatus	West Africa to Eastern English Channel & South West Wales.	Baseline data 1936, 1958, 1964. Resurvey 1993/1994, 2002, 2003.	Fischer-Piette (1936), Crisp & Southward (1958), Lewis (1964), Stubbings & Houghton (1964), Herbert et al. (2003)
Patella depressa	Senegal to Eastern English Channel & North Wales	Baseline data 1958. Resurvey 2002, 2003. Population studies 1980–2003.	Fischer-Piette (1936), Crisp & Southward (1958), Crisp & Fischer-Piette (1959), Crisp unpublished.
Patella vulgata	South Portugal to Norway 70° N	Baseline data 1954, 1958. Resurvey 2002, 2003. Population studies 1980–2003.	Fischer-Piette (1936), Crisp & Southward (1958), Bowman (1985)
Sabellaria alveolata	Morocco to Solway Firth, Scotland	Baseline data 1954, 1958, 1984. Resurvey 2002, 2003.	Gruet (1971),Wilson (1974), Cunningham et al. (1984)
Bifurcaria bifurcata	North Africa to Western English Channel and Wales	Baseline data 1936, 1954, 1958, 1955, 1959, 1952. Resurvey 2002, 2003.	Fischer-Piette (1936), Valera (1962)
Alaria esculenta	Loire-Atlantique region to Norway, limited to South & West Coasts of England and Wales, present in Scotland. Absent from most of English Channel since 1950s.	Baseline data 1936, 1954, 1955, 1958, 1970. Resurvey 2002, 2003.	Fischer-Piette (1936), Crisp & Southward (1958), Widdowson (1970)

Predicting future changes

From the historical and resurvey data outlined above, the species which have shown clear alterations in their range and population characteristics during the current period of climate warming are suggested as suitable indicator species to monitor and predict the future climate and the potential changes to marine biodiversity in Britain and Ireland (Table 1). They are also of interest as many are keystone or space dominating species whose presence or absence can determine the community composition and dynamics of a shore.

The relative abundance and distribution of the southern warm water species *Chthamalus stellatus* and the northern cold water species *Semibalanus balanoides* in Britain have been shown to be broadly controlled by temperature, with *S. balanoides* being competitively dominant during colder years and *C. stellatus* during warmer years (Southward & Crisp, 1954b; Southward, 1967; Crisp et al., 1981). The proportion of *Chthamalus* adults on shores in south-west England has also been shown to correlate with mean annual inshore sea surface temperature with a 2 year time lag, with temperature accounting for over 40% of the population variance (Southward, 1991; Southward et al., 1995). This time-series is being continued by MarClim and statistical analyses of spatial and temporal trends are ongoing.

As grazers, the trochid species *Osilinus lineatus* and *Gibbula umbilicalis* originally surveyed by Lewis and Kendall are involved in the structuring of rocky shore communities. The analyses of past and current distributions and population dynamics, coupled with experimental work on the mechanisms behind these observed changes will allow future predictions of the responses of these species to climate to be made with a high degree of confidence. Data from historical surveys and MarClim resurveys indicate that changes in climate correlate with changes in the population dynamics of southern trochid species over 3° of latitude, suggesting that a large-scale factor such as climate or hydrography is the underlying driver.

As the climate warms further it is predicted that northern cold-water species will experience a contraction at their southern range edges as environmental conditions become too warm to facilitate successful recruitment in Britain. This retreat will be compounded by increased competition from southern warm-water species, which are likely to extend their distribution northwards as their reproductive success increases. Contractions and expansions of geographic ranges will lead to species being lost from and introduced to established communities. Fluctuations in climate will also alter the relative dominance of northern and southern competitors within a community.

Research into the responses to climate by marine species using intertidal indicators has both advantages and disadvantages. Rocky shore species are amenable to research due to their accessibility, feasibility of replication in surveying and experimental design, and the ease of identification. The effects of climate change on these species also has strong consequences for community dynamics, biodiversity and the ecology of the rocky intertidal zone which can be experimentally explored (Sagarin et al., 1999; Tomanek & Helmuth, 2002). These changes concord with changes in other marine communities (see above) and may be used to predict changes offshore. There are, however, problems. These include separating the effects of air and sea temperature on the various stages of the life cycle for each species and the relative importance of air and sea temperatures to adult and larval survival. Field resurveys coupled with experimental research will provide a greater understanding of the effects of climate and assist in the correlation of climatic factors with biological responses.

The body of evidence that has been documented on the fluctuation of range and population dynamics of warm and cold water species during changes in the oceanic and air temperatures over the last century is a strong indication that these species are of use in tracking climate change (Southward et al., in press). The ability to predict future changes in the demography of these species has immense scientific, conservation and socio-economic potential. Concern is also growing among coastal zone stakeholders and environmental planners regarding the structure and functioning of marine communities in response to global environmental change. Whilst change is inevitable, it is important to incorporate current patterns and future forecasts into marine conservation and management plans.

Conclusions

This paper outlines the preliminary findings of the MarClim project to date, focussing on the responses of species at their northern limits to changes in climate ahead of completion of the resurvey in 2005. The research carried out by the MarClim Project will demonstrate the extent of changes that have occurred in the last 50 years and the results will be used to develop and test hypotheses on changes now occurring. This will allow forecasts of future change to be made with maximum confidence, based on definitive climate models and covering a wide range of temperature-sensitive species. It will also evaluate whether the climate indicator species used in this work have a wider contribution to make as part of the sustainability indicators that are needed to underpin the UK sustainable development strategy.

Acknowledgements

The MarClim Project (www.mba.ac.uk/marclim) is funded by Countryside Council for Wales, The Crown Estates, Department for Environment, Food and Rural Affairs, English Nature, Environment Agency, Joint Nature Conservation Committee, Scottish Executive, Scottish Natural Heritage, States of Jersey and Worldwide Fund for Nature. S.J.H. also received support from a small NERC research grant between 1990 and 2002 for data collection and intertidal survey, and by a Grant In Aid at the Marine Biological Association of the UK. MarClim would also like to thank Pippa Moore and Roger Herbert for assistance with fieldwork and Rosemary Bowman for information on her field observations for selected species. Data used by the authors in this study include data produced through funding from the Earth Observing System Pathfinder Program of NASA's Mission to Planet Earth in cooperation with National Oceanic and Atmospheric Administration (Fig. 1). The data was provided by the Earth Observing System Data and Information System (EOSDIS), Distributed Active Archive Center at Goddard Space Flight Center which archives, manages, and distributes this data set. The BADC are acknowledged for the use of the Met Office GISST/MOHMATN4/MOHSST6 Global Ice coverage and SST data (1856–Feb 2003).

References

Barry, J. P., C. H. Baxter, R. D. Sagarin & S. E. Gilman, 1995. Climate-related, long-term faunal changes in a California rocky intertidal community. Science 267: 672–675.

Beaugrand, G., P. C. Reid, F. Ibanez, A. J. Lindley & M. Edwards, 2002. Reorganization of North Atlantic marine copepod biodiversity and climate. Science 296 (5573): 1692–1694.

Bertness, M. D., G. H. Leonard, J. M. Levine & J. F. Bruno, 1999. Climate-driven interactions among rocky intertidal organisms caught between a rock and a hot place. Oecologia 120: 446–450.

Bode, A., I. Lombas & N. Anadon, 1986. Preliminary studies on the reproduction and population dynamics of *Monodonta lineata* and *Gibbula umbilicalis* (Mollusca Gastropoda) on the central coast of Asturias (N. Spain). Hydrobiologia 142: 31–39.

Bowman, R. S., 1985. The biology of the limpet *Patella vulgata* L. in the British Isles: spawning time as a factor determining recruitment success. The ecology of rocky coasts: essays presented to J. R. Lewis; 178–193. Hodder & Stoughton, London.

Crisp, D. J. (ed.), 1964. The effects of the severe winter of 1962–1963 on marine life in Britain. Journal of Animal Ecology 33: 179–210.

Crisp, D. J. & A. J. Southward, 1958. The distribution of intertidal organisms along the coasts of the English Channel. Journal of the Marine Biological Association of the United Kingdom 37: 157–208.

Crisp, D. J. & E. Fisher-Piette, 1959. Repartition des principales especes interastidales de la Cote Atlantique Francaise en 1954–1955. Annales Institut Oceanographique Monaco 36: 276–381.

Crisp, D. J., A. J. Southward & E. C. Southward, 1981. On the distribution of the intertidal barnacles *Chthamalus stellatus, Chthamalus montagui* and *Euraphia depressa*. Journal of the Marine Biological Association of the United Kingdom 61: 359–380.

Cunningham, P. N., S. J. Hawkins, H. D. Jones & M. T. Burrows, 1984. The geographical distribution of *Sabellaria alveolata* (L.) in England, Wales and Scotland, with investigations into the community structure of, and the effects of trampling on *Sabellaria alveolata* colonies. Report to the Nature Conservancy Council from the Department of Zoology, Manchester University, Manchester.

Fields, P. A., J. B. Graham, R. H. Rosenblatt & G. N. Somero, 1993. Effects of expected global climate change on marine faunas. Trends in Evolution and Ecology 8: 361–367.

Fischer-Piette, E., 1936. Études sur ala biogéogrpahie intercôtidale des deux rives de la Manche. Journal of the Linnean Society, Zoology 40: 181–272.

Fischer-Piette, E. & M. Prenant, 1956. Distribution des Cirripèdes intercôtidaux d'Espagne septentrionale. *Bulletin du Centrre d'Études et de Recherches Scientifiques*. Biarritz 1: 7–19.

Forbes, E., 1853. The Natural History of Europe's Seas. John van Vorst, London.

Garwood, P. R. & M. A. Kendall, 1986. The reproductive cycles of *Monodonta lineata* and *Gibbula umbilicalis* on the coast of mid-Wales. Journal of the Marine Biological Association of the United Kingdom 65: 993–1008.

Graham, R. W. & E. C. Grimm, 1990. Effects of global climate change on the patterns of terrestrial biological communities. Trends in Evolution and Ecology 5: 289–292.

Gruet, Y., 1971. The morphology, growth and associated fauna of *Sabellaria alveolata* (L.) reefs of Bernerie-en-Retz (Loire Atlantique). Tethys 3: 321–380.

Hawkins, S. J., A. J. Southward & M. J. Genner, 2003. Detection of environmental change in a marine ecosystem – evidence from the western English Channel. The Science of the Total Environment 310: 245–256.

Hawthorne, J. B., 1965. The eastern limit of distribution of *Monodonta lineata* (da Costa) in the English Channel. Journal of Conchology 25: 348–352.

Herbert, R. J., S. J. Hawkins, M. Sheader & A. J. Southward, 2003. Range extension and reproduction of the barnacle *Balanus perforatus* in the eastern English Channel. Journal of the Marine Biological Association of the United Kingdom 83: 1–10.

Houghton, J. T., Y. Ding, D. J. Griggs, M. Noguer, P. J. van der Linden, X. Dai, K. Maskell & C. A. Johnson, (eds) 2001. IPCC, 2001: Climate Change 2001: The Scientific Basis. Contribution of Working Group I to the Third Assessment Report of the Intergovernmental Panel on Climate Change. Cambridge University Press, Cambridge, United Kingdom and New York, NY, USA: 881 pp.

Hulme, M., G. J. Jenkins, X. Lu, J. R. Turnpenny, T. D. Mitchell, R. G. Jones, J. Lowe, J. M. Murphey, D. Hassell, P. Boorman, R. McDonald & S. Hill, (eds) 2002. Climate Change Scenarios for the United Kingdom: The UKCIP02 Scientific Report, Vol. Tyndall Centre for Climate Change Research, School of Environmental Sciences, University of East Anglia., Norwich. UK.

Hutchins, L. W., 1947. The bases for temperature zonation in geographical distribution. Ecological Monographs 17: 325–335.

Jeffree, C. E. & E. P. Jeffree, 1996. Redistribution of the potential geographical ranges of Mistletoe and Colarado Beetle in Europe in response to the temperature component of climate change. Functional Ecology 10: 562–577.

Jenkins, S. R., P. Aberg, G. Cervin, R. A. Coleman, J. Delany, P. Della Santina, S. J. Hawkins, E. LaCroix, A. A. Myers, M. Lindegarth, A. M. Power, M. F. Roberts & R. G. Hartnoll, 2000. Spatial and temporal variation in settlement and recruitment of the intertidal barnacle *Semibalanus balanoides* (L.) (Crustacea: Cirripedia) over a European scale. Journal of Experimental Marine Biology and Ecology 243: 209–225.

Joint Nature Conservation Committee, 1998. Marine Nature Conservation Review 1987–1998. http://www.jncc.gov.uk/mermaid.

Kendall, M. A., 1987. The age and size structure of some northern populations of the trochid gastropod *Monodonta lineata*. Journal of Molluscan Studies 53: 213–222.

Kendall, M. A. & J. R. Lewis, 1986. Temporal and spatial patterns in the recruitment of *Gibbula umbilicalis*. Hydrobiologia 142: 15–22.

Kendall, M. A., P. Williamson & P. R. Garwood, 1987. Annual variation in recruitment and population structure of *Monodonta lineata* and *Gibbula umbilicalis* populations at Aberaeron, mid-Wales. Estuarine, Coastal and Shelf Science 24: 499–511.

Lewis, J. R., 1952. The ecology of rocky shores around Anglesey. Proceedings of the Zoological Society of London 123: 482–549.

Lewis, J. R., 1964. The Ecology of Rocky Shores. English Universities Press, London.

Lewis, J. R., 1976. Long-term ecological surveillance: practical realities in the rocky littoral. Oceanography and Marine Biology Annual Review 14: 371–390.

Lewis, J. R., 1986. Latitudinal trends in reproduction, recruitment and population characteristics of some rocky littoral molluscs and cirripedes. Hydrobiologia 142: 1–13.

Lewis, J. R., R. S. Bowman, M. A. Kendall & P. Williamson, 1982. Some geographical components in population dynamics: possibilities and realities in some littoral species. Netherlands Journal of Sea Research 16: 18–28.

Murray, S. N., J. Goodson, A. Gerrard & T. Luas, 2001. Long-term changes in rocky intertidal seaweed populations in urban Southern California. Journal of Phycology 37(3): 37–38.

Parmesan, C., 1996. Climate and species range. Nature 382: 765–766.

Parmesan, C., N. Ryrholm, C. Stefanescu, J. K. Hill, C. D. Thomas, H. Descimon, B. Huntley, L. Kaila, J. Kullberg, T. Tammaru, W. J. Tennent, J. A. Thomas & M. Warren, 1999. Poleward shifts in geographical ranges of butterfly species associated with regional warming. Nature 399: 579–583.

Parmesan, C. & G. Yohe, 2003. A globally coherent fingerprint of climate change impacts across natural systems. Nature 421: 37–42.

Preece, R. C., 1993. *Monodonta lineata* (da Costa) from archeological sites in Dorset and the Isle of Wight. Journal of the Conchological Society of London 34: 339–341.

Root, T. L., J. T. Price, K. R. Hall, S. H. Schneider, C. Rosenzweig & A. Pounds, 2003. Fingerprints of global warming on wild animals and plants. Nature 421: 57–60.

Sagarin, R. D., J. P. Barry, S. E. Gilman & C. H. Baxter, 1999. Climate-related change in an intertidal community over short and long time scales. Ecological Monographs 69: 465–490.

Sims, D. W., M. J. Genner, A. J. Southward & S. J. Hawkins, 2001. Timing of squid migration reflects North Atlantic climate variability. Proceedings of the Royal Society of London – Series B 268: 2607–2611.

Southward, A. J. & D. J. Crisp, 1954a. The distribution of certain intertidal animals around the Irish coast. Proceedings of the Royal Irish Academy 57: 1–29.

Southward, A. J. & D. J. Crisp, 1954b. Recent changes in the distribution of the intertidal barnacles *Chthamalus stellatus* Poli and *Balanus balanoides* L. in the British Isles. The Journal of Animal Ecology 23: 163–177.

Southward, A. J., 1967. Recent changes in abundance of intertidal barnacles in South-west England: a possible effect of climatic deterioration. Journal of the Marine Biological Association of the United Kingdom 47: 81–95.

Southward, A. J., 1976. On the taxanomic status and distribution of *Chthamalus stellatus* (Cirripedia) in the north-east Atlantic region: with a key to the common intertidal barnacles of Britain. Journal of the Marine Biological Association of the United Kingdom 56: 1007–1028.

Southward, A. J., 1980. The western English Channel: an inconstant ecosystem?. Nature, London 285: 361–366.

Southward, A. J., 1991. Forty years of changes in species composition and population density of barnacles on a rocky shore near Plymouth. Journal of the Marine Biological Association of the United Kingdom 71: 495–513.

Southward, A. J., S. J. Hawkins & M. T. Burrows, 1995. Seventy years' observations of changes in distribution and abundance of zooplankton and intertidal organisms in the western English Channel in relation to rising sea temperature. Journal of Thermal Biology 20: 127–155.

Southward, A .J., O. Langmead, N. J. Hardman-Mountford, J. Aiken, G. T. Boalch, P. R. Dando, M. J. Genner, I. Joint, M. A. Kendall, N. C. Halliday, R. P. Harris, R. Leaper, N. Mieszkowska, R. D. Pingree, A. J. Richardson, D. W. Sims, T. Smith, A. W. Walne & S. J. Hawkins, in press. Long-term oceanographic and ecological research in the western English Channel. Advances in Marine Biology 47: 1–105.

Stebbing, A. R. D., S. M. T. Turk, A. Wheeler & K. R. Clarke, 2002. Immigration of southern fish species to south-west England linked to warming of the North Atlantic (1960–2001). Journal of the Marine Biological Association of the United Kingdom 82: 177–180.

Tomaneck, L. & B. Helmuth, 2002. Physiological ecology of rocky intertidal organisms: a synergy of concepts. Integrative and Comparative Biology 42: 771–775.

de Valera, M., 1962. Some aspects of the problem of the distribution of *Bifurcaria bifurcata* (Velley) Ross on the shores of Ireland, north of the Shannon Estuary. Proceedings of the Royal Irish Academy B 62: 77–99.

Wilson, D. P., 1974. *Sabellaria* colonies at Duckpool, north Cornwall 1971–1972 with a note from May 1973. Journal of the Marine Biological Association of the United Kingdom 54: 393–436.

Hydrobiologia (2006) 555:253–261
H. Queiroga, M.R. Cunha, A. Cunha, M.H. Moreira, V. Quintino, A.M. Rodrigues, J. Serôdio & R.M. Warwick (eds),
Marine Biodiversity: Patterns and Processes, Assessment, Threats, Management and Conservation
DOI 10.1007/s10750-005-1121-5

Effects of dredging activities on population dynamics of *Posidonia oceanica* (L.) Delile in the Mediterranean sea: the case study of Capo Feto (SW Sicily, Italy)

Fabio Badalamenti[1], Giuseppe Di Carlo[1,*], Giovanni D'Anna[1], Michele Gristina[1] & Marco Toccaceli[2]
[1]*CNR-IAMC Laboratorio di Biologia Marina, Via G. da Verrazzano 17, 91014, Castellammare del Golfo (TP)*
[2]*C.R.E.A. Soc. Coop, Via F. Guardione 30, 90139, Palermo, Italy*
*(*Author for correspondence: Tel.: +39092435013; Fax: +39092435084; E-mail: dicarlo@irma.pa.cnr.it)*

Key words: population dynamics, shoot mortality, human impact, *Posidonia oceanica*, seagrass, Mediterranean sea

Abstract

Between 1981 and 1993 a methane pipeline was deployed between Sicily (Italy) and Tunisia. This involved the construction of a pipeline trench, which damaged the *Posidonia oceanica* (L.) Delile meadow at Capo Feto (SW Sicily, Italy) and disturbed the surrounding meadow. Seagrass growth and population dynamics were examined at different depth ranges and at increasing distances from the construction site outer limit (5, 15, 30, 50 m). Results showed significant differences between the shallow (10 ± 3.3 m) and the deep (20 ± 4.6 m) meadow as well as differences among distances. The age structure of *P. oceanica* varied along the distance gradient and with depth. The mortality rate decreased with distance from the trench at all depth ranges, showing that the plants close to the excavation suffered a higher level of disturbance. Turnover and annual gross shoot recruitment rate (R_{gross}) were higher in the shallow portion of the meadow than in the deep range. Forecast of future meadow development (R_{net}) close to the trench indicates that, if present conditions are maintained, shoot density will be reduced by 50% over the next 6 to 17 yrs.

Introduction

Posidonia oceanica (L.) Delile is an endemic seagrass species of the Mediterranean Sea (Pergent-Martini & Pergent, 1995) that forms extensive meadows ranging from shallow waters to depths of about 40 m (Pérès & Picard, 1964), if sufficient light is available. In the last decade a decline of *P. oceanica* beds has been recorded, caused by both natural (Gallegos et al., 1993; Marbà & Duarte, 1994) and anthropogenic events (Short & Wyllie-Echeverria, 1996). Fishing pressure (mainly trawling) (Ardizzone & Migliuolo, 1982; Sanchez-Jerez & Ramos-Esplá, 1996), nutrient loading (Pergent-Martini & Pergent, 1995) and boat anchoring (Garcìa-Charton et al., 1993; Francour et al., 1999), are examples of anthropogenic factors which lead to a slow but constant decline of *P. oceanica* beds. In addition, there are marine operations which have caused negative impacts on seagrass beds such as the construction of industrial outfalls (Pergent-Martini et al., 1996), fish farms (Delgado et al., 1999; Ruiz & Romero, 2001; Ruiz et al., 2001) and dredge-fill operations (Guidetti & Fabiano, 2000; Short & Coles, 2001). Nowadays, coastal zones support an increasing number of pipes and cables for gas and water transport as well as communication, the maintenance of which entails damage to seagrass beds (Duarte, 2002).

In 1981, a submerged pipeline system (TRANSMED) was deployed between Cape Bon (NE Tunisia) and Capo Feto (SW Sicily). Operations included the dredging of a trench through the 30 000 ha *P. oceanica* meadow at Capo Feto. In 1993, two new pipelines were laid at the same site, enlarging the previously damaged area. This operation represented a disturbance event for the adjacent seagrass meadow, as it most probably led to an increase in sediment resuspension and deposition due to dredging and backfilling activities.

The development of reconstruction techniques, useful to reconstruct historic seagrass dynamics, allows the examination of the growth of different seagrass modules, allowing the assessment of plant health (Marbà et al., 1996; Hemminga & Duarte, 2000). Such techniques can be applied to evaluate whether a population is expanding or declining, through the study of the balance between shoot recruitment and mortality (Duarte et al., 1994; Marbà et al., 1996).

Here, we report on the case study of Capo Feto where growth and shoot demography of *P. oceanica* were assessed after the perturbation caused by the nearby pipe deployment. Plant growth and demography were estimated at different depth ranges and at increasing distances from the trench. It is hypothesised that a higher shoot mortality rate and a lower shoot net recruitment will be seen in seagrass patches closest the disturbed seabed associated with the pipe trench.

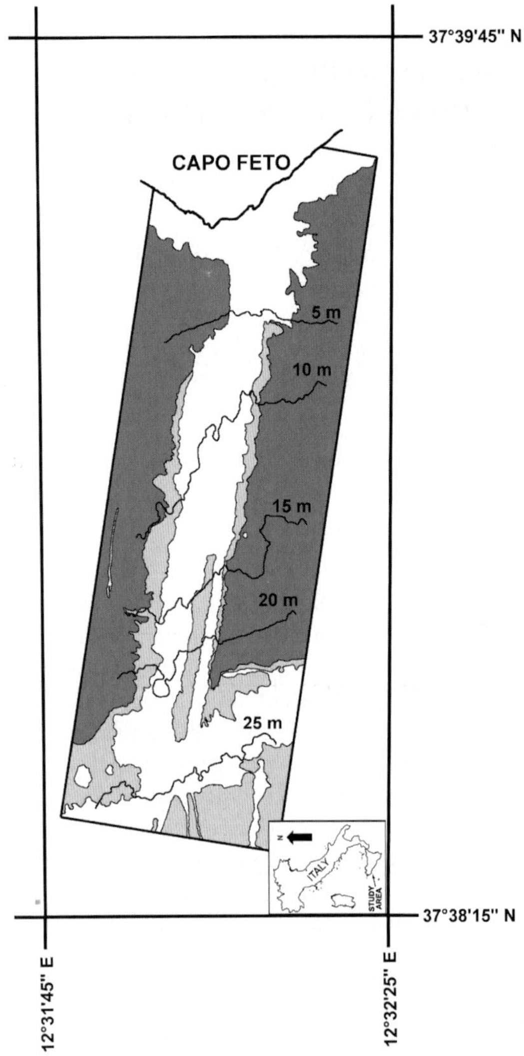

Figure 1. Side scan sonar map of study site. The area in the centre is the trench excavated for the deployment of the pipeline.

Materials and methods

The study area extends SW off Capo Feto coast, Sicily, Italy (37°39′N, 12°32′E) to a depth of 30 m (Fig. 1) (Toccaceli & Riggio, 1989; Orrù et al., 1993). Seagrass sampling was carried out between July and September 1998 at two depth ranges: shallow (10 ± 3.3 m) and deep (20 ± 4.6 m). Shoot density was recorded using 30×30 cm quadrates at four distances (5, 15, 30, 50 m) from the meadow edge/disturbed seabed interface (Fig. 1). For each distance, four sites were randomly chosen. Within each site, six random shoot counts were carried out.

For the population dynamics studies, six random samples were taken within a 10 m^2 area for each distance. Each sample contained between 50 and 100 living vertical rhizomes. Plant material was transported to the laboratory for morphological analysis (vertical rhizome length, horizontal and vertical internode length). The age structure of living shoots was estimated (Duarte et al., 1994), allowing calculation of shoot age, shoot recruitment, mortality rate and population half-life (Duarte et al., 1994; Marbà et al., 1996).

The rate of formation of horizontal rhizome internodes was estimated using a simple regression of the number of horizontal internodes, included pairs of shoots connected by a horizontal rhizome piece, against their age difference (Duarte et al., 1994). The regression slope represented the average number of rhizome internodes which are produced at each leaf plastochronic interval (P.I.). Once the leaf P.I. is converted into days, the inverse of this slope equals the time the plant takes to produce a horizontal internode. The vertical rhizome elongation rate was calculated by regressing the age of the rhizome (in years) against its total length. Shoot age was calculated by counting the number of nodes and the number of leaves present at sampling time. These two values were added and multiplied by rhizome P.I. to obtain rhizome age. The annual gross shoot recruitment rate (R_{gross}, ln units yr^{-1}) was estimated as

$$R_{gross} = \ln \sum_{t=0}^{\infty} N_t - \ln \sum_{t=1}^{\infty} N_t$$

where

$$\sum_{t=0}^{\infty} N_t$$

is the number if living shoots and

$$\sum_{t=1}^{\infty} N_t$$

is the number of living shoots with age > 1 yr. Shoot mortality rate (M, ln units yr^{-1}) was derived from the exponential decline in the abundance of living shoots (N_0) with time (t, in P.I.) as:

$$N_t = N_0 * e^{-Mt}$$

where N_0 is the number of shoots with age equal to the mode, N_t is the number of shoots older than the modal age at time t, and M is the mortality rate (ln units yr^{-1}).

The net rate of shoot population growth

$$R_{\mathrm{net}} = R_{\mathrm{gross}} - M$$

was used to make predictions about the future development of the meadow. Turnover rate and the ratio between recruitment and mortality rates (R:M) were also calculated. This ratio indicates whether a population is expanding ($R{:}M > 1$), declining ($R{:}M < 1$) or it is in equilibrium ($R{:}M = 1$) (Duarte & Sand-Jensen, 1990; Duarte et al., 1994; Marbà et al., 1996).

Data were analysed using a multifactorial multivariate design with both factors (depth and distance) fixed and orthogonal. The analysis included eight variables (vertical rhizome growth, modal, mean and median age, R_{gross}, R_{net}, shoot mortality and shoot turnover). Analysis was carried out using the FORTRAN Programme NP-MANOVA (Anderson, 2001). Shoot density data were analysed using a three-way analysis of variance. Treatment factors were: depth (two levels), distance (four levels), site (four levels). The first two factors were considered as fixed and orthogonal, while sites were taken as random and nested in the interaction of the fixed factors. Homogeneity of variance was checked for both statistical tests, using Cochran's test. When necessary, data were transformed using $\ln(X+1)$ (Underwood, 1997). When significant differences were found in the ANOVA, they were compared *a posteriori* using a Student–Newman–Keuls (SNK) test (Underwood, 1981, 1997).

Results

Posidonia oceanica shoot density was found to be significantly higher in shallow water (10 ± 3.3 m) than in deep water (20 ± 4.6 m) (Table 1), with mean values of 452.6 ± 53.62 and 254.2 ± 45.45 shoot m^{-2} respectively ($\mathrm{df}_{1,24}$; $F = 1035.94$; $p < 0.001$). A significant increase in shoot density was found at increasing distances from the trench ($\mathrm{df}_{3,24}$; $F = 4.50$; $p = 0.01$; SNK test: 5 < 15 < 30 < 50 m). In shallow water, at a distance of 5 m from the trench, density was 392.6 ± 67.8 mean shoot m^{-2}, whilst at 15 m this variable had increased to 439.8 ± 48.0 shoot m^{-2}. Further from the trench, values averaged at 483.3 ± 43.9 at 30 m distance and 494.6 ± 54.8 shoot m^{-2} at 50 m distance. At 20 m depth, shoot density showed a similar pattern with increasing values from 5 m distance (213.9 ± 50.8 shoot m^{-2}) to 50 m distance (302.2 ± 45.8 shoot m^{-2}).

Seagrass dynamics data are reported in Table 1. NP-MANOVA analysis showed significant differences between depths and distances (Table 2). Pairwise comparisons (Table 3) indicated a significant difference between depth for all variables, whilst significant differences among distances were only found for mean shoot age, shoot mortality and net recruitment rate (Table 3). Leaf P.I. attained 53.39 days for the whole meadow, a higher value than shoot P.I. which averaged at 46.8 days (Table 1). Vertical rhizome growth rates ranged between 0.67

Table 1. Average (± SD in brackets) of population dynamics variables of *Posidonia oceanica* at Capo Feto at different depths (10 and 20 m) and distances (5, 15, 30, 50 m) from the trench

Depth	Distance	Shoot density (shoot m^{-2})	Median shoot age (yr)	Maximum shoot age (yr)	Shoot recruitment rate (ln units yr^{-1})	Shoot mortality rate (ln units yr^{-1})	Net recruitment rate (ln units yr^{-1})	Shoot turnover (yr)	R_{gross}/M	Vertical rhizome growth (cm yr^{-1})	Shoot P.I. (days)
Deep	5	213.9 (50.8)	4.2 (0.8)	17.0 (2.8)	0.07 (0.02)	−0.21 (0.00)	−0.14 (0.05)	14.5 (10.4)	0.32	0.70 (0.14)	42.6
Deep	15	222.2 (35.5)	4.5 (1.6)	20.5 (4.7)	0.07 (0.02)	−0.18 (0.02)	−0.11 (0.03)	10.3 (4.3)	0.39	0.86 (0.18)	42.6
Deep	30	278.6 (49.7)	5.3 (1.0)	26.0 (11.2)	0.06 (0.04)	−0.17 (0.04)	−0.11 (0.06)	21.1 (13.4)	0.35	0.84 (0.12)	54.9
Deep	50	302.2 (45.8)	3.9 (1.4)	14.6 (4.6)	0.11 (0.10)	−0.15 (0.03)	−0.04 (0.00)	11.1 (5.3)	0.71	1.00 (0.35)	48
Shallow	5	392.6 (67.8)	1.9 (0.8)	19.3 (6.0)	0.40 (0.20)	−0.30 (0.03)	+0.10 (0.03)	2.6 (1.7)	1.32	1.00 (0.31)	40.7
Shallow	15	439.8 (48)	3.0 (1.1)	18.5. (4.1)	0.20 (0.10)	−0.21 (0.04)	−0.01 (0.00)	4.1 (1.8)	0.94	1.12 (0.2)	47.2
Shallow	30	483.3 (43.9)	3.1 (1.7)	16.7 (5.4)	0.26 (0.10)	−0.21 (0.03)	+0.05 (0.00)	2.5 (0.7)	1.25	0.96 (0.12)	49.7
Shallow	50	494.6 (54.8)	2.9 (0.7)	18.8 (5.2)	0.15 (0.06)	−0.23 (0.05)	−0.08 (0.00)	5.1 (2.0)	0.64	0.99 (0.21)	44.1

Table 2. NP-MANOVA results for the seven variables included in the statistical test

Source	Df	EMS	Perm. Units	SS	MS	F
Depth	1	RES	32	1241.20	1241.2	20.6^{***}
Dist	3	RES	32	498.55	166.18	2.75^{**}
Dexdi	3	RES	32	133.16	44.29	0.73^{ns}
RES	40			2409.57	60.23	
TOT	47					
Data not transformed				Unrestricted permutation of raw data		
No standardisation				Integer used as seed = 5		
Analysis based on Euclidean Distances				No. of permutations used = 4999		

Differences are shown both for depths (shallow and deep) and distances (5, 15, 30, 50 m).
*** $p<0.001$;** $p<0.01$; ns = not significant.

and 1.57 cm yr^{-1}, with an average value of 0.88 ± 0.22 in the deep meadow, and 1.04 ± 0.22 cm yr^{-1} in the shallow meadow. Shoot maximum age averaged between 12 and 41 yr in the deep meadow, whilst values for the shallow meadow ranged between 6 and 28 yr, indicating that the latter presents young patches with higher recruitment.

The age structure of *P. oceanica* varied along the distance gradient and with depth (Figs. 2, 3). In the shallow meadow, the population showed a significantly lower modal age (0.93 ± 0.32 yr) than in the deep meadow (2.44 ± 1.92 yr) (Table 3). Shoot recruitment (R_{gross}) rate was found to be significantly higher in the shallow meadow (on average, $0.25\pm$ 0.19 ln units yr^{-1}), than in deep meadow (0.07 ± 0.05 ln units yr^{-1}; Table 1). However, the recruitment rate recorded in shallow water was balanced by a higher mortality rate found at this depth range ($-0.23\pm$ 0.04 ln units yr^{-1}). At 10 m depth, at 5 m distance

Table 3. Pair-wise comparisons, made with NP-MANOVA and SNK tests, indicating differences within treatments for each variable

Variables	Depth	Distance
Modal shoot age	Shallow < deep	ns
Mean shoot age	Shallow < deep	5 < 15 = 30 = 50
Median shoot age	Shallow < deep	ns
R_{gross} (ln units yr^{-1})	Shallow > deep	ns
Shoot mortality (ln units yr^{-1})	Shallow > deep	5 > 15 = 30 = 50
Shoot turnover (yr)	Shallow > deep	ns
R_{net}	Shallow < deep	5 > 15 = 30 = 50
Vertical rhizome growth (cm yr^{-1})	Shallow > deep	ns

ns = not significant; < and > $p<0.05$.

from the trench, *M* averaged -0.30 ± 0.03 ln units yr^{-1}, a significantly higher value than mortality rates reported for the other three distances (15, 30 and 50 m; Table 1). Indeed, moving further from the trench, mortality rates decrease to -0.21 ± 0.04 ln units yr^{-1} (15 m distance), -0.21 ± 0.03 (at 30 m distance) and -0.23 ± 0.05 at 50 m distance. A similar trend was observed at 20 m depth ($5>15=30=50$ m distance from trench; Table 3). In this depth range, mean values of *M* at 5 m distance equal $-0.21\pm$ 0.004 ln units yr^{-1}, while average values at further distances were -0.18 ± 0.02, -0.17 ± 0.04 and -0.15 ± 0.03 ln units yr^{-1} at 15, 30 and 50 m respectively.

A significantly shorter turnover rate was reported for the shallow meadow (average 3.2 ± 1.8), than the deep meadow, mean value of 14.2 ± 9.6. The net recruitment ratio, derived from the ratio between recruitment and mortality, suggested that in the shallow meadow R_{net} is almost in equilibrium or slightly increasing (mean 0.015 ± 0.078 ln units yr^{-1}) whilst *P. oceanica* patches found in the deep meadow (20 ± 4.6 m), are declining (average -0.1 ± 0.04 ln units yr^{-1}) but at a slower rate. These findings were confirmed by the relationship between R_{gross}/M which is <1 at all distances in deep water; in shallow water, R_{gross}/M is >1 at a distance of 5 and 30 M, ~O at 10 m distance and <1 at 50 M distance.

Discussion

The population dynamics variables measured at Capo Feto fall within the range of those obtained for other Mediterranean meadows (Marbà et al., 1996). For example, shoot density varied with

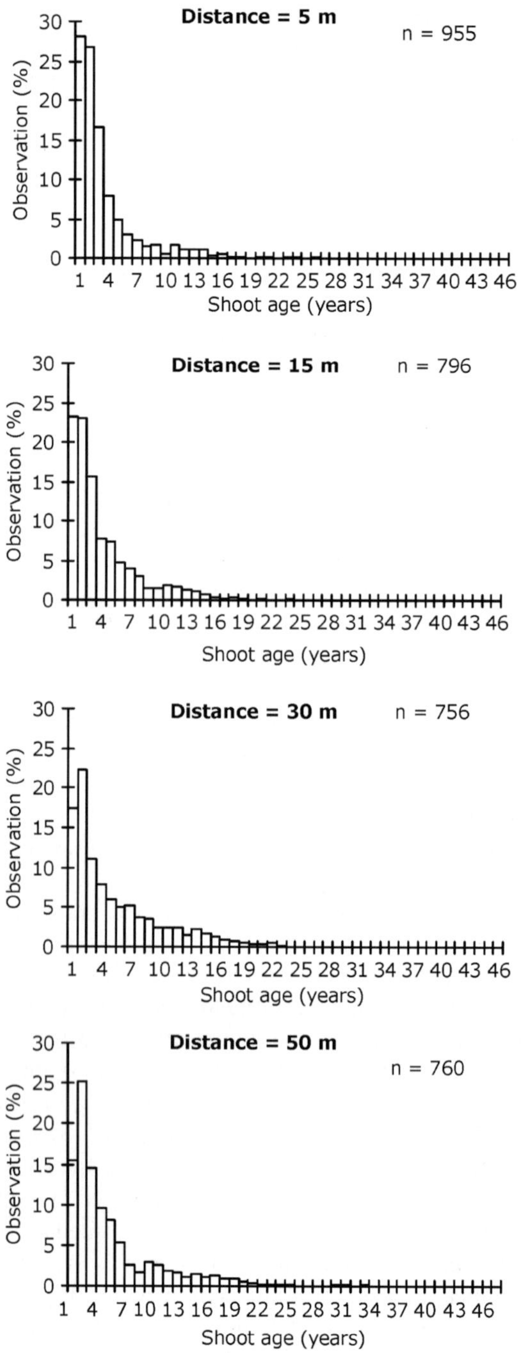

Figure 2. Shoot age structure of living shoots of *Posidonia oceanica* in the shallow (10 ± 3.3 m) meadow at increasing distances (5, 15, 30 and 50 m) from the trench.

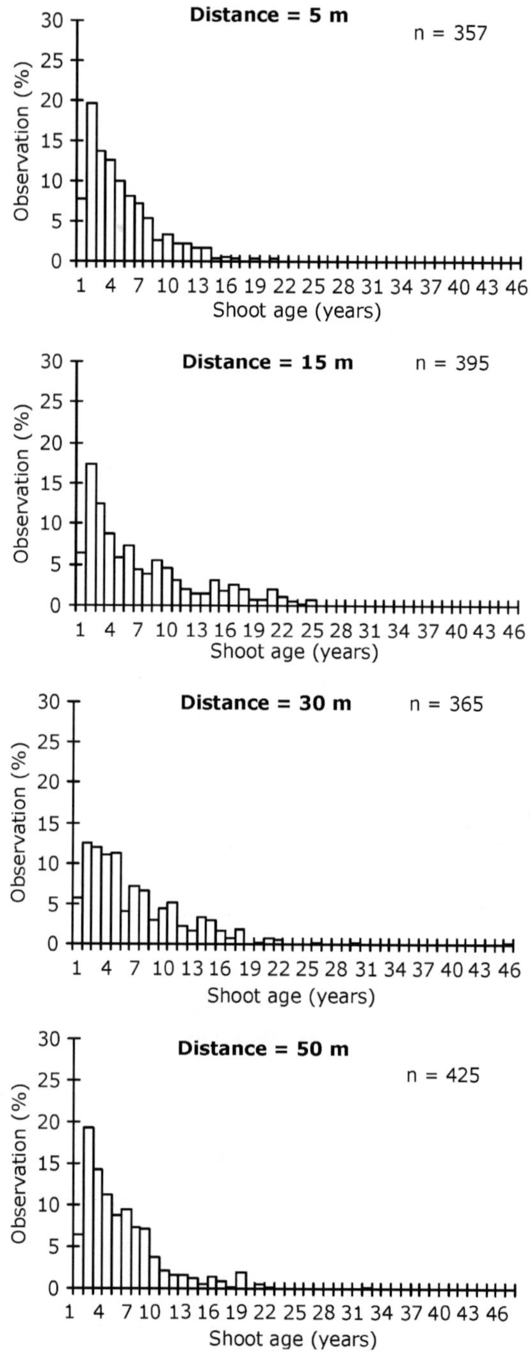

Figure 3. Shoot age structure of living shoots of *Posidonia oceanica* in the deep (20 ± 4.6 m) meadow at increasing distances (5, 15, 30 and 50 m) from the trench.

depth as in many other Mediterranean meadows (Pergent et al., 1995). However, in the area closest to the disturbed seabed associated with the trench, average shoot density, shoot age and R_{net} values decreased, while mortality rate increased, most likely as a result of the construction work. Coastal engineering and the maintenance of pipes and cable for water and gas supply often change seagrass habitat or they alter the physical characteristics of water (Hemminga & Duarte, 2000). One of the consequences of construction operations is the increase in water turbidity (Onuf, 1994), due to dredging and filling operations (Guidetti & Fabiano, 2000). Such operations do not only involve the alteration of water properties but often lead to an increase in the sedimentation rate as well as sediment movements (Orrù, pers. comm, 1994). These consequences may cause the burial of seagrass stands, with subsequent shoot death and low shoot density (Marbà & Duarte, 1994; Duarte et al., 1997; Hemminga & Duarte, 2000).

In most seagrass species, both in and outside the Mediterranean basin, rhizome growth has been demonstrated to be sensitive to sediment dynamics (Marbà & Duarte, 1994; Marbà et al., 1996; Duarte et al., 1997) and, in some cases, vertical rhizomes may overcome the burial process by increasing internodal length (Duarte & Sand-Jensen, 1990; Vidondo et al., 1997). Internodal length was demonstrated to be strictly related to the degree of burial and erosion experimentally imposed on a number of Philippine seagrass species and on the Mediterranean seagrass species *Cymodocea nodosa* (Ucria) Ashers (Duarte et al., 1997; Marbà & Duarte, 1994). Similarly, several seagrass including *P. oceanica* in natural environments display fluctuations in the rhizome elongation rate as a consequence of sediment dynamics (Marbà et al., 1996; Marbà & Duarte, 1997). At Capo Feto, vertical rhizome growth was fairly low at all depths and was found not to be affected by dredge-fill activities. This finding is in accordance with previous work by Guidetti (2001) who reported no significant differences in rhizome elongation rates of *P. oceanica* after beach replenishment. Guidetti (2001) highlighted an increased leaf production rate. On the other hand, our data shows a higher mortality and net recruitment rate in the meadow area closest to the trench. In this area, sediment deposition may have been so high that rhizome growth could not compensate for the burial process, leading to a complete disappearance of seagrass in this area. In this case, recovery, if ever possible, will require a long period of time (Talbot et al., 1990; Hemminga & Duarte, 2000).

Population variables differed between the shallow and deep meadow. A general trend was found in the age structure of the meadow, with a decline of shoot abundance with age, possibly due to shoot mortality. Plants living in deeper water tended to show a higher mortality and a lower recruitment, hence a longer turnover rate. Plants living in deep water generally showed longer recovery times as growth was reduced possibly due to light limitation. On the other hand, the shallow meadow displayed a shorter turnover rate and a lower mortality, and hence shorter recovery times may be hypothesised. At 10 m depth, a higher recruitment rate was also recorded, which may imply shorter recovery times. With regard to mortality, plants closer to the excavation site seem to be the most affected. At 5 m distance in shallow water, *P. oceanica* shoots showed the highest mortality rate, which may also be linked to low shoot density and patch size. Further from the construction site, seagrass shoots experienced a lower mortality suggesting a lower sedimentation rate hence a lower level of disturbance. The combination of increased mortality and low recruitment rates reflects an overall decline of the meadow if present conditions are maintained. Even though the source of disturbance at Capo Feto has been eliminated from the scenario, forecast of future meadow development (R_{net}) indicates that shoot density close to the trench will be reduced by 50% over the coming 6–17 yrs. *P. oceanica* has the slowest elongation rate of all seagrasses (between 1 and 6 cm yr^{-1}), with patch formation rates averaging around 3×10^{-4} $ha^{-1}\ yr^{-1}$ (Meinesz & Lefévre, 1984). This entails recovery times that generally exceed a century. Capo Feto represents a unique case study as the disturbance events were well defined and have stopped, and are not to be repeated. This cessation of disturbance should allow seagrass recovery, at least in shallow meadow areas, even if over a long decadal time period. Further studies will be necessary to evaluate the resilience of the recovering *P. oceanica* over time.

Acknowledgements

We kindly acknowledge T.M.P.C. (Trans-Mediterranean Pipeline Company) for the logistic support provided. The authors wish to thank M.C. Alessi, G. Titone and P. Di Stefano for helping during field and laboratory work. We are also grateful to T. Bahri, M.C. Chiantore, M.J. Anderson and A.C. Jensen for comments and support. This work is dedicated to Rosanna Parisi.

References

Anderson, M. J., 2001. A new method for non-parametric multivariate analysis of variance. Australian Ecology 26: 32–6.

Ardizzone, G. & A. Migliuolo, 1982. Modificazioni di una prateria di *Posidonia oceanica* (L.) Delile del medio Tirreno sottoposta ad attività di strascico. Naturalista Siciliano S. IV, VI: 509–515.

Delgado, O., J. M. Ruiz, M. Perez, J. Romero & E. Ballesteros, 1999. Effects of fish farming on seagrass (*Posidonia oceanica*) in a Mediterranean Bay: seagrass decline after organic loading cessation. Oceanologica Acta 22: 109–117.

Duarte, C. M. & K. Sand-Jensen, 1990. Seagrass colonization: patch formation and patch growth in *Cymodocea nodosa*. Marine Ecology Progress Series 65: 193–200.

Duarte, C. M., N. Marbà, N. Agawin, J. Cebrian, S. Enriquez, M. D. Fortes, M. E. Gallegos, M. Merino, B. Olesen K. Sand-Jensen, J. Uri & J. Vermaat, 1994. Reconstruction of seagrass dynamics: age determination and associated tools for the seagrass ecologist. Marine Biology Progress Series 107: 195–209.

Duarte, C. M., J. Terrados, N. S. R. Agawin, M. D. Fortes, S. Bach & W. J. Kenworthy, 1997. Response of a mixed Philippine seagrass meadow to experimental burial. Marine Ecology Progress Series 147: 285–294.

Duarte, C. M., 2002. The future of seagrass meadows. Environmental Conservation 29: 192–206.

Francour, P., A. Ganteaume & M. Poulain, 1999. Effects of boat anchoring in *Posidonia oceanica* seagrass beds in the Port-Cros National Park (north-western Mediterranean Sea). Aquatic Conservation 9: 391–400.

Gallegos, M., M. Merino, N. Marbà & C. M. Duarte, 1993. Biomass and dynamics of *Thalassia testudinum* in the Mexican Caribbean: elucidating rhizome growth. Marine Ecology Progress Series 95: 185–192.

Garcia-Charton, J. A., J. T. Bayle, J.L. Sànchez-Lizaso P. Chiesa, F. Llauradò, C. Pérez & H. Djian, 1993. Respuesta de la pradera de *Posidonia oceanica* y su ictiofauna asociada al anclaje de embarcaciones en el parque Nacional de Port-Cros, Francia. Publ Espec Inst Esp Oceanogr 11: 423–430.

Guidetti, P. & M. Fabiano, 2000. The use of lepidochronology to assess the impact of terrigenous discharges on the primary leaf production of the Mediterranean seagrass *Posidonia oceanica*. Marine Pollution Bulletin 40: 449–453.

Guidetti, P., 2001. Detecting environmental impacts on the Mediterranean seagrass *Posidonia oceanica* (L.) Delile: the use of reconstructive methods in combination with "beyond BACI" design. Journal of Experimental Marine Biology and Ecology 260: 27–39.

Hemminga, M. & C. M. Duarte, 2000. Seagrass Ecology. Cambridge University Press, Cambridge.

Marbà, N. & C. M. Duarte, 1994. Growth Response of the seagrass *Cymodocea nodosa* to experimental burial and erosion. Marine Ecology Progress Series 107: 307–311.

Marbà, N., C. M. Duarte, J. Cebrian, S. Enriquez, M. E. Gallegos, B. Olesen & K. Sand-Jensen, 1996. Growth and population dynamics of *Posidonia oceanica* on the Spanish Mediterranean coast: elucidating seagrass decline. Marine Ecology Progress Series 137: 203–213.

Marbà, N. & C. M. Duarte, 1997. Interannual changes in seagrass (*Posidonia oceanica)* growth and environmental changes in the Mediterranean littoral zone. Limnology and Oceanography 42: 800–810.

Meinesz, A. & J. R. Lefèvre, 1984. Régénération d'un herbier de *Posidonia oceanica* quarante années aprés sa destruction par une bombe dans la rade de Villefranche (Alpes-Maritimes, France). In Boudouresque, C. F., A. Jeudy de Grissac & J. Olivier (eds). International Workshop Posidonia oceanica Beds. Gis Posidonie, France, pp. 39–44.

Onuf, C. P., 1994. Seagrasses, dredging and light in Laguna Madre, Texas, U.S.A. Estuarine, Coastal and Shelf Science 39: 75–91.

Orrù, P., G. Melegari & M. Badalini, 1993. Geomorphological observations of the sea bed between Cape Bon and Cape Feto (Strait of Sicily). Geological Development of the Sicilian-Tunisian Platform. UNESCO REPORT IN MARINE SCIENCE, Paris, pp. 153–160.

Pérès, J. M. & J. Picard, 1964. Noveau manuel de bionomie benthique de la mer Méditerranée. Recueil des Travaux de la Station Marine. d'Endoume 31: 1–137.

Pergent, G., C. Pergent-Martini & C. F. Boudouresque, 1995. Utilisation de l'herbier a *Posidonia oceanica* comme indicateur biologique de la qualite du milieu littoral en mediterranee: etat des connaissances. Mésogée 54: 3–27.

Pergent-Martini, C. & G. Pergent, 1995. Impact of a sewage treatment plant on the *Posidonia oceanica* meadow: assessment criteria. Proceeding of the Second International Conference on the Mediterranean Coastal Environment MEDCOAST '95 pp. 1389–1399.

Pergent-Martini, C., V. Rico-Raimondino & G. Pergent, 1996. Impact des nutriments sur les herbiers à *Posidonia oceanica* – Données préliminaires. Journal of Recherché Océanographique 21: 35–39.

Ruiz, J. M., M. Perez & J. Romero, 2001. Effects of fish farm loadings on seagrass (*Posidonia oceanica*) distribution, growth and photosynthesis. Marine Pollution bulletin 42: 749–760.

Ruiz, J. M. & J. Romero, 2001. Effects of *in situ* experimental shading on the Mediterranean seagrass *Posidonia oceanica*. Marine Ecology Progress Series 215: 107–120.

Sanchez-Jerez, B. & A. A. Ramos-Esplà, 1996. Detection of environmental impacts by bottom trawling on *Posidonia oceanica* (L.) Delile meadows: sensitivity of fish and macrobenthic communities. Journal of Aquatic Ecosystem Health 5: 239–253.

Short, F.T. & S. Wyllie-Echeverria, 1996. Natural and human-induced disturbance of seagrasses. Environmental Conservation 23: 17–27.

Short, F. T. & R. G. Coles, 2001. Global Seagrass Research Methods. Elsevier Science Publishers B.V, Amsterdam.

Talbot, M. M. B., W. T. Knoop & G. C. Bate, 1990. The dynamics of estuarine macrophytes in relation to flood/siltation cycles. Botanica Marina 33: 159–164.

Toccaceli, M. & S. Riggio, 1989. Note bibliografiche sulle Fanerogame marine delle coste siciliane. Oebalia XV: 279–286.

Underwood, A. J., 1981. Techniques of analysis of variance in experimental marine biology and ecology. Oceanography and MarineBiology: An Annual Review 19: 513–605.

Underwood, A. J., 1997. Experiments in Ecology: Their Logical Design and Interpretation using Analysis of Variance. Cambridge University Press, Cambridge.

Vidondo, B., A. L. Middleboe, K. Stefansen, T. Ltzen, S. L. Nielsen & C. M. Duarte, 1997. Dynamics of a patchy seagrass (*Cymodocea nodosa*) landscape. Size and age distribution, growth and demography of seagrass patches. Marine Ecology Progress Series 158: 131–138.

Hydrobiologia (2006) 555:263–269

H. Queiroga, M.R. Cunha, A. Cunha, M.H. Moreira, V. Quintino, A.M. Rodrigues, J. Serôdio & R.M. Warwick (eds),
Marine Biodiversity: Patterns and Processes, Assessment, Threats, Management and Conservation
DOI 10.1007/s10750-005-1122-4

The colonisation of human-made structures by the invasive alga *Codium fragile* ssp. *tomentosoides* in the north Adriatic Sea (NE Mediterranean)

Fabio Bulleri[1,2,*], Marco Abbiati[1,2] & Laura Airoldi[1,2]
[1]*Centro Interdipartimentale di Ricerca per le Scienze Ambientali di Ravenna Università di Bologna, Via S.Alberto 163, I-48100, Ravenna, Italy*
[2]*Dipartimento di Biologia Evoluzionistica Sperimentale*
*(*Author for correspondence: E-mail: fbulleri@discau.unipi.it)*

Key words: invasive species, human-made structures, *Codium fragile*, Adriatic Sea, wave-exposure

Abstract

Human-made structures, such as groynes, breakwaters, seawalls, pier pilings and floating pontoons, are becoming common features of the landscape in urbanised coastal and estuarine areas. Despite this tendency few studies have focused on their ecology or on their potential impacts on natural assemblages of organisms. When artificial structures are introduced in areas with little or no hard substrata, they not only provide novel habitats, which enables the colonisation of sandy areas by hard-bottom dwelling species, but they can also provide suitable habitats for exotic species. Along the north-east coast of Italy, sandy shores are protected from erosion by a line of breakwaters, which runs almost uninterrupted for about 300 km. These structures provide habitat for a variety of macroalgae and invertebrates and also for the invasive green alga *Codium fragile* ssp. *tomentosoides*. The aim of this study was, therefore, to investigate patterns of distribution of this alga on breakwaters in Cesenatico. In particular, we compared the density of thalli, biomass, length and degree of branching of *C. fragile* ssp. *tomentosoides* between the landward and the seaward sides of breakwaters, to test the hypothesis that sheltered habitats (landward) represent more suitable habitats than exposed habitats (seaward). In general, the landward side of breakwaters supported greater numbers of thalli of *C. fragile* ssp. *tomentosoides* than seaward sides. Thalli grew longer and more branched in sheltered habitats, leading to an overall larger biomass of the alga on the landward side of breakwaters. The presence of sheltered human-made hard substrata in the vicinity of major trading ports and sources of eutrophication could enhance the dispersal of invasive species across regional and geographic scales. Thus, the effects of artificial structures and introduced species on coastal assemblages cannot be evaluated separately, but their synergistic nature should be considered in planning strategies for conservation of biodiversity in coastal habitats.

Introduction

The human population on Earth is rapidly expanding and about 70% of people live on or within 60 km from the coast (Hammond, 1992). This percentage is increasing (Hammond, 1992; Gray, 1997) and it is predicted that the population living on the coast will double in the next 30 years (Reilly et al., 1996; Gray, 1997). As a consequence, human-made structures, such as pier pilings, breakwaters, jetties, seawalls and floating pontoons, are becoming common features of the landscape in coastal and estuarine regions of developed countries (Glasby & Connell, 1999). Despite this tendency, few studies have focused on changes in the distribution and abundance of organisms caused by the introduction of human-made structures in shallow coastal waters, either on hard (Connell & Glasby, 1998; Glasby & Connell, 1999; Chapman & Bulleri, 2003) or soft bottoms (Davis et al., 1982; Page et al., 1999).

In the case in which human-made structures are introduced in areas with little or no hard substrata, their effects are not limited to the provision of novel habitats, but they can also alter natural patterns of distribution of marine organisms (Glasby & Connell, 1999). Human-made substrata may, in fact, function as stepping stones, enabling larvae and propagules of many marine invertebrates and algae that disperse over short distances (e.g., Jackson, 1986; Santelices, 1990) to colonise areas outside their natural ranges of distribution. A primary concern is that exotic species could take advantage of introduced human-made substrata, thus affecting species composition and biodiversity at local and regional scales (Glasby & Connell, 1999).

Along the north-east coast of Italy (Adriatic Sea), natural rocky reefs are rare and over 60% of the coastline is characterised by the presence of hard structures for protection against erosion of shores. Among these structures, breakwaters, which run parallel to the coast, provide shelter (i.e., landward side) and exposed (i.e., seaward side) hard-bottom habitats suitable for colonisation of a diverse assemblage of macroalgae and invertebrates (Bacchiocchi & Airoldi, 2003), including the invasive green alga *Codium fragile* (Sur.) Hariot ssp. *tomentosoides* (van Goor) Silva. This species, which has become a locally major component of assemblages on temperate rocky shores throughout the world (Trowbridge, 1998), was first sighted in the Mediterranean Sea in 1950, and attained its peak of abundance in the early 1960s, stabilising at lower levels afterwards (Boudouresque, 1994). The most likely vectors of introduction of *C. fragile* ssp. *tomentosoides* in the Mediterranean Sea have been identified as ship hulls and activities involving translocation of oysters (Ribera, 1994). Furthermore, the presence of sheltered habitats in close proximity to vector vessels and sources of nutrient enrichment, such as those provided along the north-east coast of Italy by harbours, coastal lagoons and breakwaters (e.g., Correggiari et al., 1992) has been suggested to increase the establishment and/or persistence of *C. fragile* ssp. *tomentosoides* (Trowbridge, 1998, 1999).

The aim of this study was to quantify patterns of distribution and abundance of *C. fragile* ssp. *tomentosoides* in low-shore habitats of breakwaters in the north Adriatic Sea and to evaluate whether the provision of sheltered habitats by breakwaters could enhance the colonisation and persistence of this invasive alga. Specifically, we tested the hypotheses that the abundance, biomass, length and degree of branching of thalli of *C. fragile* ssp. *tomentosoides* would be greater on the landward (sheltered) than on the seaward (exposed) side of breakwaters and that these patterns would be consistent among breakwaters 100s of metres apart and between times of sampling.

Materials and methods

Study site

This study was done in low-shore habitats (0–20 cm above the mean level of low water; MLLW) on the breakwaters at Cesenatico (Emilia Romagna, Italy, Fig. 1) in summer 2003. This area is subject to moderate wave-action and is characterised by a relatively large tidal amplitude (about 80 cm), in comparison to other regions of the Mediterranean Sea. Natural rocky reefs are absent in this area, the closest occurring at Gabicce (about 50 km south) and Trieste (about 300 km north). The region is characterised by the almost uninterrupted presence of groynes and offshore breakwaters, which were built to prevent erosion of sandy shores. At Cesenatico, breakwaters are made of large calcareous boulders, about 0.5–2 m in diameter. They were about 100 m long, 200 m offshore and spaced about 50 m apart. A general description of the study area and assemblages on coastal defence structures can be found in Bacchiocchi & Airoldi (2003) and references therein.

The invasive subspecies *Codium fragile* ssp. *tomentosoides* can be distinguished from the native Mediterranean species *Codium vermilara* (Olivi) Delle Chiaje by the shape of utricules (Silva, 1955). The observation of several 10 s of specimens under a microscope indicated the presence of monospecific stands of *C. fragile* ssp. *tomentosoides* in low-shore habitats of the breakwaters. During the present study (June–July 2003) reproductive structures (gametangia) were not found, while subsequent observations indicated that in the study area *C. fragile* ssp. *tomentosoides* was fertile in mid-August until the end of September. In the study area, *C. fragile* ssp. *tomentosoides* is present

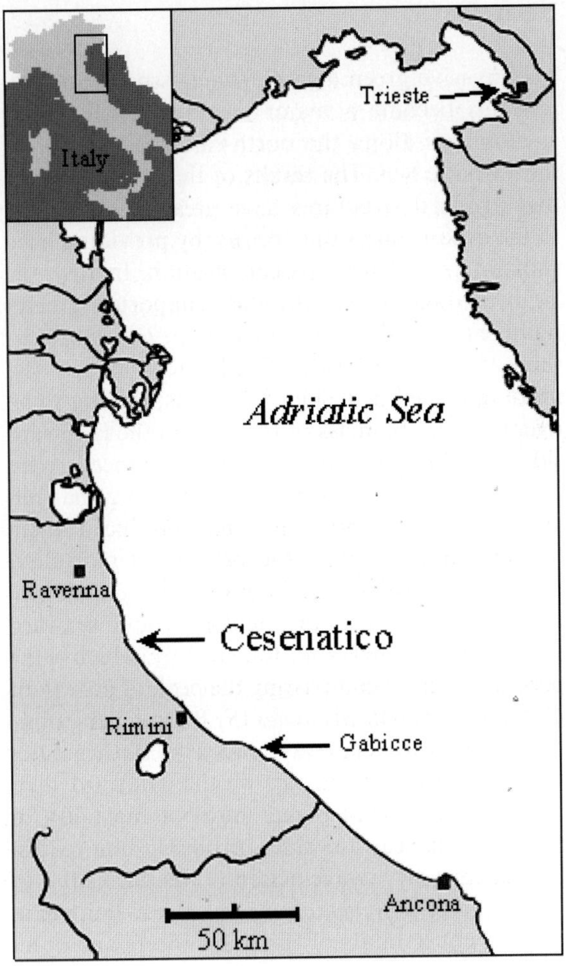

Figure 1. Map showing the study area (Cesenatico) and the closest natural rocky reefs up north (Trieste) and down south (Gabicce, Rimini).

as visible macroscopic fronds for a few months during the spring and summer (Airoldi, pers. obs.), while basal holdfasts are likely to be perennial, as reported by other authors (reviewed in Trowbridge, 1996, 1999).

Sampling design and analysis

Three breakwaters were randomly selected 100s of metres apart in spring 2003 and were sampled on the seaward and on the landward side. On each breakwater, the density (number of thalli per surface unit) of *C. fragile* ssp. *tomentosoides* was sampled in five 20 × 20 cm plots on each of three randomly chosen boulders. The size of the plots was chosen according to the results of previous pilot studies (Airoldi et al., 2000), which indicated that plots of 20 × 20 cm were the most efficient to sample the small and irregular surfaces of the blocks used to build the breakwaters. Ten thalli of *C. fragile* ssp. *tomentosoides* were also collected at random from each boulder and the biomass (wet weight), length and degree of branching (maximum number of branch dichotomies from stipe to branch tips) of each frond arising from each holdfasts were measured in the laboratory within a few hours. When more than one frond was generated by a single holdfast the average length was used in the analysis. Incipient fronds (1–5 mm) were not taken into account. Sampling was repeated in June and July 2003. Different boulders were sampled on each date, in order to obtain data independent through time.

Data were analysed by 4-way ANOVAs, including the factors Time (random), Exposure (fixed and orthogonal), Breakwater (random and orthogonal) and Boulder (random and nested within the interaction Time Exposure × Breakwater). When appropriate, pooling procedures were used following Winer et al. (1991). Cochran's test was used to test for heterogeneity of variances (Winer et al., 1991) and data were transformed when necessary. SNK tests were used for a posteriori comparison of the means (Winer et al., 1991).

Results

The density of thalli of *C. fragile* ssp. *tomentosoides* was generally greater on the landward than on the seaward side of breakwaters, but differences between exposures were significant only in July, when number of thalli greatly decreased on the seaward sides (Fig. 2a; Table 1). The density of thalli did not vary among similarly exposed sides on different breakwaters, whilst it differed among boulders within sides (Table 1).

The weight and length of thalli of *C. fragile* ssp. *tomentosoides* were greater on the landward than on the seaward side of breakwaters (Fig. 2b and c; Table 1). Differences between exposures increased from June to July, due to the growth in size of the alga on the landward sides (Fig. 2; Table 1). The weight varied among breakwaters in July, but not

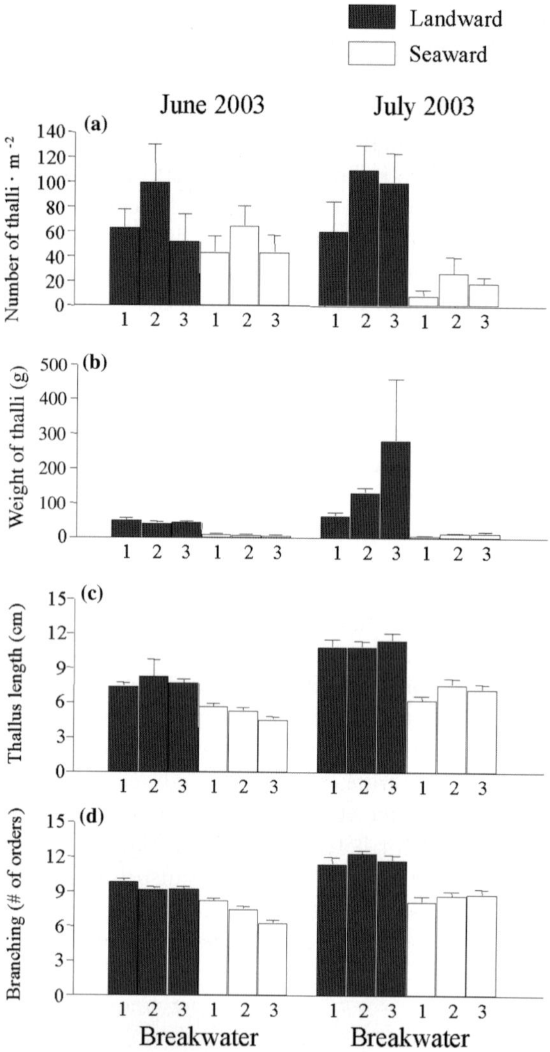

Figure 2. Mean (+SE) of density (a; n = 5), weight (b; n = 10), length (c; n = 10) and degree of branching (d; n = 10) of thalli of *Codium fragile* ssp. *tomentosoides* on the landward and the seaward side of breakwaters, in June and July 2003.

in June, resulting in a significant Time × Breakwater interaction (Table 1). The length of thalli varied among boulders (Table 1).

The order of branching of thalli of *C. fragile* ssp. *tomentosoides* was significantly greater on the landward than on the seaward side of breakwaters (Fig. 2d; Table 1). There was large variability among breakwaters in the degree of thallus branching, which changed from June to July.

Discussion

The invasive green alga *C. fragile* ssp. *tomentosoides* has become a major component of coastal assemblages along the north-east sandy coast of the Adriatic Sea. The results of this study indicate that artificial structures have greatly contributed to the expansion of this species, by providing hard substrata suitable for its colonisation. In July, the landward side of breakwaters supported greater numbers of thalli of *C. fragile* ssp. *tomentosoides* than the seaward side. Thalli grew longer and more branched in sheltered habitats, leading to an overall larger biomass of the alga on the landward side of breakwaters. These differences were consistent across breakwaters, indicating that relevant ecological processes that produced them (e.g., recruitment and post-recruitment mortality) operated consistently at the scale of 100s of metres. Differences between the landward and seaward sides also increased from June to July, which is the period of time encompassing the peak of growth of *C. fragile* ssp. *tomentosoides* (F. Bulleri, pers. obs.) This temporal trend was related to both greater rates of growth of the alga on the landward sides of breakwaters and higher rates of mortality of thalli on the seaward sides, probably due to dislodgement by wave-action. Recruitment of *C. fragile* ssp. *tomentosoides* was not limited in wave-swept habitats of the breakwaters and specialised grazers (sacoglossan slugs) were not observed on the structures. It has been, however, reported that larger thalli of *C. fragile* ssp. *tomentosoides* can be removed by waves more easily (Trowbridge, 1998, 1999). It is likely that thalli that had grown to a critical size (length) could not persist in the stressful hydrodynamic conditions on the seaward sides of breakwaters during summer swells. Overall, these results are in accordance with those of other studies (Trowbridge, 1995, 1996), which have described *C. fragile* ssp. *tomentosoides* as a species best performing in protected or moderately exposed habitats.

Besides the occurrence of suitable sheltered habitats on the breakwaters, other factors could have promoted the establishment and persistence of *C. fragile* ssp. *tomentosoides* in the study area. The successful establishment of exotic species in

Table 1. Analyses and SNK tests on the effects of Time, Exposure, Breakwater and Boulder on (A) density, (B) biomass, (C) length and (D) degree of branching of thalli of *C. fragile* ssp. *tomentosoides*. When appropriate, pooling procedures were used following Winer et al. (1991); because Time and Breakwater are random factors, some tests were only possible after elimination of terms (not significant at $P = 0.25$; Winer et al. 1991; Underwood 1997). [§]Degrees of freedom of the Residual for Biomass, Length and Branching * $P < 0.05$; ** $P < 0.01$; *** $P < 0.001$.

Source of variation		A) Density		B) Biomass		C) Length		D) Branching	
	df	MS	*F*	MS	*F*	MS	*F*	MS	*F*
Time = T	1	1.80	2.56	10.33	0.74	729.36	39.03*	2.12	4.68
Exposure = E	1	1.06	no test	333.93	38.64	775.16	no test	6.82[c]	125.19**
Breakwater = B	2	0.01	0.01	2.02	0.15	1.12	0.06	0.10	0.23
T × E	1	6.59	21.22*	8.64[a]	8.56*	86.61[b]	7.86**	0.11	eliminated
T × B	2	0.70	0.66	13.87	11.14***	18.69	1.70	0.45	6.68**
E × B	2	0.69	2.21	0.13	pooled	15.93	1.45	0.05	0.44
T × E × B	2	0.31	0.29	1.89	pooled	9.71	eliminated	0.12	1.82
Boulder(T × E × B)	24	1.07	2.30**	1.24	1.48	11.01	2.05**	0.07	1.31
Residual	144 324[§]	0.47		0.84		5.37		0.05	
Transformation		Ln (x + 1)		Ln (x + 1)		None		Ln (x + 1)	
Cochran's test		NS		NS		$P < 0.01$		NS	
SNK tests		A) Density		B) Biomass		C) Length			
		SE = 0.08		SE = 0.11		SE = 0.35			
		June: Landward = Seaward		June: Landward > Seaward		June: Landward > Seaward			
		July: Landward > Seaward		July: Landward > Seaward		July: Landward > Seaward			

[a] Tested against the pooled term B × E + T × E × B(MS = 1.01, df = 4); [b] Tested against the term Boulder (T × E × B); [c] Tested against the term E × B.

new areas has been shown to be related to their life-history attributes (Trowbridge, 1995, 1998). *C. fragile* ssp. *tomentosoides* exhibits rapid growth, and regeneration from filaments (vaucherioid stage), while fragments confer high dispersal ability on this alga. Furthermore, parthenogenetic gametes enable the persistence of populations even at low densities. *C. fragile* ssp. *tomentosoides* exhibits tolerances to salinities ranging from 12.5 to 40 ppt (reviewed in Trowbridge, 1998), inhabiting both estuarine and marine habitats. There is also evidence that *C. fragile* ssp. *tomentosoides* can be enhanced by eutrophication of waters, this alga being able to exploit different sources of nitrogen, including nitrate, nitrite, ammonia and urea (Hanisak, 1979).

The ability of *Codium fragile* ssp. *tomentosoides* to establish in new regions could also depend upon attributes of native assemblages of organisms and interactions with local environmental conditions (reviewed by Trowbridge, 1998, 1999). The establishment of an invader species, such as *C. fragile* ssp. *tomentosoides*, may be favoured by vacant niches and low diversity of species (Ribera & Boudoresque, 1995; Trowbridge, 1995; Tilman, 1997; Levin et al., 2002). Indeed, along the north-east coast of Italy, Bacchiocchi & Airoldi (2003) have shown that the diversity of species on human-made structures is generally low and they have reported a relatively large amount of unoccupied space. The causes underlying this low diversity and abundance of species are not well understood. The study area, being under the influence of the Po river plume, is subject to abrupt changes in salinity and turbidity and to high nutrient loads (Correggiari et al., 1992; Zavatarelli et al., 1998). These relatively harsh physical conditions could prevent or slow down colonisation of space on breakwaters by many taxa of macroalgae and invertebrates, thus lessening the intensity of intra-specific competition and enabling the tolerant *C. fragile* ssp. *tomentosoides* to settle and persist at low-shore levels.

Furthermore, breakwaters are subject to frequent disturbances, either natural (e.g., storms) or anthropogenic (e.g., maintenance works, harvesting of invertebrates for food). These disturbances could indirectly enhance the recruitment of *C. fragile* ssp. *tomentosoides* by removing the mussel *Mytilus galloprovincialis* Lamarck, which is the main space-occupier at low-intertidal and shallow subtidal levels on breakwaters. Although assemblages on artificial structures cannot be referred to as 'natural', experimental work is needed to identify the interactions between 'native' and exotic species on breakwaters and the environmental factors that modulate them.

Overall, these results suggest that by providing hard and sheltered substrata within sandy exposed habitats, artificial structures have the potential to alter biodiversity not only by attracting benthic species from nearby natural reefs and coastal lagoons (Bacchiocchi & Airoldi, 2003), but also by enabling the incursion of exotic species. There are numerous examples of introduced species, including algae and invertebrates, that have established in enclosed waters characterised by intense shipping activities and sources of eutrophication (Occhipinti-Ambrogi & Savini, 2003 and references therein). Indeed, the impact of human-made structures on natural assemblages would be largely underestimated without evaluating the risks of introducing exotic species. Similarly, no reliable predictions of local to regional pathways of spreading of invasive species can be formulated without embodying the influence of anthropogenic alterations to coastal landscapes. Even though human-made structures would represent sub-optimal habitats for exotic species, for their increasing presence, they could act as stepping stones (Glasby & Connell, 1999), promoting the expansion of invaders to areas that would be naturally isolated. Thus, the effects of human-made structures and introduced species on assemblages of organisms cannot be evaluated separately, but their synergistic nature should be considered in order to plan sound strategies for the conservation of biodiversity in coastal habitats.

Acknowledgements

We sincerely thank M. Carrera, S. Santin and G. Branca for help in the field and in the lab and C. Trowbridge and one anonymous referee for valuable comments on an earlier draft. This work was supported by funds from the EU project DELOS (EVK3-CT-2000–00041) and from the project COFIN (ex 40%) to M.A. L.A. was supported by an 'Assegno di Ricerca' from the Università di Bologna.

References

Airoldi, L., E. Camorani, F. Bacchiocchi & V. U. Ceccherelli, 2000. Ottimizzazione del campionamento dell'epibentos su strutture frangiflutto: confronto dell'efficienza di unità di diversa taglia. *Riassunti del 10° Congresso Nazionale S.It.E.*, 120.

Bacchiocchi, F. & L. Airoldi, 2003. Distribution and dynamics of epibiota on hard structures for coastal protection. Estuarine Coastal and Shelf Science 56: 1157–1166.

Boudoresque, C. F., 1994. Les espèces introduites dans les eaux côtieres d'Europe et de Méditerranée: etat de la question et consèquences. In Boudouresque, C. F., F. Briand, & C. Nolan (eds.) Introduced species in European coastal waters. European Commision, Luxembourg: 8–27.

Chapman, M. G. & F. Bulleri, 2003. Intertidal seawalls – new features of landscape in intertidal environments. Landscape and Urban Plannning 62: 159–172.

Connell, S. D. & T. M. Glasby, 1998. Do urban structures influence local abundances and diversity of subtidal epibiota? A case study from Sydney Harbour, Australia. Marine Environmental Research 47: 1–15.

Correggiari, A., F. Frascari, S. Miserocchi & D. Fontana, 1992. Breakwaters and eutrophication along the Emilia-Romagna coast. In Vollenweider, R. A., R. Marchetti, & R. Viviani (eds.) The response of marine transitional systems to human impact: problems and perspectives for restoration. Elsevier, Amsterdam: 277–290.

Davis, N., G. R. Vanblaricom & P. K. Dayton, 1982. Man-made structures on marine sediments: effects on adjacent benthic communities. Marine Biology 70: 295–303.

Glasby, T. M. & S. D. Connell, 1999. Urban structures as marine habitats. Ambio 28: 595–598.

Gray, J. S., 1997. Marine biodiversity: patterns, threats and conservation needs. Biodiversity and Conservation 6: 153–175.

Hammond, A., 1992. World Resources 1992–1993: Towards Sustainable Development. Oxford University Press, Oxford.

Hanisak, M. D., 1979. Growth patterns of *Codium fragile* ssp. *tomentosoides* in response to temperature, irradiance, salinity and nitrogen source. Marine Biology 50: 319–332.

Jackson, J. B. C., 1986. Modes of dispersal of clonal benthic invertebrates: consequences for species' distribution and genetic structure of local populations. Bulletin of Marine Science 39: 588–606.

Levin, P. S., J. A. Coyer, R. Petrik & T. P. Good, 2002. Community-wide effects of nonindigenous species on temperate rocky reefs. Ecology 83: 3182–3193.

Occhipinti-Ambrogi, A. & D. Savini, 2003. Biological invasions as a component of global change in stressed marine ecosystems. Marine Pollution Bulletin 46: 542–551.

Page, H. M., J. E. Dugan, D. S. Dugan, J. B. Richards & D. M. Hubbard, 1999. Effects of an offshore oil platform on the distribution and abundance of commercially important crab species. Marine Ecology Progress Series 185: 47–57.

Reilly, F. J., R. J. Spagnolo & E. Ambrogio, 1996. Marine and estuarine shallow water science and management. Estuaries 19: 165–168.

Ribera, M. A., 1994. Les macrophytes marins introduits en Méditerranée: biogéographie. In Boudouresque, C. F., F. Briand, & C. Nolan (eds.) Introduced species in Eroupean coastal waters. European Commision, Luxembourg: 37–43.

Ribera, M. A. & C. F. Boudoresque, 1995. Introduced marine plants, with spacial reference to macroalgae: mechanisms and impact. Progress in Phycological Research 11: 187–268.

Santelices, B., 1990. Patterns of reproduction, dispersal and recruitment in seaweeds. Oceanography and Marine Biology: An Annual Review 28: 177–276.

Silva, P. C., 1955. The dichotomous species of *Codium* in Britain. Journal of the Marine Biological Association of the United Kingdom 34: 565–567.

Tilman, D., 1997. Community invasibility, recruitment limitation, and grassland biodiversity. Ecology 98: 81–92.

Trowbridge, C. D., 1995. Establishment of the green alga *Codium fragile* ssp. *tomentosoides* on New Zealand rocky shores: current distribution and invertebrates herbivores. Journal of Ecology 83: 949–965.

Trowbridge, C. D., 1996. Introduced versus native subspecies of *Codium fragile*: how distinctive is the invasive subspecies *tomentosoides*?. Marine Biology 126: 193–204.

Trowbridge, C. D., 1998. Ecology of the green macroalga *Codium fragile* (Suringar) Hariot 1989: invasive and non-invasive subspecies. Oceanography and Marine Biology: An Annual Review 36: 1–64.

Trowbridge, C. D., 1999. An assessment of the potential spread and options for control of the introduced green macroalga *Codium fragile* ssp. *tomentosoides* on Australian shores. Consultancy Report. Centre for Research on Introduced Marine Pests 43.

Underwood, A. J., 1997. Experiments in Ecology: their Logical Design and Interpretation using Analysis of Variance. Cambridge University Press, Cambridge.

Winer, B. J., D. R. Brown & K. M. Michels, 1991. Statistical Principles in Experimental Design. McGraw-Hill, New York.

Zavatarelli, M., F. Raicich, D. Bregant, A. Russo & A. Artegiani, 1998. Climatological biogeochemical characteristics of the Adriatic Sea. Journal of Marine Systems 18: 227–263.

Hydrobiologia (2006) 555:271–279
H. Queiroga, M.R. Cunha, A. Cunha, M.H. Moreira, V. Quintino, A.M. Rodrigues, J. Serôdio & R.M. Warwick (eds),
Marine Biodiversity: Patterns and Processes, Assessment, Threats, Management and Conservation
DOI 10.1007/s10750-005-1123-3

Immediate effects of experimental human trampling on mid-upper intertidal benthic invertebrates at the Asinara Island MPA (NW Mediterranean)

Daniela Casu[1,3,*], Giulia Ceccherelli[2] & Alberto Castelli[3]
[1]*Dipartimento di Zoologia ed Antropologia Biologica, University of Sassari, Italy*
[2]*Dipartimento di Botanica ed Ecologia Vegetale, University of Sassari, Italy*
[3]*Dipartimento di Scienze dell'Uomo e dell'Ambiente, University of Pisa, Italy*
*(*Author for correspondence: E-mail: danicasu@uniss.it)*

Key words: Asinara Island, benthic invertebrates, MPA, trampling

Abstract

To assess potential risks of human visitation to ecological communities, the immediate effects of human trampling were investigated experimentally on small invertebrates inhabiting mid-upper intertidal hard bottoms covered by algae. Two different experimental intensities of trampling (60 and 120 footsteps) and controls (with no trampling) were applied to quadrats 20×20 cm in size (experimental area), within the two 'no-entry, no-take' zones of the Asinara Island MPA (Italy, Mediterranean Sea). One day after trampling ended, samples of benthic fauna were collected and the animals attributed to macrofaunal and meiofaunal components. Analyses of variance on the nine most common taxa of macrofauna identified significant higher abundance of bivalves, gammarid amphipods, polychaetes, isopods, oligochaetes in controls than in trampled plots. For nematodes, polychaetes, ostracods, oligochaetes, bivalves, acari, caprellid amphipods and tanaids a significant higher abundance of meiofaunal animals was found in controls than in trampled areas. Although no information on recovery is available, these results suggest that macrofaunal and meiofaunal taxa are vulnerable to this type of disturbance.

Introduction

During recent years interest in human impact on rocky intertidal communities has increased. Several studies have emphasized the direct effects on the fauna and flora such as harvesting (Duran & Castilla, 1989; Underwood & Kennelly, 1990; Murray et al., 1999) and more recently trampling (Liddle, 1975; Beauchamp & Gowing, 1982; Brosnan & Crumrine, 1994; Schiel & Taylor, 1999; Brown & Taylor, 1999; Milazzo et al., 2002). Authors have demonstrated that trampling effects on rocky shores generally cause a considerable reduction in density and cover of both sessile invertebrates and algae (Povey & Keough, 1991; Adessi, 1994; Brosnan & Crumrine, 1994; Keough & Quinn, 1998; Schiel & Taylor, 1999; Milazzo et al., 2002).

The establishment of marine parks and reserves should prevent widespread degradation and habitat loss, conserving biodiversity. However, protected areas may suffer major degradation from the effects of trampling (Brown & Taylor, 1999; Gallet & Roze, 2001; Milazzo et al., 2002). Indeed, these areas often become popular recreational destinations, leading to increased recreational pressure. Indeed, some recreational activities are not benign to the environment: collecting, although forbidden, constant foot traffic and exploratory manipulation of rocks may affect the intertidal habitat (Sousa, 1979; Murray et al., 1999).

Effective management of parks achieves goals of both public accessibility and environmental preservation. To cope with this challenge, understanding of anthropogenic impacts as well as

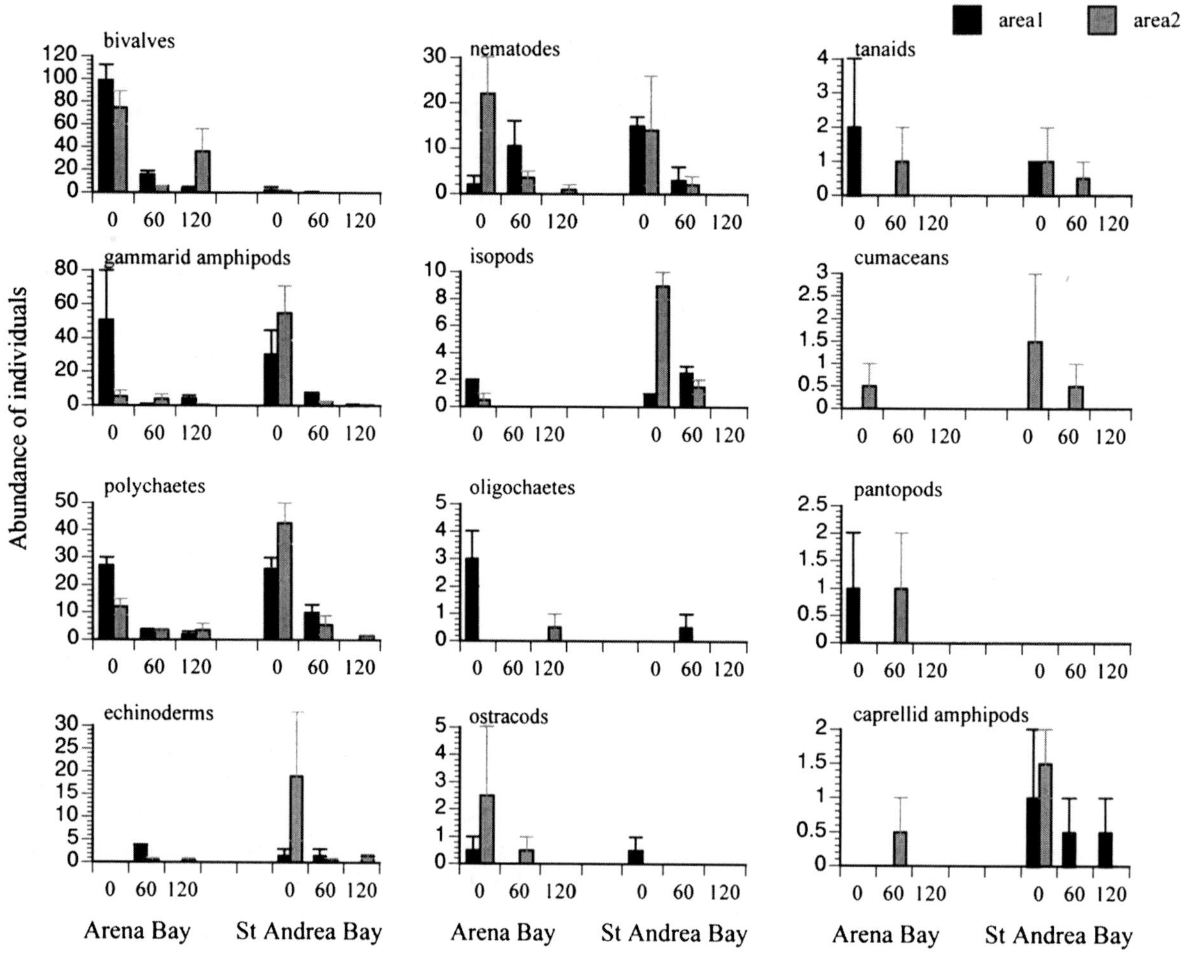

Figure 1. Mean abundance (±SE) of commonest macrofaunal taxa sampled at different trampling intensities (0, 60 and 120 steps/quadrat) and areas ($n=2$).

detailed knowledge of the systems themselves is required through experimentation.

In the Mediterranean, the effects of human trampling disturbance on benthic invertebrates of rocky shallow water have not been previously examined. Milazzo et al. (2002) have collected the only experimental data available on the effect of trampling, at Ustica Island (MPA) on macroalgae. This study has the aim of investigating the immediate effects of experimental human trampling on small invertebrates inhabiting mid-upper intertidal hard bottoms covered by algae at two 'no-entry' sites at the Asinara Island MPA. The proposed experimental design is expected to highlight the effects of trampling in the study area, advising management of other marine reserves. The establishment and the subsequent zoning of new MPAs, in particular, could have a significant benefit from a predictive experimental approach that estimates the effects of a particular habitat use and thus preventing impacts.

Materials and methods

The Asinara Island Marine Protected Area (North-western Mediterranean, 41°03′58″N 8°15′57″ E) was a prison island from 1885 to 1997, so the island has had a quieter history than the other overexploited coastlines of Sardinia, because public access and construction have been forbidden for nearly a century. This has allowed the maintenance of peculiar Mediterranean flora and fauna and has prevented serious damage to the coastal marine assemblages (Villa et al., 2002).

Table 1. Results of ANOVAs on the effects of Location, Trampling (0, 60 and 120 steps/quadrat) and Area on macrofaunal taxa

	df	Bivalves			Gammarid amphipods			Polychaetes			Echinoderms			Nematodes		
		MS	F	p	MS	F	p	MS	F	p	MS	F	p	MS	F	p
Location = L	1	8855.042	500.84	**0.0001**	2.381	0.60	0.5193	192.667	3.17	0.2169	4.660	4.420	0.1703	0.234	0.25	0.6691
Trampling = T	2	3578.792	1.11	0.4743	38.141	8.36*	**0.0370**	1463.792	10.22	0.0892	0.950	1.420	0.3422	11.124	4.63	0.0910
Area(L) = A	2	1.042	0.01	0.9923	3.962	2.69	0.1084	60.750	3.05	0.0849	1.054	3.380	0.0687	0.951	0.92	0.4241
L × T	2	3228.292	7.60	**0.0434**	2.741	0.60	0.5910	143.292	1.43	0.3398	0.950	1.42	0.3422	1.166	0.49	0.6475
T × A(L)	4	424.792	3.14	0.0553	4.565	3.10	0.0574	100.125	5.03	**0.0130**	0.670	2.14	0.1378	2.402	2.33	0.1152
Residual	12	135.208			1.473			19.917			0.312			1.031		
Cochran's test		C = 0.4931 ns			Sqrt(X + 1) ns 0.5252			C = 0.4707 ns			Ln(X + 1) ns 0.4013			Sqrt(X + 1) ns 0.4851		

	df	Isopods			Oligochaetes			Tanaids			Caprellid amphipods		
		MS	F	p	MS	F	p	MS	F	p	MS	F	p
Location = L	1	1.573	3.96	0.1847	0.161	1.50	0.3456	0.009	0.52	0.5460	0.219	15.32	0.0595
Trampling = T	2	1.558	2.39	0.2077	0.101	0.48	0.6504	0.527	2.00	0.2504	0.071	1.99	0.2510
Area (L) = A	2	0.397	21.42	**0.0000**	0.108	4.33	**0.0380**	0.018	0.08	0.9258	0.014	0.30	0.7477
L × T	2	0.398	0.61	0.5871	0.183	0.87	0.4849	0.054	0.02	0.8241	0.122	3.41	0.1369
T × A(L)	4	0.652	35.21	**0.0000**	0.210	8.43	**0.0010**	0.264	1.15	0.3785	0.036	0.74	0.5797
Residual	12	0.019			0.025			0.229			0.048		
Cochran's test		Sqrt(X + 1) ns 0.3859			Sqrt(X + 1) ns 0.4254			Ln(X + 1) ns 0.4723			Sqrt(X + 1) ns 0.4653		

*vs. T × A(L). Bold numbers indicate significance at $p < 0.05$.

Table 2. Results of SNK tests on trampling intensities for macrofaunal taxa

Bivalves		Isopods	
SE = 10.305		SE = 0.096	Arena Bay
L × T	Arena Bay	T × A(L)	0 > 60 = 120
	0 > 60 = 120		0 = 60 = 120
	St.Andrea Bay		St.Andrea Bay
	0 = 60 = 120		60 > 0 > 120
			0 > 60 > 120
Gammarid amphipods		Oligochaetes	
SE = 0.755		SE = 0.111	Arena Bay
T	0 > 60 = 120	T × A(L)	0 > 60 = 120
			0 = 60 = 120
			St. Andrea Bay
Polychaetes			0 = 60 = 120
SE = 3.155	Arena Bay		0 = 60 = 120
T × A(L)	0 > 60 = 120		
	0 = 60 = 120		
	St. Andrea Bay		
	0 > 60 > 120		
	0 > 60 = 120		

For this reason, in the last few years, this MPA has become a tourist attraction and human presence, where allowed, may conflict with conservation goals (Eckrich & Holmquist, 2000).

At the MPA two 'no-entry, no-take' locations were chosen for the experiment, Arena Bay and St. Andrea Bay, located on the east coast of the island. At each bay, between 0.2 and 0.4 m below MLLW, six experimental areas within a mixed assemblage of erect algae were randomly chosen. Two different experimental intensities of trampling (60 and 120 footsteps/quadrat) and a control (no trampling) were applied to quadrats 20 × 20 cm in size (experimental areas) and 5 m apart. Due to its size, each quadrat was entirely trampled by a trampler wearing rubber-soled shoes and 60 kg in weight.

Two replicate quadrats were randomly assigned to each experimental intensity, and two replicate samples were collected within each quadrat.

The day after trampling, samples of benthic fauna were collected by cutting through the algae to the rock surface using open-ended plastic cylinders of 40 mm in diameter corresponding to 12.56 cm^2 in surface area scraped (Brown & Taylor, 1999). All the algae and benthic invertebrates were removed down to the basal crust by using a metal scraper. In the field, samples were preserved in 4% formalin in seawater and, in the laboratory the animals were sieved at 500 μm and all passed material was filtered again on a 100 μm mesh. Organisms retained by the two meshes were attributed to macrofaunal and meiofaunal components, identified to coarse taxonomic levels and counted under a stereo microscope.

Three-way ANOVAs were performed to examine the difference in abundance of the most common taxa depending on the effects of trampling. 'Location' (Arena Bay and St. Andrea Bay) was treated as random and 'Trampling' was treated as fixed factor (0 which served as control, 60 and 120 footsteps/quadrat), they were both orthogonal, while 'area' was random and nested in location (2 levels). Cochran's test was used to check for the homogeneity of variances (Winer, 1971). Whenever necessary, data were transformed. Student-Newman-Keuls (SNK) test was used for *a posteriori* comparison of the means when appropriate (Underwood, 1997). This analysis was carried out using the GMAV 5.0 software (University of Sydney).

Results

Bivalves, gammarid amphipods, polychaetes, echinoderms, nematodes, isopods, oligochaetes, ostracods, tanaids, cumaceans, pantopods and caprellid amphipods mostly comprised the macrofauna (Fig. 1). Overall 1317 organisms were collected in the total of 24 samples. Bivalves (36.3%), gammarid amphipods (24.7%) and polychaetes (20.8%) were the dominant taxa.

Analyses of variance on the nine most common taxa of macrofauna found a significantly higher abundance of bivalves, gammarid amphipods, polychaetes, isopods, oligochaetes in controls than in trampled plots (Table 1). However, between the two trampling levels (60 and 120) an alternative hypothesis has not been identified for most comparisons (SNK Test, Table 2). For some taxa a significant heterogeneity was found between locations (bivalves) and areas (polychaetes, isopods and oligochaetes). The abundance data of ostracods, cumaceans and pantopods among the macrofauna could not be analysed by ANOVA because of variance heterogeneity, even after data

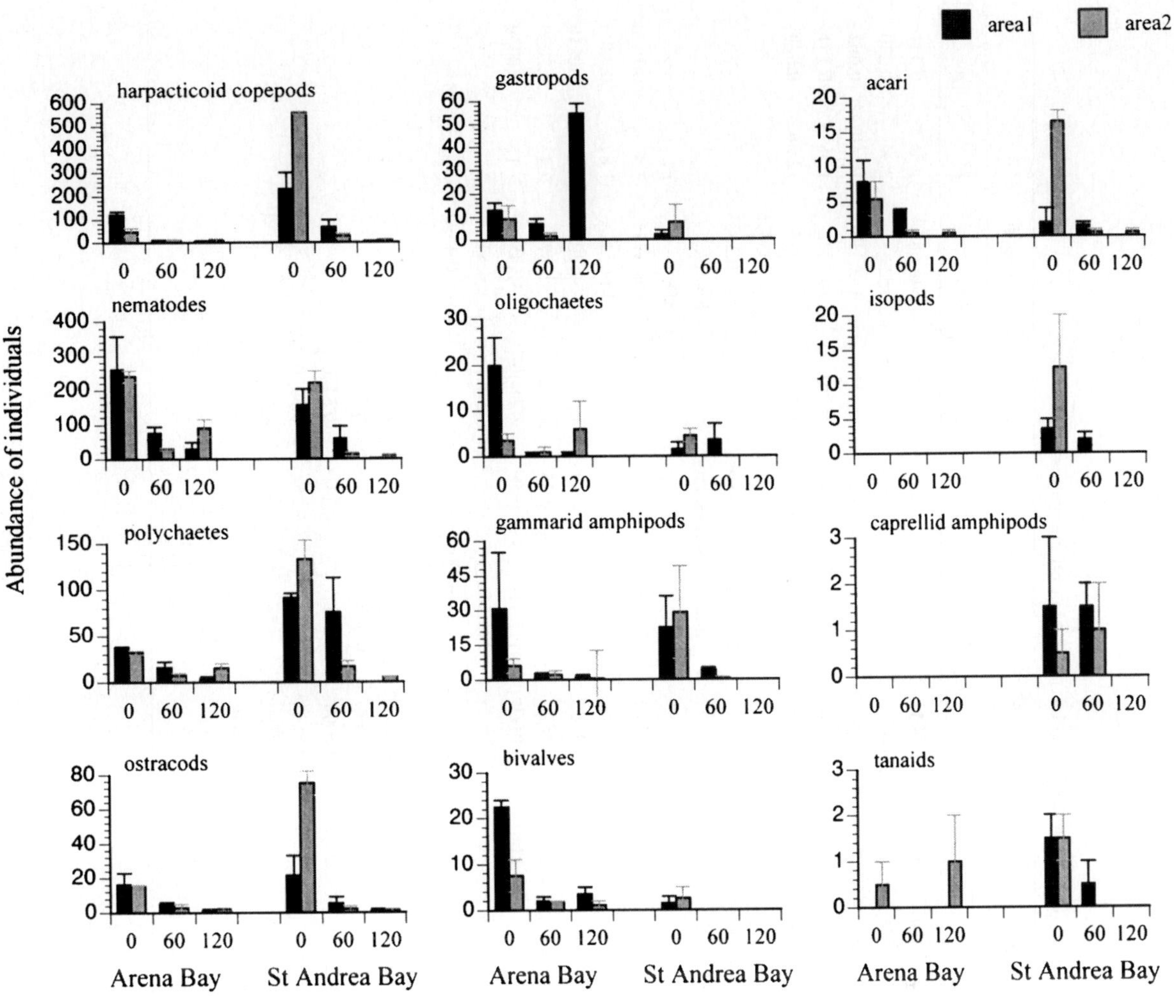

Figure 2. Mean abundance (±SE) of commonest meiofaunal taxa sampled at different trampling intensities (0, 60 and 120 steps/quadrat) and areas ($n=2$).

transformation. Results for echinoderms, nematodes, tanaids and caprellid amphipods showed no effects of trampling (Table 1).

The most common meiofaunal taxa found in samples were: harpacticoid copepods, nematodes, polychaetes, ostracods, gastropods, oligochaetes, gammarid amphipods, bivalves, acari, isopods, caprellid amphipods and tanaids (Fig. 2). A total of 6417 individuals were found: nematodes (37.1%) harpacticoid copepods (33.9%) and polychaetes (29.2%) were the dominant taxa.

For nematodes, polychaetes, ostracods, oligochaetes, bivalves, acari, caprellid amphipods and tanaids a significantly higher abundance of meiofaunal animals was found in controls than in areas trampled (Table 3). As with the macrofauna results, an alternative hypothesis between the two trampling levels (60 and 120) was not identified for most of comparisons (SNK Test, Table 4). Significant heterogeneity between locations was found for harpacticoid copepods and nematodes, while for ostracods, gastropods, bivalves and acari a significant heterogeneity was highlighted between areas. Isopod abundance in the meiofauna could not be analysed by ANOVA because of variance heterogeneity, even after data transformation. Harpacticoid copepods were not significantly affected by trampling (Table 3).

Table 3. Results of ANOVAs on the effects of Location, Trampling (0, 60 and 120 steps/quadrat) and Area on meiofaunal taxa

	df	Harpacticoid copepods			Nematodes			Polychaetes			Ostracods		
		MS	F	*P*	MS	F	*p*	MS	F	*p*	MS	F	*p*
Location = L	1	6.079	61.25	**0.015**	48.476	577.06	**0.001**	0.752	6.21	0.130	640.667	1.54	0.341
Trampling = T	2	23.637	15.27	0.062	224.398	33.42	**0.029**	13.669	3.84	0.207	2352.875	3.34	0.230
Area (L) = A	2	0.099	0.21	0.812	0.084	0.02	0.982	0.121	0.70	0.516	417.333	10.10	**0.002**
L × T	2	1.548	2.07	0.242	6.714	0.53	0.624	3.563	1.91	0.261	704.292	1.38	0.351
T × A(L)	4	0.749	1.60	0.238	12.613	2.74	0.079	1.863	10.77	**0.000**	511.083	12.36	**0.000**
Residual	12	0.469			4.601			0.173			41.333		
Cochran's test		Ln(X+1) ns 0.2492			Sqrt(X+1) ns 0.3372			Ln(X+1) ns 0.5484			ns = 0.5333		
	df	Gastropods			Oligochaetes			Gammarid amphipods			Bivalves		
		MS	F	*P*	MS	F	*p*	MS	F	*p*	MS	F	*p*
Location = L	1	950.042	1.38	0.3608	80.666	4.38	0.1710	28.166	0.25	0.6680	198.375	3.87	0.1870
Trampling = T	2	264.500	0.81	0.5511	94.291	1.97	0.3360	1162.125	29.77	**0.0320**	147.541	1.87	0.3480
Area (L) = A	2	687.708	31.32	**0.0000**	18.416	1.19	0.3360	114.166	0.58	0.5730	51.208	11.28	**0.0010**
L × T	2	324.667	0.78	0.5187	47.791	0.67	0.5620	39.041	0.35	0.7269	78.875	2.43	0.2030
T × A(L)	4	417.833	19.03	**0.0000**	71.791	4.66	**0.0160**	112.916	0.58	0.6840	32.458	7.15	**0.0030**
Residual	12	21.958			15.416			195.666			4.541		
Cochran's test		ns = 0.4269			ns = 0.3892			ns = 0.4906			ns = 0.4495		
	df	Acari			Caprellid amphipods			Tanaids					
		MS	F	*P*	MS	F	p	MS	F	P			
Location = L	1	1.041	0.03	0.8833	1.029	12.85	0.0698	0.666	1.60	0.3330			
Trampling = T	2	136.791	18.76	0.0506	0.277	1.00	0.5000	1.291	0.84	0.5440			
Area (L) = A	2	37.708	9.95	**0.0020**	0.080	0.51	0.6134	0.416	1.25	0.3210			
L × T	2	7.291	0.19	0.8350	0.278	13.88	**0.0150**	1.541	9.25	**0.0310**			
T × A(L)	4	38.708	10.21	**0.0000**	0.020	0.13	0.9696	0.166	0.5	0.7360			
Residual	12	3.791			0.157			0.333					
Cochran's test		ns = 0.3956			Ln(X+1) ns 0.5093			ns = 0.5000					

Bold numbers indicate significance at $p < 0.05$.

Table 4. Results of SNK tests on trampling intensities for meiofaunal taxa

Nematodes		Oligochaetes		Acari	
SE = 0.91		SE = 2.77	Arena Bay	SE = 1.37	Arena Bay
T	0 > 60 = 120	T × A(L)	0 > 60 = 120	T × A(L)	0 = 60 = 120
			0 = 60 = 120		0 > 60 = 120
			St. Andrea Bay		St. Andrea Bay
			0 = 60 = 120		0 = 60 = 120
			0 = 60 = 120		0 > 60 = 120
Polychaetes		Gammarid amphipods		Caprellid amphipods	
SE = 0.29	Arena Bay	SE = 2.20		SE = 0.07	Arena Bay
T × A(L)	0 = 60 > 120	T	0 > 60 = 120	L × T	0 = 60 = 120
	0 = 60 = 120				St. Andrea Bay
	St. Andrea Bay				0 = 60 > 120
	0 = 60 > 120				
	0 > 60 > 120				
Ostracods		Bivalves		Tanaids	
SE = 4.54	Arena Bay	SE = 1.50	Arena Bay	SE = 0.20	Arena Bay
T × A(L)	0 = 60 = 120	T × A(L)	0 > 60 = 120	L × T	0 = 60 = 120
	0 = 60 = 120		0 > 60 = 120		St. Andrea Bay
	St. Andrea Bay		St. Andrea Bay		0 > 60 = 120
	0 > 60 = 120		0 = 60 = 120		
	0 > 60 = 120		0 = 60 = 120		
Gastropods					
SE = 3.31	Arena Bay				
T × A(L)	0 = 60 < 120				
	0 = 60 = 120				
	St. Andrea Bay				
	0 = 60 = 120				
	0 = 60 = 120				

Discussion

Although it is known that rocky intertidal communities are sensitive to the effects of human trampling, that reduces algal coverage and sessile fauna abundance, there is little information on effects of anthropogenic disturbance on mobile fauna associated with trampled vegetation (see Keough & Quinn, 1998; Brown & Taylor, 1999). The major aim of the present experiment was to evaluate the immediate effects of three different experimental intensities of trampling on intertidal organisms. Effects of trampling were variable among organisms for both macrofaunal and meiofaunal components. Impacts of trampling caused immediate declines in density of bivalves, gammarid amphipods, polychaetes, isopods, oligochaetes among the macrofauna, while nematodes, polychaetes, gastropods, ostracods, oligochaetes, bivalves, acari, caprellid amphipods, gammarid amphipods and tanaids are the taxa most vulnerable for meiofauna.

Trampling may reduce density of animals in several ways; nevertheless the most obvious and direct effect would be the crushing impact of footsteps. However, most of the vagile taxa are highly mobile, and should be able either to escape during crushing and recolonise trampled patches after disturbance (Sherman & Coull, 1980; Billheimer & Coull, 1988). Any short-term reduction in densities could then be rapidly compensated by immigration, at least at the spatial scale of our experiment. Moreover, if the direct impact of trampling determined subsequent community composition, it would be expected that the vulnerability of individual taxa would be related to their morphology and ecology, although inconsistent data have been collected during previous investigations (Brosnan & Crumrine, 1994; Brown & Taylor, 1999). For example, in this study, hard-bodied organisms such as echinoderms, nematodes, tanaids and caprellid amphipods for the macrofauna and harpacticoid copepods for the meiofauna did not respond to trampling. Conversely, for both components there have been hard-bodied taxa significantly reduced by experimental trampling. However, the taxonomic resolution used in this paper would not be sufficient to detect all differences among experimental levels. Indeed, since species are more affected than higher taxa by both natural variability and human impact (Warwick, 1988; Terlizzi et al., 2003), the risk is that identification of taxa to higher taxonomic levels than species could lead to a significant loss of information.

Frequent trampling also breaks and crushes fronds of algae reducing cover (Milazzo et al., 2002), hence consistently reducing habitat for understory fauna. Vulnerability of the organisms that remain attached to the substrate may increase due to desiccation, solar radiation, wave action and predation (Brosnan & Crumrine, 1994; Schiel & Taylor, 1999). In such a case, it is more likely that the effects of trampling are indirect, through changes caused to the habitat itself. Algal susceptibility to be damaged by trampling is likely to depend on their morphology (Milazzo et al., 2002); indeed, quantitative habitat loss may be related to qualitative algal assemblages. This feature can also be the reason for spatial differences found at both scales (locations and areas).

In conclusion, the present experimental study indicated that small invertebrates are vulnerable to this type of disturbance although these data do not allow us to identify critical levels of trampling (i.e., threshold number of steps) sustainable within this shallow zoobenthic community. Although immediate effects of trampling are reported here, these results suggest that the correct management of some Mediterranean MPAs should take into consideration vulnerability and resilience of this community.

Acknowledgements

We would like to express our sincere gratitude to David Pala for valuable assistance in the field and the whole staff of the Asinara Island National Park for support and assistance. This study is part of a Ph.D. thesis by DC.

References

Addessi, L., 1994. Human disturbance and long term changes on a rocky intertidal community. Ecological Application 4: 786–797.

Beauchamp, K. A. & M. M. Gowing, 1982. A quantitative assessment of human trampling effects on a rocky intertidal community. Marine Environmental Research 7: 279–293.

Billheimer, L. E. & B. C. Coull, 1988. Bioturbation and recolonitazion of meiobenthos in juvenile spot (Pisces) feeding pits. Estuarine Coastal Shelf Science 27: 335–340.

Brosnan, D. M. & L. L. Crumrine, 1994. Effects of human trampling on marine rocky shore communities. Journal of Experimental Marine Biology and Ecology 177: 79–97.

Brown, P. J. & R. B. Taylor, 1999. Effects of trampling by humans on animals inhabiting *coralline* algal turf in the rocky intertidal. Journal of Experimental Marine Biology and Ecology 235: 45–53.

Duran, L. R. & J. C. Castilla, 1989. Variation and persistence of the middle rocky intertidal community of central Chile, with and without human harvesting. Marine Biology 103: 555–562.

Eckrich, C. E. & J. G. Holmquist, 2000. Trampling in a seagrass assemblage: direct effects, response of associated fauna, and the role of substrate characteristic. Marine Ecology Progress Series 201: 199–209.

Gallet, S. & F. Roze, 2001. Resistance of Atlantic heathlands to trampling in Brittany (France): influence of vegetation type, season and weather conditions. Biological Conservation 97: 189–198.

Keough, M. J. & G. P Quinn, 1998. Effects of periodic disturbances from trampling on rocky intertidal algal beds. Ecological Application 8: 141–161.

Liddle, M. J., 1975. A selective review of the ecological effects of human trampling on natural ecosystems. Biological Conservation 7: 17–36.

Milazzo, M., R. Chemello, F. Badalamenti & S. Riggio, 2002. Short-term effect of human trampling on the upper infralittoral macroalgae of Ustica Island MPA (western Mediterranean, Italy). Journal of Marine Biological Association of UK 82: 745–748.

Murray, S. N., T. G. Denis, J. S. Kido & J. R. Smith, 1999. Human visitation and the frequency and potential effects of collecting on rocky intertidal populations in southern California marine reserve. California Cooperative Oceanic Fisheries investigations reports 40: 100–106.

Povey, A. & M. J. Keough, 1991. Effects of trampling on plant and animal populations on rocky shores. Oikos 61: 355–368.

Schiel, D. R. & D. I. Taylor, 1999. Effects of trampling on a rocky intertidal algal assemblages in southern New Zealand. Journal of Experimental Marine Biology and Ecology 235: 213–235.

Sherman, K. M. & B. C. Coull, 1980. The response of meiofauna to sediment disturbance. Journal of Experimental Marine Biology and Ecology 46: 59–71.

Sousa, W. P., 1979. Experimental investigations of disturbance and ecological succession in a rocky intertidal algal community. Ecological Monographs 49: 227–254.

Terlizzi, A., S. Bevilacqua, S. Fraschetti & F. Boero, 2003. Taxonomic sufficiency and the increasing insufficiency of taxonomic expertise. Marine Pollution Bulletin 46: 556–561.

Underwood, A. J., 1997. Experiments in Ecology: Their Logical Design and Interpretation Using Analysis of Variance. Cambridge University Press, Cambridge 504 pp.

Underwood, A. J. & S. J. Kennelly, 1990. Pilot studies for designs of surveys of human disturbance of intertidal habitats in New South Wales. Australian Journal of Marine Freshwater Research 41: 165–173.

Villa, F., L. Tunesi & T. Agardy, 2002. Zoning marine protected areas through spatial multiple-criteria analysis: the case of the Asinara Island National Marine Reserve of Italy. Conservation Biology 16: 515–526.

Warwick, R. M., 1988. Analysis of community attributes of macrobenthos of Friefjiord/Langesundfjiord at taxonomic levels higher than species. Marine Ecology Progress Series 46: 167–170.

Winer, B. J., 1971. Statistical Principles in Experimental Designs (2nd ed)., McGraw-Hill, Kogakusha, Tokyo.

Hydrobiologia (2006) 555:281–287

H. Queiroga, M.R. Cunha, A. Cunha, M.H. Moreira, V. Quintino, A.M. Rodrigues, J. Serôdio & R.M. Warwick (eds),
Marine Biodiversity: Patterns and Processes, Assessment, Threats, Management and Conservation
DOI 10.1007/s10750-005-1124-2

Imposex in *Hexaplex trunculus* at some sites on the North Mediterranean Coast as a base-line for future evaluation of the effectiveness of the total ban on organotin based antifouling paints

F. Garaventa[1,*], F. Pellizzato[2], M. Faimali[1], A. Terlizzi[3], D. Medakovic[4],
S. Geraci[1] & B. Pavoni[2]
[1]*CNR – Istituto di Scienze Marine (ISMAR)–Sezione di Glenova, Via De Marini 6, 16149, Genoa, Italy*
[2]*Department of Environmental Sciences, University of Venice, Calle Larga S. Marta 2137, 30123, Venice, Italy*
[3]*Department of Biological and Environmental Science and Technology, University of Lecce, CoNISMa, Lecce, Italy*
[4]*Ruder Boskovic Institute, Center for Marine Research Rovinj, Giordano Paliaga 5, 52210, Rovinj, Croatia*
*(*Author for correspondence: Tel.: +39-010-6475426; Fax: +39-010-6475400;*
E-mail: francesca.garaventa@ismar.cnr.it)

Key words: imposex, *Hexaplex trunculus*, Mediterranean Sea, tributyltin, triphenyltin

Abstract

Imposex – the superimposition of male sexual organs (penis and vas deferens) onto female Neogastropods such as *Hexaplex trunculus* (Linné, 1758) – is used world-wide as a biomarker of ecological impact of organotin based antifouling biocides (TBT and TPhT). To limit the impact of organotin pollution, since January 1, 2003, the International Maritime Organization (IMO) has enacted a global ban on the use of organotin compounds in antifouling systems. It is important to record imposex levels and organotin contamination before the implementation of the ban, in order to assess the current situation and be able, in the future, to verify the effects of the International Protocol. In this paper, recent imposex data measured in populations of *Hexaplex trunculus* from three different Mediterranean regions are compared: the Ligurian Sea (Italy), the Lagoon of Venice (Italy) and the western coast of Istria (Croatia). In the two former locations, a partial ban on TBT has been in force for vessels less than 25 m since 1982, while in the latter region no restrictions on organotin antifouling paints have been applied yet. Gastropod samples collected from the Venice lagoon were analysed with an acid extraction followed by Grignard derivatisation, clean up and GC-MS determination, in order to relate the levels of TBT, TPhT and their metabolites with the imposex degree detected. Biological data show that the levels of imposex were very high (VDS from 4.3 to 5) in all the sampling sites considered, particularly in the Croatian coast stations. The concentrations of organotin compounds – butyltins and phenyltins – measured in the samples from the lagoon of Venice were found to partition differently in the visceral coil and in the rest of the soft body of the analysed organisms.

Introduction

Tributyltin (TBT) contamination of coastal waters is strongly related to boating, since this compound is commonly employed as a biocide in antifouling paints (Hugget et al., 1992).

TBT compounds and their derivates can cause a genital disorder in females of different species of marine snails at an ambient concentration of just a few nanograms per litre (Matthiessen & Gibbs, 1998). Such a disorder consists of the superimposition of male sexual organs onto females. This phenomenon is widely known as imposex (Smith, 1971) or pseudohermaphroditism (Jenner, 1979) and it is known to occur in 63 genera and 118 species (Fioroni et al., 1991). The effects of imposex vary, depending on the species. In some cases it does not impair reproduction, but in some others it can lead to population decline, as a consequence of sterility and reproduction failure (Bryan et al., 1986).

Imposex is a graded response and it has been used as a highly sensitive bioindicator of TBT pollution worldwide (Ellis & Pattisina, 1990).

In Mediterranean prosobranchs, imposex studies are mainly limited to populations of the following species: *Stramonita haemastoma* (Linné, 1766) (Spence et al., 1990; Rilov et al., 2000; Terlizzi, 2000), *Bolinus brandaris* (Linné, 1758) (Solé et al., 1998; Ramòn & Amor, 2001), and *Hexaplex trunculus* (Axiak et al., 1995; Terlizzi et al., 1999; Rilov et al., 2000). Along the coasts of Italy, Terlizzi et al. (1998) found this phenomenon to be widespread and related to boat density in the investigated sites. Only a few studies (Axiak et al., 1995; Rilov et al., 2000; Chiavarini et al., 2003) report the levels of organotin contamination in tissues, in addition to biomonitoring.

The negative effects of organotin compounds on non-target marine organisms (reviewed by Terlizzi et al., 2001) have forced many governments to reduce their use. Legislation in many countries banned the application of TBT-based paints to small vessels (< 25 m) and the International Maritime Organization (IMO), Marine Environment Protection Committee (MEPC), adopted a treaty for the worldwide ban of TBT that started from January 2003; the ban has been adopted by the EU since 1 July 2003 with the regulation (CE) No. 782/2003. Nevertheless in other Mediterranean countries, such as Croatia, still no restrictions whatever are applied.

This paper aims to provide data about imposex levels and organotin contamination before the

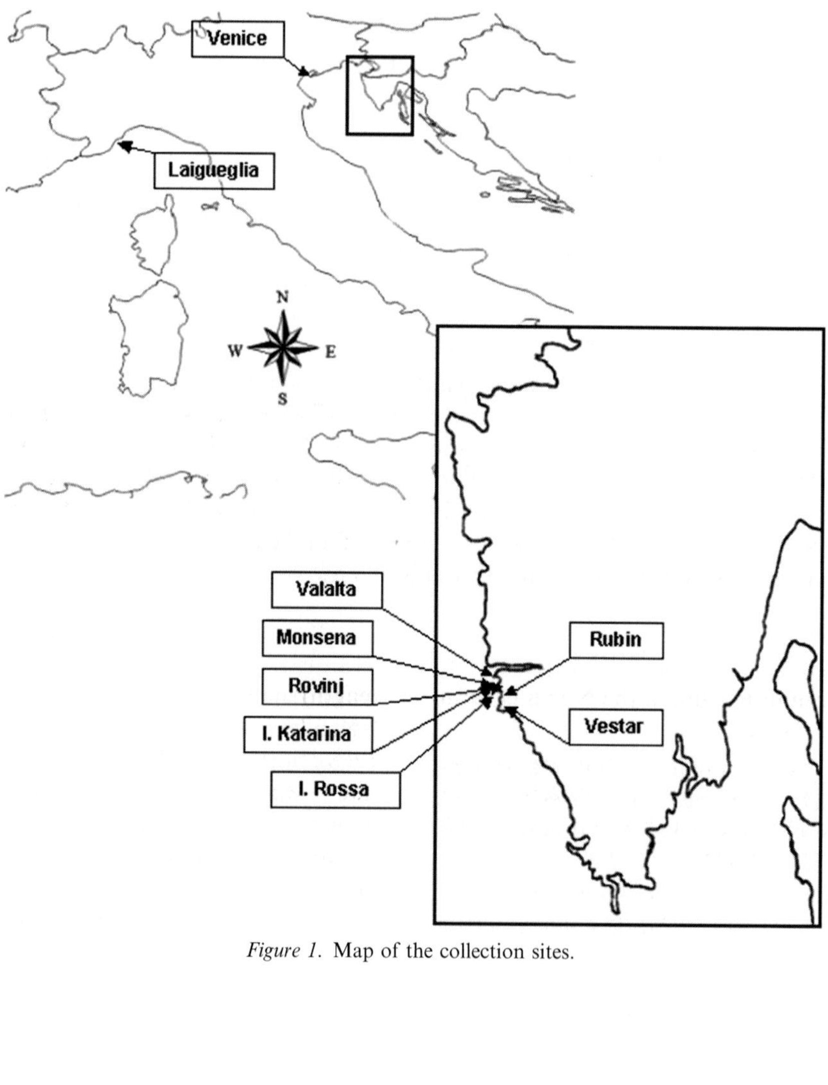

Figure 1. Map of the collection sites.

Table 1. Summary of imposex results of *Hexaplex trunculus* population

Sampling site		BA	BD	Date	Sex	No. ind.	Shell length (mm)	Imposex (%)	Avg penis length	Sterile females (%)	RPSI	RPLI	VDSI	VDS range
Monsena (HR)	NB	y	M	May-02	M	8	60.9	100	16.1	0.0	1.9	26.9	4.2	3–4.7
					F	19	60.4		4.6					
Vestar (HR)	NB	y	M	May-02	M	3	59.7	100	8.9	4.2	114.6	104.7	4.5	4.3–5
					F	24	58.7		8.5					
Villas Rubin (HR)	Marina	y	M	May-02	M	5	61.9	100	11.5	29.6	45.3	76.8	4.6	3–5
					F	27	61.2		9.2					
Is. Katarina (HR)	Marina	y	L	May-02	M	6	60.3	100	11.9	28.6	72.4	89.8	4.6	4.3–5
					F	15	63.0		10.7					
Is. Rossa (HR)	Marina	y	L	May-02	M	13	59.7	100	10.8	100.0	68.0	87.9	5.0	5
					F	14	67.7		9.5					
Rovinj (HR)	Harbour	F-Y	H	May-02	M	20	61.9	100	12.2	83.3	142.3	112.5	5.0	4.7–5
					F	6	61.9		13.8					
Vallalta (HR)	Marina	y	M	May-02	M	6	64.2	100	11.8	60.0	62.2	85.4	4.8	4.3–5
					F	25	63.7		10.0					
Venice (IT)	Outer port	C,F,N	H	Oct-02	M	28	64.5	100	19.6	20.0	9.0	44.9	4.5	4–5
					F	20	67.0		8.8					
Laiguelia (IT)	Marina	F,Y	L	Jun-02	M	9	59.7	100	12.9	3.2	12.5	50.0	4.4	4–5
					F	31	56.3		6.5					

NB = natural bay; BA = boat activity; F = fishing boat; N = navy vessels; C = commercial shipping; Y = yachting; y = yachting activity only in the summer period; BD = boat density; H = high; M = moderate; L = low; M = male; F = female; RPSI = relative penis size index; RPLI = relative penis length index; VDSI = vas deferens sequence index.

implementation of the ban, in order to assess the present situation in the Northern Mediterranean Sea and to verify, in the future, the effectiveness of the International Protocol.

Materials and methods

Organisms were collected by snorkeling from May 2002 to October 2002 from shallow waters at three sites located in the North Mediterranean Sea: one station in Laigueglia in the Ligurian sea (IT), one in the Venice lagoon in the North-western Adriatic Sea (IT) and seven in Rovinj, in the North-eastern Adriatic Sea (Istria, HR) (Fig. 1).

The level of maritime traffic differed among the study sites (Table 1). About 40 specimens of *Hexaplex trunculus* were collected from each station. After collection, the organisms were brought to the laboratory, placed in tanks with recirculating sea water and examined within 10 days.

Individuals were narcotised (7% $MgCl_2$ in seawater), the shell was measured to the nearest 0.1 mm using a calliper, cracked in a bench vice and the soft body carefully removed. Males were recognized by both the presence of a penis and a prostate and by the lack of a capsule gland. Females were recognized by certain characters present both in normal and imposexed females, namely the presence of the vagina and the capsule gland.

Three imposex indices were determined: (1) the RPSI (relative penis size index, Gibbs et al., 1987), defined by the formula (mean female penis length)^3/(mean male penis length)$^3 \times 100$; (2) the RPLI (relative penis length index), defined by the formula (mean female penis length)/(mean male penis length)$\times 100$ and (3) the Vas Deferens Sequence Index (VDSI) i.e., the mean score of the various stages of imposex developed in females (Fioroni et al., 1991). Determination of different imposex stages in females followed the general scheme proposed by Gibbs et al. (1987) and

Stroben et al. (1992) and partially modified by Axiak et al. (1995) and Terlizzi et al. (1999) for *Hexaplex trunculus*.

After imposex assessment, for each animal the visceral coil (which includes both the digestive gland and the gonad) was separated from the rest of the soft tissue. Both types of tissues were pooled from animals ($n = 2$–5) with the same degree of imposex; each pool was freeze-dried, homogenized, and stored at −20 °C in the dark in order to preserve the organotin speciation until chemical analysis.

Chemical analysis

Freeze-dried tissues (0.5 g) were twice extracted adding 13 ml of a methanolic solution of tropolone (0.03%) and 1 ml of HCl (37%) in a sonication bath for 15 min and centrifuged for 10 min. Extracts were then liquid–liquid extracted twice in separating funnels with 15 ml of dichloromethane. After dehydration through activated sodium sulphate, solution volume was reduced under nitrogen flow and the solvent exchanged to isooctane. Extracted organotins were allowed to react with 1 ml 2 M pentylmagnesiumbromide in diethylether for 5 min. The internal standard, pentylated tripropyltin, was then added and the Grignard reactant excess destroyed by carefully adding 1 ml of deionised water and 5 ml of 1 M sulphuric acid, drop by drop. Derivatised organotins were then extracted twice with 2 aliquots of hexane. After further volume reduction under nitrogen flow, the extract was purified with a 3 g Florisil column eluted with 10 ml of 1:1 hexane–toluene solution. The eluted solution was reduced to a volume of 1 ml under a gentle stream of nitrogen and analysed with a gas chromatograph coupled with a mass spectrometer. Chromatographic conditions were as follows: capillary column, HP-5 (5% phenyl methylsiloxane, i.d. 0.20 mm, length 50 m, film thickness 0.33 μm); injector temperature, 300 °C; temperature programme, 80 °C for 2 min, then 20 °C min^{-1} up to 290 °C, post run 20 min at 290 °C; transfer line temperature, 280 °C, splitless injection; carrier gas helium at 199 kPa head pressure. The MS detection was performed using electron impact ionisation (70 eV) in the selected ion monitoring mode (SIM). Detection limits of the analytical procedure, calculated as 3σ of the blanks, ranged from 0.1 to 3 ng Sn g^{-1} for butyltin and phenyltin compounds. Organotin compounds in the samples were quantified by using the response factors derived from repeated daily injections of a standard mixture of derivatized compounds. All samples were analysed in duplicate and the half variation interval of the measures was within 15%.

The accuracy of the analytical procedure was determined by analysing the Certified Reference Material (CRM) 477 (mussel tissue) which is certified only for butyltins. Recovery yields obtained were: 80% for TBT, 74% for DBT and 84% for MBT.

Results

Biometrical and imposex data for *Hexaplex trunculus* are summarised in Table 1. All collected females showed signs of imposex.

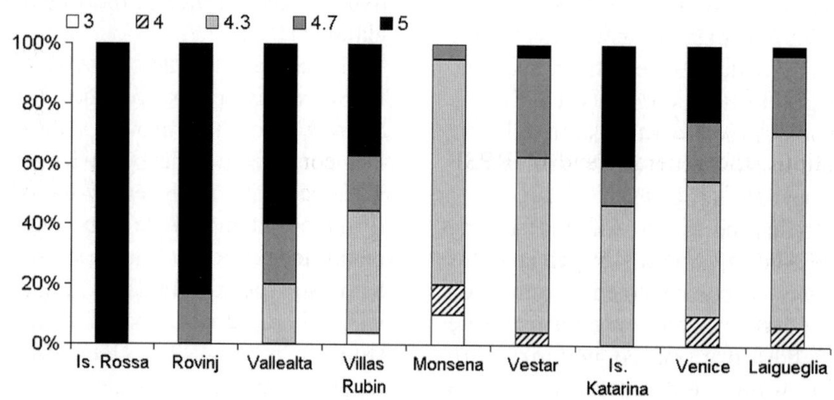

Figure 2. Percentage of imposex incidence at each of the sampling stations: 3; 4; 4.3; 4.7; 5 = imposex stages.

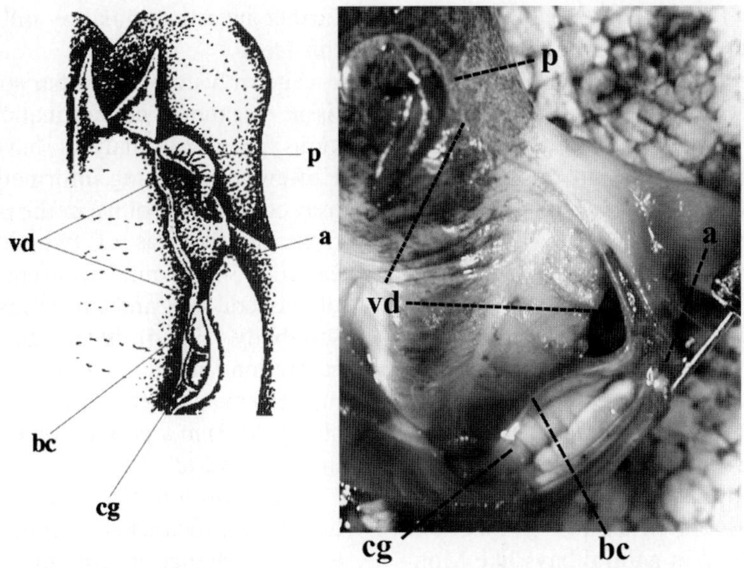

Figure 3. Particulars of capsule gland splitting at stage 5 of imposex. In the drawing (Terlizzi et al., 1999) and in the picture: cg = capsule gland; bc = bursa copulatrix; vd = vas deferens; p = penis; a = anus.

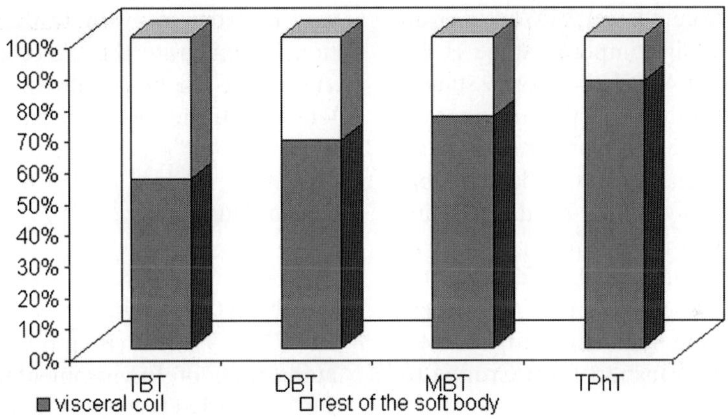

Figure 4. Average organotin partitioning in the body of organisms collected in the Venice station.

VDSI values showed a restricted range of values, from 4.2 (Monsena) to 5 (Isola Rossa and Rovinj). RPSI and RPLI values ranged from as low as 1.9 and 26.9 at Monsena to 142.3 and 112.5 in Rovinj, respectively. Both indices were over 100% in two stations; the general trend of RPSI and RPLI was not correlated to VDSI (Correlation analysis for: RPSI–R^2 = 0.4519; n = 9; RPLI–R^2 = 0.5656; n = 9). Figure 2 shows the incidence of imposex at each station; the highest degree of imposex was recorded in Isola Rossa with 100% of females at stage 5 with the split capsule gland (Fig. 3). All the Croatian stations except for Monsena and Vestar (natural bay with low yachting activity) showed a remarkable incidence of stage 5 (Isola Rossa: 100%; Rovinj: 83%; Vallealta: 60%; Villas Isola Katarina: 40% and Rubin 37%). Conversely, in Venice, a station characterised by very high ship traffic (Navy vessels, commercial ships and yachts), stage 5 was recorded in 25% of the population.

As to chemical results, average organotins contamination data in tissues of *Hexaplex trunculus* for the station of Venice were calculated. TBT (103 ± 2 ng Sn g^{-1}) prevails over its metabolites (DBT = 79 ± 18 ng Sn g^{-1}; MBT = 29 ± 8 ng Sn g^{-1}) and over TPhT (6 ± 4 ng Sn g^{-1}) according to the following sequence: TBT

> DBT > MBT > TPhT. TPhT metabolites, i.e., DPhT and MPhT were detected at concentrations below the limit of detection of the analytical procedure (< 0.4 ng Sn g^{-1}). Figure 4 shows the partitioning of the body burdens of TBT and its metabolites in Venice samples, all the congeners and in particular the metabolites (DBT, MBT and TPhT) are mainly associated with the visceral coil; e.g., 90% of TPhT is found in this tissue.

Discussion

No stages of imposex lower than 3 were found. However, differences in the incidence of imposex were detected in relation to the intensity and type of boat activity. *Hexaplex trunculus* populations sampled in sites located in natural bays like Monsena and Vestar, or located in areas with restrictions in the use of TBT based antifouling paints, like Venice and Laigueglia, were less affected by imposex. This can be seen from a lower VDSI and from the fact that the prevailing imposex stage is 4.3 (Fig. 2). Conversely, high VDSI and mostly stage 5 were found at stations without TBT restrictions.

The RPSI and RPLI overflow the limit of 100%, suggesting that these indices are less reliable markers for the analysis of TBT incidence. This observation needs further investigation, because in some samples of this survey the male/female ratio was not balanced due to low density of organisms in sampling stations and high mortality of collected specimens during transport from Croatia to the laboratory in Italy.

These preliminary data show that a slight difference between stations may result from the implementation of existing regulations. At present, it is clearly impossible to gain information about a likely decline in the severity of imposex in Italy, because data on its incidence along the coasts of Italy are quite sparse (Terlizzi et al., 1998; Chiavarini et al., 2003; Terlizzi et al., 2004). There are, however, some clues. The population from Laigueglia, for instance, had a VDSI of 4.4 (SD ± 0.2) with VDS stages ranging from 4 to 5. In a population from the same area sampled in June 1995, Terlizzi et al. (1998) measured a VDSI of 4.8 (SD ± 0.5), with the same VDS range (4–5). This comparison could indicate that there is some evidence of recovery in Italian coastal areas, but further investigations are still needed to confirm this trend.

Such a historical comparison is not possible for tissue organotin contamination, because no previous chemical analyses have been performed. However, interesting information on the metabolism of this compound in the organism is acquired from the analysis of two different tissues. Organotins were found to accumulate differently in the visceral coil and in the rest of the organisms' soft body (Fig. 4), with significantly higher concentrations in the visceral coil. This observation is in agreement with that found by Axiak et al. (1995, 2003) in a population of *Hexaplex trunculus* in the island of Malta.

In conclusion, the data provided in this paper can be regarded as a baseline for evaluating whether any change in contamination occurs before and after the ban on TBT antifouling paints enacted on July 1, 2003. Other studies concerning organotin compounds at other stations are necessary in order to evaluate the level of contamination, to integrate the biological response, and to investigate the behaviour of the compounds once taken up from the environment by the organism.

Acknowledgements

This work was partially supported by funds from the NATO Collaborative Linkage Grant N° EST.CLG.979832 (Italian-Croatian project 'Biomonitoring of Environmental Contamination by TBT in Mediterranean Coasts'). The authors thank two anonymous referees for helpful comments and revisions.

References

Axiak, V., A. J. Vella, D. Micaleff, P. Chircop & B. Mintoff, 1995. Imposex in *Hexaplex trunculus* (Gastropoda: Muricidae): first results from biomonitoring of tributyltin contamination in the Mediterranean. Marine Biology 121: 685–691.

Axiak, V., D. Micaleff, J. Muscat, A. Vella & B. Mintoff, 2003. Imposex as a biomonitoring tool for marine pollution by tributyltin: some further observations. Environmental International 28: 743–749.

Bryan, G. W., P. E. Gibbs, L. G. Hummerstone & G. R. Burt, 1986. The decline of the gastropod *Nucella lapillus* around South-West England: evidence for the effect of trybutiltin from antifouling paints. Journal of Marine Biological Association of the United Kingdom 66: 611–640.

Chiavarini, S., P. Massanisso, P. Nicolai, C. Nobili & R. Morabito, 2003. Butyltins concentration levels and imposex occurrence in snails from the Sicilian coasts (Italy). Chemosphere 50: 311–319.

Ellis, D. V. & L. A. Pattisina, 1990. Widespread neogastropods imposex: a biological indicator of global contamination? Marine Pollution Bulletin21248253.

Fioroni, P., J. Oehlmann & E. Stroben, 1991. The pseudohermaphroditism of Prosobranchs; morphological aspects. Zoologischer Anzeiger 226: 1–26.

Gibbs, P. E., G. W. Brian, P. L. Pascoe & G. L. Burt, 1987. The use of the dog-welk, *Nucella lapillus*, as an indicator of tributyltin TBT contamination. Journal of the Marine Biological Association of the United Kingdom 67: 507–523.

Huggett, R. J., M. A. Unger, P. F. Seligman & A. O. Valkirs, 1992. The marine biocide tributyltin: assessing and managing the environmental risks. Environmental Science and Technology 26: 232–237.

Jenner, M. G., 1979. Pseudohermaphroditism in *Ilyanassa obsoleta* (Mollusca: Neogastropoda). Science 205: 1407–1409.

Matthiesen, P. & P. E. Gibbs, 1998. Critical appraisal of the evidence for tributyltin mediated endocrine disruption in molluscs. Environmental Toxicology and Chemistry 17: 37–43.

Ramòn, M. & M. J. Amor, 2001. Increasing imposex in populations of *Bolinus brandaris* (Gastropoda: Muricidae) in the north-western Mediterranean. Marine Environmental Research 52: 463–475.

Rilov, G., A. Gasith, S. M. Evans & Y. Benayahu, 2000. Unregulated use of TBT-based antifouling paints in Israel (Eastern Mediterranean): high contamination and imposex levels in the two species of marine gastropods. Marine Ecology Progress Series 192: 229–238.

Smith, B. S., 1971. Sexuality in the American mud-snail, *Nassarius obsoletus* Say. Proceedings of the Malacological Society of London 39: 377–378.

Solé, M., Y. Morcillo & C. Porte, 1998. Imposex in the commercial snail *Bolinus brandaris* in the north-western Mediterranean. Environmental Pollution 99: 241–246.

Spence, S. K., S. J. Hawkins & R. S. Santos, 1990. The Mollusc *Thais haemastoma* – An exhibitor of 'Imposex' and potential biological indicator of tributyltin (TBT) contamination. P.S.Z.N. I: Marine Ecology 11: 147–156.

Stroben, E., J. Oehlmann & P. Fioroni, 1992. The morphological expression of imposex in *Hinia reticulata* (Gastropoda: Buccinidae): a potential indicator of tributyltin pollution. Marine Biology 113: 625–636.

Terlizzi, A., S. Geraci & V. Minganti, 1998. Tributyltin (TBT) pollution in the coastal waters of Italy as indicated by imposex in *Hexaplex trunculus* (Gastropoda, Muricidae). Marine Pollution Bulletin 36: 749–752.

Terlizzi, A., 2000. Imposex (Pseudoermafroditismo) in Molluschi Gasteropodi mediterranei: aspetti morfologici e considerazioni ecologiche. Bollettino Malacologico 36: 155–158.

Terlizzi, A., S. Geraci & P. E. Gibbs, 1999. Tributyltin (TBT)-induced imposex in the Neogastropod *Hexaplex trunculus* in Italian coastal waters: morphological aspects and ecological implications. Italian Journal of Zoology 66: 141–146.

Terlizzi, A., S. Fraschetti, P. Gianguzza, M. Faimali & F. Boero, 2001. Environmental impact of antifouling technologies: state of art and perspectives. Aquatic Conservation: Marine and Freshwater Ecosystems 11: 311–317.

Terlizzi, A., A. L. Delos, F. Garaventa, M. Faimali & S. Geraci, 2004. Limited effectiveness of marine protected areas: imposex in *Hexaplex trunculus* (Gastropoda, Muricidae) populations from Italian marine reserves. Marine Pollution Bulletin 48: 188–192.

Hydrobiologia (2006) 555:289–297
H. Queiroga, M.R. Cunha, A. Cunha, M.H. Moreira, V. Quintino, A.M. Rodrigues, J. Serôdio & R.M. Warwick (eds),
Marine Biodiversity: Patterns and Processes, Assessment, Threats, Management and Conservation
DOI 10.1007/s10750-005-1125-1

Environmental effects of marine fish pond culture in the Ria Formosa (Southern Portugal)

François Noël Hubert[1,*], Marc Pellaud[2] & Sofia Gamito[3]
[1]*IMAR, presently in CCMAR, Faculdade de Ciências do Mar e do Ambiente, Universidade do Algarve, Portugal*
[2]*Faculté des Sciences, Université de Genève, Switzerland*
[3]*IMAR, Faculdade de Ciências do Mar e do Ambiente, Universidade do Algarve, Portugal*
*(*Author for correspondence: Tel.: +351 289800900, Fax: +351 289818353; E-mail: fhubert@ualg.pt)*

Key words: lagoon system, semi-intensive fish farms, environmental impact

Abstract

Macrobenthic organisms, sediment and water characteristics were analysed in two fish farms in the Ria Formosa at control sites, near the outflows and inside the decantation ponds, in winter and summer 2001. Both farms operate under semi-intensive and intensive regimes. Statistical analysis reveals a very localised effect of fish farms near the outflow of the semi-intensive ponds, with an increase of density of small-sized organisms. The fish farms released high concentrations of nutrients and relatively low quantities of suspended solids. The effects of the effluents on the sediment were significant in the immediate vicinity of discharge point-source. The effects were localized mainly because all farms operate near the sea inlets, where water renewal rates are higher to ensure good fish production, thus permitting a important flushing of waste water. The effects of the effluents were more pronounced in summer because of the farms' higher productivity at that time of the year. The decantation ponds, obligatory for the intensive regime, were apparently efficient in removing part of the solid waste. However, semi-intensive fish farms represent important sources of dissolved nutrients in the Ria Formosa and the joint effect of several fish farms in a semi-enclosed coastal system may be considerable.

Introduction

The Ria Formosa is a coastal lagoon situated in the Algarve region of Southern Portugal and comprises barrier islands, barrier platforms, back barrier lagoons, inlet deltas and shoreface (Pilkey et al., 1989). This lagoon system covers a maximum area of about 63 km^2 and is characterized by a strong water renewal between tides (Sprung, 1994).

Aquaculture in the Ria Formosa is a traditional activity of great socio-economical importance. Mollusc extensive culture (mainly of *Ruditapes decussata*, Linnaeus 1758) produces up to 80% of Portugal's bivalves (Rosado & Bruxelas, 1995). Extensive multi-species fish farming is also a traditional activity being undertaken in ancient saltpan reservoirs. Semi-intensive fish farming of sea bass (*Dicentrarchus labrax*, Linnaeus 1758) and gildhead sea bream (*Sparus aurata*, Linnaeus 1758) has been developing for the past 20 years. This type of production is also undertaken in ancient saltpans or in concrete tanks. The fish density is high and the water is renewed every day through tidal gates and water pumping. The fish diet consists mostly of commercial food pellets.

Effluents from aquaculture facilities can be a major source of pollution in marine environment (Gowen et al., 1990; Wu et al., 1994; Hargreaves, 1998) and fish farming usually discharges important quantities of organic wastes in the surrounding environment (Folke & Kautsky, 1989). However, the severity and the extent of the impact depend on the kind of farming activity and on the water and sediment characteristics of the receiving system (Pillay, 1991; Wu, 1995). So far, extensive farming has been shown to cause little or no impact on the surrounding environment in the Ria Formosa (Gamito, 1994), but little is known of impact of semi-intensive fish farming in such an environment.

The objective of the study was to investigate the local effects of semi-intensive fish pond-culture on the Ria Formosa lagoon system.

Materials and methods

The present work was undertaken in collaboration with 2 semi-intensive fish farms situated in the intertidal zone of Ria Formosa. Both farms operate on the same principle. A reservoir filled on high tide supplies water to batteries of earthen and concrete ponds organized in a semi-intensive rearing section and an intensive one. Waste water from the intensive rearing section goes through a decantation process before being rejected while in the semi-intensive section waste water is directly rejected into the tidal channels. The species reared are sea-bream (*Sparus aurata*), associated with sea bass (*Dicentrarchus labrax*). Farm A is a relatively small exploitation (11 ha) producing 120 tons of fish per year with a 250 m^2 decantation pond, while farm B produces approximately 300 tons of fish annually and has a 600 m^2 decantation pond.

Four subtidal sampling stations were selected in each farm (Table 1). Sediments and benthic macrofauna were sampled at the low sea-level of spring tides on two occasions; in February 2001 and in June 2001 when the water column was less than 50 cm deep. Within each sampling zone three randomly allocated sampling units were studied. The CTRL sampling units were allocated up to 10 m outside the water reservoir gate. The DECANT units were inside the decantation pond within 10 m from the pond entrance gate. The INT units were located inside the intensive discharge channel up to 10 m outside the decantation pond exit gate. Finally, the SEMI sampling units were allocated inside the semi-intensive discharge channel up to 10 m outside the rearing ponds exit gates. In summary, stations INT and SEMI represent discharge point-source conditions from the intensive and semi-intensive rearing sections respectively, with water from the intensive section going through a decantation process before reaching station INT. Water for total suspended matter and nutrient determination was sampled at the same stations on four occasions in February 2001 and four more occasions in June 2001 (high/low sea level for spring and neap tides). Within each sampling zone, one water sampling-unit was taken on each occasion.

The studied sediment variables were grain-size analyzed following Larsonneur (1977), organic matter percentage, as well as Chlorophyll *a* and phaeopigment concentrations according to Strickland & Parsons (1968). For determination of organic matter, each sample was dried in an oven at 60 °C for 48 h. The loss of dry weight after 4 h in a furnace at 450 °C gave the amount of organic matter originally contained in the sample. Concerning the water, total suspended solids (TSS) were determined according to Strickland & Parsons (1968) and dissolved ammonia, nitrate, nitrite and phosphate concentrations were measured following methods defined by APHA (1989).

Benthic macrofauna was sampled by cylindrical cores (0.01 m^2 surface and 20 cm deep), sieved on a 500 μm sieve and identified to family taxonomic level.

In order to test for significant differences of the measured variables among zones and farms, two-way analyses of variance (ANOVA) were applied to our previously log-transformed data. The

Table 1. Characteristics of the sampling stations

Station code	Station location	Particularities
CTRL	Entrance of water reservoir	Control station in the experimental design
DECANT	Decantation pond	Decantation pond receives water from intensive section
SEMI	Semi-intensive discharge channel	Represents point-source conditions
INT	Intensive discharge channel	Represents point-source conditions

general null hypothesis was 'there is no difference of the variables among farms and/or zones'. When significant differences were observed, a Student-Newman-Keuls (SNK) *a posteriori* multiple comparison test was used to determine which means differed from one another.

A second step consisted of comparing the stations according to their benthic macrofauna using an ordination by multi-dimensional scaling (MDS) applied on double root-transformed density data and using Bray–Curtis dissimilarity index. Stations were also compared according to their sediment and water characteristics using a Principal Component Analysis (PCA) done on normalised data. The 9 variables introduced in the analysis were TSS and dissolved nutrients concentrations (phosphate, ammonia, nitrate and nitrite) for the water and the silt and organic matter percentages as well as the Chlorophyll *a* and phaeopigments concentrations for the sediment. One-way Analyses of Similarity (ANOSIM) were used to test for significant differences of macrofauna composition among seasons, farms and stations.

Software used for data analysis were Statistica (Rel. 5.1 Statsoft, Inc. 1996. Statistica for Windows), CANOCO (Ter Braak, Rel. 4.0, GLW-CPRO© 1997,98) and the PRIMER software package (Clarke, Rel. 5.0, Primer-E Ltd, 2001).

Results

Sediment variables

On the whole, in terms of the measured properties of the sediment, the two studied farms appeared very similar (Table 2). Sediments were silty-sand and sandy-silt except in the INT station of farm B where it was sand. Sediment was also sand in station CTRL in summer. The observed values of organic matter concentration were higher in summer especially in station DECANT of farm A and in station SEMI of farm B. On the other hand, values of silt and clay percentage, chlorophyll *a* and phaeopigment concentrations were generally comparable between the two seasons. There was no significant difference in the sediment variables between farms except for the pigment concentrations in summer that were significantly higher in farm B ($p < 0.05$; Table 3). In winter, station SEMI presented significantly higher percentages of fine particles and organic matter concentration than the other stations in both farms (SNK, $p < 0.05$; Table 3). This was also true in summer but only for farm B.

Water variables

The observed values of TSS and ammonia concentrations were higher in summer than in winter.

Table 2. Mean values of the sediment variables together with standard error (in parentheses)

Sediment variable	Farm A				Farm B			
	CTRL	DECANT	SEMI	INT	CTRL	DECANT	SEMI	INT
Winter								
Grain-size	Silty-sand	Silty-sand	Sandy-silt	Sandy-silt	Silty-sand	Silty-sand	Sandy-silt	Sand
Silt and clay %	8.1 (1.3)	21 (5.3)	37.5 (12.4)	27.4 (2.6)	17.7 (10)	17.4 (10.3)	34.9 (3.3)	6.2 (4)
Organic matter %	0.8 (0.1)	1.4 (0.5)	4.9 (2.3)	1 (0.2)	0.9 (0)*	0.7 (0.1)	4 (1.1)	0.6 (0.1)
[Chl *a*] ($\mu g\ g^{-1}$)	5.9 (1.3)	1 (0) *	6.6 (3.7)	1.9 (0.6)	17.8 (1.4)	2.7 (1.2)	11.9 (5)	2.4 (0.3)
[Phaeop.] ($\mu g\ g^{-1}$)	4.1 (1.7)	4.3 (0) *	18.5 (4.2)	3.4 (1.8)	7.8 (1.2)	4.3 (0.1)	44.2 (11.2)	4.1 (2.6)
Summer								
Grain-size	Silty-sand	Silty-sand	Sandy-silt	Sandy-silt	Sand	Silty-sand	Sandy-silt	Sand
Silt and clay %	10.2 (0.2)	20.1 (4.7)	40 (14.3)	36.9 (21.1)	1.5 (0.4)	12.6 (8.3)	44.2 (5.7)	3.3 (0.8)
Organic matter %	1.7 (1.5)	9.1 (5.6)	1.7 (0.7)	0.9 (0.4)	0.8 (0.1)	0.9 (0.4)	14.9 (5.2)	3.5 (2.5)
[Chl *a*] ($\mu g\ g^{-1}$)	4.3 (0) *	19.8 (9.9)	5.7 (4.1)	1.9 (0.7)	15 (1.3)	5 (3.6)	12.8 (0.9)	10.8 (6.3)
[Phaeop.] ($\mu g\ g^{-1}$)	5.1 (0) *	39.4 (14.9)	13.7 (6.7)	2 (0.6)	2.3 (1.1)	7.2 (6.9)	41 (5.7)	10.7 (7.5)

An * indicates that only one sampling unit could be analysed.

Table 3. Effects of factors farm and station on the sediment variables tested by Two-way ANOVA and Student-Newman-Keuls multiple comparison tests

Sediment variable	Source of variation			Result of Student-Newman-Keuls multiple comparison tests
	Farm	Station	Farm x Sta.	
Winter				
Silt and clay %	n.s (1)	** (3)	n.s (3)	CTRL, DECANT, INT < **SEMI**
Organic matter %	n.s (1)	** (3)	n.s (3)	CTRL, DECANT, INT < **SEMI**
[Chl *a*]	** (1)	** (2)	** (2)	Farm A: n.s (2) < Farm B: * (2)
				Farm A: n.s (2)
				Farm B: **INT** < CTRL, SEMI
[Phaeop]	* (1)	** (2)	n.s (2)	Farm A: * (2) < Farm B: ** (2)
				Farm A: INT < CTRL, SEMI
				Farm B: INT, CTRL < **SEMI**
Summer				
Silt and clay %	n.s (1)	** (3)	n.s (3)	CTRL, DECANT, INT < **SEMI**
Organic matter %	n.s (1)	* (3)	** (3)	Farm A: * (3) < Farm B: ** (3)
				Farm A: CTRL, INT, SEMI < **DECANT**
				Farm B: CTRL, DECANT, INT, < **SEMI**
[Chl *a*]	n.s (1)	n.s (2)	* (2)	Farm A: *(2) < Farm B: n.s (2)
				Farm A: type II error
[Phaeop]	n.s (1)	** (2)	** (2)	Farm A: **(2) < Farm B: n.s (2)
				Farm A: **INT** < DECANT, SEMI

An x between factors indicates interaction and degrees of freedom are in parentheses. Stations in bold differed significantly from the others. Non significant: n.s; significant at 0.05 level: *; significant at 0.01 level: **.

On the other hand, values of nitrate and phosphate concentrations were comparable between the two seasons, although station DECANT of farm B presented abnormally high phosphate concentrations in summer. Finally, nitrite concentrations were higher in summer around farm B but comparable between seasons in farm A (Table 4). Water properties did not differ significantly between the two farms except for nitrate concentration that was higher in farm B during summer ($p < 0.05$; Table 5). Stations SEMI showed the significantly highest values of TSS and nitrite concentrations in winter around both farms and of nitrate concentration around farm A in summer ($p < 0.05$; Table 5). CTRL showed the significantly lowest values of phosphate, ammonia and nitrite concentrations in summer ($p < 0.05$; Table 5).

Benthic macrofauna

In terms of total macro-invertebrate density and of *Capitellidae* density, benthic communities were only slightly different between the two farms (Table 6). In all cases, stations DECANT showed significantly higher total and *Capitellidae* densities ($p < 0.05$; Table 7), while station CTRL presented significantly lower values of total density around farm A in winter and of *Capitellidae* density in summer ($p < 0.05$; Table 7).

Ordination of stations by macrofauna composition

The MDS ordination (Fig. 1) indicates a transition in benthic macrofauna from the control stations to the decantation ponds with both semi-intensive and intensive stations in an intermediate position. The ANOSIM test indicated a significant effect of factor station on the macrofauna composition ($R = 0.332$) with stations SEMI and DECANT differing from the rest.

Ordination of stations by environmental variables

The PCA configuration shown in Figure 2 indicated that TSS and all nutrient concentrations

Table 4. Mean values of the water variables together with standard error (in parentheses)

Water variable	Farm A				Farm B			
	CTRL	DECANT	SEMI	INT	CTRL	DECANT	SEMI	INT
Winter								
TSS (mg l^{-1})	21.8 (1.1)	25 (2.8)	38 (4.8)	20.8 (3.5)	18.8 (0.6)	22.5 (1.5)	25.8 (2.8)	25.3 (4)
Phosphate (μM)	0.8 (0.2)	2.4 (0.3)	1.2 (0.3)	1.6 (0.1)	0.7 (0.1)	2.6 (0.1)	1.6 (0.3)	1.3 (0.7)
Ammonia (μM)	6.1 (3.8)	24.7 (3.9)	11.7 (5.5)	16.7 (0.2)	5.5 (1.4)	23.5 (0.3)	22.4 (7.7)	9.8 (6.6)
Nitrate (μM)	3.2 (0.6)	5.1 (0.1)	14.8 (6.7)	3.5 (0.5)	3.4 (0.5)	6.6 (0.1)	11.7 (0.2)	4.4 (0.5)
Nitrite (μM)	0.2 (0)	0.9 (0.1)	0.9 (0.3)	0.8 (0)	0.8 (0.1)	1.4 (0)	1.5 (0.4)	0.6 (0.3)
Summer								
TSS (mg l^{-1})	27.3 (2.9)	32.3 (3.6)	49.5 (10.9)	32.3 (3.4)	29.8 (3.3)	38.5 (4.4)	38.5 (7.2)	45.8 (9.8)
Phosphate (μM)	0.8 (0.2)	2.2 (0.4)	1.0 (0.3)	2.1 (0.3)	0.5 (0)	5.3 (0.7)	2.0 (0.5)	2.6 (1.2)
Ammonia (μM)	9.0 (3.5)	48.8 (10.6)	29.9 (12.8)	41.9 (9.8)	3.1 (0.6)	55 (10.3)	33.5 (9.6)	26.5 (12.8)
Nitrate (μM)	2.3 (0.1)	3.0 (0.3)	8.7 (2.8)	3.0 (0.2)	2.8 (1)	7.5 (1.4)	12 (4.1)	4.7 (1.4)
Nitrite (μM)	0.5 (0.2)	1.2 (0.3)	3.2 (1.3)	1.7 (0.4)	0.4 (0.1)	3.9 (1)	3.7 (1.5)	1.9 (1)

except nitrate were correlated. The sediment variables were also inter-correlated. The horizontal axis differentiated stations with high values of the environmental variables projected on the right side of the diagram from stations presenting low values and projected on the left side. Accordingly, the most severely impacted stations were the semi-intensive stations whatever the season or the farm as well as the intensive and decantation stations in summer. The control stations appeared relatively un-impacted. On the whole, the measured parameters did not differ significantly between farms in each of the 4 types of stations (control, decantation pond, semi-intensive and intensive) in both seasons. Only the semi-intensive discharge stations differed between farms in summer. Values of all

Table 5. Effects of factors farm and station on the water variables tested by Two-way ANOVA and Student-Newman-Keuls multiple comparison tests

Water variable	Source of variation			Result of Student-Newman-Keuls multiple comparison tests
	Farm	Station	Farm x Sta.	
Winter				
TSS	n.s (1)	** (3)	n.s (3)	CTRL, DECANT, INT < **SEMI**
Phosphate	n.s (1)	*(3)	n.s (3)	Type II error
Ammonia	n.s (1)	n.s (3)	n.s (3)	–
Nitrate	n.s (1)	** (3)	n.s (3)	CTRL, DECANT, INT < **SEMI**
Nitrite	n.s (1)	* (3)	n.s (3)	Type II error
Summer				
TSS	n.s (1)	n.s (3)	n.s (3)	–
Phosphate	n.s (1)	** (3)	n.s (3)	**CTRL** < SEMI INT < **DECANT**
Ammonia	n.s (1)	** (3)	n.s (3)	**CTRL** < DECANT, INT, SEMI
Nitrate	* (1)	** (3)	n.s (3)	Farm A: * (3) < Farm B: * (3) Farm A: CTRL, DECANT, INT < **SEMI** Farm B: type II error
Nitrite	n.s (1)	** (3)	n.s (3)	**CTRL** < DECANT, INT, SEMI

An x between factors indicates interaction and degrees of freedom are in parentheses. Stations in bold differed significantly from the others. Non significant: n.s; significant at 0.05 level: *; significant at 0.01 level: **.

Table 6. Mean values of the benthic macrofauna variables together with standard error (in parentheses)

Benthicmacro-fauna variable	Farm A				Farm B			
	CTRL	DECANT	SEMI	INT	CTRL	DECANT	SEMI	INT
Winter								
Macrofauna density (ind.m^{-2})	533 (319)	99167 (55577)	2300 (495)	7167 (1677)	4100 (1227)	57833 (25578)	15000 (6423)	8733 (3894)
Capitellidae density (ind.m^{-2})	67 (41)	20567 (15221)	867 (147)	433 (108)	367 (216)	8467 (1422)	300 (255)	267 (216)
Summer								
Macrofauna density (ind.m^{-2})	533 (178)	954000 (367000)	3967 (873)	3600 (2478)	8667 (7739)	36600 (12331)	2933 (2861)	12800 (11153)
Capitellidae density (ind.m^{-2})	33 (41)	248333 (36286)	1900 (308)	300 (245)	67 (82)	1333 (471)	300 (255)	1184 (967)

the measured parameters were higher in summer than in winter.

Discussion

In general, the studied variables showed higher values in summer than in winter in the exposed stations but were comparable between seasons in the control stations. This observation can be explained by the fact that in summer, due to high temperatures, the feeding rate in the farms is higher, thus increasing waste effluents. Therefore the effects of farming activity were stronger in summer.

There were no differences in the studied variables between the two farms, which indicate that effluent effects were similar in both farms. The difference of production between the two farms seems to be compensated for by the fact that farm B is equipped with a much larger decantation pond.

In both seasons and in both farms, the semi-intensive discharge zone showed the highest values of the environmental variables. Accordingly, the

Table 7. Effects of factors farm and station on the macrofauna variables tested by Two-way ANOVA and Student-Newman-Keuls multiple comparison tests

Macrofauna variables	Source of variation			Result of student-Newman-Keuls multiple comparison tests
	Farm	Station	Farm x Sta.	
Winter				
Macrofauna density (ind.m^{-2})	** (1)	** (3)	* (3)	Farm A: ** (3) < Farm B: * (3) Farm A: **CTRL** < SEMI, INT < **DECANT** Farm B: type II error
Capitellidae density (ind.m^{-2})	n.s (1)	** (3)	n.s (3)	CTRL, INT, SEMI < **DECANT**
Summer				
Macrofauna density (ind.m^{-2})	n.s (1)	** (3)	** (3)	Farm A: ** (3) > Farm B: n.s (3) Farm A: CTRL, INT, SEMI < **DECANT** Farm B: –
Capitellidae density (ind.m^{-2})	* (1)	** (3)	n.s (3)	Farm A: ** (3) > Farm B: n.s (3) Farm A: **CTRL** < INT, SEMI < **DECANT** Farm B: –

An x between factors indicates interaction and degrees of freedom are in parentheses. Stations in bold differed significantly from the others. Non significant: n.s; significant at 0.05 level: *; significant at 0.01 level: **.

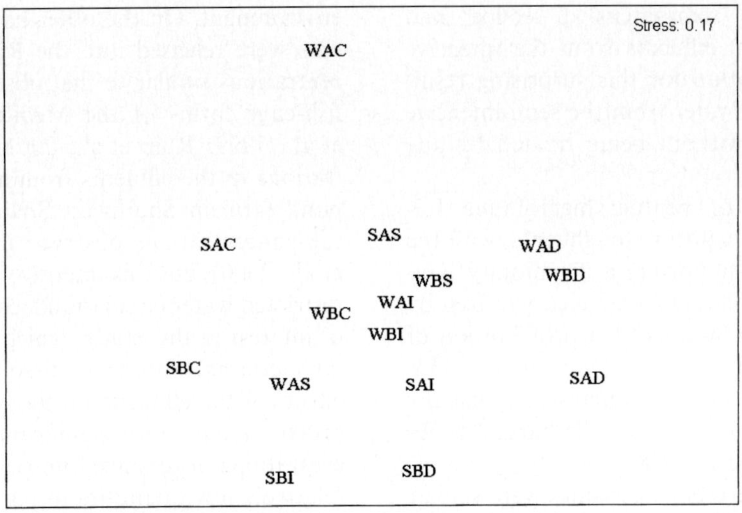

Figure 1. Two dimensional MDS configuration for macrofauna double root-transformed density; first letter refers to the season (Winter or Summer), second letter is the farm (A or B) and last is the station (Control, Semi, Intensive and Decantation). Stress = 0.17.

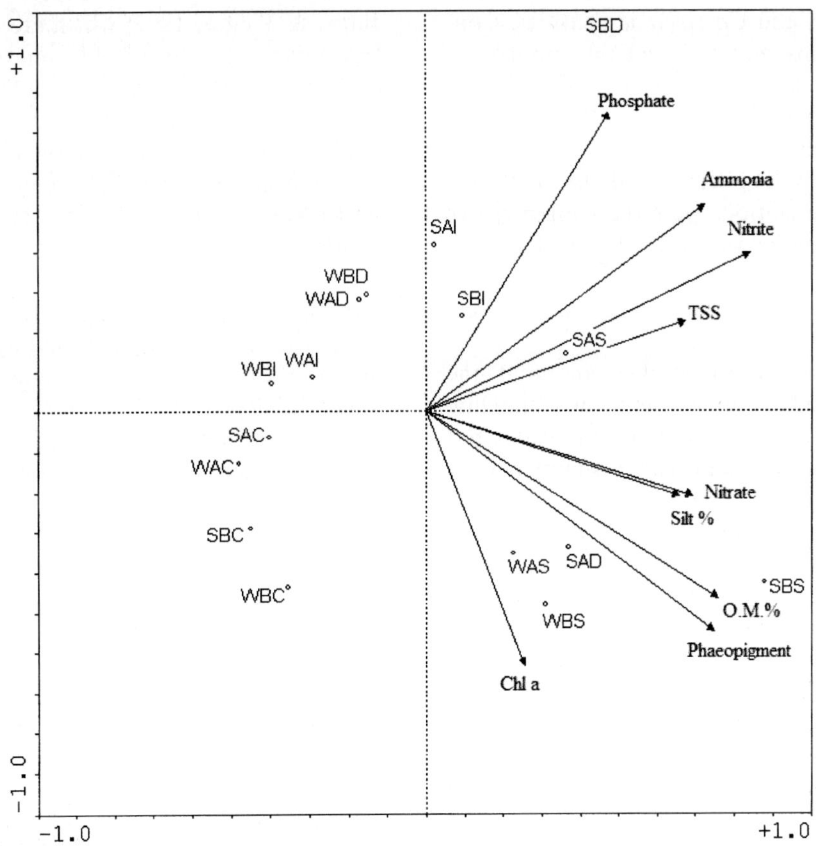

Figure 2. Two dimension PCA configuration for environmental variables; the 2 axes explained 67% of data variability. first letter refers to the season (Winter or Summer), second letter is the farm (A or B) and last is the station (Control, Semi, Intensive and Decantation).

effluents from the semi-intensive section had stronger effects than effluents from the intensive section. An explanation for this surprising result could be that waste water from the semi-intensive section is released without being treated by any decantation process.

The repartition of benthic macrofauna followed a gradient of exposure to effluents, with the decantation ponds supporting a community typical of organically polluted zones, characterized by high total density as well as by a proliferation of polychaetes of the family *Capitellidae*. This repartition of macrofauna is similar to that usually observed in organic waste discharge zones (Pearson & Rosemberg, 1978).

Total macro-invertebrate densities were similar to densities previously recorded in the Ria Formosa and in extensive fish-ponds in most stations whatever the season in both farms (Gamito, 1994; Sprung, 1994). The only exceptions were the decantation ponds that presented abnormally high total macrofauna and *Capitellidae* densities. Consequently, it seems that the peak of opportunist species characteristic of intense organic pollution (Pearson & Rosemberg, 1978) was limited to the decantation ponds.

The PCA ordination confirmed that the most severely impacted stations were the semi-intensive stations, whatever the season or the farm, as well as the intensive and decantation stations in summer. The control stations appeared relatively un-impacted.

From these observations it seems that the decantation ponds are efficient in removing organic matter and nutrients from the water but that their capacity is insufficient in summer when the farms productivity is maximal.

The nutrient concentrations in water of all stations, except the control stations, were higher than those previously observed in the Ria Formosa (Gamito, 1994; Falcão, 1996). In both seasons suspended matter concentration in water surrounding both farms was similar to values measured in tidal channels distant from discharge source (Falcão, 1996). Moreover, the TSS measured in discharge water were less than half of the values measured in extensive fish-ponds (Gamito, 1994). Accordingly, it seems that the largest part of waste feed and fish faeces settle on the pond bottoms and are not exported to the surrounding environment. On the other hand, dissolved nutrients were released into the Ria Formosa in concentrations similar to that observed near intensive fish-cage farms of the Mediterranean sea (Pitta et al., 1999; Ruiz et al., 2001). Nutrient concentrations in the effluents from an intensive earthen pond farm in Southwest Spain were higher than the concentrations observed in this study (Tovar et al., 2000), but this intensive farm was in a more restricted water circulation location than the farms of interest in this study, which could explain such high concentrations of dissolved nutrients. The effects of the effluents on the water column in the present study were significant increases in concentrations of dissolved nutrients, associated with relatively low quantities of particulate wastes.

Apart from values measured in Farm A decantation pond and in Farm B semi-intensive immediate discharge, the organic contents of sediments were comparable or even low in comparison with similar environments in the region (Fonseca, 1989; Junoy & Vieitez, 1992) which are all in a more or less advanced state of eutrophication. Such low concentrations of organic matter in sediments exposed to organic discharge could indicate that semi-intensive farms release relatively low quantities of organic wastes. This could be due to the settlement of organic matter inside the culture ponds. Another explanation would be that, due to the high water renewal rate, organic wastes are rapidly flushed to the ocean or to deposition zones inside the Ria Formosa. A final hypothesis is that most of the organic matter is rapidly consumed and mineralized by benthic fauna. However the sediment around the studied farm did show signs of effluent effects in the immediate discharge zones.

In summary, the effects of semi-intensive fish farm effluents were localized and stronger in the semi-intensive discharge zone whatever the season and the farm. The decantation ponds were relatively efficient at removing suspended matter from discharge water even if their capacity seemed insufficient in summer. Moreover, water from the semi-intensive section ought to be treated by decantation in order to diminish its impact on the surrounding environment.

This study showed that semi-intensive fish farms represent important sources of dissolved nutrients in the Ria Formosa and that the joint effects of several farms in such an enclosed system could be considerable. Finally, the Ria Formosa is subjected to the effects of other types of urban and agricultural pollution and it is important to quantify the contribution of fish farming to the overall nutrient loading of this lagoon.

Acknowledgements

The authors would like to thank the managers and the personnel of the two farms for their collaboration as well as the teachers and students of the University of Algarve for their suggestions, comments or help with field and laboratory work. Financial support was provided by S.O.C.R.A.-T.E.S European exchange program.

References

American Public Health Association (APHA), 1989. Standard Methods for the Examination of Water and Wastewater. 17th ed., APHA (eds).

Falcão, M. M., 1996. Dinamica dos Nutrientes na Ria Formosa: efeitos da interaccao da laguna com as suas interfaces na reciclagem do azoto, fosforo e silica. Ph.D thesis, Universidade do Algarve, Portugal.

Folke, C. & N. Kautsky, 1989. The role of ecosystems for a sustainable development of aquaculture. Ambio 18: 234–243.

Fonseca L. C., 1989. Estudo da influência da 'abertura ao mar' sobre um sistema lagunar costeiro: A Lagoa de Santo André. Ph.D Thesis. Universidade de Lisboa. 355 pp.

Gamito S., 1994. The benthic ecology of some Ria Formosa Lagoons, with reference to the potential for production of the gilthead seabream (Sparus aurata). Ph.D thesis. Universidade do Algarve, Portugal.

Gowen, R. J., H. Rosenthal, T. Maekinen, I. & Ezzi, 1990. Environmental impact of aquaculture activities. In Depauw, N. (ed.), Business Joins Science, Special Publication of European Aquaculture Society 12: 257–283.

Hargreaves, J. A., 1998. Nitrogen biochemistry of aquaculture ponds. Aquaculture 166: 181–212.

Junoy, J. & J. M. Viéitez, 1992. Macrofaunal abundance analyses in the Ria de Foz (Lugo, Northwest Spain). Cahiers de Biologie Marine 33: 331–345.

Larsonneur, C., 1977. La cartographie des depots meubles sur le plateau continental Francais: méthode mise au point et utilisée en Manche. Journal de Recherche Oceanographique 2: 33–39.

Pearson, T. H. & R. Rosenberg, 1978. Macrobenthic succession in relation to organic enrichment and pollution of the marine environment. Oceanography and Marine Biology, an Annual Review 16: 229–311.

Pilkey, O. H., W. J. Neal, J. H. Monteiro & J. M. A. Dias, 1989. Algarve barrier islands: a non-coastal-plain system in Portugal. Journal of Coastal Resources 5: 239–261.

Pillay, T. V. R., 1991. Aquaculture and the Environment. Blackwell Scientific, London.

Pitta, P., I. Karakassis, M. Tsapakis & S. Zivanovic, 1999. Natural vs. Mariculture induced variability in nutrients and plankton in the eastern Mediterranean. Hydrobiologia 391: 181–194.

Rosado, M. C. & A. Bruxelas, 1995. Fauna Bêntica da Ria Formosa. Estudos de Biologia e Conservação da Natureza. ICN, Lisboa.

Ruiz, J. M., M. Perez & J. Romero, 2001. Effects of fish farm loadings on seagrass (*Posidonia oceanica*) distribution, growth and photosynthesis. Marine Pollution Bulletin 42: 749–760.

Sprung, M., 1994. Macrobenthic Secondary production in the intertidal zone of the Ria Formosa – a lagoon in southern Portugal Estuarine. Coastal and Shelf Science 38: 539–558.

Strickland, J. D. H. & T. R. Parson, 1968. A Practical Handbook of Seawater Analysis. Fisheries Research Board of Canada, Ottawa 311 pp.

Tovar, A., C. Moreno, M. P. Manuel-Veiz & M. Garcia-Vargas, 2000. Environmental implications of intensive marine aquaculture in earthen ponds. Marine Pollution Bulletin 40: 981–988.

Wu, R. S. S., K. S. Lam, D. W. Mackay, T. C. Lau & V. Yam, 1994. Impact of marine fish farming on water quality and bottom sediment: a case study of the sub-tropical environment. Marine Environmental Research 38: 115–145.

Wu, R. S. S., 1995. The environmental impact of marine fish culture: toward a sustainable future. Marine Pollution Bulletin 31: 159–166.

Hydrobiologia (2006) 555:299–305
H. Queiroga, M.R. Cunha, A. Cunha, M.H. Moreira, V. Quintino, A.M. Rodrigues, J. Serôdio & R.M. Warwick (eds),
Marine Biodiversity: Patterns and Processes, Assessment, Threats, Management and Conservation
DOI 10.1007/s10750-005-1126-0

Time-series evolution of toxic organisms and related environmental factors in a brackish ecosystem of the Mediterranean Sea

M. Leonardi[1,*], F. Azzaro[1], M. Galletta[1], M. G. Giacobbe[1], M. Masò[2] & A. Penna[3]
[1]*CNR-IAMC – Consiglio Nazionale delle Ricerche, Istituto per l'Ambiente Marino Costiero, Sez. di Messina, Italy*
[2]*CSIC-ICM – Consejo Superior de Investigaciones Cientificas, Institut de Ciencies del Mar, Barcelona, Spain*
[3]*Centro di Biologia Ambientale, University of Urbino, Italy*
*(*Author for correspondence: E-mail: marcella.leonardi@iamc.cnr.it)*

Key words: toxic phytoplankton, *Alexandrium* spp., molecular probes, environmental parameters, brackish waters, Mediterranean Sea

Abstract

In the framework of the EU Project STRATEGY, a short-term study was carried out in the Marinello ecosystem, a small brackish area located on the Tyrrhenian coast of Sicily (Italy). The investigation was aimed at understanding the dynamics of phytoplankton toxic blooms in relation to other planktonic species and environmental conditions. The study started on 10 March 2003, in coincidence with the first detection of *Alexandrium minutum*, a dinoflagellate known as a producer of Paralyzing Shellfish Toxins (PST) and lasted until 4 June 2003, when the bloom collapsed. The specific identity of *A. minutum* was confirmed on field mixed samples, through the use of species-specific PCR-primers targeting the 5.8S rDNA-ITS regions. Water samples and phytoplankton net hauls were taken approximately at 10 days intervals in the Verde Pond, one of the five basins of the Marinello ecosystem, in order to evaluate the incidence of toxic and non-toxic dinoflagellate species over the whole planktonic community. The evolution of the main environmental and trophic parameters (temperature, salinity, dissolved oxygen, POC, C/N, DIN, PO_4–P) was simultaneously investigated. *Alexandrium* blooms were mostly characterized by *A. minutum* (max. 6×10^5 cells l^{-1} on April 11) and *Alexandrium tamarense* as an associated species (max. 2.5×10^4 cells l^{-1} on March 25). During the bloom, dinoflagellates or small flagellates dominated over the other taxa, with a minimum incidence of diatoms. The load of dissolved inorganic nitrogen was maximum in the pre-bloom phase (29 μM on March 19), after which it decreased sharply. An oxygen supersaturation event was registered in coincidence with the *A. minutum* bloom. The amounts of POC ranged between 266 and 658 μg l^{-1} showing a discontinuous temporal trend. A recent introduction of *A. minutum* into the Verde Pond is suggested on the basis of the absence of this species in past years.

Introduction

In recent decades a number of Harmful Algal Blooms (HABs) have been increasingly detected in many Mediterranean littoral areas as well as on other European coasts (Granéli et al.,1998). The onset of this phenomenon represents a serious risk for the economy of most European states, not only in relation to fisheries but also to human health and recreational activities. Various research projects aim at monitoring HAB events in order to manage better their effects in marine and brackish waters.

Brackish areas, as transitional zones between limnetic and marine conditions, are particularly suitable for the study of ecological processes because of the high variability of environmental parameters on small space-time scales.

In the framework of the EU Project STRATEGY (EVK3-CT-2001-00046), we carried out a short-term study, in coincidence with *Alexandrium* blooms in the Verde Pond, one of five basins of the Marinello coastal ecosystem. This is a small brackish area located on the Tyrrhenian coast of Sicily (Italy), characterized by wide fluctuations of temperature and salinity caused by continental and marine inflows. This local situation, together with the progressive confinement from the sea (Giacobbe et al., 1990), determines a high environmental variability on short spatio-temporal scales. For this reason, this ecosystem was optimal as a natural laboratory to follow the evolution of toxic and non-toxic phytoplankton populations in relation to changes in abiotic parameters.

In this paper we report the first finding of *Alexandrium minutum* Halim and *Alexandrium tamarense* (Lebour) Balech, in the Verde Pond, in spring 2003, where other HAB dinoflagellates, such as *Dinophysis sacculus* Stein and *Alexandrium margalefii* Balech, were previously detected (Giacobbe et al., 1995, 2000). We also describe the *A. minutum* and *A. tamarense* bloom dynamics in relation to other planktonic species, climatic and trophic changes. The integrated analysis of the present data can contribute to a better understanding of factors driving *Alexandrium* blooms.

Study area

The Marinello coastal system (Fig. 1) is a small littoral area located on the Tyrrhenian coast of Sicily, Italy (38.13 °N–15.05 °E), below the Tindari cliff (290 m above mean sea level). This area is characterized by strong geomorphological changes with the formation of hook-shaped littoral bars, thus delimiting impermanent small coastal ponds. At present, the system covers an area of approximately 697,000 m^2 and includes five brackish ponds, formed in different periods of the last century. The ponds do not receive either fluvial inputs or direct urban discharges; however, they are affected by various water inflows, such as runoff, groundwater, rainfall and sea water. The heterogeneity of allochthonous inputs, carrying organic and inorganic enrichment, determines the space-time differentiation of the energetic supply, trophic load, salinity (Leonardi et al., 2001), as well as sediment features and molluscan communities (Leonardi & Giacobbe, 2001).

The Verde Pond, object of this study, covers an area of 17,000 m^2, with a volume of 27,200 m^3 and a maximum depth of 2.5 m. Due to its location just under the Tindari cliff, considerable quantities of runoff water are conveyed into the pond. On the other hand, because of the permeability of the sand bank that separates it from the sea, this pond is conditioned by tide phases through periodical infiltrations of sea water. This determines wide changes in salinity and pond extension in relation to the periodical submersion of the tideway (Crisafi et al., 1981). Organic and inorganic compounds are carried by continental and runoff waters, enriched through their passage over the overhanging cliff, heavily colonized by sea-gulls. For this reason Verde Pond has a significant nutrient load (NO_3–N = 6.56 ± 5.22 μM, n = 13), being the most eutrophic among the ponds of this area (Chl a = 23.03 ± 63.56 $\mu g\ l^{-1}$, n = 13, Leonardi et al., 2000). This scenario, combined with particular climatic conditions (e.g., high summer temperatures, lack of wind, etc.), occasionally leads to dystrophic processes, associated with macroscopic, greenish water discolouration. Rare anoxia crises and fish deaths also occur in this basin (Leonardi, summer 2000, personal observation), in relation to temporary dystrophic events.

Material and methods

The study started in 10 March 2003, in coincidence with the first detection of *Alexandrium minutum*, and lasted until 4 June 2003. Water samples were collected from the surface layer (0.5 m) in a central point of the Verde pond, approximately at 10 days intervals. Phytoplankton net hauls were also carried out. Temperature, salinity and oxygen measurements were taken by an oceanographic multiparametric sensor (Heron Rocket – AMIC). Nutrient concentrations (nitrite, nitrate and orthophosphate ions) were determined according to Genovese & Magazzù (1969), except for ammonia (Aminot & Chaussepied, 1983), using a spectrophotometer VARIAN-CARY50.

Particulate Organic Carbon and Particulate Nitrogen (POC and PN) were measured on water samples (500–1000 ml) pre-screened through a 200 μm net to remove larger zooplankton and filtered on pre-combusted (450 °C, 4 h) Whatman GF/F filters (0.45 μm pore size). After removal of the inorganic carbon, by exposing to hydrochloric acid fumes (Iseki et al., 1987), the analysis was performed using acetanilide as standard at 970 °C with a Perkin Elmer 2400 CHN Autoanalizer.

Phytoplankton was analyzed in Utermöhl settling chambers using Lugol's fixed samples. Cell counts were performed by Axiovert (Giacobbe et al., 1996). Fixed specimens were stained with Calcofluor White M2R (Fritz & Triemer, 1985) and examined by epifluorescence light microscopy under UV excitation (Axiovert 200, filter set Zeiss Fs 01, 1000× magnification). Some specimens were analyzed by scanning electron microscopy. Tabular formula and morphological features of the thecal plates were studied following the Balech's criteria (1995).

The detection of *Alexandrium minutum* and other *Alexandrium* spp. was also carried out in triplicate on two field mixed samples, through the use of species-specific PCR-primers targeting the 5.8S rDNA-ITS regions (Penna & Magnani, 1999; Galluzzi et al., 2004).

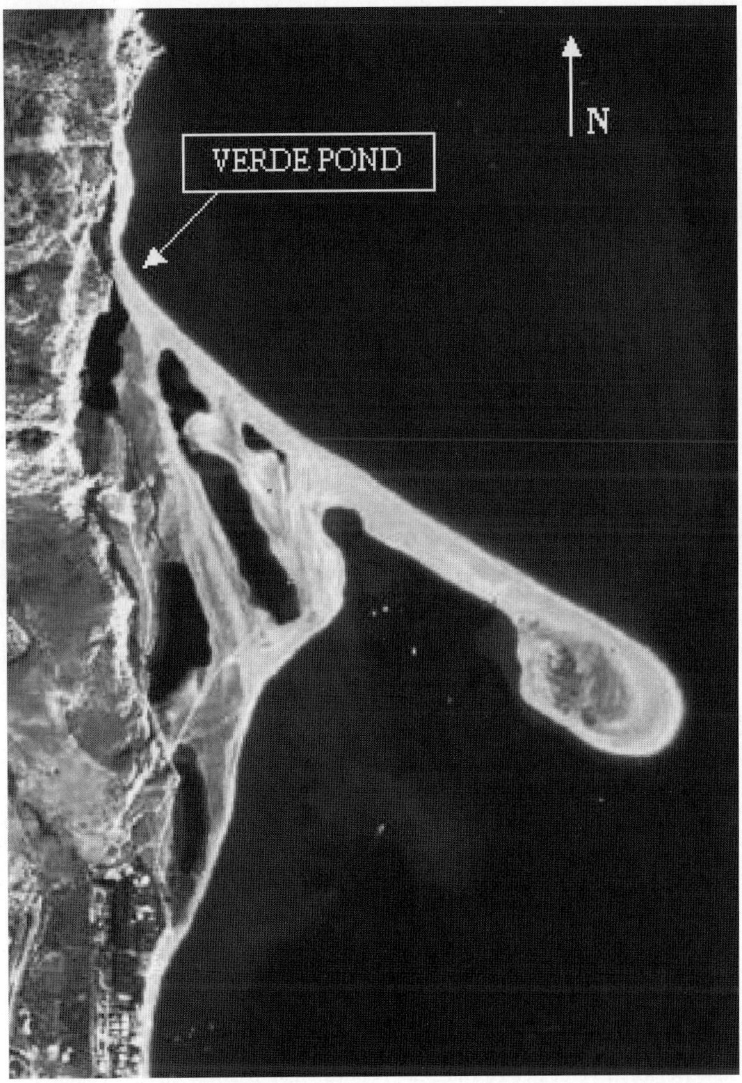

Figure 1. Remote sensing picture of the Marinello ecosystem (Sicily, Italy) with the Verde Pond.

Results

Temperature and salinity showed trends typical of Mediterranean brackish areas, with marked haline excursions and spring thermal increases (Fig. 2). During the study period, salinity ranged from 23.00 (March 10) to 27.07 psu (June 4). A rapid increase was registered on March 19 (25.72 psu) in coincidence with the syzygy tide, that determined a significant inflow of sea waters. On the same date, the temperature fell to 12.8 °C, not exceeding 15 °C until mid-April. Later, a thermal peak (23.4 °C) marked the seasonal change (April 29).

The first findings of *Alexandrium minutum* in the Verde Pond indicated cell densities of 1.8×10^5 and 2.9×10^4 cells l^{-1} (March 10 and March 19, Fig. 3). Later, the *A. minutum* incidence progressively increased up to a maximum of 6×10^5 cells l^{-1} (April 11) and the simultaneous presence of *Alexandrium tamarense* was observed, although at lower densities (max. 2.5×10^4 cells l^{-1} on March 25, Fig. 3).

The bloom of *A. minutum* on April 11 led to an oxygen maximum (12 mg l^{-1}) and water supersaturation (136%) also due to the high density of the total phytoplankton (3.8×10^6 cells l^{-1}, not shown), which was followed by a gradual O_2 decline to 7.74 mg l^{-1} (Fig. 2).

During all the study period the phytoplankton assemblage was characterized by a clear dominance of dinoflagellates or small flagellates over the other *taxa*, with a minimum incidence of diatoms (Fig. 3).

On all sampling occasions the concentration of dissolved inorganic nitrogen (DIN) was rather high, exceeding 5 μM and peaking to 29 μM on March 19 (Fig. 2). This rapid increase, coinciding with the syzygy tide, was probably due to nutrient resuspension from the bottom. Then, DIN values decreased to 7.9 and 8.7 μM during *A. tamarense* and *A. minutum* blooms, respectively (March 25

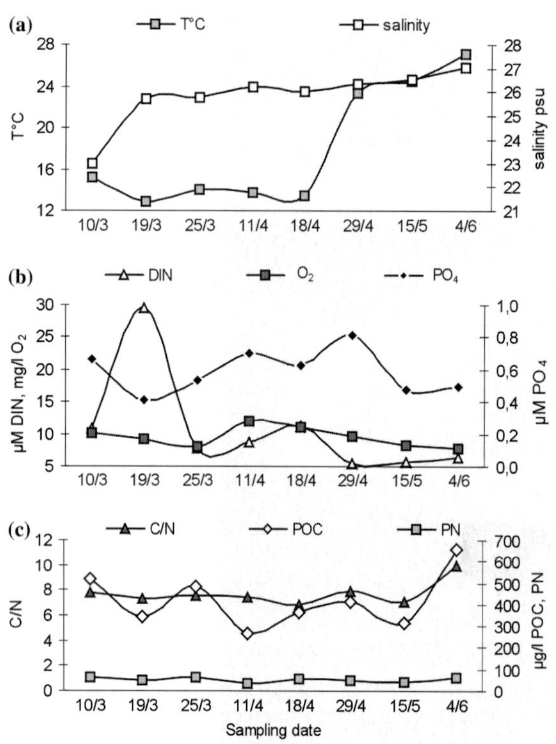

Figure 2. Trends in environmental parameters (a and b) and particulate organic compounds (c) in the Verde Pond during the study period (2003).

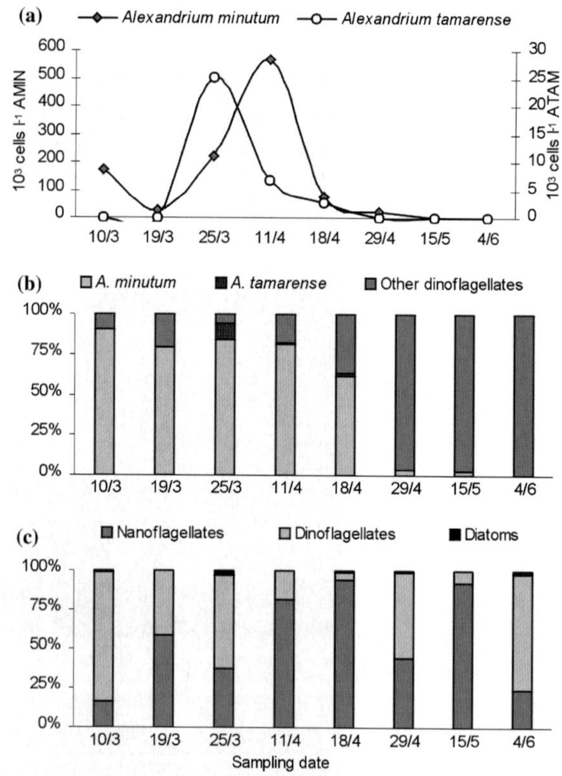

Figure 3. Trends in cell densities of *Alexandrium* species (a), their relative incidence over the whole dinoflagellate assemblage (b) and changes in the composition of the phytoplankton community (c) over the study period.

and April 11, Fig. 3), suggesting a direct uptake of nutrients (mostly nitrates) by these populations.

PO_4–P concentration was always higher than 0.4 μM, increasing to 0.8 μM on April 29 (Fig. 2), in coincidence with the total disappearance of *A. tamarense*. The N/P ratio ranged from 7.04 (April 29) to 17.93, except for a peak of 71.83 (March 19) in the pre-bloom phase. In correspondence of the *A. tamarense* and *A. minutum* peaks the N/P values were 14.9 and 12.44, close to the Redfield uptake ratio (data not shown).

The incidence of particulate organic carbon and particulate nitrogen (Fig. 2) was rather marked over the entire study period, with maxima of 658 μg of C l^{-1} and 66 μg of N l^{-1}, registered on June 4. POC and PN displayed a highly significant correlation. A statistically significant correlation was found between POC values and the relative incidence of dinoflagellates over the phytoplankton assemblage ($r = 0.83$, $p < 0.01$). Also the C/N ratio values, ranging between 6 and 8 suggested the high correspondence of the particulate organic carbon with the phytoplankton biomass (Redfield ratio).

The subsequent decline of the *Alexandrium* populations occurred in correspondence to the sharp thermal increase (April 29) and related summer weather conditions such as high pressure, fast evaporation, lack of wind (data not shown). This led to a change in the composition of the phytoplankton assemblage (Fig. 3), favouring the development of other dinoflagellate species, mostly *Prorocentrum minimum* (Pavillard) Schiller, as well as nitrogen depletion due to phytoplankton uptake. The increase of competitive pressure led to the final disappearance of *Alexandrium* species on June 4. On the same date, C/N increased to 9.91 indicating the shift towards a post-bloom phase.

The microscopical identification of *A. minutum* and *A. tamarense* (Fig. 4) was also confirmed by PCR amplifications obtained on field mixed samples (Fig. 5) – both from the Verde Pond and, for comparison, from Porto Marmoreo (Siracusa, Sicily). By using *Alexandrium* species-specific primers, designed for *A. minutum, Alexandrium catenella* (Whedon & Kofoid) Balech, *Alexandrium taylorii* Balech and *A. tamarense*, PCR amplification of the 5.8S rDNA-ITS regions yielded products of different base pair size for the field samples, containing *Alexandrium* spp. cells. Species-specific PCR amplifications for *A. minutum* revealed positive results for the presence of this species in the Verde Pond and Porto Marmoreo field samples, giving a PCR species-specific fragment of 212 bp (Fig. 5, lanes B1 and A1, respectively). When the other species-specific PCR primers were applied to

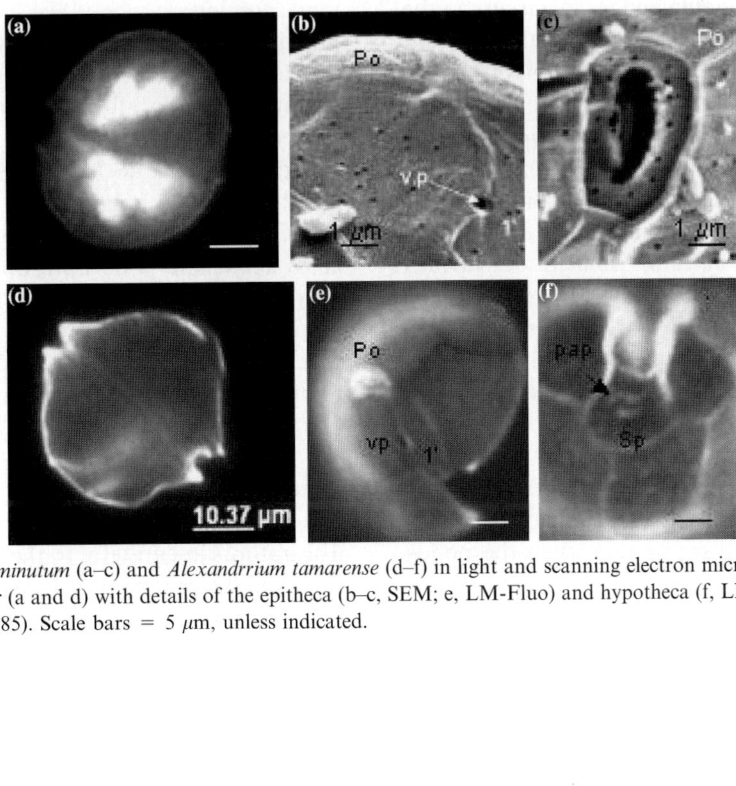

Figure 4. Alexandrium minutum (a–c) and *Alexandrrium tamarense* (d–f) in light and scanning electron microscopy. Whole cells after staining with Calcofluor (a and d) with details of the epitheca (b–c, SEM; e, LM-Fluo) and hypotheca (f, LM-Fluo). Plate tabulation according to Balech (1985). Scale bars = 5 μm, unless indicated.

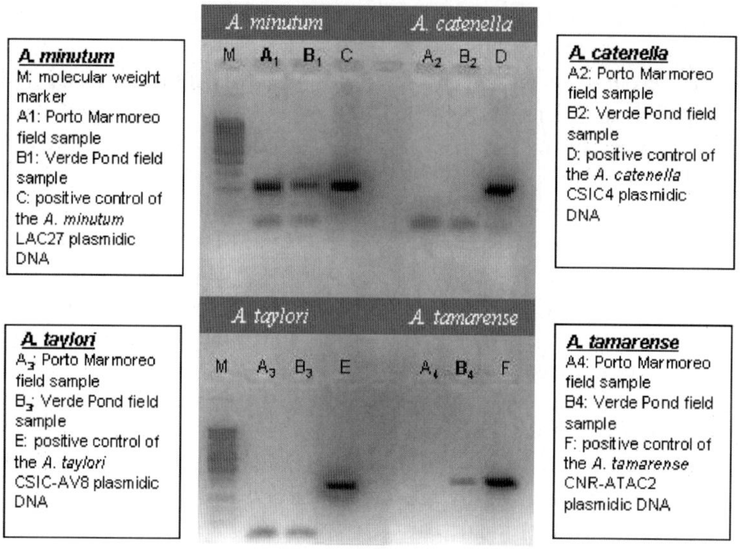

Figure 5. PCR amplification of the 5.8S rDNA-ITS regions on field samples revealing *Alexandrium minutum* (lane B1) and *Alexandrium tamarense* (lane B4). (CNR – Consiglio Nazionale delleRicerche; CSIC – Consejo Superior de Investigaciones Cientificas).

the same field samples, the presence of *A. tamarense* was detected only in the Verde Pond sample with a PCR amplified fragment of 320 bp (Fig. 5, lane B4), in agreement with the microscopic examination of the same field sample.

Discussion

The finding of *A. minutum* and *A. tamarense* in the Verde pond (Sicily, Italy) suggests their recent introduction into this area, since previous HAB surveys in the same pond only evidenced other species, such as *Alexandrium margalefii* and *Alexandrium* sp. (Giacobbe et al., 1995, 2000). The different identities were further confirmed by PCR assays on field samples and molecular analyses on *Alexandrium* clonal cultures isolated on various occasions. Thus, the 5.8S rDNA-ITS regions gene proved to be useful genetic markers at species level for *Alexandrium* spp. and offered a high resolution level for the identification of *A. minutum* and *A. tamarense*.

The existence of an *A. tamarense* strain obtained in previous years from a neighbouring pond, led us to suppose that cell inoculation may have occurred during spring tide when the ponds are subject to water inflows. Beyond this origin, as well as the possible input of cells from the nearby bay (sea storms), the existence of an *Alexandrium* seedbed cannot be ruled out, requiring further studies addressed to this specific issue.

Then, proliferations of *Alexandrium* in this ecosystem seem to be favoured by the peculiar topography of the site – limited extension, shallow waters, and consequently low hydrodynamism, as well as by the concomitance of various factors, mostly nutrient pulses of different origin and climatic forcing.

Although the *Alexandrium* blooms in the Verde Pond did not reach levels as high as in other Mediterranean localities, the pattern of dominance of associated species and *taxa* was similar to that found at Arenys de Mar in Spain (Calbet et al., 2003). However, unlike the situation reported for the Catalan coast (Garcés et al., 2000), in this ecosystem there is no anthropogenic activity that may stimulate HAB events. Thus, our study suggests that natural cues may have provided an inoculum for the later bloom development, rather than human-assisted mechanisms.

Acknowledgements

The authors are grateful to the rangers Vanella and Zucchini (Department of Protected Areas of Province Regional of Messina, Italy), who collected the monitoring data from the 'Natural Reserve of Marinello Ponds.' We also thank A. Marini and A. Cosenza for chemical analyses and technical assistance; L. Galluzzi for PCR oligonucleotide primer contribution.

References

Aminot, A. & M. Chaussepied, 1983. Manuel d'analyses chimiques en milieu marin – Centre National pour l'Exploration des Oceans, 1–395.

Balech, E., 1995. The Genus *Alexandrium* Halim (Dinoflagellata). Co. Cork, Ireland, Sherkin Island Marine Station, Sherkin Island. pp. 151.

Calbet, A., D. Vaquè, J. Felipe, M. Vila, M. M. Sala, M. Alcaraz & M. Estrada, 2003. Relative grazing impact of microzooplakton and mesozooplankton on a bloom of the toxic dinoflagellate *Alexandrium minutum*. Marine Ecology Progress Series 259: 303–309.

Crisafi, E., S. Giacobbe & M. Leonardi, 1981. Nuove ricerche idrobiologiche nell'area lagunare di Oliveri-Tindari (Messina). 1: Morfologia dei bacini e caratteristiche fisicochimiche delle acque e dei sedimenti. Memorie di Biologia Marina e di Oceanografia 11: 139–186.

Fritz, L., & R. E. Triemer, 1985. A rapid simple technique utilizing Calcofluor White M2R for the visualization of dinoflagellate thecal plates. Journal of Phycology. 21: 662–664.

Galluzzi, L., A. Penna, E. Bertozzini, M. Vila, E. Garcés & M. Magnani, 2004. Development of a real-time PCR assay for rapid detection and quantification of *Alexandrium minutum* (a dinoflagellate). Applied Environmental Microbiology 70: 1199–1206.

Garcés, E., M. Masò, M. Vila & J. Camp, 2000. Harmful algae events in the Mediterranean: are they increasing?. Harmful Algal News 20: 10–11.

Genovese, S. & G. Magazzù, 1969. Manuale d'analisi per le acque salmastre. Ed. Universitaria, Messina, pp. 134.

Giacobbe, M. G., F. Oliva, R. La Ferla, A. Puglisi, E. Crisafi & G. Maimone, 1995. Potentially toxic dinoflagellates in Mediterranean waters (Sicily) and related hydrobiological conditions. Aquatic Microbial Ecology 9: 63–68.

Giacobbe, M. G., F. D. Oliva & G. Maimone, 1996. Environmental factors and seasonal occurrence of the dinoflagellate *Alexandrium minutum*, a PSP-potential producer, in a Mediterranean lagoon. Estuarine Coastal & Shelf Science 42: 539–549.

Giacobbe, M. G., A. Penna, A. Ceredi, A. Milandri, R. Poletti & X. Yang, 2000. Toxicity and ribosomal DNA of the dinoflagellate *Dinophysis sacculus* (Dinophyta). Phycologia 39: 177–182.

Giacobbe, S., M. Leonardi, F. Azzaro & P. Rinelli, 1990. Descrizione di un esempio di "confinamento": l'area lagunare di Oliveri-Tindari (Messina). Oebalia XVI-2: 675–678.

Granéli, E., G. A. Codd, B. Dale, E. Lipiatou, S. Y. Maestrini & H. Rosenthal, 1998. Harmful Algal Blooms in European Marine and Brackish Waters. Eurohab Science Initiative, Kalmar (Sweden) November 5–7, 1998.

Iseki, K., R. W. Mac Donald & E. Carmack, 1987. Distribution of particulate matter in the south-eastern Beaufort Sea in late summer. Proceedings of NIPR Symposium on Polar Biology 1: 35–46.

Leonardi, M., F. Azzaro, M. Azzaro, F. Decembrini & L. S. Monticelli, 2000. Ciclo della sostanza organica nell'ecosistema lagunare di Tindari (ME). Biologia Marina Mediterranea 7: 222–232.

Leonardi, M., R. M. Cavalli, S. Pignatti & F. Raffa, 2001. Shapes and colours in the Tindari coastal ecosystem: new prospects for preservation and management. Proceedings 3rd International Congress on "Science and technology for the safeguard of cultural heritage in the Mediterranean basin" 2: 770–774.

Leonardi, M. & S. Giacobbe, 2001. The Oliveri-Tindari lagoon (Messina, Italy): evolution of the trophic-sedimentary environment and mollusc communities in the last twenty years. In Faranda, F. M., L. Guglielmo, & G. Spezie (eds.) Mediterranean Ecosystems: Structures and Processes. Springer-Verlag, Italia: 305–310.

Penna, A. & M. Magnani, 1999. Identification of *Alexandrium* (Dinophyceae) species using PCR and rDNA-targeted probes. Journal of Phycology 35: 615–621.

Management and Conservation

Hydrobiologia (2006) 555:309–320

H. Queiroga, M.R. Cunha, A. Cunha, M.H. Moreira, V. Quintino, A.M. Rodrigues, J. Serôdio & R.M. Warwick (eds),
Marine Biodiversity: Patterns and Processes, Assessment, Threats, Management and Conservation
DOI 10.1007/s10750-005-1127-z

Assessing the sensitivity of seabed species and biotopes – the Marine Life Information Network (*MarLIN*)

Keith Hiscock* & Harvey Tyler-Walters
Marine Life Information Network (MarLIN), Marine Biological Association of the UK, PL1 2PB, Plymouth, UK
*(*Author for correspondence: E-mail: k.hiscock@mba.ac.uk)*

Key words: biotopes, conservation, GIS, management, sensitivity, species

Abstract

Science-based approaches to support the conservation of marine biodiversity have been developed in recent years. They include measures of 'rarity', 'diversity', 'importance', biological indicators of water 'quality' and measures of 'sensitivity'. Identifying the sensitivity of species and biotopes, the main topic of this contribution, relies on accessing and interpreting available scientific data in a structured way and then making use of information technology to disseminate suitably presented information to decision makers. The Marine Life Information Network (*MarLIN*) has achieved that research for a range of environmentally critical species and biotopes over the past four years and has published the reviews on the *MarLIN* Web site (www.marlin.ac.uk). Now, by linking the sensitivity database and databases of survey information, sensitivity mapping approaches using GIS are being developed. The methods used to assess sensitivity are described and the approach is advocated for wider application in Europe.

Introduction

The need for information

Conservation biologists are being challenged to identify scientifically sound approaches to the implementation of various directives, conventions, policies and statutes that apply to the marine environment. In order to use often complex data effectively, marine scientists need to develop structures, criteria and tools to organize data and create information.

Structures to organize data and information

Structures include classifications (for instance, species dictionaries, a hierarchical classification of biotopes, categorization of human activities) and measurements (for instance, of diversity, of rarity, of degree of threat). The European Register of Marine Species (ERMS: www.marbef.org/data/erms.php) is a catalogue and dictionary of marine species recorded in European seas. In recent years, the development of a classification for seabed biotopes within the European Union Nature Information System (EUNIS) (Davies & Moss, 2002) (based on the now updated UK seabed biotopes classification: www.jncc.gov.uk/marine/biotopes) has provided a structure to compare like-with-like and to identify the diversity of biotopes in an area. There are many lists of different types of human activities although, to apply scientific information, the environmental factors that those activities influence or create need to be categorized.

Conservation of biodiversity also requires survey data to be organized to identify the number and abundances of species and habitats in different locations so that comparisons can be made between locations and so that change can be tracked. Counts of numbers of species may be difficult to achieve and diversity indices may not be suitable for advising nature conservation. Measures such as 'taxonomic distinctness', based on inter-relatedness between species rather than relative abundances, provide practical alternatives to potentially time consuming surveys (Warwick & Clarke, 2001).

Practical tools to inform the protection of biodiversity

Tools that inform the protection of biodiversity and the sustainable use of resources are likely to be of three different types:

1. criteria that identify 'importance' (of an area, a physiographic feature, a biotope or a species) for maintenance of biodiversity;
2. measurements that indicate biological quality or if a species population or habitat is degraded, and
3. information that identifies the 'sensitivity' of species and biotopes to human activities.

Methods to assess marine natural heritage importance of areas are based on well-established criteria (see, for instance, Hiscock, 1997; Hiscock et al., 2003). The most recent thinking on criteria and their application is to be seen in the work undertaken by the OSPAR Commission (the Convention on the Protection of the Marine Environment of the North-East Atlantic: www.ospar.org). Attention is also drawn to the development of criteria for assessing rarity (Sanderson, 1996). Much up-to-date information on assessing 'importance' can be found on the Web site of the UK Joint Nature Conservation Committee (www.jncc.gov.uk/marine).

'Quality' may be a comparative measure where 'best examples' of a physiographic type of feature are identified as representative locations and managed for biodiversity conservation. 'Quality' may also be measured through the presence and abundance of species indicative of environmental conditions including chemical contaminants and nutrients (see, for instance, Borja et al., 2003).

Information on the environmental preferences, growth rates, longevity and likely sensitivity to environmental change (either man-made or due to natural events) is essential background information for interpreting the likely impact of human activities and whether that impact matters for the continued survival of a species or biotope. The Marine Life Information Network (*MarLIN*) programme provides such key biology and sensitivity information and therefore supports environmental management and protection as a step towards good stewardship of the marine environment. The approach developed by *MarLIN* is described in the rest of this paper.

Developing a resource of biology and sensitivity key information: the *MarLIN* programme

Historical perspective

One of the earliest approaches to assessing wildlife sensitivity in an objective way was the 'oil spill vulnerability index' developed by Gundlach & Hayes (1978). Their index was an important starting point for preparing many sensitivity maps related to oil spills but was based mainly on likely persistence of oil in relation to shore type. Such 'surrogates' for sensitivity have been used for a long time as a 'shortcut' or 'best possible' approach. Even very recent electronic systems for identifying likely sensitivity of locations especially to prepare oil spill contingency plans use 'shore-type' mapping and the location (in Britain) of Sites of Special Scientific Interest and other conservation designations as indicators of sensitivity.

The need to make better use of scientific information and field survey data for environmental protection and management was promoted in Hiscock (1997) and updated in Hiscock et al. (2003). The approaches to assessing sensitivity developed under the work described here were designed to replace wherever possible the 'surrogates' used up to now.

Key information research

Sensitivity assessment involves the review of available literature on the life history characteristics, distribution, environmental preferences and any effects of environmental perturbation on the chosen species. In the case of biotopes, information on the community ecology and structure of the biotope (or similar community) and its associated species is collated. 'Key information' is information that is considered relevant to understanding the biology of a species or the ecology of a biotope including those environmental factors important for the survival of the species and, through characterizing, key structural and key functional species, the biotope. The categories of information researched are shown in Table 1.

Assessing 'sensitivity'

Introduction

The *MarLIN* programme has been collating, interpreting, storing and disseminating information on key features of the biology of seabed species and the likely sensitivity of species and biotopes to a range of environmental factors since August 1999. The *MarLIN* Biology and Sensitivity Key Information (BASKI) reviews are managed in a Microsoft Access database, and presented as user-friendly Web pages over the World Wide Web. During the development of the *MarLIN* programme, we were greatly aided by a technical steering group drawn from nature conservation agencies, academic institutes and fisheries laboratories. The task of that group was to determine just how 'sensitivity' could be assessed and how to overcome practical difficulties of categorizing sensitivity and of representing sensitivity in a way that could be used by decision-makers. The *MarLIN* sensitivity assessment rationale and

Table 1. Categories of information researched as a part of *MarLIN* Biology and Sensitivity Key Information assessment

Species Key Information	Biotope Key Information
Basic information	Basic information
Taxonomy and identification	Biotope classification
General biology (adult and larval/juvenile), including:	Ecology, including:
• typical abundance;	• ecological relationships;
• typical food and feeding mechanism;	• seasonal and longer term changes;
• growth rates, and	• habitat complexity;
• mobility and sociability	• productivity;
	• recruitment processes, and
	• time for community to reach maturity
Habitat preferences and distribution, including:	Habitat preferences and distribution, including:
• British, Irish and global distribution;	• physiographic preferences;
• Physiographic, substratum, wave exposure, and tidal stream preferences, and	• water clarity and temperature preferences, and
• preferred salinity range	• limiting nutrients
Reproduction and longevity, including:	Species composition, including:
• reproductive type and fecundity;	• characterizing species,
• generation time and longevity, and	• species indicative of biotope sensitivity;
• reproductive season	• species especially found within the biotope, and
	• nationally rare or scarce species associated with the biotope
Intolerance, recoverability and sensitivity	Intolerance, recoverability and sensitivity
Marine natural heritage importance, including:	Marine natural heritage importance, including:
• relevant designations for species;	• relevant designations for habitat;
• national status;	• national status;
• ecosystem importance;	• Habitat Directive Annex 1 habitat;
• Red Data Book category	• UK Biodiversity Action Plan habitat;
	• biotope importance, and
	• exploitation

definitions of terms and scales used prior to March 2003 are given by Tyler-Walters et al. (2001) (available from www.marlin.ac.uk). The *MarLIN* programme used a *sensu stricto* definition of 'sensitivity' until January 2003 together with a separate assessment of recoverability. However, those two elements are now combined to facilitate mapping of sensitivity and to correspond to the UK Government Review of Marine Nature Conservation (see Laffoley et al., 2000), which defined 'sensitivity' as:

> "A very sensitive habitat or species is one that is very easily adversely affected by external factors arising from human activities and is expected to recover over a very long period or not at all. A sensitive habitat or species is one that is easily affected by a human activity, and is expected to only recover over a long period."

The sensitivity assessment process

The procedure for assessing sensitivity of a species is illustrated in Figure 1. The assessment process involves using information from the scientific literature to identify the likely intolerance of a species to change in an environmental factor arising from human activities or natural events. The rationale then assesses the likely recoverability of the species following cessation of the human activity or natural event. Intolerance and

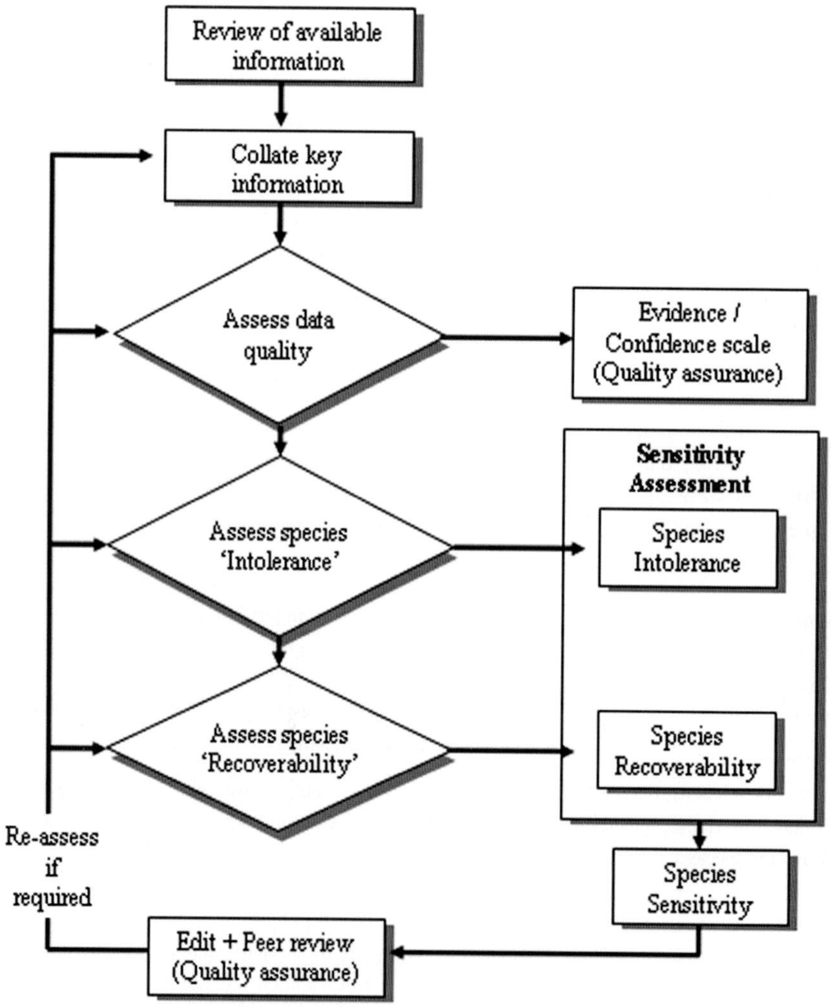

Figure 1. Protocol for researching Biology and Sensitivity Key Information for species. The protocol for biotopes is similar.

recoverability are then combined to provide a meaningful assessment of their overall sensitivity to environmental change.

Assessment of intolerance requires a specified level of environmental perturbation. Therefore, the *MarLIN* programme has developed a set of 'benchmark' levels of environmental change in the environmental factors against which to assess sensitivity. The benchmarks also allow intolerance and hence sensitivity to be compared against the predicted effects of planned projects or proposals.

The likely intolerance of the species is assessed with respect to a specified magnitude and duration of change (standard benchmarks) for 24 separate environmental factors (see Table 2).

Precedence is given to direct evidence of effect or impact. For example, information from targeted studies/experiments that looked at the effect of the specific factor on the species, or targeted work/experiments on the effects of similar factors on similar species or studies of the likely effects of a factor. The assessment of intolerance (see Table 3) is then made by reference to the reported change in environmental factors and their impact, relative to the magnitude and duration of the standard benchmarks and other relevant key information. The likely recoverability (see Table 3) of a species from disturbance or damage is dependent on its ability to regenerate, regrow, recruit or recolonize, depending on the extent of damage incurred and hence its intolerance.

The combination of intolerance and recoverability rationale shown in Table 4 takes into account the fact that, while many habitats and species will be adversely affected, even destroyed, by an activity or event, such effects 'matter' to the continued survival of that feature if it does not have the potential to recover.

Where change in a factor may affect an associated species (for instance, a grazer or a predator), the likely effect is, as far as possible, taken into account when assessing intolerance of the species being researched.

The above definitions and scenarios give rise to the sensitivity matrix shown in Table 5. The sensitivity matrix is used to automate the combination of 'intolerance' and 'recoverability' to create the sensitivity index on the *MarLIN* Web site. The sensitivity matrix shown in Table 5 is not symmetrical because the scale represents scenarios in which the potential damage to the species or habitat 'matters'. The scale is intuitively weighted towards recoverability although, in a few cases, intolerance has been given a greater weight rather than under-estimate the potential sensitivity of marine habitats and species.

The following assumptions are made in the assessment of sensitivity.

- The intolerance, recoverability, and sensitivity of a species or biotope to a specified level of environmental perturbation are dependent on

Table 2. Factors researched for sensitivity as a part of *MarLIN* Biology and Sensitivity Key Information assessment

Physical factors	Chemical factors	Biological factors
Substratum loss	Synthetic compounds	Introduction of microbial pathogens
Smothering	Heavy metals	Introduction of non-native species and translocation
Suspended sediment	Hydrocarbons	Selective extraction of this species
Desiccation	Radionuclides	Selective extraction of other species
Changes in emergence regime	Changes in nutrient levels	
Changes in water flow rate	Changes in salinity	
Changes in temperature	Changes in oxygenation	
Changes in turbidity		
Changes in wave exposure		
Noise		
Visual presence		
Physical disturbance and abrasion		
Displacement		

the biology of the species or ecology of the biotope.

- The intolerance of a hypothetical 'average' species population is assessed, representing a population in the middle of its range of habitat preferences. Populations at the limits of their environmental preferences are likely to be more intolerant of environmental perturbation.
- Recoverability assumes that the impacting factor has been removed or has been stopped and the habitat returned to a state capable of supporting the species or biotope in question.
- A species population or biotope is considered to have recovered once it has reached a state similar to that which previously existed and, in the case of a biotope, includes the main characterizing and key structural or functional species.

However, in all cases, the evidence and key information used and any judgments made in the sensitivity assessment are available on-line. The information used and evidence collated is fully referenced throughout.

Table 3. Degrees of intolerance and recoverability

Intolerance The susceptibility of a species population to damage, or death, from an external factor. Intolerance is assessed relative to change in a specific factor.	
Rank	Definition
High	The species population is likely to be killed/destroyed by the factor under consideration.
Intermediate	Some individuals of the species may be killed/destroyed by the factor under consideration and the viability of a species population may be reduced.
Low	The species population will not be killed/destroyed by the factor under consideration. However, the viability of a species population will be reduced.
Tolerant	The factor does not have a detectable effect on survival or viability of a species or structure and functioning of a biotope.
Tolerant*	Population of a species may increase in abundance or biomass as a result of the factor.
Not relevant	This rating applies to species where the factor is not relevant because they are protected from the factor (forinstance, through a burrowing habit), or can move away from the factor.

Recoverability The ability of a habitat, community, or individual (or individual colony) of species to redress damage sustained as a result of an external factor. Recoverability is only applicable if and when the impacting factor has been removed or has stopped. Ranks also only refer to the recoverability potential of a species, based on their reproductive biology etc.	
Rank	Definition
None	Recovery is not possible
Very low/ none	Partial recovery is only likely to occur after about 10 years and full recovery may take over 25 years or never occur.
Low	Only partial recovery is likely within 10 years and full recovery is likely to take up to 25 years.
Moderate	Only partial recovery is likely within 5 years and full recovery is likely to take up to 10 years.
High	Full recovery will occur but will take many months (or more likely years) but should be complete within about five years.
Very high	Full recovery is likely within a few weeks or at most 6 months.
Immediate	Recovery immediate or within a few days.
Not relevant	For when sensitivity is not relevant or cannot be assessed. Recoverability cannot have a value if there is no sensitivity and is thus 'Not relevant'.

Table 4. Defining 'sensitivity'

Sensitivity scale	Sensitivity definition or scenario
Very High	'Very high' sensitivity is indicated by the following scenario: • The habitat or species is very adversely affected by an external factor arising from human activities or natural events (either killed/destroyed, 'high' intolerance) and is expected to recover only over a prolonged period of time, i.e. >25 years or not at all (recoverability is 'very low' or 'none'). • The habitat or species is adversely affected by an external factor arising from human activities or natural events (damaged, 'intermediate' intolerance) but is not expected to recover at all (recoverability is 'none').
High	'High' sensitivity is indicated by the following scenarios: • The habitat or species is very adversely affected by an external factor arising from human activities or natural events (killed/destroyed, 'high' intolerance) and is expected to recover over a very long period of time, i.e. >10 or up to 25 years ('low' recoverability). • The habitat or species is adversely affected by an external factor arising from human activities or natural events (damaged, 'intermediate' intolerance) and is expected to recover over a very long period of time, i.e. >10 years (recoverability is 'low', or 'very low'). • The habitat or species is affected by an external factor arising from human activities or natural events (reduced viability **, 'low' intolerance) but is not expected to recover at all (recoverability is 'none'), so that the habitat or species may be vulnerable to subsequent damage.
Moderate	'Moderate' sensitivity is indicated by the following scenarios: • The habitat or species is very adversely affected by an external factor arising from human activities or natural events (killed/destroyed, 'high' intolerance) but is expected to take more than 1 year or up to 10 years to recover ('moderate' or 'high' recoverability). • The habitat or species is adversely affected by an external factor arising from human activities or natural events (damaged, 'intermediate' intolerance) and is expected to recover over a long period of time, i.e. >5 or up to 10 years ('moderate' recoverability). • The habitat or species is affected by an external factor arising from human activities or natural events (reduced viability **, 'low' intolerance) but is expected to recover over a very long period of time, i.e. >10 years (recoverability is 'low', 'very low'), during which time the habitat or species may be vulnerable to subsequent damage.
Low	'Low' sensitivity is indicated by the following scenarios: • The habitat or species is very adversely affected by an external factor arising from human activities or natural events (killed/destroyed, 'high' intolerance) but is expected to recover rapidly, i.e. within 1 year ('very high' recoverability). • The habitat or species is adversely affected by an external factor arising from human activities or natural events (damaged, 'intermediate' intolerance) but is expected to recover in a short period of time, i.e. within 1 year or up to 5 years ('very high' or 'high' recoverability). • The habitat or species is affected by an external factor arising from human activities or natural events (reduced viability*, 'low' intolerance) but is expected to take more than 1 year or up to 10 years to recover ('moderate' or 'high' recoverability).
Very low	'Very low' is indicated by the following scenarios: • The habitat or species is very adversely affected by an external factor arising from human activities or natural events (killed/destroyed, 'high' intolerance) but is expected to recover rapidly i.e. within a week ('immediate' recoverability). • The habitat or species is adversely affected by an external factor arising from human activities or natural events (damaged, 'intermediate' intolerance) but is expected to recover rapidly, i.e. within a week ('immediate' recoverability). • The habitat or species is affected by an external factor arising from human activities or natural events (reduced viability **, 'low' intolerance) but is expected to recover within a year ('very high' recoverability).

Continued on p. 316

Table 4. (Continued)

Sensitivity scale	Sensitivity definition or scenario
Not sensitive	'Not sensitive' is indicated by the following scenarios: • The habitat or species is affected by an external factor arising from human activities or natural events (reduced viability **, 'low' intolerance) but is expected to recover rapidly, i.e. within a week ('immediate' recoverability). • The habitat or species is tolerant of changes in the external factor.
Not sensitive*	• The habitat or species may benefit from the change in an external factor (intolerance has been assessed as 'tolerant*').
Not relevant	• The habitat or species is protected from changes in an external factor (i.e. through a burrowing habit or depth), or is able to avoid the external factor.

** 'Reduced viability' includes physiological stress, reduced fecundity, reduced growth, and partial death of a colonial animal or plant.

Table 5. Combining 'intolerance' and 'recoverability' assessments to determine 'sensitivity': the sensitivity matrix. 'Tolerant*' and 'Not sensitive*' are indicated for species that might benefit by change in a factor

	Recoverability						
	None	Very low (>25 yr.)	Low (>10–25 yr.)	Moderate (> 5–10 yr.)	High (1–5 yr.)	Very high (< 1 yr.)	Immediate (<1 week)
Intolerance							
High	Very high	Very high	High	Moderate	Moderate	Low	Very low
Intermediate	Very high	High	High	Moderate	Low	Low	Very Low
Low	High	Moderate	Moderate	Low	Low	Very Low	Not sensitive
Tolerant	Not sensitive	Not sensitive	Not sensitive	Not sensitive	Not sensitive	Not sensitive	Not sensitive
Tolerant*	Not sensitive*	Not sensitive*	Not sensitive*	Not sensitive*	Not sensitive*	Not sensitive*	Not sensitive*
Not relevant	Not relevant	Not relevant	Not relevant	Not relevant	Not relevant	Not relevant	Not relevant

In the case of biotopes, intolerance and recoverability are based on the intolerance and recoverability of species chosen to indicate the sensitivity of the biotope as a whole. These species include key structural or functional species; species which if lost would change the biotope, and/or species important to the function of the community within the biotope. Biotope recovery takes into account evidence of community recovery where available and/or information on the recoverability of the component species. However, where information is lacking, recoverability assessments are qualitative judgments by necessity. Although the same biotope will have survived or re-established, abundance of species within the biotope may take some time to re-establish to previous levels. If a location affected by an adverse environmental factor has been re-colonized but with different species or a different biotope (which, in the case of biotopes, may be a healthy diverse and productive system), recovery will have been considered to have not occurred.

Creating an information resource for environmental management and conservation

Searching the database

The *MarLIN* Biology and Sensitivity Key Information database allows dynamic searches of the information to be performed on-line, via a SQL (Structured Query Language) server and custom written Active Server Pages (ASP), to produce decision-support tools. Tools have been provided on the Web site to search for:

- species by phylum species name, or common name;

- species by keywords in the species description;
- species listed under UK legislation and international conventions;
- biotopes by biotope code and keywords in the biotope description;
- biotopes by species name;
- biotopes listed under UK legislation and international conventions, and
- species or biotopes sensitive to specified marine and coastal activities.

The *MarLIN* programme identifies any directives, conventions and statutes on which a species or habitat (as representative biotopes) is listed for protection. For Great Britain, the *MarLIN* Web site includes a record of whether a species is recorded as nationally rare, scarce or uncommon (see Sanderson, 1996).

Quality assurance

Reviews are placed on-line after they have been checked and edited by the Programme Director or Senior Data Researcher and clearly marked as 'Not refereed'. As many as possible of the reviews completed by *MarLIN* are refereed and the referees

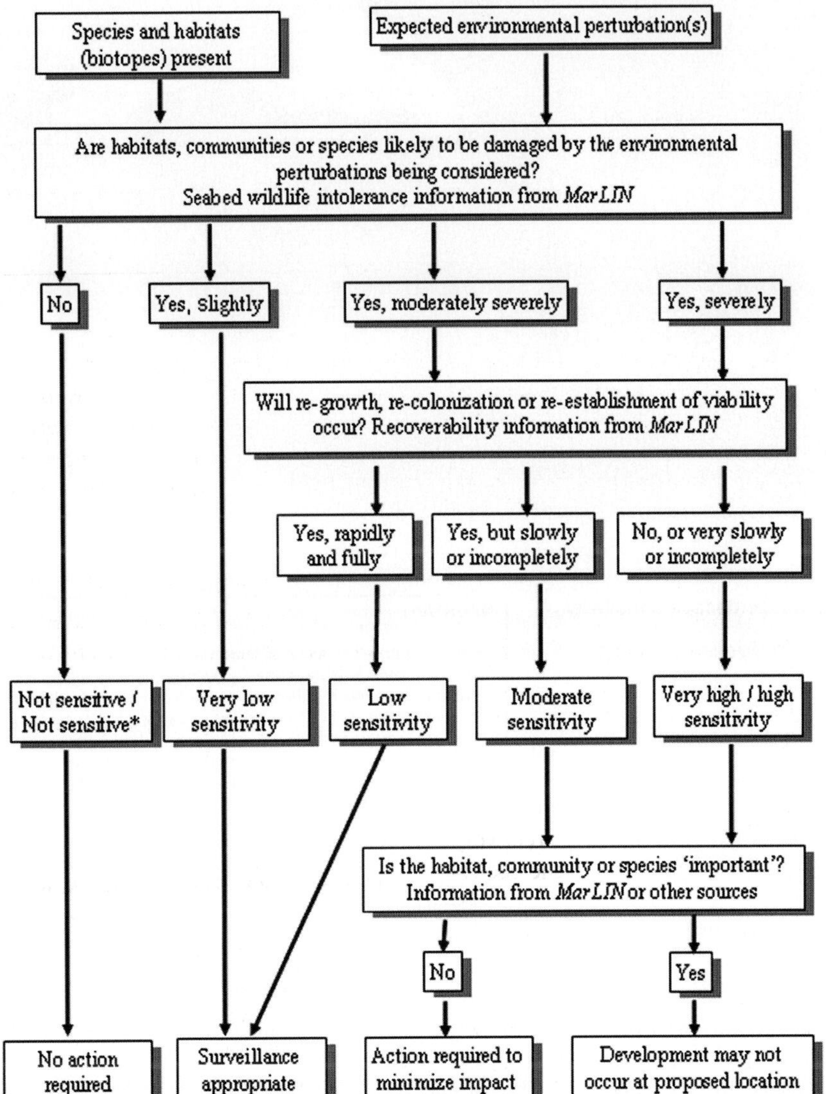

Figure 2. A 'decision tree' for environmental management incorporating concepts of sensitivity, recoverability and importance (updated from Hiscock et al., 2003). The terms and definitions used are shown in Tables 3 and 4.

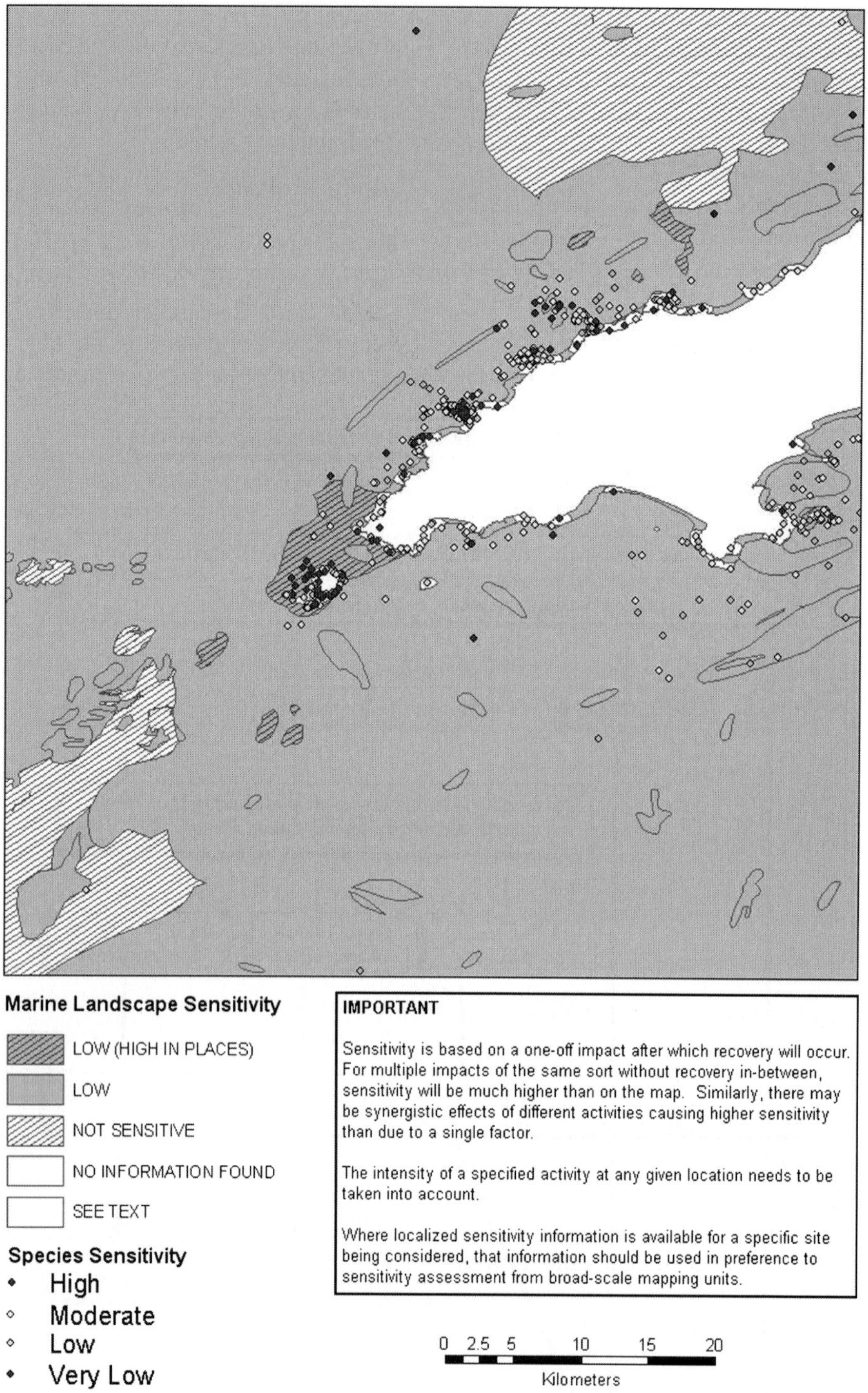

Figure 3. Sensitivity of species to physical disturbance and abrasion at specific locations and within 'marine landscapes' in the vicinity of the Lleyn Peninsula, North Wales. The map is generated by bringing together information from survey databases and the *MarLIN* Biology and Sensitivity Key Information database. The map is presented in GIS allowing interrogation of points to obtain raw data and access to different layers of the GIS holding different types of information. (A coloured map can be seen in the electronic version of *Hydrobiologia*).

are named on the Web pages. Reviews are updated where possible and the *MarLIN* programme is currently seeking resources so that updating is undertaken to take account of new literature on species and revisions of the biotopes classification.

Making decisions to protect biodiversity

Information from *MarLIN* sensitivity reviews is disseminated on the World Wide Web and can be used by anyone undertaking a review of a species or biotope, whether for academic research, an environmental assessment, or to try to understand why a change has occurred in species abundance, biotope structure etc. The information resource becomes much more powerful when accessed from the database, providing opportunities to:

1. specify the environmental factors likely to occur in relation to a specific activity and indicate levels of sensitivity to that activity, and
2. tag survey data with sensitivity assessments to map the distribution of sensitive species or biotopes.

Figure 2 illustrates how *MarLIN* information can be incorporated into decision making.

Mapping sensitivity

MarLIN has begun to link survey data and sensitivity information to map the location of sensitive species and habitats, and to trial the extension of sensitivity assessment to broad scale units for a pilot project in the Irish Sea (see Figure 3). In time, Geographical Information Systems (GIS) will allow us to map sensitivity for up to 24 separate environmental factors, and to view that information at a variety of scales from the local, to the regional or national scale. In addition, the sensitivity information could then be overlaid with the distribution of protected or threatened species and habitats, designated areas, and the location and intensity of specific activities considered damaging to the marine environment. In short, GIS could be used to produce an integrated information system for sea use management.

Discussion

The distillation of scientific literature into a Web-based decision support tool has required practical rather than perfect solutions to problems of indexing intolerance, recoverability and sensitivity. Whilst the benchmarks that have been developed are a pragmatic solution to the problem of taking one level of effect from a potentially wide range, sound judgement is needed when applying the sensitivity index to a particular proposal for development or response to an incident. Changes could be made to the research, for instance, by having several benchmark levels for a particular factor. However, the *MarLIN* approach to sensitivity assessment is both systematic and transparent and, the authors believe, a valuable asset for science-based decisions in environmental management and conservation, at both the local and regional levels.

Acknowledgements

The development of the *MarLIN* approach to sensitivity assessment and the Biology and Sensitivity Key Information reviews of species and biotopes has been undertaken in projects that were jointly funded by the UK Department for Environment, Food and Rural Affairs (Defra), English Nature (EN) and Scottish Natural Heritage (SNH). We have been especially helped in developing the approach to assessing sensitivity by *MarLIN* technical advisory groups that have included staff from nature conservation agencies, academic institutes and fisheries laboratories. The *MarLIN* Information Systems Developer, Dan Lear, has been of critical importance in creating the databases, scripts and Web pages that make the concept of a Web-based information resource on seabed sensitivity a reality.

References

Borja, A., I. Muxika & J. Franco, 2003. The application of a marine biotic index to different impact sources affecting soft-bottom benthic communities along European coasts. Marine Pollution Bulletin 46: 835–845.

Davies C. E. & D. Moss, 2002. EUNIS Habitat Classification. Final Report to the European Topic Centre on Nature

Protection and Biodiversity, European Environment Agency. February 2002. 125pp.

Gundlach, E. R. & M. O. Hayes, 1978. Classification of coastal environments in terms of potential vulnerability to oil spill damage. Marine Technology Society Journal 12: 18–27.

Hiscock, K., 1997. "Use available data". Marine Pollution Bulletin 34: 74–77.

Hiscock, K., M. Elliott, D. Laffoley & S. Rogers, 2003. Data use and information creation: challenges for marine scientists and for managers. Marine Pollution Bulletin 46: 534–541.

Laffoley D. d'A., D. W. Connor, M. L. Tasker & T. Bines, 2000. Nationally important seascapes, habitats and species. A recommended approach to their identification, conservation and protection. Prepared for the DETR Working Group on the Review of Marine Nature Conservation by English Nature and the Joint Nature Conservation Committee. Peterborough, English Nature Research Report 392, 17 pp.

Sanderson, W., 1996. Rarity of marine benthic species in Great Britain: development and application of assessment criteria. Aquatic Conservation 6: 245–256.

Tyler-Walters H., K. Hiscock, D. B. Lear & A. Jackson, 2001. Identifying species and ecosystem sensitivities. Report to the Department for Environment, Food and Rural Affairs from the Marine Life Information Network (*MarLIN*), Marine Biological Association of the United Kingdom, Plymouth. Contract CW0826.

Warwick, R.M. & K. R. Clarke, 2001. Practical measures of marine biodiversity based on relatedness of species. Oceanography and Marine Biology Annual Review 39: 207–231.

Hydrobiologia (2006) 555:321–333

H. Queiroga, M.R. Cunha, A. Cunha, M.H. Moreira, V. Quintino, A.M. Rodrigues, J. Serôdio & R.M. Warwick (eds),
Marine Biodiversity: Patterns and Processes, Assessment, Threats, Management and Conservation
DOI 10.1007/s10750-005-1128-y

Review Paper

Characterisation of the European Marine Sites in South West England: the Fal and Helford candidate Special Area of Conservation (cSAC)

W.J. Langston[1], B.S. Chesman[1,*], G.R. Burt[1], M. Taylor[2], R. Covey[3], N. Cunningham[4], P. Jonas[4] & S.J. Hawkins[1]
[1]*Marine Biological Association, Citadel Hill, PL1 2PB, Plymouth, Devon UK*
[2]*English Nature, Roughmoor, Bishop's Hull, TA1 5AA, Taunton, Somerset UK*
[3]*English Nature, Trevint House, TR1 2PA, Strangways, Truro*
[4]*Environment Agency, Manley House, Kestrel Way, EX2 7LQ, Exeter, Devon UK*
*(*Author for correspondence: E-mail: bch@mba.ac.uk)*

Key words: European Marine Sites, Habitats Directive, Fal and Helford cSAC, water quality

Abstract

The Conservation (Natural Habitats, & c.) Regulations 1994 which implement the Habitats Directive (Council Directive 92/43/EEC) and Birds Directive (79/409/EEC) in the UK include a number of provisions for the protection of European Marine Sites (marine candidate Special Areas of Conservation (cSAC) and Special Protection Areas (SPAs)), including the requirement for competent authorities to assess the effects of new and existing consents, permissions and authorisations to ensure that they are not having an adverse effect on the integrity of the site (Regulations 48 and 50). In England and Wales *site characterisation*, a process which involves the collation, integration and critical evaluation of existing site knowledge, is seen by the Environment Agency (EA) and the nature conservation bodies (English Nature – EN, and the Countryside Council for Wales – CCW) as an important step towards evaluating the effects of consented discharges and activities on the interest features of EMS, and is fundamental to the development of ecosystem-based management. An on-going project, led by the Marine Biological Association of the United Kingdom (part of the Plymouth Marine Science Partnership) in collaboration with EA, EN and CCW, sets out to characterise the impacts and risks associated with water and sediment quality in relation to the integrity and interest features of selected European Marine Sites in southwest England and Wales. Using published information and unpublished data-sets from regulatory agencies, conservation bodies, and research institutes, evidence is compiled on the links between potentially harmful activities, environmental quality and resultant biological effects, and includes an evaluation of long-term change. An overview of results is presented here for one of the UK marine sites studied, The Fal and Helford candidate SAC, highlighting the threats to biota from point-source and diffuse pollutants, together with major gaps in our understanding. Some of the priorities for new research and better integrated monitoring to address nature conservation objectives (to maintain and restore the system) are discussed.

Introduction and approach

The Habitats Directive (Council Directive 92/43/EEC) requires Member States to designate Special Areas of Conservation (SACs) for the conservation of specific interests (listed in Annexes I and II of the Directive), and stresses the need to prevent their deterioration, and the need to attain favourable conservation status of the interest features across their biogeographical ranges. Further to this, Article 6 of the Directive requires special measures to protect the ecological integrity of the designated sites. The Birds Directive (Council Directive 92/43/EEC) requires the classification of the most suitable territories as Special Protection Areas (SPAs) for the conservation of bird species contained in Annex I of the Directive as well as regularly occurring migratory species.

The Conservation (Natural Habitats, & c.) Regulations 1994 which implement the Habitats Directive and Birds Directive in the UK include a number of provisions for the protection of European Marine Sites (marine cSAC and SPAs), including the requirement for competent authorities to assess the effects of certain existing consents, permissions and authorisations to ensure that they are not having an adverse effect on the integrity of the site in view of the site's conservation objectives (Regulation 50). Similar provisions relate to the assessment of new consents, permissions and authorisations (Regulation 48). These assessments must take into account the effects of consents, permissions and authorisations in combination with each other and the cumulative effects on the site, such as background contamination. The nature conservation bodies (e.g. English Nature (EN), Countryside Council for Wales (CCW)) play a role in this process by providing advice and, in some cases, may be required to carry out assessments themselves. All competent authorities, in the exercise of their functions, need to have regard to the requirements of the Habitats Directive so far as they may be affected by the exercise of those functions. In relation to European Marine Sites (EMS) specifically the Regulations also suggest that relevant authorities should develop a management scheme for each site to ensure compliance with the requirements of the Directive.

In response to the requirements of Regulation 50, the Environment Agency (EA) is currently undertaking a project to review existing consents relevant to SACs and SPAs. In England and Wales *site characterisation*, a process which involves the collation, integration and critical evaluation of existing site knowledge, is seen by the EA and the nature conservation bodies (EN, CCW) as an important step in the review of consents. Site characterisation aims to improve understanding of the structure and function of a site, the occurrence of impacts and risks associated with anthropogenic activities, and the relative contributions for a given risk factor (e.g. relative contributions between point and diffuse inputs of a contaminant).

In the south west of England and south Wales, EA, EN and CCW commissioned the Plymouth Marine Science Partnership, led by the Marine Biological Association, to conduct initial site characterisations, focusing on water and sediment quality, for selected EMS. This paper summarises the findings of the study for the Fal and Helford candidate SAC (cSAC).[1] The objective was to characterise the site in terms of environmental quality (water, sediment and biota) using published and unpublished literature and data (up to 2002). Categories of contaminants included both toxic and non-toxic substances. Secondary objectives were to provide a synthesis of available information on vulnerability and biological impact (from bioaccumulation to community-level response); to consider activities and sources which have, or are likely to have, a significant effect on the site; and to identify areas where conditions might result in adverse effects on habitats and species for which the site was designated. Finally, gaps in our understanding of these systems are highlighted and recommendations made which will assist the relevant authorities to target future investigations, monitoring and remedial activities.

Within this study Environmental Quality Standards (EQS) and guidelines have been used as a means of identifying the main water quality issues. For many determinants this relies on summary statistics derived from EA monitoring data. However this data set is collected primarily for compliance purposes, and so there are a number of constraints in extrapolating to the type of assessment being attempted here. For example, where 'less-than' values occur in the data, a numerical value equivalent to half the detection limit has been used to derive summary statistics. Whilst this is far from ideal, with careful scrutiny of data to eliminate obvious outliers, this provides a preliminary evaluation of environmental quality. For sediments, contaminant data is generally scarce. Furthermore, statutory standards have yet to be established in the UK, therefore interim sediment quality guidelines issued by Environment Canada have been adopted as an indication of

[1] In England an amendment to legislation allowed all candidate SAC's to be treated as if they were European sites (enabling consents on these sites to be reviewed prior to the EU formally adopting the site).

the risk to biota. Available data from the EMS were compared with Threshold Effects Levels (TELs) and Probable Effect Levels (PELs) derived from published sediment toxicity data (CCME, 1999).

The following is an overview of the features of one of these sites, the Fal and Helford cSAC, together with conclusions concerning pollution sources, chemical exposure (via sediment and water) and evidence of biological impact (from bioaccumulation to community-level response).

Site description

The location and extent of the Fal and Helford cSAC is shown in Figure 1. The river estuaries are rias whose topography was determined at the end of the last glaciation when sea level rise led to flooding of river valleys and formation of present day creeks. This has led to the creation of extensive 'wetlands' in parts of the upper Fal system and, in the lower, deeper sections, to one of the largest natural harbours in the world, Carrick Roads. The

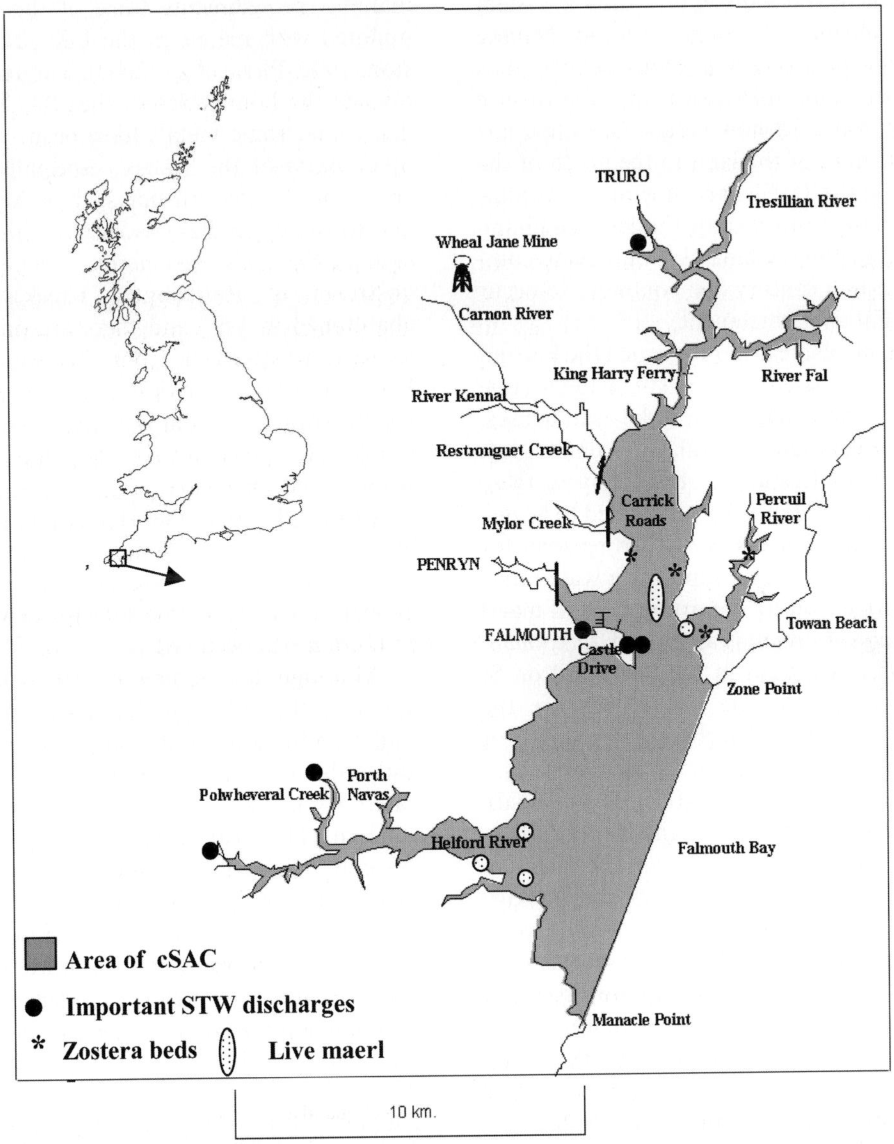

Figure 1. Fal and Helford cSAC showing boundaries of marine site, some important sewage treatment work discharges (the largest of which are Falmouth and Truro (Newham)), and the location of live maerl and seagrass beds.

catchments of the Fal and Helford are predominantly rural and support intensive mixed arable and dairy farming with associated potential threats from nutrients (plus herbicides and pesticides). There is relatively little heavy industry left to affect the site, although Falmouth Docks and a number of marinas are potential sources of disturbance (dredging activities, oil and release of antifouling and sewage). The principal urban centres with major sewage treatment works (STW) are Falmouth and Truro with a number of smaller discharges that may influence the site at a local level (Fig. 1).

The catchment and sediments of the Fal system are heavily influenced by mineralisation. Mining and processing of the metalliferous deposits has been a major feature of the area since the Bronze Age, notably in the Carnon Valley. Granite quarrying was a feature of the land to the north of the Helford during the 1800's, and operations continued into the last century until the last remaining quarry closed in 1993. China clay quarrying commenced in the 18th century and continues to occur in the upper Fal river catchment.

European interest features for which the Fal and Helford was designated as a cSAC include: large shallow inlets and bays, sub-tidal sandbanks, intertidal mudflats and sandflats, Atlantic salt meadows, reefs and estuaries. For many of these features the Fal and Helford is considered to be one of the best areas in the UK. Other reasons for selection are that the cSAC supports a wide range of communities including important beds of maerl (*Phymatolithon calcareum* (Pallas) and *Lithothamnion corallioides* (Crouan) in the lower Fal on St Mawes Bank (Fig. 1). The maerl beds are the largest in SW England and harbour a rich variety of epifaunal and infaunal species, including rare Couch's goby *Gobius couchi* (Miller & El Tawil). Near the mouths of both the Fal and Helford, there are eelgrass *Zostera marina* (Linnaeus) beds, and in the Fal/Ruan estuary, Percuil River and Passage Cove, lie beds of dwarf eelgrass *Zostera noltii* (Hornemann), with associated diverse invertebrate communities. Additionally, the site contains important sediment-dwelling species, including amphipods, polychaete worms, sea cucumber *Leptopentacta elongata* (Duben & Koren) and bivalve molluscs. The Fal and Helford have no specific designation for birds but support a nationally important population of black-tailed godwit *Limosa limosa* (Linnaeus) and many other species including some rare Grebes; *Podiceps nigricollis*, (Brehm, CL), *Podiceps auratus* (Linnaeus).

Toxic contaminants

Metals

During the years of mining activity in the Fal catchment, much of the remobilised metal and tailings were deposited in Restronguet Creek making the sediments some of the most heavily polluted with metals in the UK (Bryan & Langston, 1992; Pirrie et al., 2003). Though the creek is outside the boundaries of the cSAC, there is evidence that some metals have been transported to other parts of the system, especially to adjacent creeks on the western side such as Mylor and Pill, and to the upper Fal (Warwick et al., 1998; Pirrie et al., 2003). The last active mine (Wheal Jane, upstream of Restronguet Creek) was finally abandoned in 1991 and there are ongoing efforts to treat waste-water from this source (Hunt & Howard, 1994; Warwick et al., 1998). However, residual drainage from numerous old mines, spoil heaps and groundwater continue to influence sediment geochemistry.

Unlike the Fal, the Helford Estuary has no history of extensive mining in the immediate vicinity, although the numerous adits in the area indicate that small-scale mineral exploration and extraction has occurred.

Although the mining has ceased, sources of metals in the cSAC are dominated by the Carnon Valley and Restronguet Creek, and levels in waters and sediments of Restronguet Creek remain elevated. As, Cu, Zn, Cd and Fe are probably of most concern. The distribution of Fe in the cSAC is strongly influenced by salinity, as dissolved Fe in freshwaters entering the estuary oxidises and rapidly precipitates on mixing with seawater. This precipitation process has implications for other dissolved metals, which are scavenged by Fe oxyhydroxides produced during estuarine mixing. Since the introduction of minewater treatment at Wheal Jane, Fe inputs (and oxyhydroxide production) in Restronguet are now reduced, and scavenging of other metals in the estuary may be less pronounced. Indeed there is some evidence of

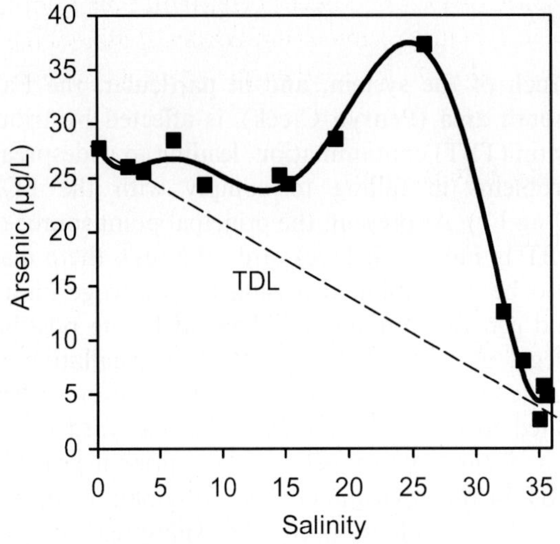

Figure 2. Plot of dissolved As against salinity, Restronguet Creek, showing marked deviation from the theoretical dilution line (TDL), indicating possible input of As from sediments in the estuary (data source: MBA).

As re-release from sediments as a result of these changes (Fig. 2). Since early life stages of some species are vulnerable to As toxicity e.g. Crustacea (Martin et al., 1981), and relatively low concentrations of As in water can inhibit growth of some important primary producers (Sanders & Vermersch, 1982), investigation of this phenomenon would seem to be important.

The possibility of re-release of other metals from sediments should also be examined. Moreover, metals may be transported out into the Carrick Roads from Restronguet Creek in the less dense surface water. Evidence of this is apparent in Figure 3, which summarises EA data for dissolved Cu in waters of the upper Fal showing high levels at the mouth of Restronguet Creek. Similar distributions apply to As and Zn (not shown). Generally, metal concentrations in water beyond the immediate influence of Restronguet Creek are below EQS and not significantly elevated. However, a secondary source of Cu and Zn in the Fal is the outfall at Falmouth Dockyard and this is reflected by elevated concentrations in the area. The significance of copper and zinc-based antifoulants and sacrificial anodes from an increasing number of yachts has still to be evaluated. There are also occasional elevated Zn levels in the upper Fal which may originate from a variety of sources including local sewage discharges and urban run-off.

Sediment metal distributions are typified by Cu in Figure 4 and illustrate the spread of contamination from Restronguet. Although concentrations decline rapidly with distance, Cu is present in sediments of all major rivers and creeks of the system, including the Helford, at levels predicted to cause effects to sensitive biota (based on Probable Effects

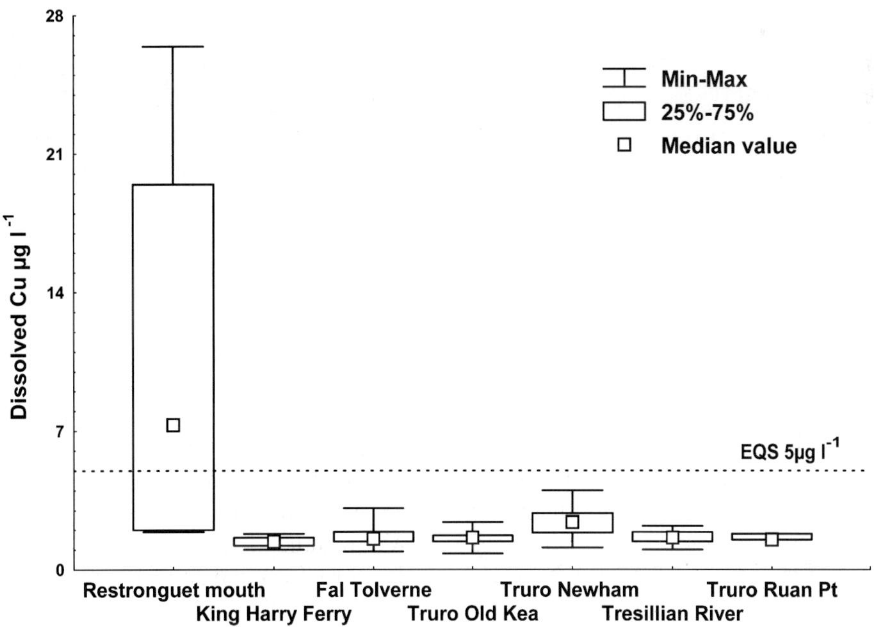

Figure 3. Dissolved Cu (μg l^{-1}) in the upper Fal Estuary (data source: EA).

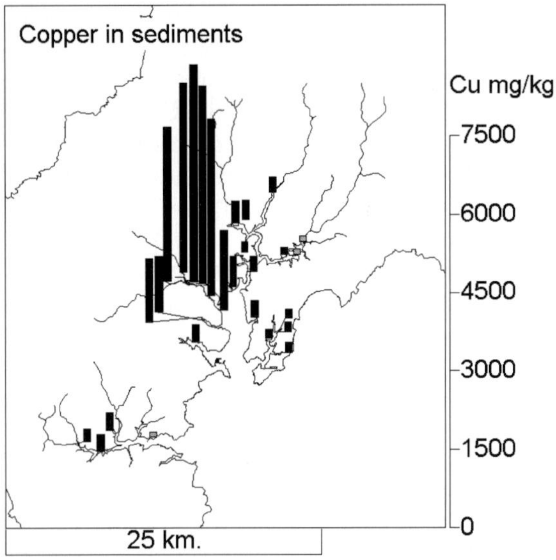

Figure 4. Copper in sediments (<100 μm) (mg kg^{-1} dry weight) Fal and Helford cSAC. Classification based on interim marine sediment quality guidelines (ISQG's) and probable effect levels (PEL's) (from CCME 1999). Black bars = effects expected; Grey = effects cannot be excluded (data source: MBA).

Levels). Distribution patterns for Zn and As are similar to Cu, though impact from these metals would not be anticipated in the Helford or Percuil because the sediment concentrations are lower.

In the more contaminated creeks there is evidence that sensitive species (e.g. some infaunal bivalves) accumulate Cu and Zn to levels that can have lethal or sub-lethal effects (Bryan & Langston, 1992). Other species have developed tolerant strains that exhibit reduced metal permeability (e.g. *Nereis diversicolor* (Müller) (Grant et al., 1989)), or show enhancement of physiological detoxification mechanisms – such as metallothionein production e.g. in oysters *Ostrea edulis* (Linnaeus) (Langston et al., 1998). There is still relatively little known of the implications for many potentially sensitive species in the Fal and Helford cSAC. Ultimately, the elimination of susceptible species and selection of tolerant forms is manifested as community-level changes (Somerfield et al., 1994). This serves as an excellent example of the benefits of a tiered approach to monitoring (involving chemistry, subcellular response and community ecology) which perhaps should be applied more widely when evaluating the quality of the marine environment.

TBT

Much of the system, and in particular, the Falmouth area (Penryn Creek), is affected by tributyltin (TBT) contamination, leading to widespread problems in failing to comply with the EQS (2 ng l^{-1}). At present, the principal point source of TBT is Falmouth Dockyard, although there may also be a component arising from sewage inputs and general shipping activities, including possible illegal usage on leisure craft. The installation of active carbon treatment plant is, however, anticipated to reduce inputs from the dockyard in the near future. A proportion of the present levels of TBT in the environment may originate from historical contamination, from its widespread use as a biocide for yachts and ships and in timber treatment. The relative contribution of sediments as a source of TBT is still to be resolved adequately but concentrations are known to range up to 10 μg g^{-1} and may penetrate to depths of >1 m. The potential for short-term sediment releases during artificial remobilisation (dredging, for example) may be considerable and is likely to contribute to ecological damage in the long term.

TBT concentrations in parts of the Fal are high by national and international standards and do not appear to have been reduced consistently by UK TBT legislation in 1987 (relating to small vessels). A treaty introduced by IMO, and now also EU law (Commission Directive 2002/62/EC) banned the application of organotin antifouling paint on all shipping from January 2003. The effect of legislation must now be carefully monitored as organotin compounds are to be removed from the hulls of all ships by 2008.

There is clear evidence of TBT impact on the reproductive capacity of dogwhelk (*Nucella lapillus* Linnaeus) populations, which have been eliminated from most of the Fal by TBT (Langston et al., 1994; Birchenough et al., 2002). Castle Drive and Towan Beach, either side of the entrance to Carrick roads, have the sole surviving populations in the area, but even these manifest advanced symptoms of imposex (83% females sterile in 1996), and therefore may be doomed (unpublished observations). There is anecdotal evidence of the decline of dogwhelk in the Helford although this is not substantiated. There is also circumstantial evidence, based on comparisons with EQS values and sediment toxicity guidelines, that other infaunal organisms, particularly molluscs, will be affected. Most faunal surveys indicate that the Fal is impoverished compared to similar systems elsewhere, and though this probably relates in part to the long history of mining impact, a significant component of the damage (and a continuing impediment to recovery) may be attributable to TBT. Unfortunately, there are constraints on providing a full assessment of TBT status caused by a lack of detailed information; notably on inputs to the system, environmental inventories (particularly for sediments), and the variables controlling partitioning. This information is needed to clarify the processes that determine organotin distributions in the estuary and to develop models that simulate the impacts of TBT in the cSAC.

Hydrocarbons

Although data are scarce and impact difficult to quantify, concentrations of hydrocarbon oils at the mouths of the Fal and the Helford (mean values for the early 1990s, 1.4 and 4 mg l^{-1}) suggest general contamination in the area, which possibly originates from the dockyard and shipping, but may also be related to run-off and/or aerial deposition. In enclosed systems such as the Fal, the potential for dispersion of hydrocarbon oils is diminished, and much of the contamination can be washed ashore. Perhaps the most vulnerable areas are intertidal habitats where sensitive shoreline seagrass and saltmarsh communities could be at risk. There are no recognised quantitative standards for total hydrocarbon. The only available standards relate to levels causing 'taints' in shellfish under the Shellfish Waters Directive (79/923/EEC, England & Wales Statutory Instrument 1997 No. 1332), and 'visible films' on waters under the Bathing Waters Directive (76/160/EEC). The latter may trigger a recommended level of 0.3 mg l^{-1} (90th percentile – this relates to mineral oils), which may be exceeded at some sites, although not at designated bathing beaches. There are no data for PAHs in tidal waters or sediments and further investigation into the sources, levels and composition of hydrocarbons in the cSAC would seem to be necessary.

Synthetic organic compounds (pesticides, herbicides, PCBs) have not been widely monitored, though the available (EA) data indicate that concentrations of the majority of these substances (which includes many endocrine disruptors) are now generally very low in the cSAC. Diffuse agricultural sources and small contributions from Sewage Treatment Works (STW) probably account for any pesticide inputs. Values for PCBs in water and oysters are largely below limits of detection at shellfish sites within the cSAC. An unpublished report from the early 1990s indicates measurable levels of 5 PCB isomers in oysters from the Helford Estuary (though these may have been re-layed from elsewhere). This highlights the need for more rigorous sampling in the cSAC for organochlorines and other synthetic organics that accumulate in sediments and may impinge on estuarine biota through food-chain magnification. There have been no investigations on the effects of endocrine disruption on biota of the Fal and Helford, other than TBT: Allen et al. (2003) have suggested that the site be ranked as medium priority for further study, which would seem a reasonable precaution.

Non-toxic contaminants

Nutrients

The effects/potential effects of nutrient enrichment are recognised in the Fal and Helford cSAC and this is reflected in the designation of the Upper Fal as a Sensitive Area (Eutrophic) and a Polluted Water (Eutrophic) under the Urban Wastewater Treatment Directive and Nitrates Directive, respectively.

In parts of the Fal/Helford system, elevated nitrogen and phosphorus species are an important factor in poor water quality, and lead to changes such as algal blooms, dissolved oxygen (DO) sags, and turbidity, which are symptomatic of eutrophication. Exaggerated DO fluctuations (which indicate a high level of biological activity in the water column), coupled with high levels of chlorophyll *a*, have been observed intermittently in several parts of the system, particularly in the Truro and Tresillian area. Indications are that the

DO status of the upper estuary is deteriorating and that concentrations in parts of the lower Fal (Carrick Roads) may also be a cause for concern.

Harmful algal blooms have been observed in the upper Fal Estuary/Truro River. An event in 1995–1996 was particularly significant because it involved the 'red tide' dinoflagellate *Alexandrium tamarense* (Lebour) which produced PSP (paralytic shellfish poisoning) toxins, resulting in a prohibition notice on the collection of shellfish during 1995 and 1996. Harmful algal blooms linked to nutrient enrichment continue to occur in the Fal and Helford cSAC and may also originate in other parts of the system. The most recent incidence, in 2002, resulted in significant mortality of marine worms as well as some shellfish. It was centred on Polwheveral Creek and Porth Navas, in the Helford Estuary, where high nutrient levels remain a cause for concern. This bloom was believed to be principally *Gyrodinium aureolum* (Hulburt), which produces toxins that can also kill fish. Calenick Creek near Truro Newham STW in the upper Fal was also affected.

The principal sources of nutrients to the cSAC as a whole are diffuse, with the agricultural component being particularly important for nitrogen, though clearly, areas of the Fal and Helford appear to be influenced locally by sewage discharges. The local influence of point-source sewage discharges is illustrated in Figure 5, which shows examples of annual median concentrations for nitrate, orthophosphate and ammonia (N) in tidal waters of the cSAC. The highest concentrations occur in the Truro River, which receives discharges from Truro Newham STW. Modelling exercises confirm that the principal source of nitrogen in the upper Fal Estuary, during summer months, is Newham STW (Fraser et al., 2000). High concentrations, especially for nitrate, also occur in the creeks of the Helford near to (smaller) STW discharges, although the relative importance of different sources has not yet been determined.

The threat of eutrophication may be exacerbated by seasonal fluctuations in the N:P ratio of waters entering the tidal estuaries, and re-release of phosphorus from sediments during periods with low oxygen levels. The role of sediments as a source/sink for nutrients is largely unknown. There may also be small inputs from mine drainage (P).

With the exception of the lower Fal Estuary, where there have been major upgrades to sewage treatment in recent years, there have been no obvious temporal reductions in nutrient levels. To the contrary, at many tidal waters sites, there is evidence of significant increases over the past 15 years. Contaminants associated with point-sources in the upper Fal are anticipated to decrease in the coming years as proposed improvement schemes to sewage treatment plants in the Truro area come on line.

The lack of water and sediment quality standards for nutrients make it difficult for consents to be judged in terms of statutory limits in receiving waters. This is identified as a significant gap in the ability to manage the system optimally. Monitoring in the upper Fal is ongoing in order to review the Sensitive Area (Eutrophic) and Polluted Water (Eutrophic) designations; monitoring is also being carried out in the lower Fal as a result of possible future designations. Hence, it would seem that an increase in nutrients should be strongly avoided as a precautionary requirement. Changes to consents (quantities and location) should therefore be considered very carefully to avoid the risk of further nutrient enrichment.

Sedimentation and turbidity have some influence on site characteristics. The large input of china clay wastes from the St Austell area through discharges to the River Fal has, in the past, had an important silting impact in the upper estuary and its saltmarshes. Though a major source of turbidity in the estuary is currently considered to be sediment resuspension, the levels of turbidity in parts of the cSAC suggest outside sources. Sewage discharges in the Truro/Tresillian area and extraction of Maerl from the lower estuary[2] could increase turbidity locally, and during blooms, it is likely that the suspended organic component includes significant amounts of microalgae. Temporally and spatially limited resuspension events also arise from dredging activity and may remobilise contaminants locally.

Impacts to biota of the Fal and Helford

As indicated above, the biological status of the site is considered to be influenced by several contami-

[2] This extraction ceased in December 2004

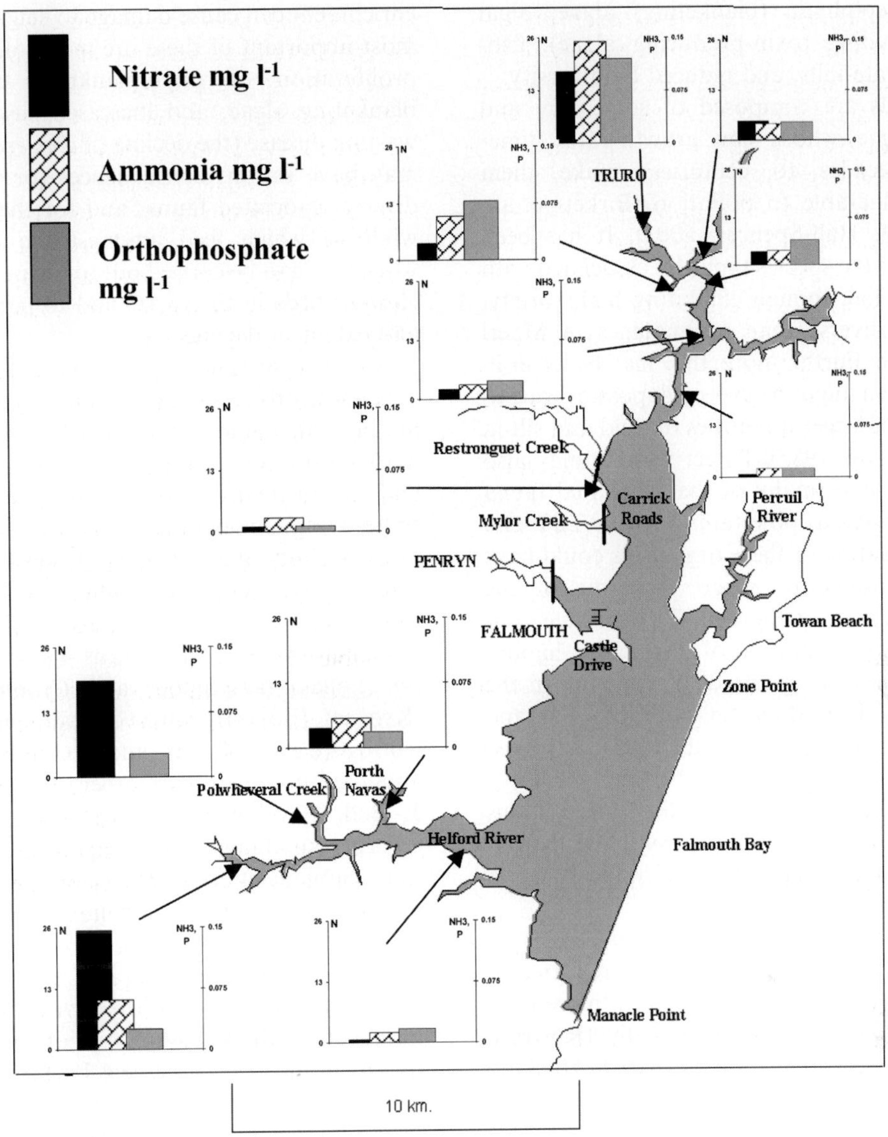

Figure 5. Annual median values for nitrate, ammonia (N) and orthophosphate in tidal waters of the Fal and Helford cSAC. Fal Nitrate data are for 2000, ammonia (N) and orthophosphate, 2001, all Helford data are for 1997 (data source: EA).

nant-related pressures including nutrients, metals and TBT. For some species, the consequence of these pressures is clear, for example *Nucella* and the very impoverished fauna of Restronguet Creek. However sub-lethal effects are more difficult to define and the precise evaluation of the threat to the conservation features of the site is therefore hampered by lack of adequate cause-effect data (a major gap in our current understanding of marine habitats generally). However, in some cases the weight of evidence, particularly if a precautionary approach is taken, suggests that adverse biological effects could well be linked to water quality. For example, even though there is no confirmation of a clear link between nutrient enrichment and ecological status, it may be argued that the weight of evidence suggests indirect links to the decline of *Zostera* and Maerl beds, an increase in epiphytic (blanketing) algae, algal blooms (including toxin-producing algae), fish- and invertebrate-kills, and reduced biodiversity.

Maerl beds are composed of some rare and important algae, whose slow growth rates (time-scales of decades to centuries) make them extremely vulnerable to pollution (Birkett et al., 1998; Grall & Hall-Spencer, 2003). It has been suggested that the species typically associated with Maerl in the Fal, though exhibiting high variety, may be less diverse than assemblages in Maerl beds elsewhere. Furthermore, they may be lower in abundance than might be expected, possibly due to the presence of larger quantities of mud and silt in the bed (Rostron, 1985). Direct toxicity may also be occurring, since many of the associated fauna could be sensitive to pollutants such as TBT and metals. Elimination of these organisms could have wider implications for maerl in terms of the structural integrity of the bed. To reinforce concerns over the conservation of this unique habitat there are subjective reports of a reduction in the proportion of live Maerl in both the Fal and Helford Estuaries (Perrins et al., 1995). A survey conducted recently by English Nature will hopefully provide a clearer picture of the current status of the Fal beds (R. Covey, *pers com*), and there is an obvious need for continued scientific study to determine impacts on this biotope.

Similarly, the diversity of species associated with *Zostera marina* beds in the Fal is reported to be lower than *Zostera* communities in 'more pristine' areas such as the Isles of Scilly (Rostron, 1985). In the Helford Estuary, Covey & Hocking (1987) indicate that *Zostera* beds have disappeared from many areas, and remaining beds are somewhat eroded, with adjacent areas of gravel and clitter becoming silted over. There is no clear evidence, as yet, to pinpoint the exact cause of the decline of *Zostera* in the Fal and Helford cSAC, although *Zostera* beds are known to be susceptible to anthropogenic disturbance (e.g. bait digging, mechanical erosion from mooring chains) and various forms of pollution, including metals and the indirect effects of nutrient enrichment: In the laboratory, metals and a number of organic contaminants have been found to reduce nitrogen fixation in the roots, and thus could affect *Zostera* viability (Williams et al., 1994). There are a variety of mechanisms by which excessive nutrient enrichment can cause damage to eelgrass beds; the most important of these are metabolic imbalance, proliferation of phytoplankton, epiphytic or blanketing algae, and increased susceptibility to wasting disease (the decline of *Zostera*, nationally, may have serious consequences for the rich and diverse associated fauna, and for the waders and wildfowl which feed amongst the beds). More work needs to be carried out to define the status of *Zostera* beds in the cSAC and to pinpoint causes and extent of decline.

In terms of general biodiversity, there is a tendency towards relative dominance by fewer species in the Fal, compared with other, less contaminated, estuarine systems in south-west Britain and the distinctive nature of the macro-invertebrate communities in certain areas (most marked in the vicinity of Restronguet Creek) is, to a large extent, due to metal contamination. For, example the Fal has a very low abundance of the sensitive sediment-dwelling crustaceans *Corophium volutator* (Pallas) (amphipod) and *Cyathura carinata* (Krøyer) (isopod), whilst certain small annelid worms (e.g. tubificids and spionids) are more abundant in the Fal than other estuaries (Holliday & Bell, 1981; Rostron, 1985; Warwick, 2001). Organic enrichment is less implicated than metals, since capitellid polychaetes (indicators of organic pollution) are relatively low in numbers in the Fal (Warwick et al., 1998).

There is some evidence that metal contamination may also influence the diversity of primary producers in the Fal system: prior to recent water treatment measures at Wheal Jane, the metal- and low pH- tolerant flagellate *Euglena mutabilis* (Schmitz) was the only representative of the phytoplankton community in the Carnon River, whereas in the less contaminated Fal and Tresillian Rivers a range of algal species were found. In more marine regions of Restronguet Creek, high metal levels were also considered responsible for a reduction in the biomass of several species, whilst Cu-tolerant *Skeletonema costatum* (Greville) and Zn-tolerant *Katodinium rotundatum* (Lohmann) were recorded in abundance (Rijstenbil et al., 1991) implying possible selection pressures due to the presence of elevated metal levels.

Such features clearly have implications for all higher trophic levels in the system. Because metal concentrations in some areas have been elevated for many years, certain populations of prey organisms have developed tolerance, accompanied by accumulation of exceptional body burdens which are potentially transferable to consumers.

The Fal does not support a salmon fishery, though anecdotal evidence suggests it may once have done so. Sea trout are present in several tributary rivers and can apparently tolerate the moderately polluted conditions. As yet there is no run of sea trout through to the Carnon River, although sporadic records are mentioned for the unpolluted Kennal, nearby. The impoverished status of these species of diadromous game fish has been linked to pollution, and it is metals that are of primary concern, although the effects of high nutrient levels (including reduced oxygen levels and high turbidity) may also be contributory factors.

The Fal Estuary supports a nationally important population of Black-tailed Godwit (a biodiversity long-list species) and Dunlin, Oystercatcher, Turnstone, Curlew, Heron, Redshank and various waterfowl are widespread. These species are often concentrated near the tops of inlets, which are subject to the influence of contaminants, including the Truro River and Restronguet Creek, but the consequences for birds have not been studied. A major discharge of metal-rich minewater from the abandoned Wheal Jane in 1992 is not thought to have had an appreciable effect on numbers of wildfowl and waders. One exception may have been the increasing mortalities and occurrences of sick mute swans observed between 1992 and 1995; however, although this has been attributed largely to heavy metal toxicity, the evidence remains equivocal. In view of the fact that various waders may overwinter in Restronguet and some of the other contaminated sites, a study to investigate the uptake of contaminants from the diet during this period might provide valuable insights into risks for birds.

Conclusions and recommendations

The available data, has allowed at least a preliminary characterisation of the condition of the Fal and Helford cSAC. Despite its importance as a designated European Marine site there are indications of some environmental degradation due to contaminants. Most faunal surveys indicate that parts of the Fal and Helford are impoverished compared to similar systems elsewhere, whilst water (and sediment) quality data identify several categories and sources of contamination which may contribute to this condition. The major limitation at present lies in connecting this chemical and biological information (applicable to most EMS). In addition to statutory requirements, future monitoring should include programmes to address the environmental issues of most concern. The introduction of appropriate and targeted biological-effects monitoring, alongside chemical determinations, should also be a priority.

Thus, a central issue for all EMS characterised in the current project is how best to monitor them to assess their ecological quality, structure and function in relation to the conservation objectives, and also to help establish cause and effect relationships. Clearly future monitoring needs to be sufficiently sensitive to detect adverse changes before they affect the achievement of conservation objectives for the site. The question is 'What approach best serves this need?'

Continuation of *relevant* time-series observations which build on established data-sets are an obvious (generic) requirement in attempting to separate human impacts from natural variation. Site-specific suggestions for prioritising chemical measurements in the Fal and Helford cSAC include long-term trends in metals (distributions and bioaccumulation). Better profiling of metals in waters is needed, paying particular attention to the distribution and dispersion of metals from Restronguet Creek as a function of river flow and tidal state (to include modelling of profiles and inputs under different tidal and hydrometerological conditions). Some consideration should be given to the possible remobilisation of metals from sediments and attempts made to characterise sediment bioavailability in relation to natural and anthropogenic components. Sub-lethal responses to metals should be investigated by, for example, examination of the extent of metallothionein induction.

TBT remains an issue despite national and international regulations. Better source apportionment, sediment distributions and modelling are needed. An extension of time-series observations on remaining *Nucella* populations and transplant studies would be valuable in establishing the prognosis for recovery of damaged parts of the ecosystem.

There should be a more comprehensive analysis of nutrient information and generation of more targetted data on distributions, sources, and seasonality, including sediments, to establish budgets and improve models. These need to be linked to biological measurements (benthic community data as well as phytoplankton production and identification) to examine cause and effect relationships and to identify conditions which stimulate harmful algal blooms.

Multivariate statistical approaches have proved to be extremely useful in detecting spatial and temporal trends in benthic communities in relation to pollutant levels (e.g. Somerfield et al., 1994). This would seem an ideal platform on which to build for future indications of biological change. These might be conducted at intervals of perhaps 5 years or so to coincide with review cycles for assessing the status of key communities and habitats on EMS (a requirement under the terms of the Habitats Directive).

Thus, an array of techniques is now available to assess biological impact and would allow screening of the EMS, in conjunction with targetted chemical analysis. By selection of an appropriate suite of indicators, sampling strategies could be tailor-made to establish with greater certainty the causes and extent of damage. These include conventional ecological survey procedures for identifying changes in the abundance and diversity of species, chemical and biomonitoring procedures for determining the concentrations and bioavailability of anthropogenic contaminants, and biochemical, physiological and behavioural biomarkers which signal exposure to, and in some cases, adverse effects of pollution. If these procedures are used in combination in well-designed survey programmes, they are expected to provide insights into which pollutants are responsible for environmental degradation and to help prioritise remediation. They are equally amenable for measuring long-term trends e.g. in the assessment of recovery. This approach would also help to address the potential risks posed by chemical contaminants for which there is no current monitoring.

Inevitably a major limitation will be the cost of application of such a scheme. It is here where the process of 'site characterisation' is perhaps most useful in terms of prioritising risks both within and between sites. To make best use of resources it is also important that those involved with investigation and management of the EMS (e.g. research organisations, and relevant authorities) establish better links, share information, and develop integrated investigative and monitoring programmes as much as possible.

Acknowledgements

The authors would like to thank David Harrison and Michael Coyle of English Nature for their comments on earlier drafts of this paper. The views represented here are those of the authors and may not necessarily reflect those of English Nature or the Environment Agency.

References

Allen, Y. T., V. Hurrell, J. Reed & P. Matthiesson, 2003. Endocrine Disruptors and European Marine Sites in England. English Nature Research Reports 531: 131 pp.

Birchenough, A. C., S. M. Evans, C. Moss & R. Welch, 2002. Re-colonisation and recovery of populations of dogwhelks *Nucella lapillus* (L.) on shores formerly subject to severe TBT contamination. Marine Pollution Bulletin 44: 652–659.

Birkett C. A., C. A. Maggs & M. J. Dring, 1998. Maerl (Volume V). An overview of dynamic and sensitivity characteristics for conservation management of marine SACs. Scottish Association of Marine Science. (UK Marine SACs Project): 116 (http://www.ukmarinesac.org.uk/pdfs/maerl.pdf).

Bryan, G. W. & W. J. Langston, 1992. Bioavailability, accumulation and effects of heavy metals in sediments with special reference to United Kingdom estuaries: a review. Environmental Pollution 76: 89–131.

CCME. 1999. Canadian sediment quality guidelines for the protection of aquatic life: Summary tables. In: Canadian environmental quality guidelines, 1999, Canadian Council of Ministers for the Environment, Winnipeg.

Covey, R. & S. Hocking, 1987. Helford River survey Heinz Guardians of the Countryside and World Wildlife Fund: 121 pp.

Fraser, A. I., D. Butterfield, R. Uncles, P. Johnes & T. R.Harrod, 2000. Fal and Helford special areas of conservation (cSAC) and the Tamar Estuaries complex cSAC/special protection area (SPA): Estimation of diffuse and point-source nutrient inputs SSLRC report to the EA: 89 pp.

Grall, J. & J. M. Hall-Spencer, 2003. Problems facing maerl conservation in Brittany. Aquatic Conservation: Marine and Freshwater Ecosystems 13: 55–64.

Grant, A., J. G. Hately & N. V. Jones, 1989. Mapping the ecological impact of heavy metals on the estuarine polychaete *Nereis diversicolor* using inherited metal tolerance. Marine Pollution Bulletin 20: 235–238.

Holliday, R. J. & R. M. Bell, 1981. The ecology of Restronguet Creek and the Fal Estuary, Cornwall, 92 pp. University of Liverpool Environmental Advisory Unit, and Billington Minerals, 1981. (unpublished report).

Hunt, L. E. & A. G. Howard, 1994. Arsenic speciation and distribution in the Carnon estuary following the acute discharge of contaminated water from a disused mine. Marine Pollution Bulletin 28: 33–38.

Langston, W. J., G. W. Bryan, G. R. Burt & N. D. Pope, 1994. Effects of sediment metals on estuarine benthic organisms, R&D Note 203. National Rivers Authority: 141 pp.

Langston, W. J., M. J. Bebianno & G. R. Burt, 1998. Metal handling strategies in molluscs. In Langston W. J. & M. J. Bebianno, (eds), Metal Metabolism in Aquatic Environments Kluwer Academic Publishers: 219–283.

Martin, M., K. E. Osborn, P. Billig & N. Glickstein, 1981. Toxicities of ten metals to *Crassostrea gigas* and *Mytilus edulis* embryos and *Cancer magister* larvae. Marine Pollution Bulletin 12: 305–308.

Perrins J. M., F. Bunker & G. M. Bishop, 1995. A comparison of the maerl beds of the Fal Estuary between 1982 and 1992. Report to English Nature.

Pirrie, D., M. R. Power, G. Rollinson, G. S. Camm, S. H. Hughes, A. R. Butcher & P. Hughes, 2003. The spatial distribution and source of arsenic, copper, tin and zinc within the surface sediments of the Fal Estuary, Cornwall, UK. Sedimentology 50: 579–595.

Rijstenbil, J. W., A. G. A. Merks, J. Peene, T. C. W. Poortvliet & J. A. Wijnholds, 1991. Phytoplankton composition and spatial distribution of copper and zinc in the Fal Estuary (Cornwall, UK). Hydrobiological Bulletin 25: 37–44.

Rostron, D., 1985. Surveys of harbours, rias and estuaries in southern Britain: Falmouth. Vol. 1. A report to the NCC by the Field Studies Council Oil Pollution Research Unit, Orielton Field Centre, Pembroke, Dyfed: 109. Peterborough: Nature Conservancy Council, 1985.

Sanders, J. G. & P. S. Vermersch, 1982. Response of marine phytoplankton to low levels of arsenate. Journal of Plankton Research 4: 881–893.

Somerfield, P. J., J. M. Gee & R. M. Warwick, 1994. Soft sediment meiofaunal community structure in relation to a long-term heavy metal gradient in the Fal Estuary system. Marine Ecology Progress Series 105: 79–88.

Warwick, R. M., 2001. Evidence for the effects of metal contamination on the intertidal macrobenthic assemblages of the Fal estuary. Marine Pollution Bulletin 42: 145–148.

Warwick, R. M., W. J. Langston, P. J. Somerfield, J. R. W. Harris, N. D. Pope, G. R. Burt & B. S. Chesman, 1998. Wheal Jane Minewater Project. Consultancy studies 1996–1999. Final biological assessment Environment Agency, 91 pp.

Williams, T. P., J. M. Bubb & J. N. Lester, 1994. Metal accumulation within saltmarsh environments: a review. Marine Pollution Bulletin 28: 277–290.

Hydrobiologia (2006) 555:335–343
H. Queiroga, M.R. Cunha, A. Cunha, M.H. Moreira, V. Quintino, A.M. Rodrigues, J. Serôdio & R.M. Warwick (eds),
Marine Biodiversity: Patterns and Processes, Assessment, Threats, Management and Conservation
DOI 10.1007/s10750-005-1133-1

Macrobenthic colonisation of artificial reefs on the southern coast of Portugal (Ancão, Algarve)

Diana Boaventura[1,2,*], Ana Moura[3], Francisco Leitão[3], Susana Carvalho[3], João Cúrdia[3], Paula Pereira, Luís Cancela da Fonseca[1,3], Miguel Neves dos Santos[3] & Carlos Costa Monteiro[3]

[1]*Laboratório Marítimo da Guia/IMAR, FCUL, University of Lisbon, Estrada do Guincho, 2750-374, Cascais, Portugal*
[2]*Escola Superior de Educação João de Deus, Avenida Álvares Cabral, 69, 1269-094 , Lisboa, Portugal*
[3]*Instituto de Investigação das Pescas e do Mar, Centro regional de Investigação das Pescas do Sul (IPIMAR-CRIPSul), Av. 5 de Outubro, 8700-305, Olhão, Portugal*
*(*Author for correspondence: E-mail: d.boaventura@netcabo.pt)*

Key words: artificial reefs, macrobenthic communities, colonisation, hard substrata, Portugal

Abstract

Artificial reef systems play an important role in the increase of natural production of biological marine resources and they have been deployed worldwide. In Portugal, seven artificial systems have been deployed along the southern coast of the Algarve. Research to date has focussed mainly on fish populations, particularly those of economical importance. The present work aims to study the macrobenthic communities of the artificial reef structures, as these will enhance the food resources and shelter, making the reefs more attractive to fish. In particular, we experimentally analysed the sequence of colonisation of macrobenthic communities of the Ancão artificial reef system, which was deployed in the summer of 2002. The study of the colonisation of benthic communities was done using samples of concrete cubic units (15×15 cm) that were suspended at the reef modules at a depth of 20 m, at the time of the reef immersion. Four replicate samples were collected by SCUBA diving from two groups of the Ancão reef every three months from the starting date. Sampling was done using essentially non-destructive methods to assess the percentage cover of macrobenthic organisms in both vertical and horizontal surfaces. The percentage cover of the taxonomic groups was compared within the different surfaces of the samples and between the two reef groups. The bottom surface of cubic samples had a significantly higher colonisation related to the dominance of barnacle cover, probably due to lower sedimentation levels. Samples from both reef groups showed a similar pattern of colonisation. Barnacles, bryozoans and serpulids dominated the samples three months immediately after the beginning of the experiment. Other invertebrates groups, such as Porifera, Hydrozoa, Anthozoa, other sessile Polychaeta, Decapoda, Gastropoda and Bivalvia, were more abundant after six months of colonisation.

Introduction

Artificial reefs are submerged structures placed on the seabed deliberately to imitate some characteristics of natural reefs, as defined by the European Artificial Reef Research Network (Jensen, 1997). Artificial reef systems are usually developed for fish exploitation, for protection of marine areas from illegal fisheries, and more recently, for preservation and rehabilitation of natural habitats (Pickering et al., 1998). These structures have been deployed worldwide and, according to Seaman & Sprague (1991), the major areas presenting this kind of activity include the Mediterranean and Caribbean Seas, South-eastern Asia, Japan, North America, Australia and some islands in the South Pacific. The expansion of this activity is related to the evolution of the structures and

Table 1. Results of ANOVA performed at months 3 and 12 for the top and bottom surfaces at the two studied reef groups

Source of variation	df	3 months		12 months	
		MS	*F*	MS	*F*
Reef group = Re	1	26.52	0.97 ns	77.88	3.22 ns
Surface Orientation = Su	1	3025.00	110.54***	1054.63	43.61***
Re × Su	1	276.39	10.10**	58.14	2.40 ns
Residual	12	27.37		24.18	
Cochran's test		*C* = 0.55 ns		*C* = 0.49 ns	
SNK tests	Re × Su, SE = 2.62				
	Top surface				
	Reef A10a (56.54) = Reef A11a (62.28 ns)				
	Bottom surface				
	Reef A10a (92.35) > Reef A11a (81.46*)				
	Reef Group A10a				
	Top (56.54) < Bottom (92.35**)				
	Reef Group A11a				
	Top (62.28) < Bottom (81.46**)				

ns = not significant; * $p<0.05$, ** $p<0.01$, *** $p<0.001$.

Table 2. Results of ANOVA performed at months 3 and 12 for the outside and inside surfaces at the two studied reef groups

Source of variation	df	3 months		12 months	
		MS	*F*	MS	*F*
Reef group = Re	1	102.52	0.62 ns	149.76	1.24 ns
Surface Orientation = Su	1	605.16	3.63 ns	5.94	0.05 ns
Re × Su	1	127.13	0.76 ns	78.10	0.65 ns
Residual	12	166.65		120.38	
Cochran's test		*C* = 0.43 ns		*C* = 0.47 ns	

ns = not significant.

was similar for both reef groups and for most of the surface orientations. The only exception was the higher mean percentage cover of barnacles found on the bottom surface on the third month with values of 76.6% and 68.1% for reef group A10a and A11a, respectively. These results conform to the significant differences obtained for the total percentage cover. Serpulids showed, in general, a small peak at the third month but mean percentage cover did not exceed 20%. The ascidians were more abundant on the bottom surface, which was less exposed to the light.

Other sessile groups such as Porifera, Hydrozoa, Anthozoa, and other sessile Polychaeta showed a generally low percentage cover which had increased after the sixth month of the experiment to 5–10%, particularly on the top surface where cover reached 20% (Fig. 5). The list of identified species to date from the scrapped surfaces of the samples is given in the Appendix A. In general, non-sessile organisms (e.g. decapods, gastropods, bivalves) were also observed to increase after the sixth month but their densities have not been processed for this article.

Discussion

Previous studies of artificial reefs have focussed mostly on aspects of fishery ecology; however, to manage and understand these artificial habitats it is essential to integrate all aspects of the hard substratum ecology (see Svane & Petersen, 2001 for a review). The present study has provided information on the patterns of macrobenthic col-

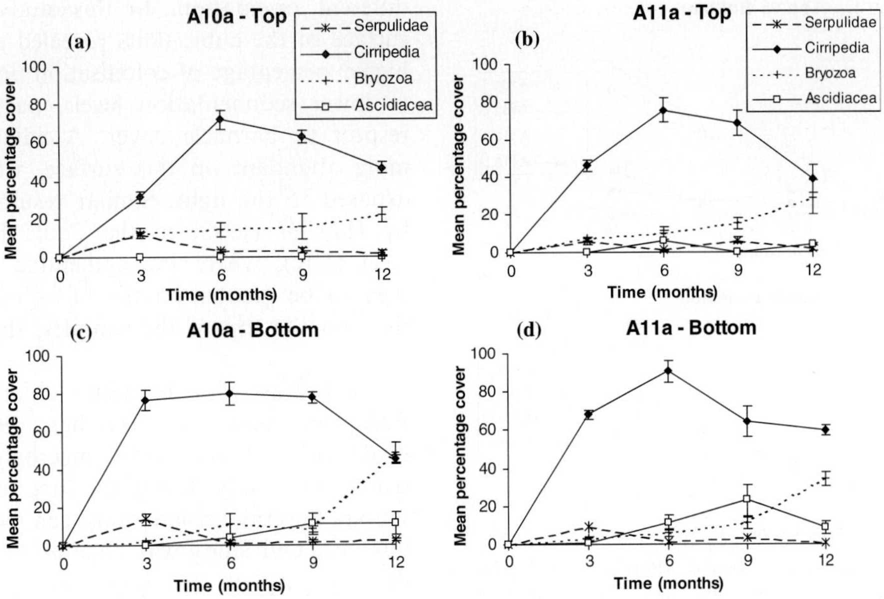

Figure 3. Major taxonomical groups at the two studied reef groups, Top vs. Bottom surfaces.

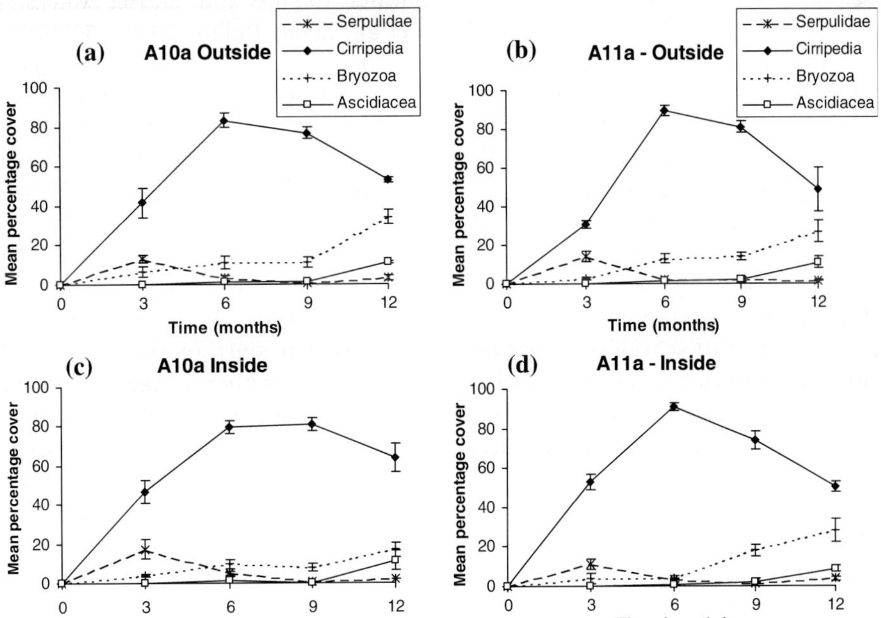

Figure 4. Major taxonomical groups at the two studied reef groups, Outside vs. Inside surfaces.

onisation of artificial reefs on the southern Portuguese coast. Two reef groups at the Ancão artificial reef system were analysed and colonisation was compared between surfaces with different orientation. The replicate cubic units used in this work were made with the same material of the reef structure (concrete) and were suspended at the reef blocks at the time of the reef immersion to ensure that the colonisation patterns would be representative of the reef structure. Three months after immersion more than half of the area of the samples was colonised by macrobenthic species and after

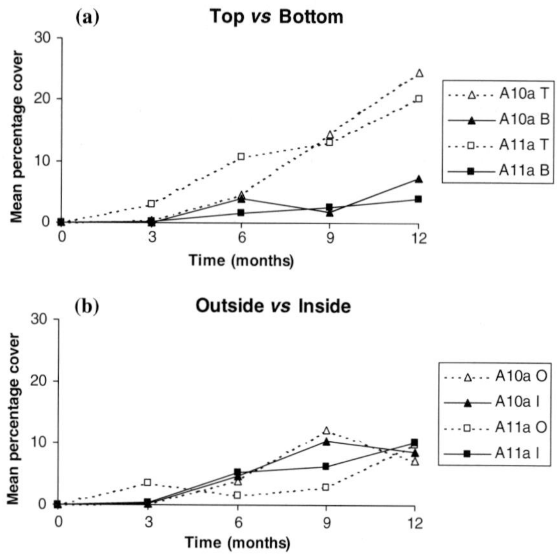

Figure 5. Other taxonomical groups (Porifera, Hydrozoa, Anthozoa, and sessile Polychaeta other than Serpulidae) at the two studied reef groups (A10a and A11a). (a) – Top (T) vs. Bottom (B) surfaces; (b) – Outside (O) vs. Inside (I) surfaces.

the sixth month the entire surface was covered. Sessile encrusting organisms such as barnacles, bryozoans, serpulids and ascidians colonised the cubic samples with a clear dominance of barnacles. These taxonomical groups have been seen to colonise artificial reefs in many other artificial reef structures (e.g. Cummings, 1994; Relini et al., 1994; Badalamenti et al., 2000; D'Anna et al., 2000; Relini, 2000). Despite the general presence of these taxonomical groups in the first year of colonisation of artificial substrata, the sequence of macrobenthic colonisation varied with seasons or places in several studies conducted in the Mediterranean (e.g. Badalamenti et al., 2000; Relini, 2000). In the Ancão reef similar patterns of colonisation were obtained at the two studied reef groups. The heavy initial settlement of barnacles matches the observations of Cummings (1994) for the colonisation of an artificial reef in Florida. The barnacle cover declined at Ancão six months after immersion and a more heterogeneous community was established.

Surfaces with different orientation showed different colonisation patterns at both reef groups of the Ancão system. This result is consistent with previous work (e.g. D'Anna et al., 2000; Glasby & Connell, 2001), which showed that different assemblages developed on surfaces of different orientation. In this study the bottom surface of the cubic units revealed a significantly higher percentage of colonisation possibly related to lower sedimentation levels, particularly with respect to barnacle cover. Ascidians were also more abundant on this surface, which was less exposed to the light. Similar results were found by Hatcher (1998) in the Poole Bay Artificial Reef (UK), where the epibenthic biomass was seen to be greater on the bases of the samples than on the top of the samples, throughout the study.

In contrast with the results obtained for other European reefs (e.g. Hatcher, 1998; D'Anna et al., 2000; Relini, 2000), macroalgae were absent in our study during the first year of colonisation. Several explanations can be offered for this fact. Our study was done at 20 m depth and the turbidity of the water was generally high. Most of the studies where macroalgae were present during the first year were done at shallower depths and clearer waters (e.g. D'Anna et al., 2000; Relini, 2000). Marked differences in colonisation stages were described by D'Anna et al. (2000) for clear and turbid waters, showing a major effect of light on algal cover. Ongoing studies on the comparison among different reefs on the southern coast, however, seem to indicate that both older artificial reef structures and artificial reefs which are closer to natural rock substrata have macroalgae (see also Monteiro & Santos, 2000). It is possible that macroalgae species will start to colonise the Ancão reef in later years. Other studies in progress at this region comparing different reefs, depths, and temporal variation will enable to clarify this and other points of colonisation patterns.

Acknowledgements

We are grateful to the INIAP (Instituto nacional de Investigação Agrária e das Pescas) scientific diving team and research vessel crew for helping in fieldwork. We would like to thank A. Silva for help in the lab. The research work was supported by the project "Implantação e estudo integrado de sistemas recifais" of MARE programme and by a Post-Doc scholarship of FCT-CFRH/BPD/3608/2000.

References

Allemand, D., E. Debernardi & W. Seaman Jr., 2000. Artificial reefs in the principality of Monaco: protection and enhancement of coastal zones. In Jensen, A. C., K. J. Collins, & A. P. M. Lockwood (eds), Artificial Reefs in European Seas. Kluwer Academics Publishers, London: 151–166.

Badalamenti, F., G. D'Anna & S. Riggio, 2000. Artificial reefs in the gulf of Castellammare (North-west Sicily): a case study. In Jensen, A. C. K. J. Collins, & A. P. M. Lockwood (eds), Artificial Reefs in European Seas. Kluwer Academics Publishers, London: 75–96.

Barnabé, G., E. Charbonnel, J.-Y. Marinaro, D. Ody & P. Francour, 2000. Artificial reefs in France: analysis, assessments and prospects. In Jensen, A. C. K. J. Collins, & A. P. M. Lockwood (eds), Artificial Reefs in European Seas. Kluwer Academics Publishers, London: 167–184.

Cummings, S. L., 1994. Colonisation of a nearshore artificial reef at Boca Raton (Palm Beach County), Florida. Bulletin of Marine Science 55: 1193–1215.

D'Anna, G., F. Badalamenti & S. Riggio, 2000. Artificial reefs in North-West Sicily: comparisons and conclusions. In Jensen, A. C. K. J. Collins, & A. P. M. Lockwood (eds), Artificial Reefs in European Seas. Kluwer Academics Publishers, London: 97–112.

Glasby, T. M. & S. D. Connell, 2001. Orientation and position of substrata have large effects on epibiotic assemblages. Marine Ecology Progress Series 214: 127–135.

Hatcher, A. M., 1998. Epibenthic colonisation patterns on slabs of stabilised coal-waste in Poole Bay, UK. Hydrobiologia 367: 153–162.

Hawkins, S. J. & H. D. Jones, 1992. Rocky Shores. Marine Field Course Guide 1. Immel Publishing, London 144 pp.

Jensen, A. C. (ed.), 1997. European artificial reef research. Proceedings of the First EARRN Conference, March 1996 Ancona, Italy. Southampton Oceanographic Centre, Southampton.

Monteiro, C. C. & M. N. Santos, 2000. Portuguese artificial reefs. In Jensen, A. C. K. J. Collins, & A. P. M. Lockwood (eds), Artificial Reefs in European Seas. Kluwer Academics Publishers, London: 249–261.

Moreno, I., 2000. Artificial reef programme in the Balearic Islands: western Mediterranean sea. In Jensen, A. C. K. J. Collins, & A. P. M. Lockwood (eds), Artificial Reefs in European Seas. Kluwer Academics Publishers, London: 219–233.

Pickering, H., D. Whitmarsh & A. Jensen, 1998. Artificial reefs as a tool to aid rehabilitation of coastal ecosystems: investigating the potential. Marine Pollution Bulletin 37: 505–514.

Relini, G., 2000. The Loano artificial reef. In Jensen, A. C. K. J. Collins, & A. P. M. Lockwood (eds), Artificial Reefs in European Seas. Kluwer Academics Publishers, London: 129–149.

Relini, G., N. Zamboni, N. Tixi & G. Torchia, 1994. Patterns of sessile macrobenthos community development on an artificial reef in the gulf of Genoa (Northwestern Mediterranean). Bulletin of Marine Science 55: 745–771.

Revenga, S., F. Fernández, J. L. González & E. Santaella, 2000. Artificial reefs in Spain: the regulatory framework. In Jensen, A. C. K. J. Collins, & A. P. M. Lockwood (eds), Artificial Reefs in European Seas. Kluwer Academics Publishers, London: 185–194.

Santos, M. N., 1997. Ichthyofauna of the artificial reefs of the Algarve coast. Exploitation strategies and management of local fisheries. PhD Thesis, University of Algarve.

Santos, M. N. & C. C. Monteiro, 1997. The Olhão artificial reef system (south Portugal): fish assemblages and fishing yield. Fisheries Research 30: 33–41.

Santos, M. N. & C. C. Monteiro, 1998. Comparison of the catch and fishing yield from an artificial reef system and neighbouring areas off Faro (Algarve, south Portugal). Fisheries Research 39: 55–65.

Seaman, Jr. W. & L. M. Sprague, 1991. Artificial habitats practices in aquatic systems. In Seaman Jr. W. & L. Sprague (eds), Artificial Habitats for Marine and Freshwater Fisheries. Academic Press Inc: 1–29.

Svane, I. & J. K. Petersen, 2001. On the problems of epibioses, fouling and artificial reefs, a review. P.S.Z.N.: Marine Ecology 22: 169–188.

Appendix A

Table A.1. List of identified *taxa* (Und. = Undetermined)

PORIFERA	
CNIDARIA	
Hydrozoa	Und. Hydrozoa
Anthozoa	Und. Anthozoa
PLATYHELMINTHES	
Turbellaria	Und. Turbellaria
NEMERTEA	
NEMATA	
ANNELIDA	
POLYCHAETA	
Polynoidae	*Malmgrenia ljungmani* (Malmgren, 1865)
	Und. Polynoidae
Pholoidae	*Pholoe synophtalmica* Claparède, 1868
Chrysopetalidae	Und. Chrysopetalidae
Phyllodocidae	Und. Phyllodocidae
Hesionidae	*Syllidia armata* Quatrefages, 1865
	Podarke sp.
	Und. Hesionidae

Continued on p. 342

Table A.1. (Continued)

Syllidae	*Autolytus* spp.
	Autolytus cf. *alexandri* Malmgren, 1874
	Autolytus cf. *brachycephalus* (Marenzeller, 1874)
	Autolytus langerhansi Gidholm, 1967
	Proceraea sp.
	Pseudobrania limbata (Claparède, 1868)
	Sphaerosyllis cryptica Ben-Eliahu, 1977
	Sphaerosyllis hystrix Claparède, 1863
	Sphaerosyllis taylori Perkins, 1981
	Exogone naidina Oersted, 1845
	Exogone verugera Claparède, 1868
	Syllides articulocirratus (Gillandt, 1979)
	Pionosyllis sp.
	Syllis spp.
	Typosyllis hyalina (Grube, 1863)
	Typosyllis cf. *variegata* (Grube, 1860)
	Typosyllis spp.
	Und. Syllidae
Nereididae	*Websterinereis glauca* (Claparède, 1870)
Lumbrineridae	*Lumbrineris* sp.
Dorvilleidae	Und. Dorvilleidae
Spionidae	*Polydora* spp.
Chaetopteridae	*Chaetopterus variopedatus* (Renier, 1804)
Ophellidae	*Aphelochaeta* sp.
Capitellidae	*Capitella* spp.
Ampharetidae	Und. Ampharetidae
Terebellidae	Und. Amphitritinae
	Und. Terebellidae
Sabellaridae	*Sabellaria* sp.
Sabellidae	Und. Sabellidae
Serpulidae	*Serpula concharum* Langerhans, 1880
	Serpula vermicularis Linnaeus, 1767
	Vermiliopsis cf. *monodiscus* Zibrowius, 1967
	Pomatocerus triqueter (Linnaeus, 1767)
	Pomatocerus lamarckii (Quatrefages, 1865)
	Hydroides cf. *norvegica* Gunnerus, 1768
	Hydroides cf. *nigra* Zibrowius, 1971
	Hydroides stoichadon Zibrowius, 1971
	Filograna sp.
	Protula sp.
	Filogranula cf. *stellata* (Southward, 1963)
	Und. Serpulidae
ARTHROPODA	
CRUSTACEA	

Table A.1. (Continued)

CIRRIPEDIA	
Balanidae	*Balanus amphitrite* Darwin, 1854
	Megabalanus tulipiformis (Ellis, 1758)
OSTRACODA	
COPEPODA	
AMPHIPODA	
Stenothoidae	*Stenothoe* cf. *valida* Dana, 1855
Aoridae	*Aora gracilis* (Bate, 1857)
	Microdeutopus versiculatus (Bate, 1856)
	Microdeutopus armatus Chevreux, 1887
Isaeidae	*Gammaropsis maculata* (Johnston, 1827)
Corophiidae	*Corophium sextonae* Crawford, 1937
	Ericthonius brasiliensis (Dana, 1855)
Ischyroceridae	*Jassa marmorata* Holmes, 1903
Caprellidae	*Caprella* sp.
	Caprella acanthifera Leach, 1814
	Phtisica marina Slabber, 1769
TANAIDACEA	
Apseudidae	*Apseudes talpa* (Montagu, 1808)
	Und. Tanaidacea
ISOPODA	
Gnathiidae	Und. Gnathiidae
Anthuridae	Und. Anthuridae
Munnidae	Und. Munnidae
Janiridae	*Janira maculosa* Leach, 1814
	Und. Isopoda
DECAPODA	
CARIDEA	
Hippolytidae	*Hippolyte longirostris* (Czerniavsky, 1868)
	Hypolite varians Leach, 1814
	Thoralus cranchii (Leach, 1817)
	Eualus occultus (Lebour, 1936)
	Eualus sp.
	Und. Caridea
ANOMURA	
Galatheidae	*Galathea intermedia* Lilljeborg, 1851
Porcellanidae	*Pisidia* cf. *bluteli* (Risso, 1816)
BRACHYURA	
Majidae	*Inachus leptochirus* Leach, 1817
Xanthidae	*Pilumnus* cf. *hirtellus* (Linnaeus, 1761)
	Und. Decapoda
PYCNOGONIDA	Und. Pycnogonida
MOLLUSCA	
GASTROPODA	

Continued on p. 343

Table A.1. (Continued)

Trochidae	*Jujubinus exasperatus* (Pennant, 1777)
Rissoidae	cf. *Pusillina sarsii* (Lovén, 1846)
	Alvania sp.
Cerithiidae	*Bittium* spp.
Muricidae	*Ocinebrina aciculata* (Lamarck, 1822)
	Muricidae sp.
Nassariidae	*Nassarius* cf. *pygmaeus* (Lamarck, 1822)
Pyramidellidae	*Chrysallida* spp.
	Odostomia spp.
Cylichnidae	cf. *Scaphander* sp.
Limapontiidae	cf. *Limapontia depressa* Alder & Hancock, 1862
	Und. Gastropoda
BIVALVIA	
Mytilidae	*Modiolus modiolus* (Linnaeus, 1758)
	Musculus costulatus (Risso, 1826)
	Musculus discors (Linnaeus, 1767)
Pinnidae	Und. Pinnidae
Pectinidae	Und. Pectinidae
Anomiidae	Und. Anomiidae
Ostreidae	Und. Ostreidae
Cardiidae	Und. Cardiidae
Tellinidae	*Tellina* sp.
Hiatellidae	*Hiatella arctica* (Linnaeus, 1767)
	Und. Bivalvia
BRYOZOA	
CYCLOSTOMATA	
Crisiidae	*Crisia ramosa* Harmer, 1891
CHEILOSTOMATA	
Scrupariidae	*Scruparia chelata* (Linnaeus, 1758)
Bugulidae	*Bugula neritina* (Linnaeus, 1758)
Umbonulidae	*Umbonula ovicellata* Hastings, 1944
Schizoporellidae	*Schizobrachiella sanguinea* (Norman, 1868)
ECHINODERMATA	
OPHIUROIDEA	
	Und. Ophiuroidea
ECHINOIDEA	
	Und. Echinoidea
CHORDATA	
UROCHORDATA	
Styelidae	*Botryllus schlosseri* (Pallas, 1766)
	Und. Urochordata

Hydrobiologia (2006) 555:345–353

H. Queiroga, M.R. Cunha, A. Cunha, M.H. Moreira, V. Quintino, A.M. Rodrigues, J. Serôdio & R.M. Warwick (eds),
Marine Biodiversity: Patterns and Processes, Assessment, Threats, Management and Conservation
DOI 10.1007/s10750-005-1130-4

The benthic impacts of the Loch Linnhe Artificial Reef

Thomas A. Wilding
Scottish Association for Marine Science, Dunbeg, Near Oban, PA37 1QA, Argyll, Scotland, UK
(E-mail: tom.wilding@sams.ac.uk)

Key words: artificial reefs, benthic impacts

Abstract

Artificial reefs are man-made structures that are placed on the seabed to mimic natural reefs. The Loch Linnhe Artificial Reef will consist of up to 40 reef modules, when completed, each constructed using concrete blocks. As part of this complex, reef modules of approximately 200 tonnes were deployed in June 2001 and January 2002 and form the basis of this study. Sediment samples were taken adjacent to-, and 5 m from, the reef perimeters and characterised in terms of sediment oxygenation (redox), particle size, organic carbon (loss on ignition) and coarseness (material >1 mm and <5 mm). Changes in molluscs and ophiuroids (>1 mm) present in the sediment were linked to chemico-physical changes occurring as a consequence of the reef. Both reefs trapped macroalgal phytodetritus at the perimeters, which subsequently decomposed. This accumulated material isolated the seabed from the overlying water column and was associated with a decrease in sediment oxygenation that was linked to changes in community structure. Loss-on-ignition (LOI) was significantly higher around one reef compared with 5 m from that reef. The reefs did not cause significant changes in the sediment texture (median particle size or coarseness) and neither LOI nor sedimentary texture changes were associated with community change.

Introduction

An artificial reef can be defined as a deliberately or accidentally deployed structure, lying on the sea- or lakebed that increases localised habitat complexity. Artificial reefs have been constructed in most coastal regions of the world, from a wide range of materials, for reasons as diverse as enhancing tourism, coastal protection and fisheries management (Baine, 2001).

Artificial reefs can impact the environment in a number of ways including the leaching of toxic contaminants, if present, from the construction material (Pickering, 1996) and/or through the interaction of the structure with the local current regime. Localised changes in the current regime around a structure such as an artificial reef have a number of implications for both the reef and the receiving environment. Increases in current speed around artificial reefs have been responsible for their undermining and subsequent burial (Mizutani et al., 2000) while decreases in current speed at the reef perimeter are likely to allow the settlement of fine material including organic particles with a subsequent decrease in mean particle size and concomitant nutrient enrichment (Guiral et al., 1995). Reefs can also alter sediments through the addition of shell fragments derived from organisms growing on the reef (Davis et al., 1982; Ambrose & Anderson, 1990; Barros et al., 2001). The accumulation of fine material, including organic material, at the perimeter of an artificial reef, has the potential to result in major changes in the nature of the sediment typical of organic enrichment (reviewed in Diaz & Rosenberg, 1995; Cloern, 2001; Gray et al., 2002).

The Loch Linnhe Artificial Reef has been designed to facilitate the scientific study of a number of reef related issues including how artificial reefs can be used in the management of fisheries

(Wilding & Sayer, 2002). On the Scottish west coast the most likely candidate for reef-based aquaculture (ranching) is the European lobster (*Homarus gammarus* L.). In lobster ranching the objective is to maximise the proportion of seeded lobsters that enter the fishery in an area over which fishing access is controlled. Lobsters are predator/scavengers that demonstrate ontogenetic shifts in habitat requirements with juveniles living in burrows until forced out by burrow collapse or insufficient food (Wickins et al., 1996). The most intuitive location for seeding juvenile lobsters is at the reef edge so that individuals leaving their burrows are more likely to move onto the reef. Understanding the near-reef environment is therefore important to ensure that juveniles are seeded in the optimal environment as close to the reef as possible.

The monitoring of impacts, such as sedimentation and/or erosion, is normally done by collecting sediment samples around the impact source and linking measured physical and biological parameters (Warwick, 1993). The identification of a particular bio-indicator group can help reduce survey costs by reducing sorting and identification time. One example of an indicator group are the molluscs as, at least in temperate waters, they contain environmental indicator species (Diaz & Rosenberg, 1995; Gray et al., 2002) and have the advantage that they are relatively easy to identify and are sufficiently robust to withstand the rigours of the sieving process.

The linkage of multivariate changes in community structure with measured changes in physical parameters poses considerable challenges (Clarke & Warwick, 2001). Redundancy analysis (RDA) (Leps & Smilauer, 2003) and the permutation based test based ANOSIM (Primer™, Plymouth Routines in Multivariate Ecological Research, Plymouth, UK; Clarke & Warwick (2001)) are methods of investigating community patterns and, in the case or RDA, the strength of relationships between measured physical factors and individual species within a multivariate data set.

The objectives of the current research were to test the hypotheses that the benthic community structure will be affected by the proximity of the reef and that this effect will be linked to changes in sediment oxygenation, particle size and the amount of organic carbon in the sediment occurring as a consequence of the accumulation of phytodetritus at the reef edge.

Materials and methods

Sampling location, dates and design

Reef modules 53 and 64 were deployed during January 2002 and June 2001, in Loch Linnhe on the west coast of Scotland, UK (Fig. 1), at 56°32.02′N, 5°27.08′W and 56°31.98′N, 5°27.24′W respectively. Each reef consisted of 5000 solid concrete blocks each 20 × 20 × 40 cm. The concrete blocks formed a scattered pile and were approximately 1–3 blocks high at the reef edge with an uneven perimeter.

Four sediment samples each from adjacent ('on'; N) and 5 m distant from each reef ('off'; F) were collected on consecutive days in July 2002 (total 16 samples) from random locations along the reef edge. Sediment samples were taken using a core tube (plastic bucket with the bottom re-

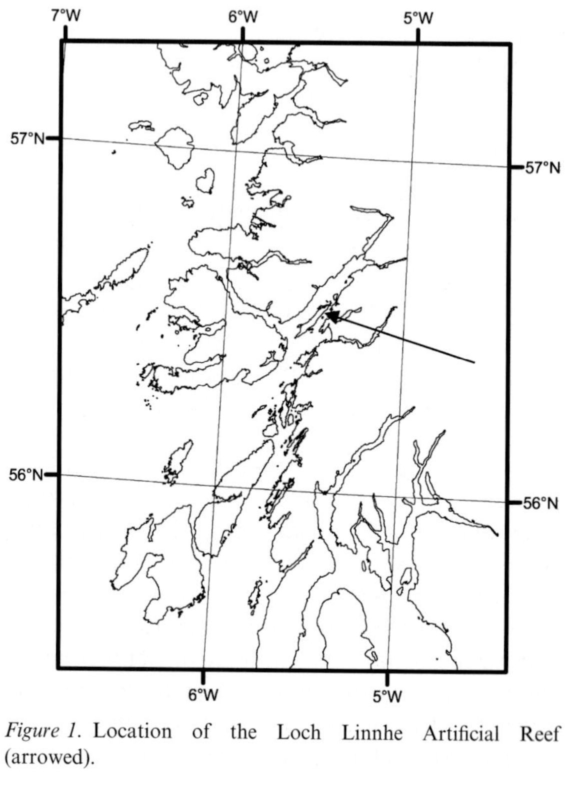

Figure 1. Location of the Loch Linnhe Artificial Reef (arrowed).

moved) with an area of 0.025 m^2. Each sample consisted of four cores combined (underwater) into a single bucket (total sample area 0.1 m^2). The samples were processed in the following order (discussed below): redox measurements, sample volume estimation, sub-sampling for organic content and particle size analysis, sieving, sediment weighing (coarseness) and preservation/staining.

Redox measurements

Redox measurements were done using redox probes (Russel pH Ltd., Auchtermuchty, Scotland, Model CMPtR106/300 mm) (Pearson & Stanley, 1979). The probe was randomly inserted to within approximately 1 cm of the bottom of the bucket containing the sample and left to stabilise for 30–60 s prior to the reading being taken. This was repeated 10 times for each replicate bucket. Between samples the probe was checked by reference to Zobell's solution (Zobell, 1946). The redox probe readings were adjusted to the hydrogen scale by the addition of a constant (198 mV, Zobell, 1946).

Sediment sub-samples from each replicate bucket were then taken for particle size and loss on ignition analysis. Each sediment sub-sample consisted of four random 50 ml aliquots taken as described in McManus (1988). The particle size analyser (Coulter LS230, Beckman – Coulter, Miami, US) outputs the median particle size, which was used as an overall indication of sediment texture. The sample volume was estimated against the internal bucket gradations.

Loss on ignition was calculated using the same sediment sub-samples described above. Approximately 7–14 g of each oven dried (65 °C, 48 h) sediment sub-sample was subject to ignition (650 °C, 12 h) and re-weighed once cool. LOI was expressed as a percentage and was arcsine transformed prior to analysis (Sokal & Rohlf, 1995).

Assessment of sample 'coarseness' and macrofaunal preservation

Each sediment sample was washed using seawater through a 5 mm then 1 mm sieve. The material retained on the 1 mm sieve (>1 mm, <5 mm) was drained and weighed. The proportion of the sample consisting of material between 1–5 mm is a granulometric assessment, here called coarseness. The two sediment fractions (>5 mm and 1–5 mm), containing the macrofauna, were then combined and preserved in borax buffered 4% formaldehyde solution containing 0.2 g/l Rose Bengal (Sigma) (Holme, 1971). The identification of macrofauna was restricted to molluscs using the keys of Tebble (1966) and Graham (1988) and to ophiuroids using the key of Mortensen (1927).

Statistical analysis

Analysis of variance (ANOVA) was used to detect differences in redox, median particle size, LOI and coarseness between treatments (where the data were normal and homoscedastic). Where a significant interaction occurred between the treatments the Tukey–Kramer multiple comparison test (Sokal & Rohlf, 1995) was used to determine which sample means were significantly different from others.

Differences in the community composition between the treatments (reef and proximity) were visualised using non-metric MDS based on the Bray–Curtis dissimilarity matrix (Primer™, Plymouth Routines in Multivariate Ecological Research, Plymouth, UK; Clarke & Warwick, 2001). Prior to analysis the species abundance data were square root transformed. The effect of both treatments (reef and proximity) were tested using the permutation based test 'two-way ANOSIM' (Primer™).

Direct ordination was done using CANOCO software (Leps & Smilauer, 2003). Prior to analysis the response of the 12 most abundant species (mean over all samples) to the environmental variables was assessed using Draftsman's plots. The outcome of the assessment was used to determine if a logarithmic transformation of the environmental variables was appropriate prior to constrained analysis. Prior to analysis, the length of the environmental gradient over which the community data was gathered was estimated using detrended correspondence analysis (DCA) (Leps & Smilauer, 2003). Species abundance data were square root transformed (Leps & Smilauer, 2003) to reduce the influence of the most abundant species and only those species represented in the top 99% of the total number of individuals were included to reduce the 'arch' effect (Leps & Smilauer, 2003). The significance of the measured environmental variables was tested using a Monte–Carlo simulation test (Leps & Smilauer, 2003) and only those variables making a significant ($p < 0.05$) contribution to the species–environment ordination were included in the final analysis and ordination diagram. Interpretation of the ordination diagram was as directed in Leps & Smilauer (2003).

Results

General observations

The sediments at the two sampled reefs (53 and 64) were visually distinct. The sediment at reef 53 was overlain with numerous stones and cobbles. Phytodetritus, consisting of rotting seaweed (mostly *Laminaria* sp. and *Fucus* sp.), accumulated around both reefs but was more pronounced at reef 53 compared with reef 64. The sediment under the rotting seaweed was occasionally black and the smell of hydrogen sulphide was noticed whilst sieving the samples from on-reef stations from reef 53. The off-reef sediment samples from both reefs, collected approximately 5 m from the reef edge, were not overlain by any phytodetritus and were visually indistinguishable from sediments further from the respective reefs.

Physical characterisation

The mean redox of the sampled sediment varied between highly oxic (>200 mV) and anoxic (<0 mV). The off-reef stations from both reefs had similar redox readings (median 130 and 160 mV from 53F and 64F respectively; Fig. 2a) and there were no overall significant differences between reef modules. However, the redox levels were significantly lower ($p < 0.001$, $n = 4$) in the on-reef samples, with the 53N redox values being particularly low (Fig. 2a). The sediment at the reef site can be classified as a silty-sand (Buchanan, 1984) and was significantly finer at reef 64 (both on- and off-reef) samples compared with reef 53 ($p = 0.004$, $n = 4$; Fig. 2b). The association between proximity to the artificial reef and the value for LOI varied between reefs. Reef 53 had the lowest and highest values for LOI from the off and on stations respectively (Fig. 2c). This contrasted with reef 64 where the values for on- and off-reef stations were similar and relatively high (Fig. 2c). The LOI value at reef 53F was significantly lower ($p < 0.01$, $n = 4$) than the other treatment combinations, which did not differ from each other (Tukey–Kramer multiple comparison test). Coarseness was not different between either the reefs or proximities (Fig. 2d).

Differences in assemblage composition

There was a clear separation of assemblage composition between both treatments (Fig. 3) where, predominantly, those samples exhibiting the greatest degree of anoxia exhibited the greatest degree of community difference. This effect was particularly pronounced at the 53N stations (Fig. 3) with the exception of station 53N3 that showed only moderate hypoxia but a very marked difference in assemblage structure compared to the 53 off-reef stations. Two-way ANOSIM showed significant changes in assemblage composition between both treatments (reef and proximity; $p = 0.03$ and 0.01 respectively).

The predominant relationship between the most abundant species and the environmental variables was either non-existent or linear. A logarithmic transformation was therefore not applied to the environmental variables prior to their inclusion in constrained ordination analysis. The length of the combined environmental gradient was short (1.6 standard deviation units) indicating that the species response to the environmental variables, over the gradient examined, was approximately linear (as opposed to unimodal). The most appropriate constrained analytical technique was therefore RDA (Leps & Smilauer, 2003).

The relative importance of the environmental variables measured in determining community structure, when considered in isolation, was (in descending order with significance in parenthesis) redox (0.008), particle size (0.008), loss on ignition (0.106) and coarseness (0.284). RDA analysis constrained only by redox and particle size accounted for 45% of the variation in species distribution (axis 1: 30%, axis 2: 15%) and can therefore be considered a good representation of species–environment relationships. The influences of redox and particle size on the community

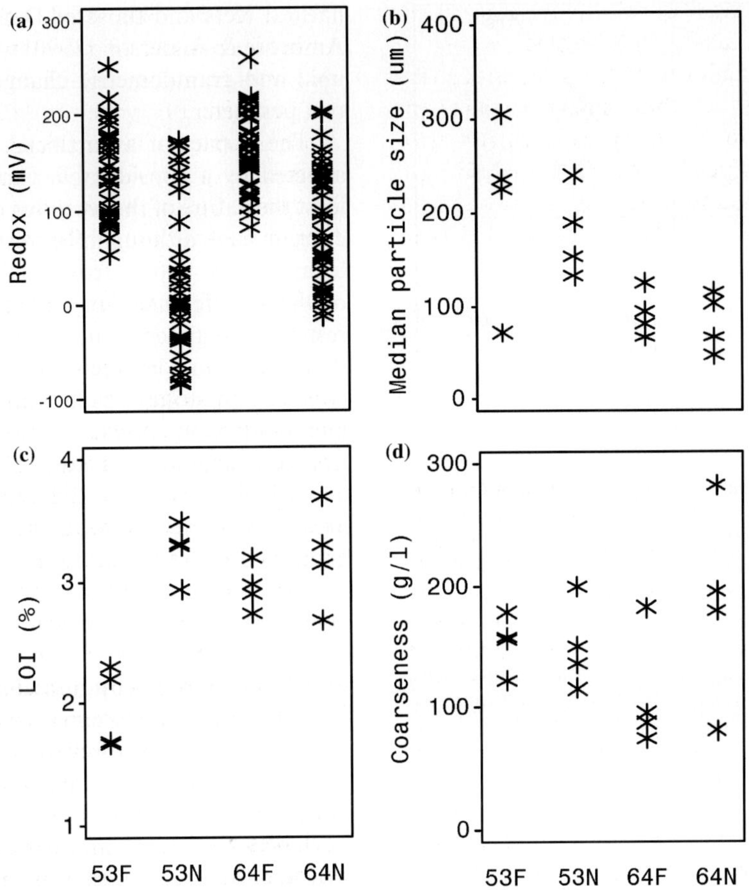

Figure 2. (a–d) Differences in redox (a), median particle size (b), percentage loss on ignition (LOI%) (c) and coarseness (d) at the four sampling stations.

structure were independent as indicated by the perpendicular arrows on the ordination diagram (Fig. 4).

RDA split the species into three broad groups depending on their response to the environmental variables. Group 1, consisting of *Hiatella artica* (L.), *Timoclea ovata* (Pennant), *Myrtea spinifera* (Montagu), *Thyasira flexuosa* (Montagu), *Corbula gibba* (Olivi) and *Ensis* sp. (L.) were more abundant in sediments with lower redox values and were not highly correlated with particle size. Group 2, containing *Mytilus edulis* (L.) and *Mysella bidentata* (Montagu), were associated with finer sediments (located opposite the Particle size arrow) but were not correlated with redox. Group 3, containing *Cylichna cylindrachae* (Pennant), *Amphiura chiajei* (Forbes) and *Amphiura filiformis* (Muller), *Abra alba* (Wood) and *Nucula* sp. (L.), were positively associated with redox (Fig. 4). No

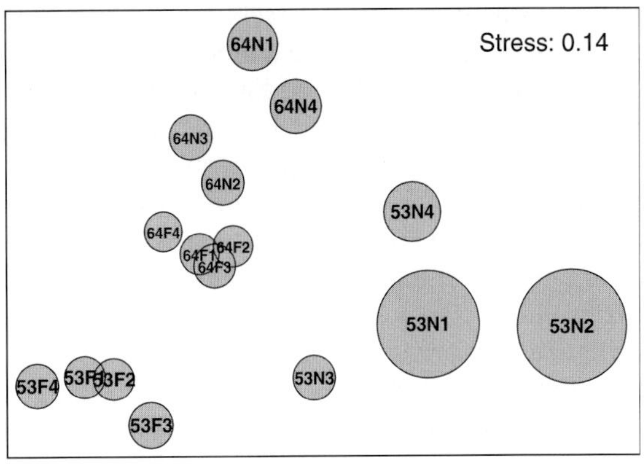

Figure 3. Relative similarity of the benthic infauna (molluscs and Ophiuroids) from different reefs (53 and 64) and proximities (on- and off-reef, N and F respectively) as shown by two-dimensional non-metric multidimensional scaling ordination. The area of the grey circles around each sample label is proportional to the degree of sediment hypoxia (larger circles denote a more hypoxic sediment).

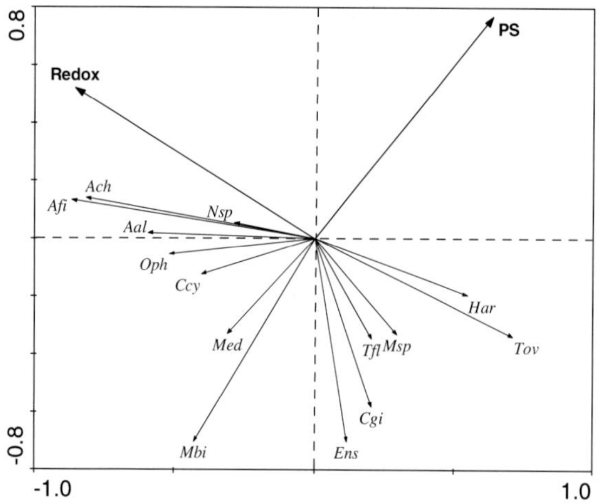

Figure 4. Redundancy analysis (RDA) correlation biplot ordination of mollusc/ophiuroid community. Key to environmental variables: redox – degree of sediment oxygenation; PS – median particle size (μm). Fourteen species showing the highest degree of association (>8%) with the canonical axes are shown in the ordination. Key to species: Aal-*Abra alba*, Ach-*Amphiura chiajei*, Afi-*Amphiura filiformis*, Cgi-*Corbula gibba*, Ccy-*Cylichna cylindrachae*, Ens-*Ensis* sp, Har-*Hiatella artica*, Mbi-*Mysella bidentata*, Med-*Mytilus edulis*, Msp-*Myrtea spinifera*, Nsp-*Nucula* sp., Oph – ophiuroids other than *A. chaijei* and *A. filiformis*, Tfl-*Thyasira flexuosa*, Tov-*Timoclea ovata*.

species were positively associated with increasing particle size (Fig. 4).

Discussion

The placement of artificial reefs made using concrete blocks in Loch Linnhe had major impacts on benthic infauna of the sediments immediately adjacent to the reef. The impacts were closely associated with a marked reduction in sediment oxygenation at the reef edge but not necessarily with an increase in organic carbon. The sediment at the different reef module locations hosted different communities that were linked with differences in sediment coarseness. However, the presence of the reef was not linked with granulometric changes. The impact of the Loch Linnhe Artificial Reef on the surrounding sediment appears to revolve around the entrapment of phytodetritus at the reef edge, which subsequently decomposes resulting in a high oxygen demand and reduced sedimentary oxygenation. The results here contrast with those of Barros et al. (2001) who reported an increase in the total number of species along the perimeters of natural reefs and those of Davis et al. (1982) and Ambrose & Anderson (1990) who linked macrofaunal and granulometric changes occurring at the reef perimeter.

The impact of an artificial reef is likely to be affected by a considerable variety of factors, not least the nature of the receiving environment. In the current case although the reefs 53 and 64 were located only 200 m from each other there were obvious differences in the appearance of the respective sediments and, by inference, the hydrodynamic regime between the sites. The presence of cobbles and stones on the sediment surface at the reef 53 location combined with the higher proportion of coarse material and lower organic carbon could be interpreted as indicating that the reef 53 area was subjected to higher current velocities compared with that at reef 64. Such an interpretation is commonly made (e.g. Davis et al., 1982). Whatever the cause of the differences between the sediments at locations 53 and 64, the off-reef stations hosted different and distinct macrofaunal communities as can be expected given the difference in the sediment type (Snelgrove & Butman, 1994).

The phytodetritus that accumulated around both reef modules had routinely been seen drifting in the water column and saltating just above the seabed in the area prior to reef deployment (Wilding & Sayer, 2002) but never accumulating. The accumulation of such material around the reef perimeter is likely to have occurred as a consequence of localised decreases in current velocity around the reef perimeter. The accumulation, and by inference the decrease in current speed, was particularly pronounced at reef 53, but not associated with any discernable change in sediment particle size. The biotically mediated alterations in sediment coarseness, consisting of the addition of shell fragments, that has been reported (Davis et al., 1982; Ambrose & Anderson, 1990; Barros et al., 2001) was also not found. The absence of significant granulometric changes around the reef is counter-intuitive given the extent of the phytodetrital accumulation (and, by inference, change in current regime) but may be explained by the relative immaturity of the reefs as granulometric changes may take longer than the 6–12 months the reefs had been in place to be detectable.

The accumulation of organic matter above normal levels, such as occurred around 53 and to a lesser degree 64, can be expected to cause effects similar to that of localised organic enrichment (Gray et al., 2002). The extent of the impact of the accumulation of phytodetritus is proportional to a number of factors including the nature of the deposited material (Gray et al., 2002). The phytodetritus that accumulated at the reef edge consisted mostly of intertidal and sub-tidal macroalgae, including fucoids and *Laminaria* sp. The relatively high carbon: nitrogen ratio in such material is generally considered to result in moderate hypoxia where it accumulates (Gray et al., 2002) presumably as the rate of decomposition is relatively slow. The degree of impact at reef 53 is, therefore, surprising and indicates that the development and severity of sedimentary hypoxia involves more than just the nature of the material deposited. In the present case, the major change in sediment oxygenation was associated with the accumulation of phytodetritus at depths of up to 30 cm. The presence of large accumulations of such material, as well as being a source of biological oxygen demand, will effectively isolate the benthos from the overlying oxygenated water and, in addition, indicates localised decreases in current velocity. These related processes are likely to act synergistically to cause the drastic reduction in sediment oxygenation that was observed around reef 53 and initiate a number of processes that radically change the nature of the sediment. The changes that occur under anoxic conditions include the growth of sulphate reducing bacteria that may be directly pathogenic to bivalves (de Zwaan et al., 2002) and which also release toxic compounds that cause mortality in the benthos (Snelgrove & Butman, 1994). This was observed in sediments adjacent to reef 53, which smelt of hydrogen sulphide and where the greatest reduction in macrofaunal abundance occurred. The differences in the degree of impact at the different reefs (higher impact at reef 53) may have occurred not only because of the development of more severe hypoxia at reef 53 but also because communities living in coarser sediments tend to be more sensitive to such perturbations (Diaz & Rosenberg, 1995). Reef 64, which trapped less organic matter, and which is likely to host a community more able to capitalise on a flux of organic matter, showed less of a community impact as a consequence of reef placement. The accumulation of phytodetritus at the reef edge and its subsequent decay and associated hypoxia may vary temporally, particularly in areas such as Loch Linnhe that are subject to spring- and neap tidal variations. The recovery of the benthos to a non-impacted status will take time and results in a lag phase between the chemical and biotic recovery of marine sediments (Macleod et al., 2004). The delay in recovery of the biotic component following a short-term impact, such as periodic hypoxia, may explain why station 53N4 was relatively oxic yet still demonstrated a radically altered benthic assemblage.

In the current case the molluscs appeared to be an excellent indicator group as they were found over the full range of environmental conditions encountered including the highly impacted stations. Species such as the opportunistic and hypoxia resistant *C. gibba* (Diaz & Rosenberg, 1995) were closely associated with the most impacted 53N stations and showed a positive correlation with organic carbon (data not shown). Also common within some of the stations at the reef site was *A. alba*, a species considered moderately sensitive to hypoxia (Diaz & Rosenberg, 1995) but which was found in the highest abundances in highly oxic, fine sediments such as the 64F stations suggesting factors other than redox and granulometry affect its distribution. The brittle stars *A. chiajei* and *A. filiformis* were more abundant at the reef 64 location and decreased slightly in abundance at the reef edge. Surface filter/deposit feeding organisms such as *A. alba* (Diaz & Rosenberg, 1995) and motile ophiuroids are likely to move away from less favourable areas and, while reef 64 did not suffer severe hypoxia, organisms such as the ophiuroids may have suffered reduced feeding efficiency as a consequence of being smothered or subject to reduced water currents hence their apparent dispersion away from the reef edge. Species including *M. edulis* and *M. bidentata* did not demonstrate any correlation with sediment hypoxia but were more dominant in fine sediments. *Myrtea spinifera, H. artica, T. ovata, T. flexuosa, C. gibba and Ensis* sp. should be considered to be either tolerant to, or show a preference for, hypoxic conditions. Of those species *T. ovata* showed a clear increase in abundance at the perimeter of reef 53 and can, therefore, be considered an indicator of severely hypoxia-impacted muddy sediments.

One of the main objectives of the Loch Linnhe Artificial Reef is to investigate the potential for artificial reefs as a tool in fisheries enhancement. In the case of the Loch Linnhe Artificial Reef the most likely candidate species on the Loch Linnhe Artificial Reef is the European Lobster (*Homarus gammarus* L.) (Sayer & Wilding, 2002). Although lobsters are considered relatively resistant to hypoxia (Diaz & Rosenberg, 1995) they are likely to grow fastest in areas where there is a good food and oxygen supply. The results presented here suggest that the survivorship and growth of juvenile lobsters will be greatest around artificial reefs that trap some organic material, resulting in an increase in benthic production (food in the form of macrofauna), but in insufficient quantities to cause a catastrophic decrease in sediment oxygenation.

The entrapment of phytodetritus and subsequent changes in the near-reef sediments reported here has implications that extend beyond stock enhancement. The development of near-reef hypoxia may both increase the leaching of contaminants from the construction material and increase the susceptibility of the benthos to those contaminants (Skei et al., 1996; Gunnarsson et al., 2000). The potential and implications of such changes must be evaluated prior to reef deployment particularly where waste disposal is one of the reef objectives or where materials of convenience are being used.

Acknowledgments

My thanks to Douglas MacAlpine, Steve Douglas, Simon Thurston and Hugh Brown for their support. This research was supported by NERC grant NER/D/S/2000/01307. We are grateful to Foster Yeoman Limited for their long-term commitment to the ongoing manufacture and deployment of the Loch Linnhe Artificial Reef.

References

Ambrose, R. F. & T. W. Anderson, 1990. Influence of an artificial reef on the surrounding infaunal community. Marine Biology 107: 41–52.

Baine, M., 2001. Artificial reefs: a review of their design, application, management and performance. Ocean & Coastal Management 44: 241–259.

Barros, F., A. J. Underwood & M. Lindegarth, 2001. The influence of rocky reefs on structure of benthic macrofauna in nearby soft-sediments. Estuarine Coastal and Shelf Science 52: 191–199.

Buchanan, J. B. 1984. Sediment analysis. In Holme, N. A. & McIntyre, A. D. (eds), Methods for the Study of Marine Benthos. Blackwell Scientific Publications: 41–65.

Clarke, K. R. & R. M. Warwick, 2001. Change in Marine Communities: An Approach to Statistical Analysis and Interpretation (2nd ed.). Primer-E Ltd, Plymouth.

Cloern, J. E., 2001. Our evolving conceptual model of the coastal eutrophication problem. Marine Ecology-Progress Series 210: 223–253.

Davis, N., G. R. Vanblaricom & P. K. Dayton, 1982. Man made structures on marine sediments: effects on adjacent benthic communities. Marine Biology 70: 295–303.

de Zwaan, A., J. M. F. Babarro, M. Monari & O. Cattani, 2002. Anoxic survival potential of bivalves: (arte)facts. Comparative Biochemistry and Physiology – a Molecular and Integrative Physiology 131: 615–624.

Diaz, R. J. & R. Rosenberg, 1995. Marine benthic hypoxia: a review of its ecological effects and the behavioural responses of benthic macrofauna. Oceanography and Marine Biology – an Annual Review 33: 245–303.

Graham, A., 1988. Molluscs: Prosobranch and Pyramidellid Gastropods. 2nd edn. The Linnean Society of London, 662 pp.

Gray, J. S., R. S. S. Wu & Y. Y. Or, 2002. Effects of hypoxia and organic enrichment on the coastal marine environment. Marine Ecology-Progress Series 238: 249–279.

Guiral, D., N. Gourbault & M. N. Helleouet, 1995. Sediment nature and meiobenthos of an artificial reef (Acadja) used for extensive aquaculture. Oceanologica Acta 18: 543–555.

Gunnarsson, J., M. Bjork, M. Gilek, M. Granberg & R. Rosenberg, 2000. Effects of eutrophication on contaminant cycling in marine benthic systems. Ambio 29: 252–259.

Holme, N. A., 1971. Macrofauna sampling. In Holme, N.A., & McIntyre, A.D. (eds), Methods for the Study of Marine Benthos. Blackwell Scientific Publications: 80–131.

Leps, J. & P. Smilauer, 2003. Multivariate analysis of ecological data using CANOCO. Cambridge University Press, Cambridge, UK 269 pp.

Macleod, C. K., C. M. Crawford & N. A. Moltschaniwskyj, 2004. Assessment of long term change in sediment condition after organic enrichment: defining recovery. Marine Pollution Bulletin 49: 79–88.

McManus, J., 1988. Grain size determination and interpretation. In Tucker, M. (ed.), Techniques in Sedimentology. Blackwell Science Ltd.: 63–85.

Mizutani, N., K. Iwata & N. Jinno, 2000. Velocity field and local seabed scour around submerged structure under wave action. Proceedings of the Coastal Structures '99, Volumes 1 & 2, Rotterdam, 755–762.

Mortensen, T. H., 1927. Echinoderms of the British Isles. Dr Backhuys W., Uitgever, Rotterdam.

Pearson, T. H. & S. O. Stanley, 1979. Comparative measurement of the redox potential of marine sediments as a rapid means of assessing the effect of organic pollution. Marine Biology 53: 371–379.

Pickering, H., 1996. Artificial reefs of bulk waste materials: a scientific and legal review of the suitability of using the cement stabilised by-products of coal-fired power stations. Marine Policy 20: 483–497.

Sayer, M. D. J. & T. A. Wilding, 2002. Planning, licensing, and stakeholder consultation in an artificial reef development: the Loch Linnhe reef, a case study. ICES Journal of Marine Science 59: S178–S185.

Skei, J., K. Hylland, M.T. Schaanning, J. A. Berge, J. S. Gunnarsson, M. Skold & D.O. Eriksen, 1996. Interactions between eutrophication and contaminants 1. Principles, experimental design and synthesis. Marine Pollution Bulletin 33: 64–70.

Snelgrove, P. V. R. & C. A. Butman, 1994. Animal-sediment relationships revisited – cause versus effect. Oceanography and Marine Biology 32: 111–177.

Sokal, R. R. & F. J. Rohlf, 1995. Biometry: The Principles and Practice of Statistics in Biological Research (3rd ed.). Freeman W.H., and Company, New York 887 pp.

Tebble, N., 1966. British Bivalve Seashells. A Handbook for Identification (2nd ed.). Her Majesty's Stationary Office, Edinburgh 212 pp.

Warwick, R. M., 1993. Environmental impact studies on marine communities – pragmatical considerations. Australian Journal of Ecology 18: 63–80.

Wickins, J. F., J. C. Roberts & M. S. Heasman, 1996. Within-burrow behaviour of juvenile European lobsters *Homarus gammarus* (L.). Marine and Freshwater Behaviour and Physiology 28: 229–253.

Wilding, T. A. & M. D. J. Sayer, 2002. Evaluating artificial reef performance: approaches to pre- and post-deployment research. ICES Journal of Marine Science 59: S222–S230.

Zobell, C. E., 1946. Studies on redox potential of marine sediments. Bulletin of the American Association of Petrology and Geology 30: 477–513.